Outstanding Contributions to Logic

Volume 15

More information about this series at http://www.springer.com/series/10033

Sergei Odintsov
Editor

Larisa Maksimova on Implication, Interpolation, and Definability

Springer

Editor
Sergei Odintsov
Sobolev Institute of Mathematics
Novosibirsk
Russia

ISSN 2211-2758 ISSN 2211-2766 (electronic)
Outstanding Contributions to Logic
ISBN 978-3-319-69916-5 ISBN 978-3-319-69917-2 (eBook)
https://doi.org/10.1007/978-3-319-69917-2

Library of Congress Control Number: 2017956323

Printed on acid-free paper

This Springer imprint is published by Springer Nature
The registered company is Springer International Publishing AG
The registered company address is: Gewerbestrasse 11, 6330 Cham, Switzerland

Preface

Several scientific affairs in 2012 and 2013 were dedicated to Prof. Larisa Maksimova who celebrated her anniversary in November 2013. Among them are the special session on the conference Advances in Modal Logic 2012 (22–25 August 2012, Copenhagen, Denmark) and the conference Mal'tsev Meeting 2013 (12–15 August 2013, Novosibirsk, Russia). Mal'tsev Meeting is an annual conference of Novosibirk algebraic and logical school, an outstanding representative of which is Larisa Maksimova. Professor Heinrich Wansing, a member of the editorial board of the series Outstanding Contributions to Logic, participated Mal'tsev Meeting 2013, where he presented an invited talk. The idea to publish a volume of this series devoted to Larisa Maksimova belongs namely to Heinrich Wansing. The field of interests of Larisa Maksimova can be characterized as applied algebraic logic. She intensively used in her investigations various algebraic methods; established algebraic analogs of numerous logical properties such as different kinds of interpolation and definability properties, disjunction property; and carefully studied the relationship between relational and algebraic semantics for various kinds of logics. However, her investigations were aimed all the time at obtaining concrete logical results. In many cases, it was an exhaustive description of logics with one or another remarkable property in a big class of logics. The present book collects original contributions written by colleagues, collaborators, and former Ph. D. students of Maksimova that are related to one or another topic of her research. It contains a "historical" part (Chaps. 1–3) including a scientific autobiography of Professor Maksimova, a survey of her works on relevant logics, and a survey of her first famous results on classes of superintuitionistic logics and of normal modal logics extending **S4**. The rest of the book constitutes research papers as well as a few surveys.

Novosibirsk, Russia
September 2017

Sergei Odintsov

Acknowledgements

I would like to thank Prof. Maksimova for her kind attention to my work on this book. As a customer of the seminar "Non-standard logics" in Novosibirsk State University, I am grateful to Prof. Maksimova for creating and supporting this form of intellectual communication, which caused, in particular, the change of my field of scientific interests from the classical computability theory to non-classical logics. I would like to thank all authors of this volume for valuable contributions and all anonymous referees for useful reviews. My special thanks to Veta Yun for comprising the list of publications by Larisa Maksimova and to Sergei Drobyshevich for his essential help in the work on the camera-ready copy on the book. Finally, I am grateful to Heinrich Wansing for the suggestion to edit this volume.

Contents

Contributors

Lev Beklemishev Steklov Mathematical Institute, RAS, Moscow, Russia; Moscow M.V. Lomonosov State University, Moscow, Russia; National Research University Higher School of Economics, Moscow, Russia

Katalin Bimbó Department of Philosophy, University of Alberta, Edmonton, AB, Canada

Alex Citkin Metropolitan Telecommunications, Warren, NJ, USA; Metropolitan Telecommunications, New York, NY, USA

Janusz Czelakowski Insitute of Mathematics and Informatics, Opole University, Opole, Poland

Giovanna D'Agostino Department of Mathematics and Computer Science, Udine, Italy

J. Michael Dunn School of Informatics, Computing, and Engineering and Department of Philosophy, Indiana University, Bloomington, IN, USA

Anastasia Karpenko Novosibirsk State University, Novosibirsk, Russia

Larisa Maksimova Sobolev Institute of Mathematics, Novosibirsk, Russia

Alexei Muravitsky Louisiana Scholars' College, Northwestern State University, Natchitoches, LA, USA

Sergei Odintsov Sobolev Institute of Mathematics, Novosibirsk, Russia

Vladimir V. Rybakov Institute of Mathematics and Informatics, Siberian Federal University Krasnoyarsk, and (part time) Institute of Informatics Systems (Siberian Division RAS), Krasnoyarsk, Russia

Valentin Shehtman Kharkevich Institute for Information Transmission Problems, RAS, Moscow, Russia; National Research University Higher School of Economics, Moscow, Russia; Moscow State University, Moscow, Russia

Dmitry Tishkovsky School of Computer Science, The University of Manchester, Manchester, UK

Alexander Yashin Moscow State University of Psychology and Education, Moscow, Russia

Veta F. Yun Sobolev Institute of Mathematics, Novosibirsk, Russia

Chapter 1
Maksimova, Relevance and the Study of Lattices of Non-classical Logics

Sergei Odintsov

Abstract We outline the main stages of Maksimova's investigation and present in details her results published between 1972 and 1979 and concerning the study of pretabularity and interpolation properties in superintuitionisitc logics and in normal extensions of the logic $S4$.

1.1 Introduction

Larisa Maksimova was one of the first students of the Algebra and Logic Department of Novosibirsk State University (NSU). This department was founded by Analoly Maltsev at the Faculty of Mechanics and Mathematics of NSU in 1960, the year when Maksimova entered the University. A very special attention to the application of logical methods in the field of algebra and vise versa was typical for Anatoly Maltsev since his fist logical work (see Maltsev (1936)). Larisa Maksimova became a very bright representative of Maltsev's logical and algebraic school known first of all for effective application of the universal algebra methods to the study of the lattice of intermediate logics and the lattice of normal modal logics. The first period of her research covering undergraduate and post-graduate studies under supervision of Anatoly Maltsev was devoted to relevance logic. This period starts with her first article Maksimova (1964) devoted to the question of independence for axioms of Ackermann's calculus for rigorous implication and finished with the defence of PhD Dissertation "Logical calculi of rigorous implication" (Maksimova 1968). Notice that Maksimova (1964) was published in the beginning of fifth year of her university studies. Starting with Ackermann's calculus for rigorous implication Maximova then extended her work to the systems R of Relevant Implication and E of Entailment suggested by Anderson and Belnap. In particular, she developed algebraic semantics

S. Odintsov (✉)
Sobolev Institute of Mathematics, Novosibirsk, Russia
e-mail: odintsov@math.nsc.ru

S. Odintsov (ed.), *Larisa Maksimova on Implication, Interpolation, and Definability*, Outstanding Contributions to Logic 15, https://doi.org/10.1007/978-3-319-69917-2_1

for E and suggested for the first time a semantic definition of entailment via ternary accessibility relation. The results of this period remain however not very well known to the community of logicians. Fortunately, our volume includes a chapter[1] written by Katalin Bimbó and J. Michael Dunn devoted to Maximova's contribution to the field of relevant logic. Bimbó and Dunn provide an excellent survey of Maximova's results and carefully emerge them in the setting of contemporary investigations in the area. They finish the survey with the words:"She is definitely a pioneer in the area of research on rigorous and relevant implications."

After the defence of PhD thesis Larisa Maksimova switched her attention to the investigation of superintuitionistic and normal modal logics. In the next few years she obtained strong positive results, which bring her the worldwide recognition and were considered by contemporary researchers not only as unexpected, but also as unbelievable. In (1972) Maximova proved that among the continuum of superintuitionistic logics there are exactly 3 pretabular logics. Recall that a logic is *tabular* if it can be determined via a finite-valued matrix or, equivalently, via a frame with a finite number of possible worlds. A logic L is *pretabular* if it is not tabular, but every logic extending L is tabular. In Maksimova (1975) a similar result was proved for normal $S4$-extensions: exactly 5 normal modal logics containing $S4$ are pretabular. Two years later (1977) Maximova completely described superintuitionistic logics with the Craig interpolation property, the number of such logics is finite again — 8 (including the trivial one). The results on pretabular logics were obtained via description of pretabular varieties of pseudo-Boolean and topological Boolean algebras. An algebraic analog of the Craig interpolation property is not so obvious. It was proved in Maksimova (1977) that a superintuitionistic logic posesses the Craig interpolation property if and only if the respective variety of pseudo-Boolean algebras has the amalgamation property. Thus, to obtain the desired result it would be enough to find out all varieties of pseudo-Boolean algebras with amalgamation property, which actually was done in Maksimova (1977). In (1979a) Maksimova investigated normal $S4$-extensions satisfying the Craig interpolation theorem. The situation is more complicated in this case. For a normal modal logic extending $S4$, the Craig interpolation theorem is equivalent to the so called superamalgamation property of the respective variety of topological Boolean algebras, whereas the amalgamation property is equivalent to a weak form of interpolation theorem (for necessitated formulas). Maksimova (1979a) has found the list of 50 varieties of topological Boolean algebras containing all varieties with the amalgamation property and the list of 38 varieties of topological Boolean algebras containing all varieties with the superamalgamation property. Thus, there is only a finite number (≤ 37) of non-trivial normal $S4$-extensions with the Craig interpolation theorem and a finite number (≤ 49) of non-trivial normal $S4$-extensions satisfying the interpolation theorem for necessitated formulas. Curious that up to this moment the exact number of non-trivial normal $S4$-extensions with the Craig interpolation theorem remains unknown. In subsequent sections we survey these results and the setting in which they arise with more details.

[1] See Chap. 3.

1.2 Pretabular Logics

In 1972, the year of publication of the result on pretabular superintuitionistic logics (Maksimova 1972), the class of superintuitionistic logics was recognised as an important object of investigation. For example, Tsutomu Hosoi (1967a) wrote: "So we shall not discuss such problems as *the* mathematics built on an intermediate logic or the application of the intermediate logics, though we do not deny the usefulness of them, especially, that of many valued logics. In a word, we regard the intermediate logics merely as *objects* of mathematical study." Hosoi (1967a) defines a logic as a set of formulas (constructed from a fixed set $Prop$ of propositional variables with the help of connectives $\vee$, &, $\supset$, and $\neg$) closed under the rules of substitution and *modus ponens*. In this case, a superintuitionistic logic is a logic containing the set of theorems of intuitionistic propositional logic Int, and an intermediate logic is a superintuitionistic logic contained in the set of classical tautologies.[2] If a logic is given axiomatically, the above definition identifies it with the set of its theorems. However, as emphasizes Hosoi (1967a) the given definition of intermediate logics allows not to restrict the studies to logics defined by axiomatic methods, but it also covers the logics defined semantically, with the help of models. By a *model* of a logic L Hosoi means a logical matrix, which is a characteristic model of L. It is remarkable that despite such models of intermediate logics as pseudo-Boolean (p.-B.) algebras[3] were well studied in 1967 (*The Mathematics of Metamathematics* by H. Rasiowa and R. Sikorski was published in 1963[4]) Hosoi prefer to work with the equivalent notion of a regular model. A *regular model* $\mathsf{M} = \langle \mathsf{A}, \{1_M\} \rangle$, where $\mathsf{A} = \langle A, \&, \vee, \supset, \neg \rangle$, is a matrix with a single distinguished value 1_M and such that: (i) $\mathsf{M} \models \Phi$ for all $\Phi \in Int$; (ii) if $a = 1_M$ and $a \supset b = 1_M$, then $b = 1_M$; (iii) if $a = 1_M$, then $b \supset a = 1_M$ for all $b \in A$; (iv) if $a \supset b = 1_M$ and $b \supset a = 1_M$, then $a = b$. It was proved in Hosoi (1967) that every intermediate logic can be characterized by a regular model (its Lindenbaum algebra) and that $\mathsf{M} = \langle \mathsf{A}, \{1_M\} \rangle$ is a regular model if and only if A is a p.-B. algebra.

Hosoi (1967a) suggests the classification of intermediate logics by slices. Let $Z = ((p \supset q) \supset r) \supset (((q \supset p) \supset r) \supset r)$, let for an intermediate logic L and a formula Φ, $L + \Phi$ be the least intermediate logic containing $L \cup \{\Phi\}$. Finally, for a regular model M, we denote $L\mathsf{M} = \{\Phi \mid \mathsf{M} \models \Phi\}$. The set of formulas $L\mathsf{M}$ is an intermediate logic, which is called a *logic of* M. The *n-th slice* ($n = 1, 2, \ldots \omega$) is defined as the following set of intermediate logics:

---

[2]Recall that the trivial logic coinciding with the set of all formulas is the only superintuitionistic logic, which is not intermediate.

[3]Recall that $\langle A, \&, \vee, \supset, \neg, 1 \rangle$ is a *pseudo-Boolean algebra* if $\langle A, \&, \vee \rangle$ is a bounded lattice with the greatest element 1 with restpect to its lattice ordering $\leq$ and the operations $\supset$ and $\neg$ are such that:

$$c \leq a \supset b \text{ iff } a\&c \leq b; \quad b \leq \neg a \text{ iff } a\&b = 0,$$

where 0 is the least element with respect to $\leq$. Every pseudo-Boolean algebra is a distributive lattice.

[4]Rasiowa and Sikorski (1963) was translated into Russian by V.A. Yankov and published in 1972.

$$\mathscr{S}_n = \{L \mid L + Z = L\mathsf{S}_n\},$$

where S_n is a regular model based on a linearly ordered p.-B. algebra with $n+1$ elements for $n \neq \omega$ and with countable number of elements for $n = \omega$. Hosoi proved that every intermediate logic belongs to a unique slice and that every slice has the least and the greatest elements. It follows readily from the definition that the greatest element of $\mathscr{S}_n$ is $L\mathsf{S}_n, n \leq \omega$. For $n < \omega$, the least element of $\mathscr{S}_n$ is $LP_n = Int + P_n$, where

$$P_1 = ((p_1 \supset p_0) \supset p_1) \supset p_1, \quad P_{n+1} = ((p_{n+1} \supset P_n) \supset p_{n+1}) \supset p_{n+1}.$$

The least element of $\mathscr{S}_\omega$ coincides with Int.

Notice that $L\mathsf{S}_\omega$ coincides with Dummett's logic $LC = Int + (p \supset q) \vee (q \supset p)$ introduced in Dummett (1959), where it was proved that LC can not be determined via a finite matrix. Dunn and Meyer (1971) proved that every proper extension of LC coincides with one of logics $L\mathsf{S}_n$. It follows readily from the results of (Dummett 1959) and Dunn and Meyer (1971) that LC is a pretabular logic. Two other examples of pretabular logics can be found in Hosoi and Ono (1970), the sequel of Hosoi (1967a), where all logics of second slice were characterized semantically and axiomatized. They proved that $L \in \mathscr{S}_2$ iff $L = L(\mathbf{B}_0^n + \mathbf{B}_0)$[5] for some $1 \leq n \leq \omega$, where $\mathbf{B}_0$ denotes the two-element Boolean algebra, A^n is the direct power of an algebra A, and $\mathsf{A} + \mathsf{B}$ denotes a p.-B. algebra obtained from p.-B. algebras A and B via identifying the least element of A and the greatest element of B. Moreover, it was proved in Hosoi and Ono (1970) that every logic L of the third slice containing axiom $\neg p \vee \neg\neg p$ is of the form $L(\mathbf{B}_0 + \mathbf{B}_0^n + \mathbf{B}_0)$. The proofs of the above results essentially use the results of McKay (1967), who proved that every subdirectly irreducible p.-B. algebra has the form $\mathbf{B}_0 + \mathsf{A}$, where A is some other p.-B. algebra and that every intermediate logic is completely determined by its subdirectly irreducible models. It follows from the results of (Hosoi and Ono 1970) that the logics LP_2, the least element of $\mathscr{S}_2$, and $LQ_2 = LP_2 + \neg p \vee \neg\neg p$, the least logic of $\mathscr{S}_3$ containing $\neg p \vee \neg\neg p$, are pretabular.

Thus, the classes of extensions of the logics LC, LP_2 and LQ_3 were described in the literature before the publication of Maksimova (1972). It was known that every proper extension of each of these logics is tabular. Moreover, it was proved that neither of LC, LP_2, or LQ_3 is tabular. Thus, all three logics are pretabular, it remains to prove that there are no other pretabular logic. To solve this problem it is not enough to know that every superintuitionistic logic can be determined by a p.-B. algebra, it is necessary to take into account the fact that there is a one-to-one correspondence between the class of superintuitionistic logics and the class of subvarieties of the variety of p.-B. algebras. Maksimova (1972) mentioned this fact giving a reference to Gerchiu and Kuznetsov (1968). This ono-to-one correspodence can be given by the following rule: to every superintuitionistic logic L we assign the variety $\mathfrak{M}_L$ determined by the identities $\{\Phi = 1 \mid \Phi \in L\}$. Following (Kuznetsov 1971) Maksimova

[5]The logic $L\mathsf{A}$ of a p.-B. algebra is defined in the same way as a logic of the regular model $\langle \mathsf{A}, \{1_\mathsf{A}\}\rangle$.

says that a variety $\mathfrak{M}$ is *tabular*, if it is generated by a finite algebra, and *pretabular* if it is not tabular but any its proper subvariety is tabular. Respectively, a superintuitionistic logic L is called *tabular* (*pretabular*) if the corresponding variety $\mathfrak{M}_L$ is tabular (pretabular). One more important notion introduced by Kuznetsov (1963) and used in Maksimova (1972) is that of finite approximability. Maksimova (1972) says that a superintuitionistic logic is *finitely approximable* if the variety $\mathfrak{M}_L$ is generated by its finite algebras. The term "finite approximability" has algebraic roots (Maltsev 1958), the content of this notion is synonymous to Harrop's finite model property (Harrop 1958). Kuznetsov's motivation of the notion of finite approximability is more or less the same as Harrop's for finite model property: it helps to prove decidability of different problems for propositional logics (see Kuznetsov 1963 and Citkin 2008 for details). Subtle distinctions between these two notions are discussed in Citkin (2008). Moreover, Citkin (2008) claims that Kuznetsov was unaware of Harrop's publication in 1963, Kuznetsov (1963) contains no reference to Harrop (1958). Kuznetsov (1971) proved that every pretabular variety of p.-B. algebras is finitely approximable.

Thus, it remains to describe finitely appoximable pretabular varieties of p.-B. algebras. To this end Maksimova (1972) defines three varieties: $\mathfrak{M}_1$ is a variety generated by algebras L_n, $2 \leq n < \omega$, where L_n is a linearly ordered p.-B. algebra with n elements; $\mathfrak{M}_2$ is a variety generated by algebras $\mathsf{B}_n = \mathbf{B}_0 + \mathbf{B}_0^n + \mathbf{B}_0$, $0 \leq n < \omega$; finally, $\mathfrak{M}_3$ is a variety generated by algebras $\mathsf{C}_n = \mathbf{B}_0 + \mathbf{B}_0^n$, $0 \leq n < \omega$. In the concluding section of (Maksimova 1972) the logics of these varieties were axiomatized and its was proved that these varieties are different and that

$$\mathfrak{M}_1 = \mathfrak{M}_{LC}, \quad \mathfrak{M}_2 = \mathfrak{M}_{LQ_3}, \quad \text{and} \quad \mathfrak{M}_3 = \mathfrak{M}_{LP_2}.$$

To prove that $\mathfrak{M}_1$, $\mathfrak{M}_2$, and $\mathfrak{M}_3$ are the only pretabular varieties among finitely approximated varieties of p.-B. algebras Maksimova (1972) uses the presentation of p.-B. algebras via partially ordered (p.o.) sets. If $\mathsf{S} = \langle S, \leq \rangle$ is a p.o. set, then the family $B(\mathsf{S})$ of its cones (upward closed sets w.r.t. $\leq$) forms a topology on S with the interior operator $\mathscr{I}$. The algebra $\mathsf{B}(\mathsf{S}) = \langle B(\mathsf{S}), \&, \vee, \supset, \neg, 1 \rangle$, where $\&$ and $\vee$ are set theoretical intersection and union, $X \supset Y = \mathscr{I}((S \setminus X) \vee Y)$ and $\neg X = X \supset \varnothing$ for $X, Y \in B(\mathsf{S})$, and finally $1 = S$, is a p.-B. algebra. For a cone S' of S, the mapping $h(X) = X \& S'$ is homomorphism of $\mathsf{B}(\mathsf{S})$ onto $\mathsf{B}(\mathsf{S}')$, where $\mathsf{S}' = \langle S', \leq \cap (S')^2 \rangle$. Mappings of p.o. sets corresponding to embeddings of p.-B. algebras are characterized as follows.

Lemma 1 (Lemma 2 in Maksimova (1972)) *Let mapping θ from p.o. set S onto p.o. set $\mathsf{S}' = \langle S', \leq' \rangle$ satisfy the conditions:*

(1) $x \leq y \implies \theta(x) \leq' \theta(y)$,
(2) $\theta(x) \leq' \theta(y) \implies (\exists y')(\theta(y) = \theta(y')$ *and* $x \leq y')$.

Then, $\mathsf{B}(\mathsf{S}')$ *is isomorphically embedded into* $\mathsf{B}(\mathsf{S}')$.

If S and S' are considered as Kripke frames for intuitionistic logic, mappings satisfying conditions (1) and (2) from this lemma are usually called p-morphisms.

Maksimova (1972) however does not work with Kripke semantics, all what she needs is a kind of Stone representation for p.-B. algebras. Every p.-B. algebra A can be embedded into $\mathsf{B}(\mathsf{S_A})$, where $\mathsf{S_A}$ is a p.o. set of prime filters on A,[6] via the mapping $\varphi(a) = \{P \mid P \in \mathsf{S_A},\ a \in P\}$. If A is finite, then φ is an isomorphism of A and $\mathsf{B}(\mathsf{S_A})$.

For a p.o. set S Maksimova (1972) introduces the following characteristics: its length $l(\mathsf{S})$, its exterior branching $D_1(\mathsf{S})$, and its interior branching $D_2(\mathsf{S})$. The length $l(\mathsf{S})$ is defined as the supremum of cardinalities of chains in S. For $a \in \mathsf{S}$, define $d_1(a)$ as the cardinality of the set of all immediate successors of a, which are maximal in S, and $d_2(a)$ as the cardinality of the set of all immediate successors of a, which are not maximal in S. The exterior and interior branchings of S are defined as follows:

$$d_1(\mathsf{S}) = \sup_{a\in\mathsf{S}} d_1(a), \quad d_2(\mathsf{S}) = \sup_{a\in\mathsf{S}} d_2(a).$$

Theorem 1 (Theorem 1 on Maksimova (1972)) *If a variety $\mathfrak{M}$ of p.-B. algebras is generated by a system of algebras such that the lengths, exterior and interior branchings of their representing p.o. sets are bounded by some natural number k, then $\mathfrak{M}$ is tabular.*

Further, Maksimova (1972) makes the following observations on generators of varieties $\mathfrak{M}_1$–$\mathfrak{M}_3$:

1. If $l(\mathsf{S_A}) \geq n \geq 1$ for a p.-B. algebra A, then A contains a subalgebra isomorphic to L_{n+1}.
2. If a variety $\mathfrak{M}$ contains a finite p.-B. algebra A with $d_2(\mathsf{S_A}) \geq n \geq 1$, then B_n belongs to $\mathfrak{M}$.
3. If a variety $\mathfrak{M}$ contains a finite p.-B. algebra A with $d_1(\mathsf{S_A}) \geq n \geq 1$, then C_n belongs to $\mathfrak{M}$.

Let $\mathfrak{M}$ be a finitely approximated variety of p.-B. algebras. From the above observations we immediately infer the following statements:

1. If lengths of representing p.o. sets for algebras in $\mathfrak{M}$ are not bounded, then $\mathfrak{M}$ contains $\mathfrak{M}_1$.
2. If interior branchings of representing p.o. sets for finite algebras in $\mathfrak{M}$ are not bounded, then $\mathfrak{M}$ contains $\mathfrak{M}_2$.
3. If exterior branchings of representing p.o. sets for finite algebras in $\mathfrak{M}$ are not bounded, then $\mathfrak{M}$ contains $\mathfrak{M}_3$.

The next theorem readily follows from the above statements, Theorem 1 and the result of Kuznetsov (1971) on finite approximability of pretabular varieties.

Theorem 2 (Theorem 4 of Maksimova (1972)) *Any pretabular variety of p.-B. algebras contains one of the varieties $\mathfrak{M}_1$, $\mathfrak{M}_2$, and $\mathfrak{M}_3$.*

[6]Recall that $P \supseteq A$ is a prime filter on a p.-B. algebra A if it satifies the conditions: a) $P \neq A$; b) $(x\&y) \in P$ whenever $x, y \in P$; c) $(x \vee y) \in P$ iff $x \in P$ or $y \in P$.

As a consequence we obtain that any pretabular superintuitionistic logic is contained in one of LC, LQ_3, and LP_2. Since logics LC, LQ_3, and LP_2 are not tabular, we obtain the description of all pretabular superintuitionistic logics.

The technique and results of Maksimova (1972) were used later to describe pretabular normal extensions of $S4$ in Maksimova (1975). Three of five pretabular normal modal logics over $S4$ are the greatest modal companions of LC, LQ_3, and LP_2. However, prior to discuss (Maksimova 1975) we say a few words on a preceding work (Maksimova and Rybakov 1974) written in coauthorship with her first student Vladimir Rybakov, the author of a chapter in this book. Despite this is a joint paper the results are devided between authors. It is noticed in the introduction of Maksimova and Rybakov (1974) that the results of Sects. 1,2,4 belong to Rybakov, while the results of Sect. 3 belong to Maksimova. The article is devoted to the study of interrelations between superintuitionistic logics and normal modal logics extending $S4$. The authors take into account the lattice structure of two classes of logics. Superintuitionistic logics are understood as above, i.e., as sets of propositional formulas in the language $\{\&, \vee, \supset, \neg\}$[7] containing Int and closed under the rules of substitution and *modus ponens*. The trivial logic or the set of all such formulas we denote as $For_{\mathcal{I}}$. Formulas from $For_{\mathcal{I}}$ are denoted as $A, B, C, \ldots$. The lattice of all superintuitionistic logics is denoted $\mathcal{L}$, its meet operation coincides with the set-theoretical intersection $\cap$, where as the join $L_1 + L_2$ is the least logic containing logics L_1 and L_2.

Modal formulas are considered in the language $\{\&, \vee, \rightarrow, \sim, \Box\}$[8] and denoted as $\alpha, \beta, \gamma, \ldots$, the set of all such formulas is denoted $For_{\mathcal{M}}$. A *normal modal logic* is defined in Maksimova and Rybakov (1974) as a subset of $For_{\mathcal{M}}$ containing all axioms of $S4$ and closed under the rules of substitution, *modus ponens*, and of normalization $\frac{\alpha}{\Box\alpha}$. For a normal modal logic M and $X \subseteq For_{\mathcal{M}}$, the least normal modal logic containing M and X is denoted as $M + X$, and we write $M + \alpha$ for $M + \{\alpha\}$. The lattice of all normal modal logics with lattice operations $\cap$ and $+$ is denoted as $\mathcal{M}$.

Gödel (1933) defined the translation T from $For_{\mathcal{I}}$ to $For_{\mathcal{M}}$ and conjectured that T is a faithful embedding of Int into $S4$. The conjecture was confirmed by McKinsey and Tarski (1948). The relation between $L \in \mathcal{L}$ and $M \in \mathcal{M}$ defined with the help of Gödel translation as

$$\forall A \in For_{\mathcal{I}}(A \in L \iff T(A) \in M)$$

[7] $A \equiv B$ is an abbreviation for $(A \supset B)\&(B \supset A)$.

[8] $\alpha \leftrightarrow \beta$ is an abbreviation for $(\alpha \rightarrow \beta)\&(\beta \rightarrow \alpha)$ and $\Diamond\alpha$ for $\sim\Box\sim\alpha$.

plays the central part in Maksimova and Rybakov (1974).[9] Later logics L and M related as above received the following names: L is a *superintuitionistic fragment of* M, and M is a *modal companion of* L. Both names were suggested by Leo Esakia (1976).

Dummett and Lemmon (1959) introduced a mapping $\tau : \mathscr{L} \to \mathscr{M}$ such that $\tau(L) = S4 + T(L)$ and proved, in fact, that $\tau(L)$ is the least modal companion of L. Rybakov proved in Maksimova and Rybakov (1974) that τ is a lattice embedding preserving infinite joins as well as that the lattice $\mathscr{M}$ is distributive. Maksimova introduced the mapping $\rho : \mathscr{M} \to \mathscr{L}$ sending every normal $S4$-extension to its superintuitionistic fragment and proved that for every $L \in \mathscr{L}$, the set $\rho^{-1}(L)$ forms an interval in the lattice $\mathscr{M}$, i.e., that every superintuitionistic logic L has the greatest modal companion, which Maksimova denotes $\sigma(L)$.

Prior to define $\sigma(L)$ we recall some facts on the relations between algebraic semantics for Int and algebraic semantics for $S4$, i.e., between p.-B. algebras and topological Boolean (t.-B.) algebras. An algebra $\mathsf{A} = \langle A, \&, \vee, \to, \sim, \Box, 1\rangle$ is a *t.-B. algebra* if its reduct $\langle A, \&, \vee, \to, \sim, 1\rangle$ is a Boolean algebra, and the operation $\Box$ satisfies the identities:

$$\Box x \leq x, \quad \Box\Box x = \Box x, \quad \Box(x\&y) = \Box x\&\Box y, \quad \Box 1 = 1.$$

For every t.-B. algebra A, the set of its open elements $\{a \mid a \in A,\ \Box a = a\}$ has a natural structure of a p.-B. algebra, which is denoted as $G(\mathsf{A})$. The operations & and $\vee$ of $G(\mathsf{A})$ are induced by the respective operations of A, whereas the pseudo-complement and negation operations are defined as follows: $a \supset b = \Box(a \to b)$ and $\neg a = \Box(\sim a)$. A formula α is *true in* a t.-B. algebra A, $\mathsf{A} \models \alpha$, if the identity $\alpha = 1$ holds on A. The logic $M(\mathsf{A})$ of A is the set of all formulas that are true in A. The logic of a family of t.-B. algebras is defined similarly. The lattice of subvarieties of the variety of p.-B. algebras is dually isomorphic to the lattice $\mathscr{M}$.

To every p.-B. algebra A Maksimova assigns a unique t.-B. algebra $\mathscr{S}(\mathsf{A})$ such that for every t.-B. algebra B, if $\mathsf{A} = G(\mathsf{B})$, then the subalgebra of B generated by the set of its open elements (the universe of A) is isomorphic to $\mathscr{S}(\mathsf{A})$. Further, she defines a mapping $\sigma : \mathscr{L} \to \mathscr{M}$ as follows:

$$\sigma(L) = \{\alpha \mid \forall \mathsf{A}(L\mathsf{A} \supseteq L \ \Rightarrow\ \mathscr{S}(\mathsf{A}) \models \alpha)\},$$

and proves that $\sigma(L)$ is the greatest modal companion of $L \in \mathscr{L}$ in $\mathscr{M}$.

In this way, she obtain the following interrelations between mappings ρ, τ, and σ for $L \in \mathscr{L}$ and $M \in \mathscr{M}$:

---

[9] Maksimova and Rybakov (1974) use the generally accepted version of Gödel translation such that $T(p) = \Box p$ for a propositional variable p and

$$\begin{aligned} T(A\&B) &= T(A)\&T(B), \quad T(A \supset B) = \Box(T(A) \to T(B)),\\ T(A \vee B) &= T(A) \vee T(B), \quad T(\neg A) = \Box(\sim T(A)), \end{aligned}$$

where $A, B \in For_{\mathscr{I}}$.

$$\tau\rho(M) \supseteq M \supseteq \sigma\rho(M), \quad \rho\tau(L) = L, \quad \rho\sigma(L) = L.$$

Maksimova also pointed out that if $M \in \mathcal{M}$ is finitely approximable, then its superintuitionistic fragment $\rho(M)$ is finitely appoximable too, and Rybakov proved in the last section of Maksimova and Rybakov (1974) that every pretabular normal $S4$-extension is finitely approximated. Finally, Rybakov proved the following tabularity criterium.

Lemma 2 (Lemma 4.5 of Maksimova and Rybakov (1974)) *Logic M from $\mathcal{M}$ is tabular if and only if there is $n \in \omega$ such that $\alpha(n) \in M$, where*

$$\alpha(n) = \bigvee_{1 \leq i < j \leq n+1} \Box(p_i \leftrightarrow p_j)$$

Concluding the discussion of this work we notice that the syntactic characterization of σ is known as Blok-Esakia theorem, which was proved independently by Blok (1976) and Esakia (1976), and says that $\sigma(L) = Grz + T(L)$. Here $Grz = S4 + \Box(\Box(p \to \Box p) \to p) \to p$ is a logic discovered by Grzegorczyk (1967) as a modal companion of Int different from $S4$.

Now we come back to the discussion of Maksimova (1975), which starts with the definition of logics $PM1$–$PM5$. Three of these logics are the greatest modal counterparts of pretabular superintuitionistic logics:

$$PM1 = \sigma(LC), \quad PM2 = \sigma(LP_3), \quad PM3 = \sigma(LQ_2).$$

Other logics are defined semantically:

$$PM4 = L(\{\mathsf{U}_n \mid n \in \omega\}), \quad PM5 = L(\{\mathsf{V}_n \mid n \in \omega\}),$$

where for $n \in \omega$, finite t.-B. algebras

$$\mathsf{U}_n = \langle U_n, \&, \vee, \to, \sim, \Box, 1\rangle \text{ and } \mathsf{V}_n = \langle V_n, \&, \vee, \to, \sim, \Box, 1\rangle$$

are such that their reducts $\langle U_n, \&, \vee, \to, \sim, 1\rangle$ and $\langle V_n, \&, \vee, \to, \sim, 1\rangle$ are finite Boolean algebras with n atoms, the $\Box$-operation on V_n is defined as follows: $\Box 1 = 1$ and $\Box a = 0$ for $a \neq 1$. The $\Box$-operation of U_n has more complex definition. Assume that $a_1, \ldots, a_n$ are all atoms of U_n, then

$$\Box b = \begin{cases} 1, & \text{if } b = 1, \\ a_n, & \text{if } a_n \leq b < 1, \\ 0, & \text{if } a_n \not\leq x. \end{cases}$$

The following theorem is a crucial point of Maksimova (1975).

Theorem 3 (Theorem 2 in Maksimova (1975)) *Every non-tabular $M \in \mathcal{M}$ is contained in one of $PM1$–$PM5$.*

We give a short sketch of the proof. Let M be non-tabular. Two cases are possible, either $\rho(M)$ is non-tabular, or tabular.

In the first case, one can use the classification of pretabular superintuitionistic logics from Maksimova (1972) and Kuznetsov's result that every non-tabular superintuitionistic logic is contained in a pretabular one to conclude that $\rho(M)$ is contained in one of LC, LQ_3, LP_2. If $\rho(M) \subseteq LC$, then $\sigma\rho(M) \subseteq \sigma(LC)$ by monotonicity of σ established in Maksimova and Rybakov (1974). Taking into account $M \subseteq \sigma\rho(M)$, we obtain $M \subseteq \sigma(LC)$. If $\rho(M)$ is contained in LQ_3 or in LP_2, we infer in a similar way that M is contained in $\sigma(LQ_3)$ or in $\sigma(LP_2)$.

The case of tabular $\rho(M)$ is more complicated and uses the duality between t.-B. algebras and quasiorderings. If $\mathsf{Q} = \langle Q, R\rangle$ is a quasi-ordered set, i.e., R is a reflexive and transitive relation on Q, then the algebra $\mathscr{T}(\mathsf{Q}) = \langle P(Q), \&, \vee, \to, \sim, \Box, 1\rangle$, where $\langle P(Q), \&, \vee, \to, \sim, 1\rangle$ is a Boolean algebra of all subsets of Q and

$$\Box X = \{a \mid \forall b(aRb \Rightarrow b \in X)\} \text{ for } X \subseteq Q,$$

is a t.-B. algebra. To a t.-B. algebra B one can assign its representing set $\mathsf{Q}_\mathsf{B} = \langle Q_\mathsf{B}, R\rangle$, where Q_B is the set of all prime filters on B and R is such that

$$F_1 R F_2 \Leftrightarrow \forall a \in B(\Box a \in F_1 \Rightarrow \Box a \in F_2)$$

for $F_1, F_2 \in Q_\mathsf{B}$. In this case R is a quasi-ordering on Q_B and it is well known that algebra B can be embedded into $\mathscr{T}(\mathsf{Q}_\mathsf{B})$ via the mapping φ given by the rule $\varphi(a) = \{F \mid F \in Q_\mathsf{B} \,\&\, a \in F\}$.

For a quasi-ordering $\mathsf{Q} = \langle Q, R\rangle$, Maksimova (1975) introduces the following characteristics. Let $[a]$ be coset of $a \in Q$ w.r.t. equivalence $\sim_R = R \cap R^{-1}$. Denote by $\sigma(\mathsf{Q})$ the *sceleton of* Q, i.e., the partial ordering obtained by taking the quotient of Q w.r.t. $\sim_R$. Put

$$\mu_1(\mathsf{Q}) = \sup\{|[a]| \mid [a] \text{ is maximal in } \sigma(\mathsf{Q})\},$$
$$\mu_2(\mathsf{Q}) = \sup\{|[a]| \mid [a] \text{ is not maximal in } \sigma(\mathsf{Q})\},$$

where $|A|$ is the cardinality of the set A. For $a \in Q$, we define $\nu(a) = |\{[b] \mid [a] \leq [b]\}|$ and

$$\nu(\mathsf{Q}) = \sup\{\nu(a) \mid a \in Q\}.$$

Further, Maximova observes the following facts.

(F1) If a family of t.-B. algebras $\{\mathsf{B}_i \mid i \in I\}$ is such that there is $k \in \omega$ such that $\nu(\mathsf{Q}_{\mathsf{B}_i}) \leq k$, $\mu_1(\mathsf{Q}_{\mathsf{B}_i}) \leq k$, and $\mu_2(\mathsf{Q}_{\mathsf{B}_i}) \leq k$ for every $i \in I$, then logic $M(\{\mathsf{B}_i \mid i \in I\})$ defined by this family is tabular (Theorem 1 in Maksimova (1975)).

(F2) If t.-B. algebra B is such that $\mu_1(\mathsf{Q}_\mathsf{B}) \geq n$, then $M\mathsf{B} \subseteq M(\mathsf{U}_n)$ (Lemma 11 in Maksimova (1975)).

(F3) If t.-B. algebra B is such that $\mu_2(\mathsf{Q_B}) \geq n$, then $M\mathsf{B} \subseteq M(\mathsf{V}_n)$ (Lemma 12 in Maksimova (1975)).

Now we come back to the case of tabular $\rho(M)$. It was proved in Maksimova and Rybakov (1974) that M is finitely approximable in this case. From tabularity of $\rho(M)$ one can infer (we omit here the details) that $M = M(\{\mathsf{B}_i \mid i \in I\})$ for a family of finite t.-B. algebras $\{\mathsf{B}_i \mid i \in I\}$ such that all numbers $\nu(\mathsf{B}_i)$ are restricted by some natural number k. Since M is not tabular, we obtain by (F1) that either $\sup\{\mu_1(\mathsf{Q}_{\mathsf{B}_i} \mid i \in I\}) = \omega$ or $\sup\{\mu_2(\mathsf{Q}_{\mathsf{B}_i} \mid i \in I\}) = \omega$. In the first case, (F2) implies $M \subseteq PM4$. In the second case, $M \subseteq PM5$ by (F3). Thus, every non-tabular $M \in \mathscr{M}$ is contained in one of $PM1$–$PM5$.

It follows from Scroggs (1951) that $PM5 = S5$ and that $S5$ is pretabular. For other logics $PM1$–$PM4$ Maksimova provides the following finite axiomatizations:

$$PM1 = S4 + \{E, T(Z)\}, \quad PM2 = S4 + \{T(A_2), \alpha_2\}, \quad PM3 = S4 + \alpha_3,$$

$$PM4 = S4 + \{T(Z), T(D_2), \alpha_4\},$$

where

$$\begin{aligned}
Z &= (p \supset q) \vee (q \supset p),\\
A_2 &= (\neg p \vee \neg\neg p)\&((\neg\neg p\&((q \supset p) \supset (r \supset q))\&((r \supset q) \supset r)) \supset r),\\
D_2 &= (\neg\neg p\&((q \supset p) \supset q)) \supset p,\\
E &= \Box p \vee \Diamond((\Box(q \vee \Box p)\&{\sim}\Box p) \vee (\Box({\sim}\, q \vee \Box p)\&{\sim}\Box p)),\\
\alpha_2 &= \Box(p \leftrightarrow (\Diamond p\&{\sim}\Diamond(\Diamond p\&{\sim}p))) \vee \Box(p \leftrightarrow (\Box p \vee \Diamond(\Diamond{\sim}p\&p))),\\
\alpha_3 &= \Box(p \leftrightarrow \Box p) \vee \Box(p \leftrightarrow \Diamond p),\\
\alpha_4 &= \Box\Diamond p \rightarrow \Diamond\Box p.
\end{aligned}$$

Finally, she formulates the main result of this work.

Theorem 4 (Fundamental Theorem of Maksimova (1975)) *There exists exactly five pretabular normal modal logics containing* $S4$*:* $PM1$–$PM5$.

To prove this statement it is enough to check that logics $PM1$–$PM5$ are pairwise incomparable, to verify that $PM1$–$PM5$ are not tabular using the tabularity criterium of Maksimova and Rybakov (1974) (see Lemma 2 above), and finally to apply Theorem 3.

Important consequence of this theorem is the decidability of the property "to be tabular" for normal $S4$-extensions: there is an algorithm checking given $\alpha \in For_{\mathscr{M}}$ whether $S4 + \alpha$ is tabular. Indeed, $S4 + \alpha$ is tabular if and only if α does not belong to any of $PM1$–$PM5$, and each of this logics is finitely axiomatizable and finitely approximable. In a similar way the property "to be tabular" is decidable for superintuitionistic logics. According to Chagrov and Zakharyaschev (1997) the idea of using pretabular logics to recognise the tabularity property belongs to Kuznetsov.

1.3 Logics Satisfying the Craig Interpolation Theorem

Gabbay and Maksimova (2005) wrote in the preface to their monograph: "Interpolation has also become a traditional question to ask of a logical system along with other traditional questions such as decidability/complexity and axiomatizability". In the implicational form the interpolation theorem is formulated as follows: if a formula $A \to B$ holds in a logic L, then there is a formula C (so called *interpolant*) such that both implications $A \to C$ and $C \to B$ hold in L and only common non-logical terms have occurences in C.[10] Namely in this form the interplation theorem was proved by Craig (1957) for first order logic. Later Schütte (1962) proved this theorem for first order intuitionistic logic, and Gabbay (1971) for some of its extensions. For several modal logics the interpolation theorem was proved in Gabbay (1972) and Czermak (1975). Rasiowa (1972) investigated interpolation in many-valued predicate logics.

What is more important Jonsson (1965) related the interpolation principle for equalities (IPE) with the so called amalgamation property. He proved that under some additional conditions on a variety of algebras the amalgamation property implies that the equational theory of the variety satisfies IPE. Bacsich (1975) proved that the inverse implication does not hold in general case. For varieties of p.-B. algebras Maksimova (1977) proved that the amalgamation property is equivalent to IPE, moreover, it is equivalent to the fact that the Craig interpolation theorem holds for the supserintuitionistic logic corresponding this variety. Following this way and describing all amalgamable varieties of p.-B. algebras Maksimova (1977) has found all superintuitionistic logics satifying the Craig interpolation theorem.

Now we pass to exact definitions. By the *Craig interpolation theorem* (CIT) *for* $L \in \mathscr{L}$ Maksimova (1975) means the following statement:

- For any $A, B \in For_{\mathscr{I}}$, if $A \supset B \in L$, then there is $C \in For_{\mathscr{I}}$ such that $A \supset C, C \supset B \in L$ and C contains only those propositional variables that which occur simultaneously in both A and B.

Let $\mathfrak{K}$ be an arbitrary class of algebras. The *interpolation principle for equalities* (IPE) in $\mathfrak{K}$ is the following statement:

- For any pairwise disjoint tuples of variables $\mathbf{x}$, $\mathbf{y}$, $\mathbf{z}$ and equalities $p_1(\mathbf{x}, \mathbf{y})$, ..., $p_n(\mathbf{x}, \mathbf{y}), q(\mathbf{x}, \mathbf{z})$, if the implication $\bigwedge_{i=1}^{n} p_i(\mathbf{x}, \mathbf{y}) \Rightarrow q(\mathbf{x}, \mathbf{z})$ holds in $\mathfrak{K}$, then there are m and equalities $r_1(\mathbf{x})$, ..., $r_m(\mathbf{x})$ such that both

$$\bigwedge_{i=1}^{n} p_i(\mathbf{x}, \mathbf{y}) \Rightarrow \bigwedge_{i=1}^{m} r_i(\mathbf{x}) \text{ and } \bigwedge_{i=1}^{m} r_i(\mathbf{x}) \Rightarrow q(\mathbf{x}, \mathbf{z})$$

  hold in $\mathfrak{K}$.

If all algebras in $\mathfrak{K}$ are partially ordered (via $\leq$), Maksimova (1977) defines also the *interpolation principle for inequalities* (IPI):

[10] By non-logical terms we mean elements of the signature in case of first order logic and propositional variables in case of propositional logics.

- For all terms $t(\mathbf{x}, \mathbf{y})$ and $u(\mathbf{x}, \mathbf{z})$ if $t(\mathbf{x}, \mathbf{y}) \le u(\mathbf{x}, \mathbf{z})$ holds in $\mathfrak{K}$, then there is a term $v(\mathbf{x})$ such that both inequalities $t(\mathbf{x}, \mathbf{y}) \le v(\mathbf{x})$ and $v(\mathbf{x}) \le u(\mathbf{x}, \mathbf{z})$ hold in $\mathfrak{K}$.

The class of algebras $\mathfrak{K}$ is *amalgamable* (see Jonsson (1965)) if for any A_0, A_1, $\mathsf{A}_2 \in \mathfrak{K}$ we have:

(A) For any monomorphisms $i_1 : \mathsf{A}_0 \to \mathsf{A}_1$, $i_2 : \mathsf{A}_0 \to \mathsf{A}_2$ there are $\mathsf{A} \in \mathfrak{K}$ and monomorphisms $\varepsilon_1 : \mathsf{A}_1 \to \mathsf{A}$, $\varepsilon_2 : \mathsf{A}_2 \to \mathsf{A}$ such that $\varepsilon_1 i_1 = \varepsilon_2 i_2$, i.e., such that the following diagram commutes:

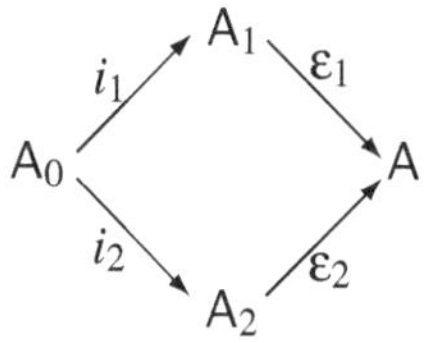

A class $\mathfrak{K}$ is *strongly amalgamable* if for any A_0, A_1, $\mathsf{A}_2 \in \mathfrak{K}$ condition (A) is satisfied and $\varepsilon_1(\mathsf{A}_1) \cap \varepsilon_2(\mathsf{A}_2) = \varepsilon_1 i_1(\mathsf{A}_0)$. A class $\mathfrak{K}$ of partially ordered algebras is *superamalgamable* if condition (A) is satisfied for A_0, A_1, $\mathsf{A}_2 \in \mathfrak{K}$, and if

$$\varepsilon_j(a) \le \varepsilon_k(b) \Rightarrow (\exists c \in \mathsf{A}_0)(a \le_j i_j(c) \wedge i_k(c) \le_j b),$$

where $\{i, j\} = \{1, 2\}$, $\le_j$ is the order in A_j, $a \in \mathsf{A}_j$, $b \in \mathsf{A}_k$.

It turns out that for a variety of p.-B. algebras each of the above properties is equivalent to CIT for the respective logic.

Theorem 5 (Theorem 1 of Maksimova (1977)) *For every $L \in \mathscr{L}$ the following conditions are equivalent:*

1. *CIT holds for L;*
2. *The variety $\mathfrak{M}_L$ satisfies IPI;*
3. *$\mathfrak{M}_L$ satisfies IPE;*
4. *$\mathfrak{M}_L$ is superamalgamable;*
5. *$\mathfrak{M}_L$ is strogly amalgamable;*
6. *$\mathfrak{M}_L$ is amalgamable;*
7. *Condition* (A) *holds for any well-connected* A_0, A_1, $\mathsf{A}_2 \in \mathfrak{M}_L$.[11]

Further, Maksimova (1977) listed eight varieties of p.-B. algebras defined by the following identities:

$H_1 : x = x;$
$H_2 : \neg x \vee \neg\neg x = 1;$
$H_3 : x \vee (x \supset (y \vee \neg y)) = 1;$

[11] Recall that a p.-B.algebra A is well-connected if $a \vee b = 1$ implies $a = 1$ or $b = 1$ for $a, b \in \mathsf{A}$.

$H_4: x \vee (x \supset (y \vee \neg y)) = 1; (x \supset y) \vee (y \supset x) \vee (x \equiv \neg y) = 1$
$H_5: x \vee (x \supset (y \vee \neg y)) = 1, \neg x \vee \neg\neg x = 1;$
$H_6: (x \supset y) \vee (y \supset x) = 1;$
$H_7: x \vee \neg x = 1;$
$H_8: x = 1.$

It is clear that H_1 is the variety of all p.-B. algebras, H_8 is a variety consisting of one-element algebras corresponding to the trivial logic $For_{\mathscr{I}}$ and amalgamable in a trivial way, and H_7 is a variety of Boolean algebras, which is known to be amalgamable.

To prove that varieties H_1–H_6 are amalgamable Maximova use the following construction. Let A_0, A_1, and A_2 be t.-B. algebras such that A_0 is a subalgebra of A_1 and A_2. Recall that $\mathsf{S}_{\mathsf{A}_i}$ is a representing p.o. set for A_i, whose elements are prime filters on A_i. Consider the set

$$S(\mathsf{A}_0, \mathsf{A}_1, \mathsf{A}_2) = \{\langle F_1, F_2\rangle \mid F_1 \in \mathsf{S}_{\mathsf{A}_1} \wedge F_2 \in \mathsf{S}_{\mathsf{A}_2} \wedge F_1 \cap \mathsf{A}_1 = F_2 \cap \mathsf{A}_2\}.$$

If we order $S(\mathsf{A}_0, \mathsf{A}_1, \mathsf{A}_2)$ componentwise, its algebra $\mathsf{B}(S(\mathsf{A}_0, \mathsf{A}_1, \mathsf{A}_2))$ of cones satisfies condition (A) for A_0, A_1, and A_2. This fact immediately implies that variety H_1 of all p.-B. algebras is amalgamable. Other cases require additional considerations.

More complicated problem is to prove that every amalgamable variety of p.-B. algebras is one of H_1–H_8. To solve this problem Maksimova find out suitable families of algebras generating varieties H_1–H_8. It is also necessary to characterize the complements of these varieties proving for each of H_2–H_8 statements like this:

Lemma 3 (Lemma 15 in Maksimova (1977)) *A p.-B. algebra* A *does not belong to* H_2 *if and only if algebra* $\mathsf{C}_2 = \mathbf{B}_0^2 + \mathbf{B}_0$ *can be embedded into* A.

The next series of statements proved in Maksimova (1977) concerns the possibility to construct from given algebras more complex algebras in amalgamable varieties. The following lemma is an example of such statement. A variety $\mathfrak{M}$ is *weakly amalgamable* if condition (A) holds for finite elements of $\mathfrak{M}$.

Lemma 4 (Lemma 24 of Maksimova (1977)) *Let* $\mathfrak{M}$ *be a (weakly) amalgamable variety of p.-B. algebras. If* B *is an arbitrary (a finite) p.-B. algebra and* $\mathsf{B} + \mathbf{B}_0^n + \mathbf{B}_0 \in \mathfrak{M}$ *for some* $n \geq 3$, *then* $\mathsf{B} + \mathbf{B}_0^m + \mathbf{B}_0 \in \mathfrak{M}$ *for all* $m \geq 1$.

The result of all this hard work is the following:

Theorem 6 (Theorem 2 of Maksimova (1977)) *For every variety* $\mathfrak{M}$ *of p.-B. algebras the following conditions are equivalent:*

a) $\mathfrak{M}$ *is amalgamable;*
b) $\mathfrak{M}$ *is weakly amalgamable;*
c) $\mathfrak{M}$ *coincides with one of* H_1–H_8*;*

Corollary 1 *The amalgamation problem for varieties of p.-B. algebras is decidable: given a finite basis of identities one can effectively check whether a variety defined by this basis is amalgamable.*

Given a finite basis of identities for $\mathfrak{M}$ the inclusions $\mathfrak{M} \subseteq H_i$ one can check using statements like Lemma 3. To check inverse inclusions one can use the families of algebras generating H_1–H_8.

Passing from varieties to their logics one can obtain the decidability of the problem: given a formula A to determine whether the Craig interpolation theorem holds for $Int + A$. Finally, Theorem 6 implies:

Theorem 7 (Theorem 3 of Maksimova (1977)) *There exists exactly seven non-trivial superintuitionistic logics in which the Craig theorem is true. These logics can be axiomatized by formulas:*

(1) $p \supset p$;
(2) $\neg p \vee \neg\neg p$;
(3) $p \vee (p \supset (q \vee \neg q))$;
(4) $p \vee (p \supset (q \vee \neg q))$, $(p \supset q) \vee (q \supset p) \vee (p \equiv \neg q)$;
(5) $p \vee (p \supset (q \vee \neg q))$, $\neg p \vee \neg\neg p$;
(6) $(p \supset q) \vee (q \supset p)$;
(7) $p \vee \neg p$.

Notice that 3rd and 6th logics in this list coincide with pretabular LP_2 and LC. Concluding this paper Maximova provides also the description of all logics with CIT extending the positive fragment Int^+ of intuitionistic logic. There are only three logics of this kind: Int^+, the positive fragment of Dummett's logic LC, and the positive fragment of classical logic.

Again these results on superintuitionistic logics were intensively used in the study of normal $S4$-extensions satisfying CIT (Maksimova 1979a). In this case CIT has another algebraic counterpart, the superamalgamation property, whereas amalgamation corresponds to a weaker form of interpolation theorem.

A logic $M \in \mathcal{M}$ satisfies the Craig interpolation theorem for necessitated formulas (CITN) if for every implication of the form $\Box\alpha \to \beta \in M$ there is $\gamma \in For_{\mathcal{M}}$ such that both $\Box\alpha \to \gamma$ and $\gamma \to \beta$ are in M, and C contains only those propositional variables that occur in both α and β.

In subsequent works CITN was replaced by the interpolation property for deducibility (IPD): for any α and β the condition $\alpha \vdash_M \beta$ implies $\alpha \vdash_M \gamma$ and $\gamma \vdash_M \beta$ for some γ whose propositional variables are in both α and β. Here $\alpha \vdash_M \beta$ means that β can be obtained from α and elements of M via *modus ponens* and necessitation rules.

It is clear that CITN and IPD are equivalent in normal $S4$-extensions.

Maksimova (1979a)(Theorems 1 and 2) proved the two chains of equivalences for every $M \in \mathcal{M}$:

- (M satisfies CIT) $\Leftrightarrow$ ($\mathfrak{M}_M$ satisfies (IPI)) $\Leftrightarrow$ ($\mathfrak{M}_M$ is superamalgamable)
- (M satisfies CITN) $\Leftrightarrow$ ($\mathfrak{M}_M$ satisfies (IPE)) $\Leftrightarrow$ ($\mathfrak{M}_M$ is amalgamable)

Recall that $\mathfrak{M}_M = \{\mathsf{A} \mid \mathsf{A} \models M\}$ is the variety of t.-B. algebras corresponding to logic M.

If $\mathfrak{M}$ is a variety of t.-B. algebras, then algebras $G(\mathsf{A})$ of open elements constructed from algebras $\mathsf{A} \in \mathfrak{M}$ also form a variety (of p.-B. algebras). We put $\mathfrak{M}^\circ = \{G(\mathsf{A}) \mid \mathsf{A} \in \mathfrak{M}\}$. It is clear that the logic of this variety coincides with the superintuitionistic fragment of $M(\mathfrak{M})$. Moreover, this variety inherits the amalgamation property.

Proposition 1 (Proposition 1 in Maksimova (1979a)) *If a variety $\mathfrak{M}$ of t.-B. algebras is amalgamable, then $\mathfrak{M}^\circ$ is amalgamable too.*

This statement essentially restricts the set of logics which can satisfy CIT or CITN. If a non-trivial $M \in \mathscr{M}$ satisfies one of these theorems, then $\rho(M)$ is one of 7 non-trivial superintuitionistic logics satisfying CIT.

Among a number of properties of amalgamable varieties of t.-B. algebras proved in Maksimova (1979a) we distinguish the following.

Lemma 5 (Lemma 10 in Maksimova (1979a)) *Let a variety $\mathfrak{M}$ of t.-B. algebras be amalgamable and contain algebras U_{n+1} and V_m. If A is a finite t.-B. algebra such that $\mathscr{S}(G(\mathsf{A})) \in \mathfrak{M}$, $\mu_1(\mathsf{Q}_\mathsf{A}) \leq n$, and $\mu_2(\mathsf{Q}_\mathsf{A}) \leq m$, then $\mathsf{A} \in \mathfrak{M}$.*[12]

Proposition 1 and Lemma 5 explain the choice of parameters for the following classification of varieties of t.-B. algebras. To every non-trivial variety $\mathfrak{M}$ of t.-B. algebras Maksimova (1979a) assigns the following *characteristic*

$$Ch(\mathfrak{M}) = (\mathfrak{M}^\circ, \mu_1(\mathfrak{M}), \mu_2(\mathfrak{M})),$$

where the variety $\mathfrak{M}^\circ$ of p.-B. algebras was defined above and

$$\mu_i(\mathfrak{M}) = \sup\{\mu_i(\mathsf{Q}_\mathsf{A}) \mid \mathsf{A} \in \mathfrak{M} \wedge |\mathsf{A}| < \omega\},\ i = 1, 2.$$

For a variety $\mathfrak{N}$ of p.-B. algebras and $m, n \leq \omega$ we denote by $\mathscr{T}(\mathfrak{N}, m, n)$ the greatest variety with characteristic $(\mathfrak{N}, m, n)$. It is clear that $\mathscr{T}(\mathfrak{N}, m, 0)$ is defined only if $\mathfrak{N}$ is the variety of Boolean algebras.

Studying restrictions which amalgamation and superamalgamation properties impose on characteristics Maksimova (1979a) proved the following theorems.

Theorem 8 (Theorem 3 in Maksimova (1979a)) *Every amalgamable variety of t.-B. algebras belongs to the following list:*

$\mathscr{T}(H_i, m, n)$, where $1 \leq i \leq 5$, $m, n \in \{1, 2, \omega\}$;
$\mathscr{T}(H_6, 1, 1)$;

---

[12]See the previous section for the definition of t.-B. algebra $\mathscr{S}(\mathsf{A})$ assigned to a p.-B. algebra A and for the definition of characteristics $\mu_1(\mathsf{Q}_\mathsf{A})$ and $\mu_2(\mathsf{Q}_\mathsf{A})$ of the representing quasi-ordering for A.

$\mathcal{T}(H_7, n, 1)$, *where* $n \in \{1, 2, \omega\}$;
$\mathcal{E}$ — *the variety of one-element algebras.*

Theorem 9 (Theorem 4 in Maksimova (1979a)) *Every superamalgamable variety of t.-B. algebras coincides with one of the following 38 varieties:*

$\mathcal{T}(H_i, m, n)$, *where* $1 \leq i \leq 2$, $m, n \in \{1, 2, \omega\}$;
$\mathcal{T}(H_i, 1, 1)$, $\mathcal{T}(H_i, 1, 2)$, $\mathcal{T}(H_i, 1, \omega)$, $\mathcal{T}(H_i, 2, 1)$, $\mathcal{T}(H_i, \omega, 1)$, *where* $i \in \{3, 4, 5\}$;
$\mathcal{T}(H_6, 1, 1)$;
$\mathcal{T}(H_7, n, 1)$, *where* $n \in \{1, 2, \omega\}$;
$\mathcal{E}$ — *the variety of one-element algebras.*

It does not follow from these two theorems that there is at least one variety of t.-B. algebras, which is amalgamable but not superamalgamable. Due to this reason Maksimova pointed out an example of such variety. This is $\mathcal{T}(H_3, \omega, \omega)$, which can be defined by the identity

$$\Box x \vee \Box(\Box x \to (\Box y \vee \Box{\sim}\Box y)) = 1.$$

The logic of this variety equals $\tau(LP_2)$.

Further, Maksimova noticed that for each of varieties $\mathcal{E}$, $\mathcal{T}(H_i, m, n)$, where $i \leq 7$, $m, n \leq \omega$, there is an algorithm deciding whether a given finite system of identities forms a base of the variety. This follows from the series of statements exemplified by Lemma 3 (similar lemmas were proved in Maksimova (1977) for each of varieties H_2–H_8), from the equivalence $\mathsf{A} \notin H_i \Rightarrow \mathcal{S}(\mathsf{A}) \notin \mathcal{T}(H_i, m, n)$ and from the following statement.

Lemma 6 (Lemma 15 in Maksimova (1979a)) *For a variety* $\mathfrak{M}$ *of t.-B. algebras and* $n \in \omega$ *the following equivalences are valid:*

$$\mu_1(\mathfrak{M}) \geq n \Leftrightarrow \mathsf{V}_n \in \mathfrak{M}, \quad \mu_2(\mathfrak{M}) \geq n \Leftrightarrow \mathsf{U}_{n+1} \in \mathfrak{M}.$$

In this way, for every finite set T of varieties of the form $\mathcal{E}$, $\mathcal{T}(H_i, m, n)$, where $i \leq 7$, $m, n \leq \omega$, there is an algorithm deciding whether a given finite system of identities form a base of a variety from T. In particular, amalgamation and super amalgamation properties are decidable for varieties of p.-B. algebras.

To transfer the above results from varieties to logics one can uses the results of Maksimova (1979). The logic of $\mathcal{T}(\mathfrak{N}, m, n)$ coincides with logic $\Gamma(L(\mathfrak{N}), m, n)$ introduced in Maksimova (1979). All logics of this form were axiomatized in Maksimova (1979) modulo $L(\mathfrak{N})$. Thus, in $\mathcal{M}$ there are not more than 49 non-trivial logics satisfying CITN and not more than 37 non-trivial logics satisfying CIT.

The exact number of normal $S4$-extensions with CIT or CITN remains unknown. It was mentioned in Gabbay and Maksimova (2005) CIT was established for 29 logics in $\mathcal{M}$ and CITN for 41 logics in $\mathcal{M}$. Thus, recognizing CIT and CITN for normal $S4$-extensions are nice examples of decidable problems such that algorithms deciding these problems remain unknown.

1.4 Conclusion

In the previous two sections we surveyed only a few results of Maksimova published between 1972 and 1979, her fist results on classes of superintuitionistic and normal modal logics. This was a conscious choice. On one hand, discussing all results of the author published after 1972 needs a book-length review. On the other hand, namely these first articles demonstrated that so general positive results on classes of logics are possible, they strongly influenced on a general trend of investigations in the field of non-classical logics: passing from the study of concrete systems to the study of big classes of logics. These articles vividly demonstrate the main features of Maksimova's approach to the study of lattices of non-classical logics: using the powerful methods of universal algebra, the importance of relational representations of algebraic models and of the duality between algebraic and relational (Kripke) semantics, the special attention to the decidability questions, the ability to suggest a system of notions which allows to deal with a variety of cases, which looks initially as unobservable.

In subsequent papers Maksimova proved the Craig interpolation theorem and the interpolation property for deducibility for many of logics from the lists found in Maksimova (1979a), in particular she completely described the normal modal logics with CIT extending Grzegorczyk's logic Grz. She investigated different kinds of interpolation properties in different classes of logics as well as various definability properties and its algebraic counterparts. The most well known properties of this kind are different versions the Beth property saying that the implicit definability is equivalent to the explicit definability. The state of investigations of interpolation and definability properties corresponding to first years of this century is presented in the impressive monograph written by Gabbay and Maksimova (2005). In recent years Maksimova written a series of papers devoted to the class of extensions of Johansson's minimal logic, which is much more complicated than the class of superintuitionistic logics.

Acknowledgements As a customer of the seminar "Non-standard logics" in Novosibirsk State University I am gratefull to Prof. Maksimova for creating and supporting this form of intellectual communication, which caused, in particular, the change of my field of scientific interests from the classical computability theory to non-classical logics.

References

Bacsich, P. D. (1975). Amalgamation properties and interpolation theorem for equational theories. *Algebra Universalis*, *5*, 45–55.

Blok, W. J. (1976). *Varieties of interior algebras*, Ph.D. thesis, University of Amsterdam.

Chagrov, A., & Zakharyaschev, M. (1997). *Modal Logic*. Oxford: Clarendon Press.

Citkin, A. (2008). A mind of non-countable set of ideas. *Logic and Logical Philosophy*, *17*, 23–39.

Craig, W. (1957). Three uses of Herbrand-Gentzen theorem in relating model theory and proof theory. *Journal of Symbolic Logic*, *22*, 269–285.

Czermak, J. (1975). Interpolation theorem for some modal logics. In H. E. Rose & J. C. Shepherdson (Eds.), *Logic Colloquium'73* (pp. 381–393). Amsterdam: North-Holland.
Dummett, M. (1959). A propositional calculus with denumerable matrix. *The Journal of Symbolic Logic, 24*, 97–106.
Dummett, M. A., & Lemmon, E. J. (1959). Modal logics between *S*4 and *S*5. *Zeitschrift für Mathematische Logik und Grundlagen der Mathematik.*
Dunn, J. M., & Meyer, R. K. (1971). Algebraic completeness results for Dummett's LC and its extensions. *Zeitschrift für Mathematische Logic und Grundlagen der Mathematik, 17*, 225–230.
Gabbay, D. M., & Maksimova, L. L. (2005). *Interpolation and Definability*. Oxford: Clarendon Press.
Esakia, L. L. (1976). On modal companions of superintuitionistic logics. *In VIIth Soviet Symposium on Logic, Kiev*. [On modal companions of superintuitionistic logics].
Esakia, L. L., & Meskhi, V Yu. (1974). O pyati 'kriticheskikh' modalnykh systemakh. *In Theory of Logical Inference (Summaries of Reports of the All-Union Symposium, Moscow, 1974). Part I.* Moscow. [On five 'critical' modal systems].
Esakia, L. L., & Meskhi, V Yu. (1977). Five critical modal systems. *Theoria, 40*, 52–60.
Gabbay, D. (1971). Semantic proof of the Craig interpolation theorem for intuitionistic logic and its extensions, part I, part II, In R. O. Gandy & C. M. E. Yates (eds.) *Logic Colloquium'69* (pp. 391–410)., North-Holland, Amsterdam.
Gabbay, D. (1972). Craig's interpolation theorem for modal logics. In W. Hodges (Ed.), *Lecture Notes in Mathematics* (pp. 111–127)., Conference in mathematical logic, London 1970 Berlin: Springer.
Gerchiu, V Ya., & Kuznetsov, A. V. (1968). O mnogoobraziyakh psevdo-bulevykh algebr, zadannykh tozhdestvami ogranichennoj dliny. *In 9th All-Union Algebraic Colloquium, Resume of Communications and Papers* (pp. 54–65). Gomel. [On the varieties of pseudo-Boolean algebras, specified by idenities of bounded length].
Gödel, K. (1933). Eine interpretation des intuitionistischen aussagenkalkuls. *Ergebnisse eines mathematischen Kolloquiums, 4*, 39–40. [An interpretation of intuitionistic propositional calculus].
Grzegorczyk, A. (1967). Some relational systems and the associated topological spaces. *Fundamenta Mathematicae, 60*, 223–231.
Harrop, R. (1958). On the existence of finite models and decision procedures for propoiional calculi. *Proceedings of Cambridge Philosophical Society, 54*, 1–13.
Hosoi, T. (1967). On the axiomatic method and the algebraic method for dealing with propositional logics. *Journal of the Faculty of Science, Iniversity of Tokio, Setion I, 14*, 131–169.
Hosoi, T. (1967a). On intermediate logics I, *Journal of the Faculty of Science, Iniversity of Tokio, Setion I, 14*, 293–312.
Hosoi, T., & Ono, H. (1970). The intermediate logics of the second slice. *Journal of the Faculty of Science, Iniversity of Tokio, Section I, 17*, 457–461.
Jónsson, B. (1965). Extensions of relational sctructures. In J. W. Addison, L. Henkin, & A. Tarski (Eds.), *Theory of Models* (pp. 146–157). Amsterdam: North-Holland.
Kuznetsov, A. V. (1963). O nerazreshimosti obschikh problem polnoty, razresheniya i ekvivalentnosti dla ischisleniĭ vyskazyvaniĭ. *Algebra i logika, 2*(4), 47–66. [On non-decidability of general problems of completeness, decidability and equivalency for propositional calculi].
Kuznetsov, A. V. (1971). Nekotorye svoĭstva reshetki mnogoobraziĭ psevdo-bulevykh algebr. In *11th All-Union Algebraic Colloquium, Resume of Communications and Papers* (pp. 255–256). Kishinev. [Certain properties of the lattice of varieties of pseudo-Boolean algebras].
Maksimova, L. L. (1964). O sisteme aksiom ischisleniya strogoĭ implikatsii. *Algebra i logika, 3*(5), 59–68. [On the system of axioms of the calculus of rigorous implication].
Maksimova, L.L. (1968) . *Logicheskie ischisleniya strogoĭ implikatsii*, Ph.D. thesis, SO AN SSSR, Novosibirsk. [Logical calculi of rigorous implication].
Maksimova, L. L. (1972). Predtablichnye superintuicionistskie logiki. *Algebra i logika, 11*(5), 558–570. [Pretabular superintuitionistic logics].

Maksimova, L.L. (1975). Predtablichnye rasshireniya sistemy Lewisa S4, *Algebra i logika, 14*(1), 28–55. [Pretabular extensions of Lewis' S4].

Maksimova, L. L. (1977). Teorema Kreĭga v supserintuicionistskikh logikakh i amalgamiruemyje mnogoobrazja psevdobulevykh algebr. *Algebra i Logika, 16*(6), 643–681. [Craig's theorem in superintuitionistic logics and amalgamable varieties of pseudo-Boolean algebras].

Maksimova, L. L. (1979). Ob odnoj klassifikacii modalnylh logik, *Algebra i Logika, 18*(3), 328–340. [On one classification of modal logics].

Maksimova, L. L. (1979a). Interpolacionnye teoremy v modalnykh logikalh i amalgamiruemye mnogoobrazija topobulevykh algebr, *Algebra i Logika, 18*(5), 556–586. [Interpolation theorems in modal logics and amalgamable varieties of topoboolean algebras].

Maksimova, L. L. & Rybakov, V. V. (1974). O reshetke normalnykh modalnykh logik, *Algebra i logika, 13*(2), 188–216. [On the lattice of normal modal logics].

Maltsev, A. I. (1936). Untersuchungen aus dem Gebiete der mathematischen Logik, *Mathematicheskiĭ Sbornik, 1*(3), 323–335. [Investigations from the field of mathematical logic].

Maltsev, A. I. (1958). O gomomorfizmakh konechnykh grupp, Ivanovskiĭ Gosudarstvennyĭ Pedagogicheskiĭ Institut. Uchenye Zapiski *18*, 49–60. [On homomorphisms of finite groups].

McKay, C. G. (1967). On finite logics. *Indagationes Mathematicae, 29*, 363–365.

McKinsey, J. C. C., & Tarski, A. (1948). Some theorems about the sentential calculi of Lewis and Heyting. *Journal of Symbolic Logic, 13*, 1–15.

Rasiowa, H. (1972). The craig interpolation theorem for *m*-valued predicate calculi. *Bulletin de là Académie Polonaise des Science, 20*, 341–346.

Rasiowa, H., & Sikorski, R. (1963). Monografie Matematyczne. *The mathematics of metamathematics*. Warszawa.

Schütte, K. (1962). Der interpolationsatz der intuitionistischen predikatenlogik, *Mathematische Annalen, 148*, 192–200. [Interpolation theorem of intuitionistic predicate logic].

Scroggs, J. (1951). Extensions of S5. *Journal of Symbolic Logic, 16*, 112–120.

Chapter 2
A Short Scientific Autobiography

Larisa Maksimova

Abstract These short autobiographic notes highlight main stages of my scientific and pedagogical activities.

2.1 My Family

I was born in Novosibirsk region, November 5, 1943. My father Lev D. Maksimov was a geobotanist, PhD in Biology. He was born in 1906 in the Yaroslavl region in a family of peasants. In 1932 he graduated from Leningrad State University. He defended his PhD thesis in 1937.

My mother, Taisia M. Maslennikova was born in 1920 in the city of Kursk. Shortly after the graduation from high school she went to work in the Nature Reserve situated near Kursk, where my father was a director at that time. There they got married.

In 1940 my parents moved to Siberia, where my father took up a position at Tomsk State University. My sister Natalia was born there. In 1941 the war broke out. Living in the city had become very difficult, and the family moved to a village near to Novosibirsk. In 1944, after I was born, all four of us moved to Novosibirsk. My father got the position of associate professor at Novosibirsk Pedagogical Institute, where he became the dean of the Geography faculty, as well as the head of department. He passed away in 1950 after a serious illness. My mother became a widow with two children.

By that time, my mother had graduated from the Pedagogical Institute and taught geography classes at school. In 1965 she got married again and moved back to Kursk, where she spent the rest of her life.

After Graduation from School in 1960, I entered the Faculty of Mechanics and Mathematics of Novosibirsk State University (NSU), which I graduated from in 1965. Since 1964 I have been working at Sobolev Institute of Mathematics SB RAS.

In 1973 I got married Oleg E. Trofimov, a mathematician. At NSU we were classmates. After graduating from NSU, he worked at the Institute of Automation

L. Maksimova (✉)
Sobolev Institute of Mathematics, Novosibirsk, Russia
e-mail: lmaksi@math.nsc.ru

S. Odintsov (ed.), *Larisa Maksimova on Implication, Interpolation, and Definability*, Outstanding Contributions to Logic 15, https://doi.org/10.1007/978-3-319-69917-2_2

and Electrometry SB RAS. There he defended his PhD thesis and then habilitation thesis as well. He passed away in 2008.

In 1975 our daughter Anna was born. Anna graduated from NSU, got her PhD degree. Now she works at the Institute of Philology SB RAS. In 2010 she got married, and in 2011 gave birth to her daughter Olya.

2.2 Studies at Novosibirsk State University

In 1960, I left school with honors and entered the Faculty of Mechanics and Mathematics of NSU. To do so I had to, like all other applicants, pass five exams: mathematics (both written and oral), physics, Russian language and literature (in the form of an essay) and a foreign language (German in my case).

It was a new university that has just been opened in 1959, at which point a number of new students was admitted as freshmen and some, those who transferred from other universities and constituted the majority of students, were admitted as sophomores. Thus I studied with the second wave of students. Among the first wave of students to graduate from NSU in January 1964 was Yu.L. Ershov.

NSU has arisen almost simultaneously with the Siberian Branch of the USSR Academy of Sciences, whose first chair was M.A. Lavrentiev. The university was built to train scientific personnel for the eastern part of the country and had been in a close cooperation with the Academy of Sciences. Both then and now most professors of the university conduct their scientific research at the institutes of SB RAS, while also teaching courses at NSU and supervising students and PhD-students.

Among people who gave lectures to us was a number of outstanding scientists, such as: S.L. Sobolev, A.I. Mal'tsev, A.M. Budker, A.I. Shirshov, M.I. Kargapolov, G.I. Marchuk, A.A. Lyapunov, A.P. Ershov, and many others.

When Siberian Branch of the Academy of Sciences was founded, almost all its employees came to Siberia from Moscow, Leningrad and other cities. 30 kilometers away from Novosibirsk, in forest, a new township called Akademgorodok was built, where both the university and various institutes of SB RAS have been situated.

Almost all students stayed in a dormitory. Like other students, I shared a room with three other girls.

Some students were from Novosibirsk, but most of them came from different parts of USSR.

At that time it was hard to get from Akademgorodok to the center of Novosibirsk. I visited my mother and sister every Saturday and went back on Sunday.

Starting with my fourth term I had attended all courses and special seminars, which dealt with logic: Algebra and Logic, Model Theory, Theory of Algorithms as well as lectures on the theory of algebraic systems and on iterative Post algebras taught by A.I. Maltsev.

In 1963, I had to choose my specialization and I chose the Algebra and Logic Department led by Professor Anatoly Ivanovich Maltsev. Then he asked me to read a small article by V. Donchenko on the calculus of strict implication which was formulated by W. Ackermann in 1956, and to make a talk on this paper at the "Algebra and Logic" seminar. After this talk Maltsev has formulated a research problem for my term thesis, which was to investigate the independence of Ackermann's list of axioms. On behalf of Anatoly Ivanovich this work was supervised by M.A. Taitslin, who has recently defended his PhD thesis. At that time nobody in Novosibirsk studied non-classical logics. Evidently, Anatoly Ivanovich was interested in algebraic approach to logical calculi, which was suggested by A. Tarski in his papers and developed in the book "Mathematics of Metamathematics" by H. Rasiowa and R. Sikorski, which was published in Warsaw in 1963.

I managed thoroughly to solve the problem: it turned out that two axioms of Ackermann's calculus were derivable from the others, while the rest were independent. These results were published in my first article in "Algebra and Logic" journal in 1964. In the beginning of my fifth year of studies I got a position at the Sobolev Institute of Mathematics.

2.3 After Graduating from NSU

My whole life is connected with both Sobolev Institute of Mathematics and Novosibirsk State University.

In Sobolev Institute of Mathematics I worked under supervision of Professor A.I. Maltsev, and Dr. N.V. Belyakin was overseeing my work as well.

In July 1967, A.I. Maltsev suddenly passed away. It was great blow for all of us. When I defended my PhD thesis, he was no longer with us.

I defended my PhD thesis at NSU in May 1968.

During a year (1972–1973) I had Postdoctoral studies under Professor Helena Rasiowa at Mathematical Institute of Warsaw University.

In 1986, I defended my habilitation thesis at Sobolev Institute of Mathematics.

In 1972, I received the title of Associate Professor, and in 1993 the title of Professor.

In 2009, the Russian Academy of Sciences has awarded me the A.I. Maltsev Prize.

As a member of a team of Novosibirsk logicians, in 2010 I became a winner of the Russian Federation Government Prize in Education.

2.4 Research

My main area of interest is Mathematical Logic, especially Algebraic Logic and Non-Classical Logics, including modal and superintuitionistic logics, relevance logics, many-valued logics, temporal and dynamic logics, and Theory of Algebraic Structures.

Here I give a brief summary of my main research results in chronological order. Most of these results were published in journals "Algebra i Logika", "Sibirskii

Matematicheskii Zhurnal" originally in Russian. English versions of these journals are published by Springer. Also there are publications in "Doklady Akademii Nauk SSSR" (English version is available too), in international journals "Studia Logica", "Journal of Symbolic Logic", "Logic Journal of the IGPL", "Annals of Pure and Applied Logic", "Journal of Applied Non-Classical Logic", "Theoretical Computer Science", and others.

At first, I studied relevance logics. In my Master's thesis (1965), I investigated Ackermann's system of rigorous implication and proved that two of its axioms are dependent and the others independent.

Since 1964 I work at Sobolev Institute of Mathematics, Siberian Branch of Russian Academy of Sciences, Novosibirsk.

When graduated from Novosibirsk State University (1965), I continued my investigations in relevance logics and proved a number of results for the system E of entailment, R of relevance and their fragments. In particular, a decidable relevant calculus was constructed, whose first degree fragment coincides with one of E and R. These results were collected in my Ph.D.Thesis "Logical calculi of rigorous implication" defended in Novosibirsk State University in 1968 under supervision of Prof. Anatolii I. Maltsev.

From September 1972 to June 1973 I had Postdoctoral studies under Professor Helena Rasiowa at Mathematical Institute of Warsaw University, was a participant of First Logical Semester (February–May 1973) at International Mathematical Banach Centre in Warsaw.

Relevant Logics. My first paper (1964) was in area of relevant logics. It investigated axiom system of Ackermanns calculus for rigorous implication. In my later research, I found an adequate algebraic semantics for relevant logics, adequate relational semantics for the logic E of entailment, and proved a number of separation theorems for relevant logics and representation theorems for structures with implication, found a number of principles of variable separation.

Pretabular Logics. At the same time I started a study of superintuitionistic logics. This area was much more popular than relevant logics.

In 1971 I had proved that there are exactly three pretabular superintuitionistic logics (SIL) having the finite model property (FMP). At the same time an abstract by A.V. Kuznetsov was published which stated that every pretabular SIL has the finite model property. Thus there are exactly three pretabular superintuitionistic logics. A logic is called tabular if it can be characterized by finitely many finite algebras; A pretabular logic is a maximal non-tabular logic. To obtain this result the duality theory between varieties of Heyting algebras and categories of partially ordered frames was developed.

It should be noted that there are fruitful inter-connections between the families of superintuitionistic logics and of normal extensions of the modal S4 logic found by M.Dummett and E. Lemmon, Maksimova and Rybakov (1974), L. Esakia, W. Blok and other authors and based on Gödel-McKinsey-Tarski translation of the intuitionistic logic Int into S4. This is a base of many parallels between the two families.

The number of pretabular normal extensions of the well-known modal S4 logic is equal to five, it was proved independently in my paper of 1975 and by L.Esakia and S.Meskhi (announced in 1974 and published in 1977).

As a corollary, the decidability of the tabularity problems for superintuitionistic logics and for extensions of S4 was obtained. In 1999 I proved in coauthorship with Voronkov that these problems are NP-complete.

Notice, for comparison, that W. Block proved the existence of a continuum of pretabular logics over the modal K4 logic.

Interpolation in Superintuitionistic Logics and in Extensions of S4. Many of my papers are devoted to the problem of interpolation in non-classical logics. Interpolation property is one of the most significant and interesting properties of logical theories. Classification of theories having the interpolation property has been the subject of research of many people, starting from 1957, when William Craig proved his interpolation theorem for the classical predicate logic. It was a difficult problem, how to prove or disprove interpolation even for one particular theory.

In my paper of 1977 the problem of interpolation in propositional superintuitionistic logics was solved. It was proved that in the continuum of superintuitionistic logics only eight logics possess the Craig interpolation property CIP.

When this result was announced, some people did not believe it. A bit earlier a paper was published where infinitely many superintuitionistic logics with CIP were found, but the proof turned out to be incorrect.

Each of the logics L with CIP is recognizable over the intuitionistic logic Int, i.e. there is an algorithm which, for any finite system Ax of axiom schemes, recognizes if the equality $\mathrm{Int} + Ax = L$ is true. As a corollary, the problem of interpolation is decidable for superintuitionistic calculi. Simultaneously the exhaustive list of amalgamable varieties of Heyting algebras was obtained, and their amalgamability problem was solved. This result was very surprising, and was cited in many papers and monographs on non-classical logics and universal algebra.

Also the problem of interpolation for extensions of the modal logic S4 was solved in a series of my papers of 1979–1982. In 1979 I proved that there are only finitely many normal logics above S4 which have the interpolation property, in 1982 all extensions of the Grzegorczyk logic with interpolation were described. As a corollary, the problem of interpolation for extensions of S4 and the amalgamability problem for varieties of closure algebras are decidable.

Again, these results disproved a paper which was to that time published in the Journal of Symbolic Logic.

The above-mentioned results on superintuitionistic logics and on extensions of the modal S4 logic were summarized in my habilitation thesis "Decidable properties of superintuitionistic and modal logics" (Sobolev Institute of Mathematics, 1986).

Later the complexity bounds for interpolation problems over Int and S4 were found. In particular, the interpolation problem is PSPACE-complete over Int and PSPACE-hard over S4.

In addition, the interpolation problems over both Int and S4 are strongly decidable (2000). It means that this property is recognizable even in the case where a logic is given by finitely many axiom schemes and rules of inference.

These results can not be extended to all modal logics. For example, in 1989 I had found a continuum of logics with interpolation above the modal provability logic.

Although the number of propositional superintuitionistic logics with the interpolation property is finite, it appeared that the family of superintuitionistic predicate logics with equality, which possess interpolation, has the continuum cardinality. In 1996 I proved this result by constructing an infinite independent sequence of axioms preserving interpolation and the Beth property. An analogous result was obtained for modal logics.

Algebraization. A general algebraic approach was worked out in my investigations. Duality theory for modal and Heyting algebras was developed. Algebraic equivalents of many important properties of logical theories were found. The methods were proposed that combine algebraic and semantical tools and allow to solve logical and algebraic problems at the same time.

It was established that CIP and other variants of interpolation in arbitrary modal logics are equivalent to appropriate variants of amalgamation in varieties of modal algebras. Moreover, studying these properties can be reduced to the study of finitely indecomposable algebras (1979, 1992).

Having in mind the equivalence of amalgamation and interpolation properties, I investigated some variants of the joint embedding property for varieties of modal and Heyting algebras and found their logical equivalents. They may be formulated in terms of various principles of variable separation and Hallden- reasonability (1995).

Interpolation and Beth Definability. Later various versions of interpolation and of the Beth property (related to interpolation) were considered for modal and temporal logics. A logic L has the Beth property, whenever every object implicitly definable in L is also explicitly definable. I proved that the strong interpolation is equivalent to the strong Beth definability but the weak variants of these properties are independent. Algebraic equivalents of the mentioned properties were found (1991). Some necessary conditions of interpolation in monomodal logics were obtained, it allows to disprove interpolation theorems in large families of modal logics. On the other hand, it was proved in my paper of 1992 that all modal logics satisfying the transitivity axiom have the Beth definability.

Failure of Interpolation in Temporal Logics. Once one of my colleagues told me that there is a paper which states the interpolation theorem for a temporal logic with linear time, and asked me if it is correct. I succeeded to find a counter-example. In fact, the temporal logics with discrete moments of time have neither interpolation nor the Beth property. It is valid for logics of linear time (1990) and of branching time (1991).

Projective Beth Property. The behavior of the projective Beth property, a stronger analog of Beth's Theorem on implicit definability, turned out to be similar to that of interpolation. It is proved that there are exactly 16 propositional logics extending the intuitionistic logic which possess the projective Beth property. These logics are fully described; they are recognizable over Int (2000–2001). Simultaneously, all varieties of Heyting algebras with Strong Epimorphism Surjectivity SES are found. It is stated that SES is base-decidable on the class of varieties of Heyting algebras.

It was established in 2011 that the restricted interpolation property IPR is equivalent over Int to the projective Beth property PBP. Thus all main versions of the interpolation property are decidable over Int.

Similar results are obtained for modal logics over S4 but the interpolation problem over S4 is not completely solved. It was proved in 2013 that there are only finitely many extensions of S4 with IPR or with PBP, and all of them are recognizable over S4. This implies decidability of IPR and PBP over S4. We have a finite list of logics which may have IPR or PBP. But not all logics of this list really have the property under examination, and for some logics of this list we do not know if they possess some interpolation property or not. Fortunately, all these problems are completely solved over the Grzegorczyk logic (1992–2010).

Tabular Logics. Applying my methods to the class of varieties of boolean algebras with operators generated by finitely many finite algebras, I proved that there exist uniform decision procedures for interpolation, various versions of the Beth property, Strong epimorphisms surjectivity, amalgamation and joint embedding problems for such varieties.

Positive Logics and Extensions of the Johansson Minimal Logic J (Since 2003). The picture of interpolation and definability in positive logics extending the positive fragment Int^+ of Int is quite similar to that in superintuitionistic logics. There are only finitely many positive logics with CIP, IPR or PBP, and all of them are recognizable over Int^+. All these properties are are decidable and PSPACE-complete over Int^+ (2003).

The situation in extensions of J is much more complicated. In 2011 the decidability of the weak interpolation property over J was obtained. Most of the results on interpolation and definability over Int are extended to so-called well-composed logics (2010–2015). But still we do not know whether the number of logics with CIP over J is finite or infinite. The same problem is open for IPR and PBP.

Conferences. Most of these results were presented at international logical and algebraic conferences. In 1966 I had a 20-minute talk at the International Congress of Mathematicians in Moscow. I participated in International Congresses of Logic, Methodology and Philosophy of Sciences (Bucharest 1971, Hannover 1979, Salzburg 1983, Moscow 1987, Uppsala 1991, Florence 1995, Krakow 1999, Beijing 2007), in many Logic Colloquia (Berlin 1989, University of Keele 1993, Clermon-Ferrand 1994, San Sebastian 1996. Prague 1998, Utrecht 1999, Paris 2000 and 2010, Vienna 2001, Münster 2002, Helsinki 2003, Athens 2005, Nijmegen 2006, Bern 2008, Sofia 2009), in Workshops Advancies in Modal Logic (Berlin 1996, Nancy 2008, Moscow 2010, Copenhagen 2012) and in many other conferences. During many years, we organise the annual Maltsev Conference in Novosibirsk.

I was an invited speaker of conferences in Moscow, Novosibirsk, Kishinev, Warsaw, Barcelona, Kanazawa, Varna, Clermont-Ferrand, Helsinki, Copenhagen, Kazan, had invited lectures and seminars in universities of Tokyo, Warsaw, Berlin, Vienna, Uppsala.

2.5 Teaching

Although I mainly keep research position as a principal researcher at Sobolev Institute of Mathematics, Siberian Branch of Russian academy of Sciences, I also have part-time teaching at Novosibirsk State University as Professor of Department of Algebra and Logic (Mathematical Faculty).

I started teaching in 1965, after graduating from Novosibirsk State University. First I had practical studies on logic and theory of algorithms in two student groups. It was very useful for me because I could learn this course in detail.

During many years I give lecture courses and seminars on Mathematical Logic and Recursion Theory for undergraduate students of the first and second years. This course is one of the main lecture courses at Mathematical Faculty, it takes two semesters: from February to December, and includes basic notions and theorems of Set Theory, Mathematical Logic and Recursion Theory. Other courses (so-called "special courses") correspond to graduate courses in European universities, but they are also attended by students at all levels. From 1972 I taught such lecture courses in modal and tense logics, temporal and dynamic logics, many-valued and other non-classical logics.

In 60-th and later years, nobody of my older colleagues in Novosibirsk worked in Non-Classical Logic. When I defended my PhD thesis, Professor A.D. Taimanov told me: Now you must develop your area. Since 1970 I organized a weekly seminar on Non-Standard Logics for students of all levels and other people interested in the subject, this seminar works until now.

Besides that, in 1981–1985 I taught courses in Algebra and Mathematical Analysis at the Faculty of Natural Sciences of Novosibirsk University. During several years, from 1993 I taught a course of modal logic for MSc students of Philosophical Faculty. From Spring 1997 I had given a half-year course of Logical Foundations of Programming for MSc students of Mathematical Faculty.

In 1993 I gave a tutorial on modal logics at Logic Colloquium '93 (Klermont-Ferrand, France), in 1996 a lecture course on interrelation of algebraic, semantical and logical properties of modal logics for participants of Mini-Semester "Logic, Algebra and Computer Science" at the International Banach Centre (Warsaw, Poland).

At Uppsala University I taught a graduate course "Modal, Intuitionistic and Temporal Logics" (1999).

Text-book on Logic. When I started teaching courses on Mathematical Logic, we had no russian text-books on this subject. There were only some translated books by A. Church, S. Kleene, E.Mendelson, and a few others. Once Prof. Igor Lavrov proposed to collect exercises which might be used by students for training in logic. This short collection was published by Novosibirsk State University:

I.A. Lavrov, L.L. Maksimova. Problems in Logic. Novosibirsk State University, 1970, 112 p.

This book contained main definitions and exercises on Set Theory and Mathematical Logic. Later we continued our work, and added exercises on Theory of Algorithms and also some solutions and comments. So we published the book

I.A. Lavrov, L.L. Maksimova. Problems on Set Theory, Mathematical Logic and Recursion Theory, which is widely used in teaching Mathematical Logic in our universities. It was published in 1975 (2nd edition - 1986, 3nd edition - 1995) by "Nauka", Moscow; 4th and 5th editions 2000 and 2001. This book is translated into Hungarian (Budapest 1988), English (New York 2002) and Polish (2004).

My Students. I supervised undergraduate, MSc and PhD students in Novosibirsk State University, practical studies in research and PhD students in Sobolev Institute of Mathematics.

Below is the list of persons those who obtained PhD under my supervision.

- I was a supervisor of MSc (1973) and PhD (1979) theses of *Vladimir Rybakov*, Professor, Doctor of Physical and Mathematical Sciences (1987). During many years he was Head of Department of Algebra and Logic of Krasnoyarsk State University. He is well-known with his investigations of admissible rules and unification in non-classical logics. Now he is Professor of Manchester Metropolian University.
- My PhD student *Margarita Verhosina* prepared her PhD thesis on separation problem of intermediate logics in 1983. She was a Docent of Irkutsk State University; died in 2014.
- Under my supervising *Sergei Mardaev* prepared his MSc (1984) and PhD (1987) theses on superintuitionistic logics satisfying various finiteness conditions. He defended Habilitation thesis in 2001. Sergei was a leading researcher of Sobolev Institute of Mathematics, he obtained interesting results on fixed points problem for modal logics. Died in 2013.
- My PhD student *Pavel Shreiner* graduated from Novosibirsk University with MSc degree in 1995. In 1998 he defended his PhD thesis "Interpolation and Definability in Superintuitionistic Logics of Finite Domains". He was a Docent of Siberian University of cooperation, moved to Industry.
- My PhD student *Dmitry Tishkovsky* also graduated from Novosibirsk State University with MSc degree in 1995. He defended his PhD thesis "Algebraization of Superintuitionistic Predicate Logics" in 1999. He is a researcher of the University of Manchester.
- My PhD student *Mikhail Sheremet* graduated from Novosibirsk State University with MSc degree in 1998. He was a student of Prof. Viktor Gorbunov who died in 1999. Then Mikhail finished the work on his PhD thesis devoted to quasivarieties in 2001 under my supervising. He is a senior researcher of Sobolev Institute of Mathematics, Novosibirsk.

- My PhD student *Veta Yun (Murzina)* graduated from Novosibirsk State University with MSc degree in 2000, defended her PhD thesis On modal logics of domains in 2003. She is a senior researcher of Sobolev Institute of Mathematics.
- My PhD student *Anastasia Karpenko* graduated from Novosibirsk State University with MSc degree in 2007, defended her PhD thesis on weakly transitive modal logics in 2010. She is a reader of Novosibirsk State University..

2.6 Other Activities

Since 1989 I am a member of the Dissertation Council for Awarding Scientific Degrees at Sobolev Institute of Mathematics, Novosibirsk. This Council takes into consideration PhD and habilitation theses on algebra and mathematical logic, and makes decisions on them.

I was an official referee and gave official reports on many PhD theses and habilitation dissertations in mathematical logic, among them theses of well-known russian logicians S. Artemov, L. Beklemishev, V. Shehtman, S. Odintsov and some of their PhD students, V. Orevkov, M. Zakharyaschev, and many others, F. Wolter (Free University of Berlin, 1993). I was a member of Commission for defense of PhD thesis of M. Veanes (Uppsala University, 1997). Now I cannot remember all the names.

On request of appointment boards, I had given official reports on applications for professorship of K. Bowen (Syracuse University, USA, 1974) and J.M. Dunn (Indiana University, USA, 1976), and some others.

I was a member of Honorary Consulting Board of International Journal "Studia Logica" (Poland) and a collecting editor of "Journal of Applied Non-Classical Logic" (France). I was a referee of a lot of mathematical papers submitted to Russian and international journals and conferences.

I translated (together with E. Palyutin) into Russian a well-known book by G. Sacks "Saturated Model Theory" (this translation was published in Moscow, 1976), the paper by A.Maltsev with his compactness theorem (originally published in German, 1936) for Collection of Maltsev's Selected Papers (Novosibirsk, 1979) and other papers.

Being a Professor of Department of Algebra and Logic of Novosibirsk State Universty, I was a chair of Subsection of Mathematical Logic at many International Scientific Student Conferences which have been held in Spring of every year at Novosibirsk State University. Since 1970, I organized weekly Seminar on Non-Standard Logics at Novosibirsk State University.

Since 1974 I was a Chair of Section of Non-Classical Logics at almost all Soviet conferences on Mathematical Logic (Novosibirsk 1974, 1979, 1984, Moscow 1986, Leningrad 1988) and at some International conferences on Mathematical Logic and on Algebra (Novosibirsk 1989, 1997, 1999). I was a member of Organizing and Program Committees of a number of algebraic and logical Russian and International conferences.

JAIST, Uppsala University, and King's College London. Together with Professor Hiroakira Ono we organized two Workshops on Non-Standard Logics and Logical Aspects of Computer Science (Kanazawa, Japan 1994, and Irkutsk, Russia 1995) as activities of Joint project on Non-Standard Logics of JAIST and Novosibirsk State University (1994–1996, coordinators H. Ono and L. Maksimova).

In 1995 Prof. Hiroakira Ono invited me to JAIST, this visit was supported by a grant of Japan Society for Promotion of Science. I was happy to stay in Japan from April to May. There I met many colleagues in JAIST, Tokyo and other cities.

From November 1998 to October 1999 I was a Visiting Professor at Computing Science Department, Uppsala University, supported by a grant of TFR (Swedish Research Council for Engineering Science). This visit was organized by Prof. Andrey Voronkov. It was an excellent time for research. At home, in Novosibirsk, I always had a lot of other duties.

In 2003, Prof. Dov Gabbay invited me to King's College, London. The purpose of the visit was preparing a book on interpolation. In fact, I planned to write such a book since 1990, and Dov supported this idea. But the work went too slowly. I needed files in Latex but most of my articles existed only in hard copies. At last Dov suggested his collaboration, and I was happy to accept it. It is worth to mention that Dov Gabbay had many results on interpolation in intuitionistic and modal logics, he created semantic methods for proving this property.

The joint work started during my three-weeks visits in March and July of 2003 and finished in March-August of 2004. The manuscript in the camera-ready form was prepared due to Jane Spurr. Of course, I am greatly thankful to Dov and Jane. Our book was published by Oxford University Press in 2005. The book contains several chapters written by Dov Gabbay and the most of my results on interpolation and definability published until 1992. We planned to prepare the second volume but, unfortunately, it was not done.

Grants. I received a research grant of American Mathematical Society for work in Mathematics (1993–1996). I was awarded with Stipend of Russian Federation for distinguished scientists (1994–1996, 1997–1999, 2000–2003).

I was a leader of Research Projects "Non-Standard Logics and Their Applications in Knowledge Representation" (1993–1995) and "Non-Standard Logics and Logical Aspects of Computer Science" (1996–1997) supported by grants of Russian State Committee for High Education. Since 1997 I was the leader of many research projects supported by Russian Foundation of Humanities or Russian Foundation for Basic Research, participated in international projects.

In 2005–2007 I participated in European INTAS project "Algebraic and deductive methods in non-classical logics and their applications in Computer Science" (led by Alberto Policriti). It was collaboration of four teams: from Italy (team leader – Alberto Policriti), from Poland (Ewa Orlowska), from Georgia (Leo Esakia) and from Russia (L. Maksimova).

In conclusion I would like to say that I am happy to be a member of Siberian School of Algebra and Logic based in Sobolev Institute of Mathematics and Novosibirsk State University. I am grateful to my teachers, my friends and colleagues in

Novosibirsk, Moscow, St-Petersburg and other cities of Russia, in Bulgaria, England, France, Germany, Italy, Japan, Poland, Sweden, USA and in many other countries.

I wish to thank the authors of the papers in this book, and I am indebted to Prof. Sergei Odintsov, the Editor of the volume, for his great efforts in preparing the book.

Chapter 3
Larisa Maksimova's Early Contributions to Relevance Logic

Katalin Bimbó and J. Michael Dunn

Abstract This paper presents an overview of the pioneering contributions of Larisa Maksimova to relevance logics. She is one of the first researchers who set out to methodically study systems of relevance logics, initially, focusing on Ackermann's Π' of "Rigorous Implication," and then extending her work to Anderson and Belnap's systems E of Entailment and R of Relevant Implication, and other related logics. Not only did she develop an algebraic semantics for E, but also we find that a semantic definition of entailment via a ternary accessibility relation appears — the very first time — in an abstract by her.

Introduction

Larisa Maksimova wrote her Ph.D. thesis on relevance logic. With the famous algebraist Anatoliĭ Ivanovich Mal'tsev as her advisor, it is no mystery that Maksimova should work on the algebra of logic, but what is more mysterious is how she would choose to write on the algebra of relevance logic (often called relevant logic). The answer seems to us to lie in the fact that she originally did not label what she was doing as connected with relevance logic, but rather with the work of Wilhelm Ackermann on his "strenge Implikation" (often translated into English as "rigorous implication," to avoid confusion with Lewis's "strict implication"). Ackermann's work was a strong influence on A. R. Anderson and N. D. Belnap, and their famous (1975) book *Entailment: The Logic of Relevance and Necessity*, vol. 1, was dedicated to his memory. By the time she wrote her dissertation, she was well aware of the work by Anderson and Belnap and lists seven of their works in her references (as opposed to just the one of Ackermann). But of course this leaves the question as to how she came to become interested in Ackermann's work.

K. Bimbó
Department of Philosophy, University of Alberta, 2–40 Assiniboia Hall,
Edmonton, AB T6G 2E7, Canada
e-mail: bimbo@ualberta.ca
URL: https://www.ualberta.ca/~bimbo

J. M. Dunn (✉)
Department of Philosophy and School of Informatics, Computing, and Engineering,
Indiana University, 901 East 10th Street, Bloomington, IN 47408–3912, USA
e-mail: dunn@indiana.edu

S. Odintsov (ed.), *Larisa Maksimova on Implication, Interpolation, and Definability*,
Outstanding Contributions to Logic 15, https://doi.org/10.1007/978-3-319-69917-2_3

We have learned from Larisa Maksimova that Anatoliĭ Mal'tsev asked her to review a paper by V. V. Donchenko (1963) devoted to Ackermann's calculus, which she did, and she presented this review at one of the seminars in the Sobolev Institute of Mathematics. After her talk Mal'tsev posed a question on the independence of axioms in Ackermann's calculus, which was not addressed in Donchenko's work. This question of Mal'tsev initiated Larisa's investigations of Ackermann's calculus.

Despite the obstacles to scholarly communication raised by the "Cold War," the journal *Algebra i Logika*, founded by Maksimova's teacher Mal'tsev, has been routinely translated into English and republished since 1968 as *Algebra and Logic*, first, by the Consultants Bureau, and then by other publishers.[1] Since this appears to have been Maksimova's main publication venue of choice, especially, early in her career, it let her work become well-known to "relevant" (pun intended) researchers in the West.

We will not attempt to summarize in this introduction her many achievements and contributions in the area of relevance logic. Even our paper by itself is too short for that. But we will mention her restricted theorem regarding the admissibility of Ackermann's rule (γ) for the system E, her algebraic semantics for the system E, and most importantly, a semantic definition of entailment via a ternary accessibility relation. This definition — later independently discovered or invented by others too, most notably by Routley and Meyer — is original with her.

3.1 From Syntax to Algebras

Logics are often formulated as axiomatic calculi. The main systems of relevance logics are no exception — initially, they were conceived as axiomatic systems. Maksimova started to investigate *syntactic properties* of calculi of rigorous implication in her (1964) paper.

A perennial question about axiomatic systems is the *independence of the axioms*. This is the topic of Maksimova (1964) with respect to Ackermann's calculus of rigorous implication. The axioms and rules of Π' are the following. (We use $p_1, p_2, p_3, \ldots$ for propositional variables, A, B, C, ... for formulas, $\sim$ for negation, $\wedge$ for conjunction, $\vee$ for disjunction and $\rightarrow$ for implication.)

(A1) $\mathrm{A} \rightarrow \mathrm{A}$
(A2) $(\mathrm{A} \rightarrow \mathrm{B}) \rightarrow ((\mathrm{B} \rightarrow \mathrm{C}) \rightarrow (\mathrm{A} \rightarrow \mathrm{C}))$
(A3) $(\mathrm{A} \rightarrow \mathrm{B}) \rightarrow ((\mathrm{C} \rightarrow \mathrm{A}) \rightarrow (\mathrm{C} \rightarrow \mathrm{B}))$
(A4) $(\mathrm{A} \rightarrow (\mathrm{A} \rightarrow \mathrm{B})) \rightarrow (\mathrm{A} \rightarrow \mathrm{B})$
(A5) $(\mathrm{A} \wedge \mathrm{B}) \rightarrow \mathrm{A}$
(A6) $(\mathrm{A} \wedge \mathrm{B}) \rightarrow \mathrm{B}$
(A7) $((\mathrm{A} \rightarrow \mathrm{B}) \wedge (\mathrm{A} \rightarrow \mathrm{C})) \rightarrow (\mathrm{A} \rightarrow (\mathrm{B} \wedge \mathrm{C}))$
(A8) $\mathrm{A} \rightarrow (\mathrm{A} \vee \mathrm{B})$

[1] See Hollings (2014).

(A9) $B \to (A \vee B)$
(A10) $((A \to C) \wedge (B \to C)) \to ((A \vee B) \to C)$
(A11) $(A \wedge (B \vee C)) \to ((A \wedge B) \vee C)$
(A12) $(A \to B) \to ({\sim}B \to {\sim}A)$
(A13) $(A \wedge B) \to {\sim}(A \to B)$
(A14) $A \to {\sim}{\sim}A$
(A15) ${\sim}{\sim}A \to A$
(α) A and $A \to B$ imply B
(β) A and B imply $A \wedge B$
(γ) A and ${\sim}A \vee B$ imply B
(δ) $A \to (B \to C)$ and $\vdash B$ imply $A \to C$

The last rule slightly differs from the three other rules in Π', because B, as indicated by the turnstile, has to be a theorem.

Maksimova shows that two of the axioms, (A1) and (A3) are provable from the rest of the system. It is easy to prove that $A \to A$ is a theorem of this calculus with the double negation axioms and a rule of transitivity which is derivable in the system. From $A \to {\sim}{\sim}A$ and ${\sim}{\sim}A \to A$, $A \to A$ follows in one step.

The proof that the principal type schema of B (i.e., the axiom called prefixing) is a theorem is more complicated and lengthier. The proof given by her relies on all the negation axioms, as well as on suffixing and Ackermann's (α) and (δ) rules.

The second part of the paper utilizes *many-valued matrices* to prove that the other axioms are independent. The only exception is the double negation elimination axiom (i.e., ${\sim}{\sim}A \to A$), the independence of which is a straightforward consequence of the other axioms being theorems of intuitionistic logic — under an obvious translation that matches connectives denoted by similar symbols. The 12 finite algebras that are created in this process are interesting, because they are all models of proper sublogics of Π'.

For example, the contraction axiom does not take a distinguished value in the following matrix, when $v(A) = 2$ and $v(B) = 1$. The matrix is $\mathfrak{M}_4 = \langle\{1, 2, 3, 4\}; \{3, 4\}, {\sim}, \wedge, \vee, \to\rangle$, where the operations are defined by (1)–(4). (The elements of the algebra are four positive integers, with their usual ordering.) We also include the result of the calculations for the contraction axiom starting with the values for A and B as above.

(1) ${\sim}a = 5 - a$
(2) $a \wedge b = \min(a, b)$
(3) $a \vee b = \max(a, b)$

(4)

$\to$	1	2	3	4
1	3	3	3	3
2	2	3	3	3
3	1	2	3	3
4	1	1	2	3

$$\begin{array}{ccccccccc} (A & \to & A & \to & B) & \to & (A & \to & B) \\ 2 & 3 & 2 & 2 & 1 & 2 & 2 & 2 & 1 \end{array}$$

The lion's share of the work — beyond inventing the matrices — is to verify that the other axioms take a distinguished value under any assignment of the values, and that the four rules lead from formulas with distinguished values to formulas with distinguished values.

In 1964, it was not known that the rule (γ) is admissible in E. Furthermore, E has been traditionally axiomatized by Anderson and Belnap via omitting not only the rule (γ), but also the rule (δ) in lieu of several axioms such as specialized assertion $((\mathrm{A} \to \mathrm{A}) \to \mathrm{B}) \to \mathrm{B}$, and necessity encasement $(\Box\mathrm{A} \wedge \Box\mathrm{B}) \to \Box(\mathrm{A} \wedge \mathrm{B})$, where $\Box\mathrm{A}$ is $(\mathrm{A} \to \mathrm{A}) \to \mathrm{A}$. (The negation axioms are often varied too.[2]) Thus, if we take into consideration the admissibility of (γ) (which was proved in Meyer and Dunn 1969), Masimova's result showing the independence of the axioms (when (A1) and (A3) are excluded) is the first proof that an *axiomatization of* E comprises independent axioms. A similar result for the (δ)-free formulation of E has not been obtained until some 10 years later by Chidgey.[3]

In her next paper of (1966), Maksimova continued to investigate Ackermann's Π' together with E, Anderson and Belnap's calculus of entailment. She immediately noted that Π' and E do not possess a deduction theorem in the usual sense. Then, she went on to formulate suitable versions of the deduction theorem that can be proved to hold in Π' and E, respectively.

The two core lemmas that facilitate the proofs of the deduction theorems reveal the role of shared propositional variables in theorems.

Definition 3.1.1 Let $p_1, \ldots, p_n$ be propositional variables. Then P_n is the formula $(p_1 \to p_1) \wedge \ldots \wedge (p_n \to p_n)$.

The conjunction connective is associative (and commutative) both in Π' and E, hence, we omitted some parentheses (as usual).

Lemma 3.1.2 *If the propositional variables occurring in* A *are* $p_1, \ldots, p_n$*, then* $\mathrm{P}_n \to (\mathrm{A} \to \mathrm{A})$ *is a theorem of* Π' *(of* E*).*

The lemma is proved by induction on the length of the formula A.

Both Π' and E contain the (δ) rule (as a primitive or as an admissible rule), hence, it is immediate that if A is a theorem, then $\mathrm{P}_n \to \mathrm{A}$ is a theorem too.

Lemma 3.1.3 *If* $p_1, \ldots, p_n$ *are all the propositional variables occurring in* A *and* B*, then* $\mathrm{P}_n \to (\mathrm{A} \to \mathrm{B})$ *follows from* $\mathrm{A} \to \mathrm{B}$.

The proof of this lemma relies on the previous lemma together with instances of the suffixing axiom, which are chained into a proof by applications of the rule (α).

Theorem 3.1.4 (Deduction theorem for E) *Let* Γ *comprise implicational formulas, and let* $p_1, \ldots, p_n$ *be all the propositional variables occurring in formulas in* Γ, A *and* B*. If* $\Gamma, \mathrm{A} \vdash_E \mathrm{B}$*, then* $\Gamma \vdash_E (\mathrm{A} \wedge \mathrm{P}_n) \to \mathrm{B}$.

A deduction theorem for $E_{\to}$, the *implicational fragment* of E, was formulated and proved by Anderson and Belnap (1975, Sect. 4.1). However, that theorem does not extend in a straightforward step to the whole of E, because of the presence of

[2]See, e.g., (Anderson and Belnap (1975), Sect. 21.1) and (Anderson et al. (1992), Sect. R2)

[3]See Anderson and Belnap (1975, Sect. 26.2 and 29.2).

the extensional connectives ($\wedge$ and $\vee$). Indeed, they do not seem to have formulated a version of the deduction theorem for E ever. Maksimova restricts the shape of the premises (with respect to their main connective), and adds a new formula P_n into the implication that absorbs a premise. P_n may be thought of as a concrete substitute for the *intensional truth constant*, which is often denoted by $\boldsymbol{t}$. At about the same time, Dunn in his thesis (1966) defined $\boldsymbol{t}$ for the n-variable fragment of R in the same way as Maksimova defined P_n.[4] Later on, Meyer et al. (1974) formulated a deduction theorem for R that had $\boldsymbol{t}$ in it, hence, they called the theorem *enthymematic deduction theorem*. It says that if $\Gamma, \mathrm{A} \vdash_R \mathrm{B}$, then $\Gamma \vdash_R (\mathrm{A} \wedge \boldsymbol{t}) \to \mathrm{B}$. This is a slightly more general theorem than Maksimova's, because the elements of Γ are arbitrary formulas of R (and it is for R, not for E).

Theorem 3.1.5 (Deduction theorem for Π') *Let Γ comprise implicational formulas, and let $p_1, \ldots, p_n$ be all the propositional variables occurring in formulas in Γ,* A *and* B. *If $\Gamma, \mathrm{A} \vdash_{\Pi'} \mathrm{B}$, then $\Gamma \vdash_{\Pi'} (\mathrm{A} \wedge \mathrm{P}_n) \to ({\sim}\,\mathrm{P}_n \vee \mathrm{B})$.*

The proof of both theorems has a similar structure to certain proofs of the deduction theorem in classical (or intuitionistic) logic, namely, the induction proceeds on the length of the proof.

The difference between Π' and E is highlighted by how a defined material implication behaves with respect to deducibility. Let $\supset$ be defined by $\sim$ and $\vee$, that is, $\mathrm{A} \supset \mathrm{B}$ is an abbreviation for ${\sim}\mathrm{A} \vee \mathrm{B}$. Then, Maksimova shows that in Π', $\Gamma, \mathrm{A} \vdash_{\Pi'} \mathrm{B}$ iff $\Gamma \vdash_{\Pi'} \mathrm{A} \supset \mathrm{B}$, where Γ is an arbitrary set of formulas of Π'. On the other hand, $\vdash_E {\sim}(\mathrm{A} \wedge ({\sim}\mathrm{A} \vee \mathrm{B})) \vee \mathrm{B}$ (i.e., $\vdash_E (\mathrm{A} \wedge ({\sim}\mathrm{A} \vee \mathrm{B})) \supset \mathrm{B}$), but $\mathrm{A} \wedge ({\sim}\mathrm{A} \vee \mathrm{B}) \nvdash_E \mathrm{B}$.

3.2 Algebraic Semantics for *E*

After studying some essential syntactic features of an axiomatic calculus (such as the replacement property), the next straightforward step is to investigate the *Lindenbaum algebra* and *algebraic models* of the logic.

Ackermann showed that Π' theorems cannot have a certain peculiar shape. Namely, there is no theorem of Π' having the form $p \to (\mathrm{A} \to \mathrm{B})$. Maksimova (1967a) extended this result to other classes of formulas using algebraic models. We will not include details of her findings here, because they are included in Anderson and Belnap (1975, Sect. 22.1.1).[5]

---

[4]This portion of Dunn (1966) was published as Sect. 28.2.2 in the 1st volume of *Entailment*, Anderson and Belnap (1975).

[5]We should point out that Maksimova (1967a) is missing from the references in Anderson and Belnap (1975) and the content of the mentioned section is attributed to Maksimova (1967b), which does not contain those results or their proofs — as should become obvious from our account of the content of the latter paper.

Maksimova in her (1967b) paper carried out a thorough investigation of *algebraic models of* E, and also of models of some related logics. The latter are EI, which is E without the double negation elimination axiom, E_+, which is the negation-free fragment, $E_\to$, which is the implicational fragment, and $E_{\underset{\to}{\sim}}$, which is the negation–implication fragment. In this paper, she swapped the earlier axiomatization which contained the (δ) rule for one in which the only rules are modus ponens (α) and adjunction (β), and certain axioms are added, in particular, specialized assertion and the encasement of necessity.

Definition 3.2.1 *(E models)* Let $\mathfrak{A} = \langle A;\, D, \wedge, \vee, \sim, \to\rangle$ be an algebra with D, a distinguished subset of elements. $\mathfrak{A}$ is an E *model* when every homomorphism from the set of formulas of E into A evaluates every theorem of E into an element of D.

The notion of an EI model is similar, except that theorems of EI should take a distinguished value. The models for the three other fragments also differ in that the operation that corresponds to a connective that is omitted from the logic is omitted from the algebra.

Lemma 3.2.2 *In every E (or EI) model* $\mathfrak{A} = \langle A;\, D, \wedge, \vee, \sim, \to\rangle$, *the reduct* $\mathfrak{A}' = \langle A;\, \wedge, \vee\rangle$ *is a distributive lattice. Similarly, in any* E_+ *model* $\mathfrak{A} = \langle A;\, D, \wedge, \vee, \to\rangle$, *the reduct* $\mathfrak{A}' = \langle A;\, \wedge, \vee\rangle$ *is a distributive lattice.*

Given the lemma, it is natural to ask whether any of these *logics is prime*, that is, the set of its theorems is a prime filter on the Lindenbaum algebra of the logic. (We parenthesize the operation omitted in the parenthesized sort of model.)

Theorem 3.2.3 *Every* EI (E_+) *model* $\mathfrak{A} = \langle A;\, D, \wedge, \vee, (\sim,) \to\rangle$ *is a homomorphic image of an* EI (E_+) *model* $\mathfrak{B} = \langle B;\, P, \wedge, \vee, (\sim,) \to\rangle$, *in which* P *is a prime filter.*

Given this theorem the next claim is practically obvious.

Theorem 3.2.4 (Disjunction property for EI and E_+) *If* $\vdash_{EI} \mathrm{A} \vee \mathrm{B}$, *then* $\vdash_{EI} \mathrm{A}$ *or* $\vdash_{EI} \mathrm{B}$; *similarly, for* E_+, *if* $\vdash_{E_+} \mathrm{A} \vee \mathrm{B}$, *then* $\vdash_{E_+} \mathrm{A}$ *or* $\vdash_{E_+} \mathrm{B}$.

E, the whole logic of entailment is not prime, hence, Maksimova proves a *restricted disjunction property* for E.

Definition 3.2.5 A filter F is *i-prime* (or prime for implications), if $(a_1 \to b_1) \vee \ldots \vee (a_n \to b_n) \in F$ (where $n \in \mathbb{N}$) implies $a_i \to b_i \in F$, for some i (where $1 \leq i \leq n$).

The definition clearly excludes a formula of the form $p \vee \sim p$, which would be worrisome together with the disjunction property. (Incidentally, $p \vee \sim p$ is a theorem of E.)

Theorem 3.2.6 *Every E model* $\mathfrak{A} = \langle A;\, D, \wedge, \vee, \sim, \to\rangle$ *is a homomorphic image of an E model* $\mathfrak{B} = \langle B;\, P, \wedge, \vee, \sim, \to\rangle$, *in which* P *is an i-prime filter.*

Theorem 3.2.7 (Admissibility of γ in E for implicational formulas) *If* $\vdash_E \sim A \vee B$ *and* $\vdash_E A$, *where* B *is* $(C_1 \to C_2) \vee \ldots \vee (C_n \to C_{n+1})$, *then* $\vdash_E B$.

The first proof of the admissibility of the rule (γ) in E (as well as in R, RM and EM) is in Meyer and Dunn (1969). Their proof is algebraic too, however, it does not impose constraints on the shape of B (or A). The rule (γ) gained a lot of notoriety since it was dropped from Π'. There is a whole section, Sect. 25 in Anderson and Belnap (1975), which is devoted to this rule and two proofs of its admissibility; the second is also by Meyer and Dunn, but differs from the first (which is their earlier proof). We thought that none of the authors or contributors of (the first volume of) *Entailment* was aware of the proof of the admissibility of (γ) in E for a special subclass of formulas in Maksimova's (1967b) — until we recently discovered a review of her paper by Anderson from (1971).

In the remainder of her paper, Maksimova is concerned with establishing that $E_{\overset{\sim}{\to}}$ is a *conservative extension* of $E_{\to}$, and EI is a *conservative extension* of E_{+}.

The semantical interpretations of modal logics, obviously and inevitably, influenced how logicians thought about a semantics for relevance logics. The connections between topological interpretations and the modal logic $S4$ may suggest that a closure or an interior operator could be used in the interpretation of other connectives beyond $\Diamond$ or $\Box$. Maksimova in her (1968b) uses an *interior operator* in algebraic models of E.

The problem of the decidability of E was at the top of Anderson's list in Anderson (1963), and decidability was an open problem for other relevance logics such as R and T too. Maksimova introduced SE, a *new calculus* of rigorous implication, which is a subsystem of E formulated with the constant $\boldsymbol{t}$ in the language, and which she proved decidable.[6] The axioms and rules of SE are as follows. (We reuse some of the axioms and rules that we listed in Sect. 3.1.) SE contains (A4)–(A12), (A14)–(A15), and the rules (α) and (β) from Π', together with axioms (A16)–(A18) and rules (R3)–(R5). ($\boldsymbol{t}$ is intensional truth, as we already mentioned above, and it is in the language of SE.)

(A16) $((A \to B) \wedge (B \to C)) \to (A \to C)$
(A17) $(A \to \sim A) \to \sim A$
(A18) $(\boldsymbol{t} \to A) \to A$
(R3) $A \to B$ implies $(B \to C) \to (A \to C)$
(R4) $A \to B$ implies $(C \to A) \to (C \to B)$
(R5) A implies $\boldsymbol{t} \to A$

[6] She seems to use the term "rigorous implication" not only to denote Ackermann's Π' calculus, but other closely related calculi too, each of which avoids the paradoxes of material implication. The "S" in "SE" may stand for "strenge" or for "(sub)system."

Maksimova stresses that (A → B) → ((B → C) → (A → C)) is *not a theorem of* SE, and she proves this later in her paper. However, first, she establishes some properties of SE and its relationship to E.

Lemma 3.2.8 *If* $\vdash_{SE}$ A *and the propositional variables occurring in* A *are* $p_1, \ldots, p_n$, *then* $\vdash_E \mathrm{A}^{\mathrm{P}_n}$, *where* $\mathrm{A}^{\mathrm{P}_n}$ *is the formula* A *except that every occurrence of* $\boldsymbol{t}$ *in* A *is replaced by* P_n *(from Definition 3.1.1).*

In the proof of this lemma, the $\boldsymbol{t}$ in (A18) and (R5) is replaced by P_n, which is a suitable definition of $\boldsymbol{t}$ in the n-variable fragment of the logic. Then the proof is a straightforward induction on the proof of A in SE.

Another feature that Maksimova considers to be important for her calculus of rigorous implication SE is that it coincides with E on first-degree entailments. The logic of first-degree entailments is often denoted by **fde**, and it was first isolated from E. However, R and T are also extensions of **fde**, which makes the **e** in the acronym (and the "entailment" in the name) of this logic slightly misleading.

Definition 3.2.9 *(First-degree entailments)* The formula A → B is a *first-degree entailment* if A and B are →-free formulas.

Maksimova proves that if A is provable in the calculus for **fde** in Belnap (1967), then $\vdash_{SE}$ A.

We conjecture that the third property that Maksimova considered to be important for a logic of rigorous implication is its *decidability*, and in fact, it appears to have motivated the formulation of the calculus SE. Taking a quick informal look at SE, we see a logic in which entailment has few properties. The central remaining property of → is *contraction*, which is surprising, because the presence of the contraction axiom in a logic is often associated with difficulties in proving decidability, in particular, in proving decidability through constructing a finite proof-search tree using a sequent calculus. Axiom (A16) is a version of transitivity for entailment, and (R3) and (R4) are rule forms of (A2) and (A3), respectively, from Π' or E. These rules are necessary for the logic to turn out to be structural. (A17) is a theorem of Π' and E, but it is not provable in SE from the rest of axioms and rules.

In the second part of her (1968b) paper, Maksimova, essentially, proves the *finite-model property for* SE, though she does not use this term. First, the notion of a *normal* SE *model* is defined as a certain algebra $\mathfrak{A} = \langle A; \boldsymbol{t}, \wedge, \vee, \sim, \rightarrow \rangle$.[7] The similarity type of $\mathfrak{A}$ is $\langle 0, 2, 2, 1, 2 \rangle$, and it is a model when $\boldsymbol{t} \wedge \mathrm{A} = \boldsymbol{t}$ in $\mathfrak{A}$ (with the propositional variables of the theorem A interpreted as elements of A). (In effect, a single distinguished element of A replaces what was D in some earlier definitions.) Normality means that for any $a, b \in A, a = b$ iff $\boldsymbol{t} \wedge (a \rightarrow b) \wedge (b \rightarrow a) = \boldsymbol{t}$. Then the following lemma is proved.

Lemma 3.2.10 *If* $\mathfrak{A}$ *is a normal* SE *model, then* $\langle A; \wedge, \vee, \sim \rangle$ *is a De Morgan lattice, and* $\boldsymbol{t} \leq a \rightarrow b$ *just in case* $a \leq b$.

[7]Maksimova is careful to distinguish between the components of the language of the logic SE and the particles of a model — as she did earlier too. We continue to re-use the same symbols for the connectives in order to shorten our presentation.

A full characterization of normal SE models is obtained by combining the previous two properties with a series of quasi-inequations, which are directly obtainable from the axioms and rules of SE by replacing the main connective by $\leq$.

Lemma 3.2.11 *Any normal SE model $\mathfrak{A} = \langle A; \boldsymbol{t}, \wedge, \vee, \sim, \rightarrow \rangle$ can be embedded into a normal SE model $\mathfrak{A}' = \langle A'; \boldsymbol{t}, \wedge', \vee', \sim', \rightarrow' \rangle$, in which $A' = A \cup \{\top, \bot\}$ (where $\top, \bot \notin A$) and $\langle A'; \wedge', \vee', \sim' \rangle$ is a bounded De Morgan lattice.*

$\top$ and $\bot$ are added to the De Morgan lattice reduct of $\mathfrak{A}$ as expected, that is, $\top$ is the greatest and $\bot$ is the least element in A'. We recall the definition of $\rightarrow'$ though.

$$a \rightarrow' b = \begin{cases} a \rightarrow b & \text{if } a, b \in A, \\ \top & \text{if } a = \bot \text{ or } b = \top, \\ \bot & \text{if either } b = \bot \text{ and } a \neq \bot, \text{ or } a = \top \text{ and } b \neq \top. \end{cases}$$

The notions of a closure and an interior operator originated in topology, but these concepts had been placed into a more abstract setting in universal algebra and they have been used in a range of applications.[8]

Definition 3.2.12 Let $\mathfrak{A} = \langle A; \wedge, \vee, \sim, \top, \bot \rangle$ be a bounded De Morgan lattice. $\mathfrak{I}$ is an *interior operator* on A when it satisfies (1)–(4).

(1) $\mathfrak{I}\top = \top$
(2) $\mathfrak{I}(a \wedge b) = \mathfrak{I}a \wedge \mathfrak{I}b$
(3) $\mathfrak{I}a \leq a$
(4) $\mathfrak{I}\mathfrak{I}a = \mathfrak{I}a$

An element $a \in A$ is *open* if $\mathfrak{I}a = a$.

The next lemma is a core component in the decidability proof for SE.

Lemma 3.2.13 *Let $\mathfrak{A} = \langle A; \boldsymbol{t}, \wedge, \vee, \sim, \rightarrow, \top, \bot \rangle$ be a normal SE model with an interior operator $\mathfrak{I}$ such that $\mathfrak{I}$ satisfies* (5) *and* (6).

(5) $\mathfrak{I}\boldsymbol{t} = \boldsymbol{t}$
(6) $\sim \mathfrak{I}a = \mathfrak{I} \sim \mathfrak{I}a$

Then the algebra $\mathfrak{B} = \langle B; \boldsymbol{t}, \wedge, \vee, \sim, \rightarrow_{\mathfrak{B}} \rangle$, where $B = \{a \colon a \in A \;\&\; a \leq \mathfrak{I}a\}$ and $a \rightarrow_{\mathfrak{B}} b = \mathfrak{I}(a \rightarrow b)$ is a normal SE model.

The use of the interior operator reflects a remarkable insight about the nature of $\rightarrow$. In a modal logic context, $\mathfrak{I}$ is easily linked to $\Box$, that is, to necessity. On the other hand, the motivations and explanations surrounding E often involve necessity. Not only the title of Anderson and Belnap (1975) and Anderson et al. (1992) talk about the *logic of necessity*, but the connection between $\rightarrow$ and $\Box$ is programmatic.

[8] Some earlier work of Maksimova (e.g., her 1967c) concerned connections between quasi-ordered sets and closure operators.

> The foregoing seems to us to be a strong and natural list of valid entailments, all of which are *necessarily* true, and in each of which the antecedent is *relevant* to the consequent. ... as regards necessity, we observe here that, as might be expected, a theory of *logical necessity* is forthcoming in $\mathbf{E}_{\rightarrow}$. — Anderson and Belnap (1975, Sect. 4.3)

The elements of $\mathfrak{B}$ are the *open elements* from $\mathfrak{A}$. The conditions (5) and (6) ensure that $\boldsymbol{t}$ is open, and that the negation of an open element is open. (The set of opens is closed under meet and join then.) Thus the crux is the definition of $\rightarrow_{\mathfrak{B}}$, which ensures that an implication is necessary, so to speak.

The next step toward decidability is to guarantee *finiteness*. First, if M is a finite subset of the carrier set A in a normal SE model, then there is a normal SE model $\mathfrak{B} = \langle B; \boldsymbol{t}, \wedge', \vee', {\sim}', \rightarrow' \rangle$, where $M \subseteq B$, and the primed operations are extensions of their counterpart in $\mathfrak{A}$ on the subset $M \cup \{\, \boldsymbol{t} \,\}$. B is taken to be the set of elements in $\mathfrak{A}$ generated by $\wedge$ and $\vee$ from $M \cup \{\, {\sim}a \colon a \in M \,\} \cup \{\, \top, \bot, \boldsymbol{t}, {\sim}\boldsymbol{t} \,\}$. If M is finite, then so is B, because a De Morgan lattice is distributive. An interior operation $\mathfrak{I}$ on $\mathfrak{B}$ is defined by $\mathfrak{I}a = \inf\{\, b \colon b \in B \;\&\; b \leq a \,\}$. $\mathfrak{I}$ is well-defined and has its requisite properties. Thus, in $\mathfrak{B}$, $\rightarrow$ is definable as the interior of $\rightarrow$ in $\mathfrak{A}$, whereas the other operations are restrictions of their matching pair in $\mathfrak{A}$.

The accumulated results lead to a proof of the next theorem.

Theorem 3.2.14 (Decidability of SE) *If* $\nvdash_{SE}$ A, *then there is a finite normal* SE *model* $\mathfrak{B}$ *such that* $\boldsymbol{t} \wedge \mathrm{A} \neq \boldsymbol{t}$. *The number of elements of* B *is* $\leq 2^{2^{2r+4}}$, *where* r *is the number of subformulas of* A.

Finally, Maksimova shows — and this kind of "explains" the choice of SE — that **4** (the four-element Boolean algebra, which is also a De Morgan lattice) refutes suffixing, when one of the atoms plays the role of $\boldsymbol{t}$. Implication is defined as follows.

$$a \rightarrow b = \begin{cases} ({\sim}a \wedge b) \vee \boldsymbol{t} & \text{if } a \leq b, \\ ({\sim}a \wedge b) \wedge {\sim}\boldsymbol{t} & \text{if } a \nleq b. \end{cases}$$

Let a valuation assign $\bot$ to A, $\top$ to B and $\boldsymbol{t}$ to C. The values calculated for all the subformulas are as shown below.

$$\begin{array}{ccccccccccc} (\mathrm{A} & \rightarrow & \mathrm{B}\,) & \rightarrow & ((\mathrm{B} & \rightarrow & \mathrm{C}\,) & \rightarrow & (\mathrm{A} & \rightarrow & \mathrm{C})) \\ \bot & \top & \top & \bot & \top & \bot & \boldsymbol{t} & \boldsymbol{t} & \bot & \boldsymbol{t} & \boldsymbol{t} \end{array}$$

Maksimova defended her Ph.D. thesis (a so-called candidacy thesis) at the Siberian Division of the Academy of Sciences of the USSR in Novosibirsk, in 1968. Her thesis, (1968a) is based on — though written anew — the sequence of her papers (1964), (1966), (1967a), (1967b) and (1968b). As we tried to intimate, the papers constitute "logical steps" in a research program culminating in the proof of the decidability of a calculus of rigorous implication, with the proof using, essentially, algebraic techniques. We should note also that although, in the 1960s, there were several people in Pittsburgh — including Dunn — who hoped for a quick resolution of the decidability problem of E and R, it took another 26 years or so before a proof of the undecidability of these calculi was published.[9] This makes even more remarkable that Maksimova carved out a part of E (as her SE calculus), which retains both contraction and the distributivity of conjunction and disjunction, that she was able to prove decidable.

In 1969, Maksimova published a short, but truly remarkable abstract. She makes a move from algebraic models of E (SE and R) to set-theoretic representations with topologies. In fact, in this abstract, she states the following isomorphic representation theorem for SE, E and R models.

Theorem 3.2.15 *Let* $\mathfrak{A} = \langle A; D, \wedge, \vee, \sim, \rightarrow \rangle$ *be an* E *(*R *or* SE*) model. Then there is a compact topological space* $\mathcal{S} = \langle S; \leq, g, \tau \rangle$*, where* $\leq$ *is a partial order,* g *is an involution and* τ *is a ternary relation, such that the algebra of clopen sets is isomorphic to* $\mathfrak{A}$*.*

The abstract gives the definition of the four operations in the algebra of clopen sets, however, for the result to hold, $\mathcal{S}$ must have further properties. Maksimova states that "we provide axioms that characterize the classes of $\langle S; \leq, g, \tau \rangle$ systems on which the algebras of clopen sets are E (R or SE) models." We describe below a potential set of conditions (alias "axioms") and some small additions to the claim that yield the stated result. (The set of conditions is not unique — exactly, as a logic may be axiomatized by different sets of axioms, hence, we cannot claim that Maksimova meant exactly this set of conditions.) It is completely plausible to assume that the postulates were left out of the abstract due to a limitation on its length.

Before we undertake the reconstruction, we have to point out also that a topological representation of distributive lattices existed for some 30 years by then; indeed, Maksimova (1967b) used the representation from Stone (1936). The idea of adding a partial order to the topological space is well-suited for semantics of logics, because a partial order relation naturally emerges in the algebra of a logic on sets of filters, ideals, cones, etc. A representation of distributive lattices using *ordered topological spaces* was published a year later in Priestley (1970).

The modeling of negation from a function g that is an involution has become customary in relevance logic, though g is usually denoted by *, and sometimes is called the Routley star (with reference to Routley and Meyer 1972). While the use of an involution for this purpose might not be original with Maksimova (as it is not original with the Routleys either), she seems to have found the application of g in a semantics for relevance logic earlier than the Routleys did.[10]

In sum, this abstract is groundbreaking with respect to giving the first definition of $\rightarrow$ from a ternary relation, and also in using a partially ordered topological space in a representation of a distributive lattice.

---

[9] See Urquhart (1984), Anderson et al. (1992, Sect. 65) and Urquhart (2007).

[10] See Dunn (1986, Sect. 4.3–4.4) for more on De Morgan lattices and their representation including the fact that Białynicki-Birula and Rasiowa (1957) first originated the use of an involution in the representation of De Morgan lattices. The latter are the algebraic counterparts of first-degree entailments. (Cf. Lemma 3.2.10.).

We turn to filling in a few details now. For the sake of simplicity (which is a relative notion, as the details will make it plain), we consider only R models.

Definition 3.2.16 *(R models)* Let $\mathfrak{A} = \langle A; D, \wedge, \vee, \sim, \rightarrow\rangle$ be an algebra of similarity type $\langle 2, 2, 1, 2\rangle$, where $D \subseteq A$ (i.e., D is a distinguished subset of elements) and D is a filter. $\mathfrak{A}$ is an *R model* if it satisfies (1)–(8). ($d, d_1, d_2 \in D$.)

(1) $\langle A; \wedge, \vee, \sim\rangle$ is a De Morgan lattice;
(2) $a \rightarrow (b \rightarrow c) \leq b \rightarrow (a \rightarrow c)$;
(3) $a \rightarrow b \leq (c \rightarrow a) \rightarrow (c \rightarrow b)$;
(4) $a \rightarrow (a \rightarrow b) \leq a \rightarrow b$;
(5) $a \rightarrow (b \rightarrow c) \leq a \rightarrow ({\sim}c \rightarrow {\sim}b)$;
(6) $d \leq a \rightarrow b$ implies $a \leq b$;
(7) $d_1 \leq a \rightarrow b$ and $d_2 \leq a$ imply $d_1 \wedge d_2 \leq b$;
(8) $d_1 \leq a \rightarrow b$ and $d_2 \leq b \rightarrow a$ imply $a = b$.

R models are closely related to De Morgan monoids.[11]

Definition 3.2.17 *(De Morgan monoids)* Let $\mathfrak{A} = \langle A; \boldsymbol{t}, \vee, \sim, \circ\rangle$ be an algebra of similarity type $\langle 0, 2, 1, 2\rangle$. $\mathfrak{A}$ is a *De Morgan monoid* if (1)–(11) hold. ($a \wedge b$ is ${\sim}({\sim}a \vee {\sim}b)$, $a \rightarrow b$ is ${\sim}(a \circ {\sim}b)$ and $a \leq b$ iff $a \vee b = b$.)

(1) $a \circ (b \circ c) = (a \circ b) \circ c$
(2) $a \circ b = b \circ a$
(3) $\boldsymbol{t} \circ a = a$
(4) $a \vee b = b \vee a$
(5) $(a \vee b) \vee c = a \vee (b \vee c)$
(6) $a = a \vee (a \wedge b)$
(7) $a \wedge (b \vee c) = (a \wedge b) \vee (a \wedge c)$
(8) ${\sim}{\sim}a = a$
(9) $a \circ (b \vee c) = (a \circ b) \vee (a \circ c)$
(10) $(a \rightarrow b) \circ a \leq b$
(11) $a \leq a \circ a$

The logic $R^{\boldsymbol{t}}$ algebraizes into a De Morgan monoid. The theorems of relevance logics do not imply each other, in general, which is a consequence of provable implications being relevant. $\boldsymbol{t}$ provides the best of the classical/intuitionistic and of the relevant worlds (or logics), because it implies all the theorems of the logic, but it neither destroys the differences between the theorems nor violates variable sharing when it is viewed as containing all the propositional variables (as if it would be an infinite conjunction of self-implications). De Morgan monoids, by the above definition, are a variety. $R^{\boldsymbol{t}}$ models may be defined from De Morgan monoids by adding D, where $D = [\,\boldsymbol{t}\,)$ (the principal filter generated by $\boldsymbol{t}$), and then every $R^{\boldsymbol{t}}$ theorem has to evaluate to an element of D. Alternatively, we could use $\boldsymbol{t}$ directly

[11]These were introduced in Dunn's (1966) dissertation, where he independently investigated the algebra of R. See also Anderson and Belnap (1975, Sect. 28.2).

(without turning $\mathfrak{A}$ into a matrix) by requiring that if A is a theorem of R^t, then under any valuation, $\boldsymbol{t} \leq$ A.

We have already seen that in a finite-variable context $\boldsymbol{t}$ is definable, but D or $\boldsymbol{t}$ cannot be omitted, in general, from the definition of $R^{(t)}$ models.[12] $\boldsymbol{t}$ has been added to other relevance logics or their fragments — beyond SE or R^t — and it is a preeminently useful little constant.

A soundness theorem (with respect to algebraic R models) for the logic R would comprise of showing that no matter how the propositional variables of R are mapped into elements of $\mathfrak{A}$, a theorem of R turns into an element of D. (We do not go into the details of the proof.)

Maksimova gives the following definition of $\rightarrow$, and we will assume this definition of $\rightarrow$ below. ($S_1, S_2 \subseteq S$.)

$$S_1 \rightarrow S_2 = \{ z \colon \forall x, y\, ((x \in S_1 \;\&\; \tau(x, y, z)) \Rightarrow y \in S_2) \}$$

Definition 3.2.18 *(Topological spaces for R)* Let $\mathcal{S} = \langle S; \mathcal{O}, \leq, I, g, \tau \rangle$ be a compact, totally order-disconnected topological space. $\mathcal{S}$ is a *topological space for R* when it satisfies (1)–(12). (O, O_1 and O_2 are clopen cones.)

(1) $\langle S; \leq \rangle$ is a poset, and $\emptyset \neq I \subseteq S$;
(2) if $x \leq y$, then $g(y) \leq g(x)$;
(3) $g(g(x)) = x$;
(4) $\tau \subseteq S \times S \times S$;
(5) $x \leq y$ iff $\exists z\, (z \in I \;\&\; \tau(x, y, z))$;
(6) I is a cone (i.e., $I^{\uparrow} = I$);
(7) $\tau(y, w, x) \;\&\; \tau(z, v, w)$ implies $\exists w\, (\tau(w, v, x) \;\&\; \tau(z, w, y))$;
(8) $\tau(y, w, x) \;\&\; \tau(z, v, w)$ implies $\exists w\, (\tau(z, w, x) \;\&\; \tau(y, v, w))$;
(9) $\tau(y, z, x)$ implies $\exists w\, (\tau(y, w, x) \;\&\; \tau(y, z, w))$;
(10) $\tau(y, z, x)$ implies $\tau(g(z), g(y), x)$;
(11) $O_1 \rightarrow O_2$ is a clopen cone;
(12) $\{ x \colon g(x) \notin O \}$ is a clopen cone.

We made explicit the set of open sets $\mathcal{O}$ — for easy reference. *Compactness* means that if $\bigcup_{j \in J} O_j \supseteq S$ (where $O_{j \in J} \in \mathcal{O}$), then there is a finite J' ($J' \subseteq J$) such that $\bigcup_{j \in J'} O_j \supseteq S$. In words, each open cover has a finite subcover.

Total disconnectedness means that two elements x, y (i.e., $x \neq y$) can be distinguished by some clopen set Q, that is, $x \in Q$, but $y \notin Q$. With the order, we can limit our consideration to *cones* and to pairs of elements such that one is not less than the other. $\mathcal{S}$ is *totally order-disconnected* when for any x and y, if $x \nleq y$, then there is a clopen cone O such that $x \in O$, but $y \notin O$.

Definition 3.2.19 Let $\mathcal{S} = \langle S; \mathcal{O}, \leq, I, g, \tau \rangle$ be a topological space for R (as in Definition 3.2.18). The *R model of* $\mathcal{S}$ is the algebra $\mathfrak{A}_{\mathcal{S}} = \langle A_{\mathcal{S}}; D_{\mathcal{S}}, \cap, \cup, \sim_{\mathcal{S}}, \rightarrow_{\mathcal{S}} \rangle$, where the components are defined by (1)–(6). (O, O_1 and O_2 are clopen cones.)

[12] For a detailed algebraic explanation of this point, see Anderson and Belnap (1975, Sect. 28.2).

(1) $A_{\mathfrak{s}}$ is the set of clopen cones of $\mathcal{S}$;
(2) $D_{\mathfrak{s}} = \{ O \colon I \subseteq O \}$;
(3) $\cap$ is intersection;
(4) $\cup$ is union;
(5) $\sim_{\mathfrak{s}} O = \{ x \colon g(x) \notin O \}$;
(6) $O_1 \rightarrow_{\mathfrak{s}} O_2 = \{ z \colon \forall x, y\, (x \in O_1 \;\&\; \tau(x, y, z)) \Rightarrow y \in O_2) \}$.

Theorem 3.2.20 *Given a topological space $\mathcal{S}$ for R (as in Definition 3.2.18), $\mathfrak{A}_{\mathfrak{s}}$, the R model of $\mathcal{S}$ (as in Definition 3.2.19) is an R model.*

In order to prove that for any R model $\mathfrak{A}$, there is an isomorphic algebra $\mathfrak{A}_{\mathfrak{s}}$, we have to do some further work.

Lemma 3.2.21 *If $\mathfrak{A} = \langle A; D, \wedge, \vee, \sim, \rightarrow \rangle$ is an R model (as in Definition 3.2.16), then there is an R model $\mathfrak{A}^{\top\bot} = \langle A'; D', \wedge', \vee', \sim', \rightarrow', \top, \bot \rangle$, where $A' = A \cup \{ \top, \bot \}$ while $\top, \bot \notin A$, $\langle A'; \wedge', \vee' \rangle$ is a bounded distributive lattice, $D' = D \cup \{ \top \}$ and the operations of $\mathfrak{A}$ and $\mathfrak{A}^{\top\bot}$ coincide on A.*

The proof of this lemma is straightforward. The lemma itself may be viewed as the algebraic analog of the conservativeness of the addition of $\boldsymbol{T}$ and $\boldsymbol{F}$ to R.

Definition 3.2.22 Let $\mathfrak{A} = \langle A; D, \wedge, \vee, \sim, \rightarrow, \top, \bot \rangle$ be a bounded R model. The *topological space of* $\mathfrak{A}$ is $\mathcal{S}_{\mathfrak{a}} = \langle S_{\mathfrak{a}}; \mathcal{O}_{a}, \subseteq, I_{\mathfrak{a}}, g_{\mathfrak{a}}, \tau_{\mathfrak{a}} \rangle$, where the components are defined by (1)–(6).

(1) $S_{\mathfrak{a}}$ is the set of prime filters on $\langle A; \wedge, \vee, \top, \bot \rangle$;
(2) $\mathcal{O}$ is the topology generated from the subbase $\{ ha \colon a \in A \} \cup \{ -ha \colon a \in A \}$, where $ha = \{ x \in S_{\mathfrak{a}} \colon a \in x \}$;
(3) $\subseteq$ is the subset relation;
(4) $I_{\mathfrak{a}} = \{ x \in S_{\mathfrak{a}} \colon D \subseteq x \}$;
(5) $g_{\mathfrak{a}}(x) = \{ a \colon \sim a \notin x \}$;
(6) $\tau_{\mathfrak{a}}(x, y, z)$ iff $\forall a, b\, ((a \rightarrow b \in z \;\&\; a \in x) \Rightarrow b \in y)$.

Theorem 3.2.23 *Given an R model $\mathfrak{A}$ (as in Definition 3.2.16) where $\mathfrak{A}$ is bounded, $\mathcal{S}_{\mathfrak{a}}$, the topological space of $\mathfrak{A}$ (as in Definition 3.2.22) is a topological space for R.*

The proof of this lemma is fairly straightforward — except, perhaps, the properties that involve topological concepts. The other conditions are proved as in a proof establishing that the canonical frame for R^T falls into the class of frames for the logic. The topological properties (11) and (12) ensure that the set of clopen cones is closed under the defined $\sim$ and $\rightarrow$ operations — in addition to (finite) intersection and (finite) union, which stand in for meet and join.

The requirement that the topological space is totally order disconnected implies that it is Hausdorff. Completeness amounts to finding a topological space for R, with a suitable homomorphism into its algebra being an isomorphism. If $\mathfrak{A}_R$ is the Lindenbaum algebra of R (which is a model for R with D, the equivalence classes of theorems of R), then it can be embedded into $\mathfrak{A}_R^{\top\bot}$. Then the suitable map is the h that appears in (2) in Definition 3.2.22. ha is a clopen cone in $\mathcal{S}_{\mathfrak{a}}$, hence, an element of $\mathfrak{A}_{\mathfrak{s}}$. Any two a and b in A are separated by some $x \in S_{\mathfrak{a}}$, hence, $ha \neq hb$. Furthermore, $\mathcal{S}_{\mathfrak{a}}$ is Hausdorff, and so $\mathfrak{A}_{\mathfrak{s}} \cong \mathfrak{A}_R^{\top\bot}$, that is, $\mathfrak{A}_{\mathfrak{s}}$ is isomorphic to $\mathfrak{A}_R^{\top\bot}$.[13]

3.3 Theories of Π', E and R

A topological representation theorem for the algebra of a logic can be conceptualized in terms of theories of the logic. Maximally consistent sets of formulas (or maximally consistent theories) are the ultrafilters of the non-algebraist. For relevance logics, in general, the appropriate objects are not ultrafilters, of which there might be too few. However, there are suitable substitutes such as *filters* and *prime filters*, which we had mentioned already.

Maksimova (1970) is concerned with extensions of E, as well as of Π' and R, on one hand, and with classical logic, on the other.

Definition 3.3.1 (*Theory*) A set of formulas T is an E *theory* if all the axioms of E are elements of T, and T is deductively closed, that is, if $T \vdash_E \mathrm{A}$, then $\mathrm{A} \in T$.

Definition 3.3.2 Let T be an E theory. T is an E-*logic* when the rule of substitution is admissible for T.

The easiest example of an E-logic is E itself; it is also the least E-logic (under inclusion). An E theory, however, may contain formulas that are not instances of a theorem of E. For example, $(p \to\sim q) \to ((p \to q) \to (\sim p \to q))$ is not a theorem of E. Adding this formula to E generates an E theory, which is not an E-logic, because $(r \to\sim s) \to ((r \to s) \to (\sim r \to s))$ is not an element of that theory. (As often, we tacitly assume that p, q, r and s are pairwise distinct propositional variables.) On the other hand, if we add the schema $(\mathrm{A} \to\sim \mathrm{B}) \to ((\mathrm{A} \to \mathrm{B}) \to (\sim \mathrm{A} \to \mathrm{B}))$ to E and generate an E theory, then we obtain an E-logic, because substitution for propositional variables is — obviously — admissible.

Given a pair of theories, we might consider their *common* and their *cumulative* consequences. (As usual, two sets of formulas on the left-hand side of the turnstile are viewed as the union of those sets; that is, T_1, T_2 is a shorthand for $T_1 \cup T_2$ in $T_1, T_2 \vdash_E \mathrm{A}$.)

Definition 3.3.3 Let T_1 and T_2 be E theories. Then the operations $\sqcap$ (tapering) and $\sqcup$ (cumulation) are defined by (1) and (2).

(1) $T_1 \sqcap T_2 = \{\mathrm{A}\colon T_1 \vdash_E \mathrm{A} \text{ and } T_2 \vdash_E \mathrm{A}\}$
(2) $T_1 \sqcup T_2 = \{\mathrm{A}\colon T_1, T_2 \vdash_E \mathrm{A}\}$

[13] A detailed proof would be even longer than this two-page outline, and we do not include that here. Somewhat similar claims and proofs may be found in Bimbó and Dunn (2008, Chap. 9), as well as in Urquhart (1996), Bimbó (2007a) and Bimbó (2009).

It is easy to show that if $\mathcal{T}$ is the set of all E theories, then $\sqcap$ and $\sqcup$ are, indeed, operations on $\mathcal{T}$.

Theorem 3.3.4 $\langle \mathcal{T}; \sqcap, \sqcup \rangle$, *the set of all E theories, is a distributive lattice.*

Corollary 3.3.5 *The set of E-logics with the operations $\sqcap$ and $\sqcup$ is a distributive lattice.*

The proof of Theorem 3.3.4 uses the notion of a *disjunctive part* of a formula from Belnap (1960). We denote by $\mathrm{B}/\mathrm{A}/$ the formula B of which A is a disjunctive part.

Maksimova observes that the proof of the theorem uses four properties of deducibility. (We omit the subscript $_E$ from $\vdash$ so that the next corollary reads smoothly.)

(1) If $T \vdash \mathrm{A}/\mathrm{B}/$ and $T \vdash \mathrm{A}/\mathrm{B} \rightarrow \mathrm{C}/$, then $T \vdash \mathrm{A}/\mathrm{C}/$.
(2) If $T \vdash \mathrm{A}/\mathrm{B}/$ and $T \vdash \mathrm{A}/\mathrm{C}/$, then $T \vdash \mathrm{A}/\mathrm{B} \wedge \mathrm{C}/$.
(3) If $T \vdash \mathrm{A} \vee \mathrm{A}$, then $T \vdash \mathrm{A}$.
(4) If $T \vdash \mathrm{B}$, then $T \vdash \mathrm{A}/\mathrm{B}/$.

Corollary 3.3.6 *Let L be a logic and let $\mathcal{T}_L$ be the set of L theories. If for any $T \in \mathcal{T}_L$,* (1)–(4) *are true, then $\langle \mathcal{T}_L; \sqcap, \sqcup \rangle$ is a distributive lattice.*

An ultrafilter never contains $\bot$, which is $a \wedge -a$ in a Boolean algebra. In a De Morgan lattice, however, $\bot$ is not definable as $a \wedge \sim a$, which is why a De Morgan lattice (sometimes) has to be rounded out by $\top$ and $\bot$. (Cf. Lemmas 3.2.11 and 3.2.21.) It is easy to construct a De Morgan lattice in which some prime filters contain neither a nor $\sim a$, and some prime filters contain both a and $\sim a$.

Definition 3.3.7 An E theory T_1 is *complete* when $T_1 \vdash \mathrm{A}$ iff $T_1 \nvdash \sim \mathrm{A}$. An E theory T_2 is *consistent* iff for each A, either $T_2 \nvdash \mathrm{A}$ or $T_2 \nvdash \sim \mathrm{A}$.

The definitions of this section transfer to Π', with the obvious modifications.

Lemma 3.3.8 *A Π' theory T_1 is the intersection of all complete Π' theories that contain T_1. An E theory T_2 is the intersection of all complete E theories that contain T_2 when T_2 is a Π' theory.*

Maksimova establishes a difference between Π' and E, on one hand, and R, on the other, with respect to their relationship to classical logic. (The obvious translations between the languages of these logics are assumed again. In particular, beyond conjunction and disjunction, $\sim \mapsto \neg$ and $\rightarrow \mapsto \supset$, where $\neg$ and $\supset$ are the classical connectives, which are called "negation" and "conditional.")

Lemma 3.3.9 *There is a consistent E-logic, which is not contained in classical logic.*

The addition of the schema $\sim(\sim(\mathrm{A} \rightarrow \mathrm{B}) \rightarrow (\mathrm{C} \rightarrow \mathrm{D}))$ to E results in a consistent E-logic. The corresponding classical formula $\neg(\neg(\mathrm{A} \supset \mathrm{B}) \supset (\mathrm{C} \supset \mathrm{D}))$ is easily falsifiable simply by setting D to true. Then, by Post completeness, this schema trivializes classical logic.

Theorem 3.3.10 *Every consistent R-logic is contained in classical logic.*

3.4 Set-Theoretic Semantics for Relevance Logic

Maksimova (1971) returns to the representation of E and R models. She proves *isomorphic representation theorems*, which can be interpreted as completeness theorems for logics when their Lindenbaum algebra is in the class of those models. Philosophers might find it easier to interpret a relation than an operation (which is also a relation, but a single-valued one). However, operational structures, by and large, are better behaved and understood mathematically. Algebras are operational structures, and so it should not be a surprise that Maksimova designed representations built out of sets of operational structures.

Definition 3.4.1 Let $\mathfrak{A} = \langle A;\, D, \rightarrow\rangle$ be an $E_{\rightarrow}$ model, and let $\mathfrak{G} = \langle G;\; \cdot\,\rangle$ be the free groupoid generated from the set S, where $S = \{\, [a) \colon a \in A \,\} \cup \{\, D \,\}$. $h \colon A \longrightarrow \mathrm{P}(G)$ such that h satisfies (1)–(2). ($s \in S$ and $u, v \in G$.)

(1) $s \in h(a)$ iff $a \in s$;
(2) $u \cdot v \in h(a)$ iff $\exists b \in A\, (u \in h(b \rightarrow a)\ \&\ v \in h(b))$.

$[a)$ is the principal cone generated by a in A, where $a \leq b$ holds (by definition) iff $a \rightarrow b \in D$.

Lemma 3.4.2 *The map h is* 1–1 *(injective) from A into* $\mathrm{P}(G)$*; moreover,* (1)–(3) *are true too.*

(1) $a \leq b$ *iff* $h(a) \subseteq h(b)$;
(2) $D \in h(a)$ *iff* $a \in D$;
(3) $h(a \rightarrow b) = \{\, u \colon u \cdot h(a) \subseteq h(b) \,\}$.

The groupoid operation of $\mathfrak{G}$ is lifted to subsets of G as usual. That is, $u \cdot h(a) = \{\, u \,\} \cdot h(a)$, and further, $\{\, u \cdot v \colon v \in h(a) \,\}$.

Lemma 3.4.3 *If $\mathfrak{A}$ is an $E_{\rightarrow}$ model with $\mathfrak{G}$ and h defined as above, then h has the following properties.*

(1) $u \in h(a)$ *iff* $D \cdot u \in h(a)$;
(2) $u \cdot D \in h(a)$ *implies* $u \in h(a)$;
(3) $w \cdot (u \cdot v) \in h(a)$ *implies* $(u \cdot w) \cdot v \in h(a)$;
(4) $(u \cdot v) \cdot v \in h(a)$ *implies* $u \cdot v \in h(a)$.

Next, the free groupoid is normalized by taking its quotient. The order relation $\prec$ is defined on G as follows: $u \prec v$ iff $\forall a \in A\, (v \in h(a) \Rightarrow u \in h(a))$.

Lemma 3.4.4 *The relation $\prec$ is a pre-order, and $\cdot$ is monotone with respect to $\prec$ in both arguments, that is,*

(i) $u \prec v$ *implies* $w \cdot u \prec w \cdot v$ *and* $u \cdot w \prec v \cdot w$.

Definition 3.4.5 Let $\mathfrak{G}$ be as defined above. Then $\mathfrak{G}' = \langle G_{/\equiv}, \cdot\rangle$, where $u \equiv v$ iff $u \prec v\ \&\ v \prec u$ and $[u] \cdot [v] = [u \cdot v]$.

The pre-order has an obvious counterpart on the quotient algebra, namely, $[u] \prec [v]$ iff $u \prec v$.

Lemma 3.4.6 *Let* $\mathfrak{G}'$ *be as defined above, and let* j *be the type-lifted version of* h*; that is,* $j(a) = \{ [u] \colon u \in h(a) \}$. *Then* (1)–(4) *are true of* j. *(From now on,* $u, v, w \in G'$*, where* $G' = G_{/\equiv}$.*)*

(1) $j(a)$ *is a co-cone (i.e., a downward closed set with respect to* $\prec$*);*
(2) $a \leq b$ *implies* $j(a) \subseteq j(b)$*;*
(3) $a \in D$ *iff* $[D] \in j(a)$*;*
(4) $j(a \to b) = \{ u \colon u \cdot j(a) \subseteq j(b) \}$.

The construction of $\mathfrak{G}$ does not depend on $\mathfrak{A}$ not containing other operations beyond $\to$. Thus, for an $E_{\overset{\wedge}{\to}}$ model, we obtain the corresponding $\mathfrak{G}$, h, $\mathfrak{G}'$ and j.

Lemma 3.4.7 *If* $\mathfrak{A}$ *is an* $E_{\overset{\wedge}{\to}}$ *model, then* $\mathfrak{G}'$ *satisfies* (1)–(4).

(1) *There is an* $e \in G'$ *such that* $e \cdot u = u$*;*
(2) $u \prec u \cdot e$*;*
(3) $(u \cdot w) \cdot v \prec w \cdot (u \cdot v)$*;*
(4) $u \cdot v \prec (u \cdot v) \cdot v$.

The above conditions, in particular, the last three ones correspond to an axiomatization of $E_{\to}$ with axioms that are principal type schemas of the combinators CII, B′ and W. (The addition of $\wedge$ does not disturb this picture, because the interpretation of $\wedge$ comes for free.)

Theorem 3.4.8 *If* $\mathfrak{A} = \langle A;\, D, \to, \wedge \rangle$ *is an* $E_{\overset{\wedge}{\to}}$ *model, then* $\mathfrak{A} \cong \mathfrak{B}$*, where* $\mathfrak{B} = \langle B;\, D_B, \to_B, \wedge_B \rangle$*,* B *is the set of co-cones on* $\vec{\mathfrak{G}}'$*, which satisfies conditions* (1)–(4) *from Lemma 3.4.7, and* D_B*,* $\to_B$ *and* $\wedge_B$ *are as follows.*

(1) $D_B = \{ X \colon X \in B \;\&\; e \in X \}$
(2) $X \to_B Y = \{ u \colon u \cdot X \subseteq Y \}$
(3) $X \wedge_B Y = X \cap Y$

Theorem 3.4.9 (Representation theorem) *The algebra* $\mathfrak{A} = \langle A;\, D, \to \wedge \rangle$ *is an* $E_{\overset{\wedge}{\to}}$ *model iff* $\mathfrak{A} \cong \mathfrak{B}$*, where* $\mathfrak{B}$ *is defined as in the previous theorem.*

The previous constructions can be adapted to $R_{\to}$ and $R_{\overset{\wedge}{\to}}$ models too. In the corresponding $\mathfrak{G}'$, $u \cdot v \prec v \cdot u$ holds, which leads to analogues of the previous two theorems for $R_{\to}$ and $R_{\overset{\wedge}{\to}}$.

Theorem 3.4.10 *Any* $E_{\to}$ *model can be embedded into an* $E_{\overset{\wedge}{\to}}$ *model; similarly, for* $R_{\to}$ *models with respect to* $R_{\overset{\wedge}{\to}}$ *models. Therefore,* $E_{\overset{\wedge}{\to}}$ *is a conservative extension of* $E_{\to}$*, and* $R_{\overset{\wedge}{\to}}$ *is a conservative extension of* $R_{\to}$.

In her (1973b) paper, Maksimova continues her investigations into representations of models of various logics. She introduces a class of logics, which includes E and R. (Again, we re-use some of the axioms and rules that we listed earlier.)

Definition 3.4.11 The *positive assertive logic*, that we denote by A_+, contains the following axioms and rules. (A1), (A5)–(A11), (α), (β), (R3)–(R4) and the rule (R6).

(R6) A implies (A → B) → B

Actually, Maksimova calls these logics *regular* positive logics. We use "assertive" instead of "regular," because the latter term has a different use in the (English-language) literature on relevance logics, and (R6) is the rule form of the axiom called *assertion*.

Further axioms that will be considered for addition to A_+ are (A2)–(A4) and (A19)–(A25).

(A19) ((A → B) ∧ A) → B
(A20) (A → (B → C)) → (B → (A → C))
(A21) ((A → B) ∧ (B → C)) → (A → C)
(A22) A → (B → B)
(A23) (A → B) → (C → (A → B))
(A24) A → (B → A)
(A25) A → (A → A)

The models for all these logics contain at least a distributive lattice. In particular, models for E_+ and R_+ fall into the following class of algebras, but they have some further properties too.

Definition 3.4.12 Let $\mathfrak{A} = \langle A; D, \wedge, \vee, \rightarrow\rangle$ be an algebra, where $\emptyset \neq D \subseteq A$, and there are three binary operations. $\mathfrak{A}$ is a *lattice with rigorous implication* (a "strimpla" or an A_+-lattice, for short) iff (1)–(7) hold.

(1) $\langle A; D, \wedge, \vee\rangle$ is a distributive lattice, where D is a filter;
(2) $a \rightarrow b \in D$ iff $a \leq b$;
(3) $a \in D$ iff $a \rightarrow b \leq b$;
(4) $a \leq b$ iff $b \rightarrow c \leq a \rightarrow c$;
(5) $a \leq b$ iff $c \rightarrow a \leq c \rightarrow b$;
(6) $(a \rightarrow b) \wedge (b \rightarrow c) \leq a \rightarrow (b \wedge c)$;
(7) $(a \rightarrow b) \wedge (c \rightarrow b) \leq (a \vee c) \rightarrow b$.

Next Maksimova defines a relational structure, which is reminiscent of her definition in Maksimova (1969) and of a definition in Routley and Meyer (1972), but differs from both.

Definition 3.4.13 A *structure* for an A_+-lattice is $\mathcal{S} = \langle S; R, I\rangle$, where $\emptyset \neq I \subseteq S$, $R \subseteq S \times S \times S$, and (1)–(4) are true.

(1) $u \geq v$ iff $\exists i\, (i \in I \;\&\; Riuv)$;
(2) $\exists i\, (i \in I \;\&\; Riuu)$;
(3) $\exists i\, (i \in I \;\&\; Ruiu)$;
(4) $u \leq u' \;\&\; v \leq v' \;\&\; w' \leq w \;\&\; Ruvw$ implies $Ru'v'w'$.

The ternary relation that she denoted by τ before is R now, and $\leq$, which was explicitly postulated, now defined from R. She also includes not only a special element of S, but a *subset of distinguished elements* of S, which we denoted by I, because a set like this functions as an *identity* in some contexts. The set algebra associated to a structure is the one in the following definition.

Definition 3.4.14 Let $\mathcal{S}$ be a structure for an A_+-lattice. $\mathfrak{A}_\mathcal{S} = \langle A_\mathcal{S};\, D_\mathcal{S}, \wedge, \vee, \rightarrow \rangle$ is an *algebra*, where the components are defined by (1)–(5).

(1) $A_\mathcal{S}$ is the set of cones on A (with respect to $\leq$);
(2) $D_\mathcal{S} = \{\, a\colon a \in A_\mathcal{S} \;\&\; I \subseteq a \,\}$;
(3) $a \wedge b = a \cap b$;
(4) $a \vee b = a \cup b$;
(5) $a \rightarrow b = \{\, u\colon \forall v, w\, ((Ruvw \;\&\; v \in a) \Rightarrow w \in b) \,\}$.

The groupoid representation that Maksimova started to build in her (1971) can be refined to yield a representation for A_+-lattices.

Definition 3.4.15 A *special structure for an A_+-lattice* is $\mathcal{S} = \langle S;\, R, I \rangle$, which is constructed from $\mathfrak{G} = \langle G;\, \leq, \cdot \rangle$, a partially ordered groupoid, according to (1)–(5). (We use x, y, z for elements of G and u, v, w for elements of S.)

(1) $x \cdot y$ implies $z \cdot x \leq z \cdot y$ and $x \cdot z \leq y \cdot z$
(2) $S \subseteq G$ such that $\forall x, y, u\, (u \leq x \cdot y \Rightarrow (\exists v\, u \leq v \cdot y \;\&\; \exists v\, u \leq x \cdot v))$
(3) $\exists e\, (e \in G \;\&\; e \cdot x = x)$
(4) $Ruvw$ iff $u \cdot v \geq w$
(5) $I = \{\, u\colon u \leq e \,\}$

Condition (2) characterizes a relation between G and its subset S, which in the intended representation amounts to the so-called *squeeze lemma* in the Routley–Meyer semantics. Informally speaking, (2) then states that if the product of two filters x and y is a subset of a prime filter u, then there is a prime filter v in place of one or the other filter (i.e., x or y) such that the product with v is (still) the subset of the same prime filter u.

Theorem 3.4.16 *For any A_+-lattice $\mathfrak{A}$, there is an $\mathcal{S}$, which is a special structure for an A_+-lattice, such that $\mathfrak{A} \cong \mathfrak{A}_\mathcal{S}$.*

The components of the special structure deserve some mention. G is taken to be the set of filters on A, including A and $\emptyset$. The order relation on G is defined as $F_1 \geq F_2$ iff $F_2 \supseteq F_1$. The groupoid operation on filters is $F_1 \cdot F_2 = \{\, c\colon \exists b\, (b \in F_2 \;\&\; b \rightarrow c \in F_1) \,\}$. Filters or their sets do not give a suitable representation of distributive lattices (with the obvious operations $\cap$ and $\cup$); therefore, S is taken to be the set of *prime filters*, including A and $\emptyset$. Condition (4) in the Definition 3.4.15 reveals the reason why Maksimova has to single out a special subset S of G. Namely, the natural operation $\cdot$ (as above) on (principal) filters is *not an operation* on prime filters. Either the objects (principal cones or filters) are unsuitable, or the operation is partial. Having a (total) operation as well as prime filters should keep everybody happy, so to speak, for the price of having two kinds of objects and a ternary relation too.

Maksimova also proves that the representation result can be extended to A_+-lattices to which a top element is added. Furthermore, she shows how to replace D with a single element, and vice versa, how to define D, if a single distinguished element is included.

The addition of axioms from Definition 3.4.11 requires additional conditions in the algebraic models, in the partially ordered groupoids and in the structures. Maksimova gives three lists of conditions and proves that they are necessary and sufficient conditions. In other words, especially, speaking of conditions for structures, this means that all the logics that she considers are *canonical*. We denote the conditions by (m#), (g#) and (s#) with the number matching the number of the added axiom.

The conditions on models are the following.

(m2) $a \to b \leq (b \to c) \to (a \to c)$
(m3) $a \to b \leq (c \to a) \to (c \to b)$
(m4) $a \to (a \to b) \leq a \to b$
(m19) $(a \to b) \wedge a \leq b$
(m20) $a \to (b \to c) \leq b \to (a \to c)$
(m21) $(a \to b) \wedge (b \to c) \leq a \to c$
(m22) $a \leq b \to b$
(m23) $a \to b \leq c \to (a \to b)$
(m24) $a \leq b \to a$
(m25) $a \leq a \to a$

The groupoid conditions are as follows.

(g2) $(x \cdot y) \cdot z \leq y \cdot (x \cdot z)$
(g3) $(x \cdot y) \cdot z \leq x \cdot (y \cdot z)$
(g4) $x \cdot y \leq (x \cdot y) \cdot y$
(g19) $x \leq x \cdot x$
(g20) $(x \cdot y) \cdot z \leq (x \cdot z) \cdot y$
(g21) $x \cdot y \leq x \cdot (x \cdot y)$
(g22) $x \cdot y \leq y$
(g23) $(x \cdot y) \cdot z \leq x \cdot z$
(g24) $x \cdot y \leq x$
(g25) $x \cdot x \leq x$

The conditions on structures are the following.

(s2) $Ru_1u_2v \,\&\, Rvu_3w$ implies $\exists v\,(Ru_1u_3v \,\&\, Ru_2vw)$
(s3) $Ru_1u_2v \,\&\, Rvu_3w$ implies $\exists v\,(Ru_2u_3v \,\&\, Ru_1vw)$
(s4) Ru_1u_2w implies $\exists v\,(Ru_1u_2v \,\&\, Rvu_2w)$
(s19) $Ruuu$
(s20) $Ru_1u_2v \,\&\, Rvu_3w$ implies $\exists v\,(Ru_1u_3v \,\&\, Rvu_2w)$
(s21) Ru_1u_2w implies $\exists v\,(Ru_1u_2v \,\&\, Ru_1vw)$

(s22) $Ruvw$ implies $\exists i\,(i \in I \,\&\, Rivw)$
(s23) $Ru_1u_2v \,\&\, Rvu_3w$ implies Ru_1u_3w
(s24) $Ruvw$ implies $\exists i\,(i \in I \,\&\, Riuw)$
(s25) $Ruvw$ implies $\exists i\,(i \in I \,\&\, Riuw) \vee \exists i\,(i \in I \,\&\, Rivw)$

In the last part of her paper, Maksimova extends the positive logics with a negation operation, which results in a De Morgan lattice reduct in the algebras of the logics. She uses the function g on the frame to model the negation operation. The corresponding conditions are those we mentioned earlier.

A short paper, Maksimova (1973a) caused ambivalent emotions among relevance logicians.[14] She proved a conjecture of Mints, and thereby she disproved a conjecture that she credits to Prawitz and Meyer. The latter pertains to the relationship between R and E, namely, it states that $A \to B$ in E is the same as $\Box(A \to B)$ in $R^\Box$. If true, this conjecture would have completed elegantly the intuitions about entailments being necessary (relevant) implications — what we already mentioned in Sect. 3.2. But Mints thought he had found a putative counterexample to the translation. Mints conjectured that $\nvdash_E ((p \to (q \to r)) \wedge (q \to (p \vee r))) \to (q \to r)$. The importance of his formula was that he had proved that the formula that is the translation of this E formula in $R^\Box$ is provable, $\vdash_{R^\Box} \Box(\Box(p \to \Box(q \to r)) \wedge \Box(q \to (p \vee r)) \to \Box(q \to r))$, and so there would be a counterexample to the Prawitz and Meyer conjecture. Maksimova's paper took for granted Mints's proof that the last formula is provable in $R^\Box$, and so she only explicitly shows — using her relational semantics — that the formula without the $\Box$'s is not valid in structures for E.

In a recent email from Maksimova, she told us that Mints told her about his formula and its proof when they both were attending the same conference. She reconstructed a proof for us which she included in her email, and said we could share. We will not give the whole proof here, but since the proof has been unpublished we point out for the record that the key lemma is proving $\vdash_R ((p \to (q \to r)) \wedge (q \to (p \vee r))) \to (q \to r)$. In $R^\Box$ we can then put a box in front of it using the rule of necessitation, move it across the arrows using the axioms for $\Box$, and then "rebox" it using the rule of necessitation again. It is important that we use R here, because of its relative freedom of permutation versus E. In particular it is important that we have permutation (A20), because the trick is to first prove $\vdash_R ((q \to (p \to r)) \wedge (q \to (p \vee r))) \to (q \to r)$ (mind your p's and q's, as the old saying goes). Proving this is relatively straightforward relying critically on contraction and $\vdash_R ((p \to r) \wedge p) \to r$.

Maksimova notes in the conclusion to her paper that unlike for the positive fragment of E, a structure for E cannot have merely one distinguished element. While in Kripke's semantics for various normal modal logics there is a single "actual world," which, in fact, can be selected as any possible world in the canonical structure, in the semantics of E, there is no single situation that can play a similar role. Maksimova

[14] See Anderson and Belnap (1975, pp. 351–352).

investigated the algebraic models of E (see Sect. 3.2). Since E is not a prime logic, and its theorems are "not equal," so to speak, the set of the equivalence classes of theorems is not a prime theory.[15]

3.5 Variable Sharing

Maksimova's research interests shifted to other logics, but she turned back to relevance logics later in the 1970s to further investigate the variable sharing property. (Sometimes, this property is called the *relevance principle*.) A simple formulation for R is R82 from Anderson and Belnap (1975, p. 417).

R82. Suppose $\vdash_{\mathbf{R}} A \rightarrow B$. Then A and B share a sentential variable.

Maksimova in her papers of (1976, 1989) sharpened and extended the above principle — first considering a related principle. A logic is *Halldén reasonable* when $\vdash$ A $\vee$ B implies $\vdash$ A or $\vdash$ B, provided that A and B do not have common variables. This property is similar to the disjunction property, and it is immediate that any prime logic is Halldén reasonable. Not the other way around though.

Definition 3.5.1 Let α, β be sets of propositional variables such that $\alpha \cap \beta = \emptyset$. Furthermore, let A's be formulas built from elements of α by the connectives $\vee$, $\wedge$, $\rightarrow$, $\circ$ and $\sim$. Similarly, for B's and β. The *classes of formulas* denoted by K_1 and K_2 are defined by (1)–(7).

(1) $\mathrm{A} \vee \mathrm{B} \in K_1$ and $\mathrm{A} \wedge \mathrm{B} \in K_2$;
(2) if $\mathrm{C}, \mathrm{D} \in K_1$, then $\mathrm{C} \vee \mathrm{D} \in K_1$;
(3) if $\mathrm{C}, \mathrm{D} \in K_2$, then $\mathrm{C} \wedge \mathrm{D} \in K_2$;
(4) if $\mathrm{C} \in K_1$ and $\mathrm{D} \in K_2$, then $\mathrm{D} \rightarrow \mathrm{C} \in K_1$;
(5) if $\mathrm{C}, \mathrm{D} \in K_2$, then so is $\mathrm{C} \circ \mathrm{D}$;
(6) if $\mathrm{C} \in K_1$, then $\sim\mathrm{C} \in K_2$;
(7) if $\mathrm{C} \in K_2$, then $\sim\mathrm{C} \in K_1$.

Maksimova considers again a whole class of logics, that she introduced earlier. She states five versions of the relevance principle about A_+ and its extensions. Accordingly, a *structure* is defined to be a quadruple $\langle S, R, I, g\rangle$, where $I \subseteq S$ is a non-empty set; $R \subseteq S^3$; $g \colon S \longrightarrow S$ and (1)–(6) are satisfied.

(1) $x \leq x$, where $x \leq y$ iff $\exists i\ (i \in I\ \&\ Rixx)$;
(2) $\exists i\ (i \in I\ \&\ Rxix)$;
(3) $x' \leq x\ \&\ y' \leq y\ \&\ z \leq z'\ \&\ Rxyz$ implies $Rx'y'z'$;
(4) $g(g(u)) = u$;
(5) $Rxyz$ implies $Rxg(z)g(y)$;
(6) $Rxg(x)x$.

[15]For later discussions of the same point in the Routley–Meyer semantics, see Dunn (1993) and Dunn (2001).

The definition of the valuation and of the truth conditions for $\sim$, $\wedge$, $\vee$, $\circ$ and $\rightarrow$ is as usual. Maksimova also lists conditions involving R that correspond to the axioms (A2)–(A4) and (A19)–(A25).

In her earlier papers, Maksimova showed her deftness not only in finding small finite algebraic models, but also in modifying them so that they acquire some desirable feature while retaining the property of being models. Now she shows the usefulness of some small structures and how to combine them with other structures.

Example 3.5.2 Let $\mathcal{S}_0$ comprise the following components. $S = \{0, 1\}$; $I = \{1\}$; $g(0) = 1$ and $g(1) = 0$, and finally, $Rxyz$ holds iff $x = 0$ or $y = 0$ or $z = 1$.

$\mathcal{S}_0$ can be viewed as **2**, the two-element Boolean algebra, where $Rxyz$ is $x \wedge y \leq y$. Indeed, for any formula A that is a theorem of A_+, $1 \models$ A and $0 \not\models$ A.

Example 3.5.3 Let $\mathcal{S}_e$ be the trivial structure, that is, $S = \{1\} = I$ and $R111$. ($g(1) = 1$, inevitably, however, g will not be required for what $\mathcal{S}_1$ is used.)

Every negation-free formula is true in $\mathcal{S}_e$, since R is the total relation. The basic construction that is used to prove a series of theorems for formulas in which the variables are segregated to particular subformulas is the *product of models*.

Definition 3.5.4 Let $\mathcal{S}_1 = \langle S_1, I_1, R_1, g_1, \models_1\rangle$ and $\mathcal{S}_2 = \langle S_2, I_2, R_2, g_2, \models_2\rangle$ be two models, where the $\models$'s are the satisfiability relations induced by the valuation functions v_1 and v_2, respectively. $\mathcal{S} = \langle \mathcal{S}_1, \mathcal{S}_2\rangle$ is the *$\alpha\beta$-product model* of $\mathcal{S}_1$ and $\mathcal{S}_2$, with its components defined by (1)–(5).

(1) $S = S_1 \times S_2$;
(2) $I = I_1 \times I_2$;
(3) $R\langle x_1, x_2\rangle\langle y_1, y_2\rangle\langle z_1, z_2\rangle$ iff $R_1x_1y_1z_1$ & $R_2x_2y_2z_2$;
(4) $g(\langle x_1, x_2\rangle) = \langle g_1(x_1), g_2(x_2)\rangle$;
(5) $\langle x_1, x_2\rangle \models p$ iff $x_1 \models_1 p$ & $p \in \alpha$ or $x_2 \models_2 p$ & $p \in \beta$.

The construction builds on the Cartesian product of the sets of situations, with the valuation functions being restricted to one or the other set of propositional variables. The role of the classes K_1 and K_2 becomes clear in light of the following lemma.

Lemma 3.5.5 *Let $\mathcal{S}$ be an $\alpha\beta$-product model. For any formula* C $\in K_1$, $\langle x_1, x_2\rangle \models$ C *iff* $x_1 \models_1$ C *or* $x_2 \models_2$ C. *For any formula* D $\in K_2$, $\langle x_1, x_2\rangle \models$ D *iff* $x_1 \models_1$ D *and* $x_2 \models_2$ D.

Theorem 3.5.6 *Let* C $\in K_1$ *and let* $\mathbb{M}$ *be a class of models closed under forming $\alpha\beta$-product models.* C *is true in* $\mathbb{M}$ *when either* $\mathrm{C}^{\setminus\alpha}$ *or* $\mathrm{C}^{\setminus\beta}$ *is true in* $\mathbb{M}$.

The superscripts on C indicate that C is modified, namely, the subformulas containing variables from the subtracted set are recursively omitted. (E.g., if $p, p_0, p_1 \in \alpha$ but $q \in \beta$, then $\sim p$ is replaced by p, $p \wedge q$ is replaced by q, $p_1 \wedge p_1$ is replaced by p_1, etc.)

Theorem 3.5.7 *Let* D *be a formula generated by the connectives from formulas of the form in* (1)–(3).

(1) $\mathrm{A} \to \mathrm{C}$, *where* $\mathrm{C} \in K_1$;
(2) $\mathrm{C} \to \mathrm{A}$, *where* $\mathrm{C} \in K_2$;
(3) A.

Let $\mathbb{M}$ *be a class of models such that* $\mathbb{S}_0 \in \mathbb{M}$, *and* $\mathbb{M}$ *is closed under forming* $\alpha\beta$*-product models. Then,* $\mathbb{M} \models \mathrm{D}$ *implies that* $\mathbb{M} \models \mathrm{D}^{\backslash\alpha}$.

The results obtained from the $\alpha\beta$-product model constructions lead to the following series of results, which are stated in terms of provability in the logics (each of which is either A_+ or one of its extension considered above).

Theorem 3.5.8 *Let* L_1 *be a logic, which may contain in addition to* A_+ *some of* (A19)–(A21), (A2)–(A4). *If* $\vdash_{L_1} \mathrm{A}_1 \to \ldots (\mathrm{A}_{n-1} \to (\mathrm{A}_n \vee \mathrm{B}))$ *and the* A*'s have no variable in common with* B, *then* $\vdash_{L_1} \mathrm{A}_1 \to \ldots (\mathrm{A}_{n_1} \to \mathrm{A}_n)$.

The calculus of rigorous implication SE that was introduced by Maksimova in 1968, does not quite fit the definition of the a positive assertive logic, because $\boldsymbol{t}$ is in the language of SE. However, Theorem 3.5.8 is true for SE too, provided that the A's and B do not contain occurrences of $\boldsymbol{t}$.

Theorem 3.5.9 *Let* L_2 *be a logic, which may contain in addition to the axioms and rules of* L_1 *(from the previous theorem) some of the axioms* (A22)–(A25). *If* $\vdash_{L_2} \mathrm{A}_1 \to \ldots (\mathrm{A}_{n-1} \to (\mathrm{A}_n \vee \mathrm{B}))$ *and the* A*'s have no variable in common with* B, *then* $\vdash_{L_2} \mathrm{A}_1 \to \ldots (\mathrm{A}_{n-1} \to \mathrm{A}_n)$ *or* $\vdash_{L_2} \mathrm{B}$.

The last four axioms in the list after Definition 3.4.11 introduce the possibility of irrelevance. This is very clearly the case if one of (A22)–(A24) is included. Less obviously, (A25), the so-called *mingle axiom* (which is a special instance of (A24), the so-called positive paradox) harbors special prospects for irrelevance.

Theorem 3.5.10 *Let* L_2 *be a logic (from the previous theorem). If* $\vdash_{L_2} \mathrm{A}_1 \to \ldots ((\mathrm{A}_{n-1} \wedge \mathrm{B}) \to \mathrm{A}_n)$ *or* $\vdash_{L_2} (\mathrm{A}_1 \wedge \mathrm{B}_1) \to \ldots ((\mathrm{A}_{n-1} \wedge \mathrm{B}_{n-1}) \to \mathrm{A}_n)$, *then* $\mathrm{A}_1 \to \ldots (\mathrm{A}_{n-1} \to \mathrm{A}_n)$, *provided that the* A*'s and* B*'s have no variable in common.*

Theorem 3.5.11 *Let* L_2 *be a logic (from Theorem 3.5.9). If* $\vdash_{L_2} (\mathrm{A}_1 \wedge \mathrm{B}_1) \to \ldots ((\mathrm{A}_{n-1} \wedge \mathrm{B}_{n-1}) \to (\mathrm{A}_n \vee \mathrm{B}_n))$, *then* $\vdash_{L_2} \mathrm{A}_1 \to \ldots (\mathrm{A}_{n-1} \to \mathrm{A}_n)$ *or* $\vdash_{L_2} \mathrm{B}_1 \to \ldots (\mathrm{B}_{n-1} \to \mathrm{B}_n)$, *if the* A*'s and the* B*'s do not have a variable in common.*

Next Maksimova considers extensions of E and R, with axioms. E_+ is an extension of A_+ with (A2) and (A4), whereas R_+ is an extension of E_+ with (A20). Negation may be added by axioms (A12), (A14)–(A15) and (A17). We denote by L_3 an extension of E or R by some of the axioms (A22)–(A25). The logic that extends R with (A25) is called *R-mingle* or RM.

Theorem 3.5.9 holds for any L_3 logics. The following theorem is an appropriately modified version of Theorem 3.5.10 for L_3 logics.

Theorem 3.5.12 *If* $\vdash_{L_3} \mathrm{A}_1 \to \ldots ((\mathrm{A}_{n-1} \wedge \mathrm{B}) \to \mathrm{A}_n)$, *then* $\vdash_{L_3} {\sim}\mathrm{B}$ *or* $\vdash_{L_3} \mathrm{A}_1 \to \ldots (\mathrm{A}_{n-1} \to \mathrm{A}_n)$, *provided that the* A*'s and* B *have no variable in common.*

Theorem 3.5.11 is true for all L_3 logics save those in which (A25) is an axiom, such as for example RM. Let L_4 stand for one of E_+, R_+, E or R.

Theorem 3.5.13 *If* $\Gamma, \mathrm{A} \vdash_{L_4} \mathrm{B}$, *where no variable of* A *occurs in the formulas in* Γ *or in* B, *then* $\Gamma \vdash_{L_4} \mathrm{B}$.

The theorems of Π' are theorems of E; however, the two calculi differ in their consequence relation, because of Π''s (γ) rule.

Theorem 3.5.14 *If* $\Gamma, \mathrm{A} \vdash_{\Pi'} \mathrm{B}$, *where no variable of* A *occurs in the formulas in* Γ *or in* B, *then* $\vdash_{\Pi'} {\sim}\,\mathrm{A}$ *or* $\Gamma \vdash_{\Pi'} \mathrm{B}$.

3.6 Summary

We have examined the early work of Larisa Maksimova on relevance logic. It was early in two senses. First, it began to be published in the mid 1960s and early 1970s, just when relevance logic came into its fore under the leadership of Alan R. Anderson and Nuel D. Belnap. Second, it was very early in her career — originating in her Ph.D. dissertation with her work inspired by Wilhelm Ackermann, one of the inspirations also of Anderson and Belnap and their system E of entailment. Her work contains a variety of detailed results, but two important themes have to do with looking at these systems algebraically, and through that lens, by proving a representation theorem and developing a semantics with a ternary accessibility relation. (That relation plays a role similar to Kripke's binary accessibility relation in his semantics for modal logic.) We think that she was probably the first to do this. Maksimova used her semantics to establish a number of important properties of E and other relevance logics, as we have discussed in our paper. She is definitely a pioneer in the area of research on rigorous and relevant implications.

Acknowledgements We would like to thank Larisa Maksimova for providing us with e-copies of some of her publications that were otherwise inaccessible to us. We are thankful to Sergeĭ Odintsov for his help in this matter, and to the anonymous referees for their comments.
Parts of this paper belong to our larger project concerning the relational semantics of intensional logics, which is supported by an *Insight Grant* (IG #435–2014–0127), awarded by the *Social Sciences and Humanities Research Council of Canada*.

References

Anderson, A. R. (1963). Some open problems concerning the system E of entailment. *Acta Philosophica Fennica, 16*, 9–18.
Anderson, A. R. (1971). Review: L. L. Maksimova, on models of the calculus E. *Journal of Symbolic Logic, 36*(3), 521.
Anderson, A. R., & Belnap, N. D. (1975). *Entailment: The logic of relevance and necessity* (Vol. I). Princeton: Princeton University Press.
Anderson, A. R., Belnap, N. D., & Dunn, J. M. (1992). *Entailment: The logic of relevance and necessity* (Vol. II). Princeton: Princeton University Press.
Belnap, N. D. (1960). EQ and the first order functional calculus. *Zeitschrift für mathematische Logik und Grundlagen der Mathematik, 6*, 217–218.
Belnap, N. D. (1967). Intensional models for first degree formulas. *Journal of Symbolic Logic, 32*, 1–22.
Białynicki-Birula, A., & Rasiowa, H. (1957). On the representation of quasi-Boolean algebras. *Bulletin de l'Académie Polonaise des Sciences, 5*, 259–261.
Bimbó, K. (2007a). Functorial duality for ortholattices and De Morgan lattices. *Logica Universalis, 1*, 311–333.
Bimbó, K. (2007b). Relevance logics. In D. Jacquette (Ed.), *Philosophy of logic*, D. Gabbay, P. Thagard & J. Woods, (Eds.), Handbook of the philosophy of science (Vol. 5, pp. 723–789). Amsterdam: Elsevier (North-Holland).
Bimbó, K. (2009). Dual gaggle semantics for entailment. *Notre Dame Journal of Formal Logic, 50*, 23–41.
Bimbó, K. (2015). *Proof theory: Sequent calculi and related formalisms*. Discrete Mathematics and its Applications. Boca Raton: CRC Press.
Bimbó, K., & Dunn, J. M. (2008). *Generalized Galois logics. Relational semantics of nonclassical logical calculi. CSLI Lecture Notes*, (Vol. 188). Stanford: CSLI Publications.
Bimbó, K., & Dunn, J. M. (2012). New consecution calculi for $R^t_{\rightarrow}$. *Notre Dame Journal of Formal Logic, 53*, 491–509.
Bimbó, K., & Dunn, J. M. (2013). On the decidability of implicational ticket entailment. *Journal of Symbolic Logic, 78*, 214–236.
Donchenko, V. V. (1963). Nekotorye voprosy, svyazannye s problemoĭ rasresheniya dlya ischisleniya strogoĭ implikatsii, *Problemy Logiki*, Akademiya Nauk SSSR, Moscow, pp.18–24. [Some questions concerning the decision problem for Ackermann's calculus of rigorous implication].
Dunn, J. M. (1966). *The Algebra of Intensional Logics*, Ph.D. thesis, University of Pittsburgh, Ann Arbor (UMI).
Dunn, J. M. (1973). A 'Gentzen system' for positive relevant implication, (abstract). *Journal of Symbolic Logic, 38*, 356–357.
Dunn, J. M. (1986). Relevance logic and entailment. In D. Gabbay & F. Guenthner (Eds.), *Handbook of philosophical logic* (1st edn., Vol. 3, pp. 117–224). Dordrecht: D. Reidel.
Dunn, J. M. (1993). Partial gaggles applied to logics with restricted structural rules. In K. Došen & P. Schroeder-Heister (Eds.), *Substructural logics* (pp. 63–108). Oxford: Clarendon.
Dunn, J. M. (2001). A representation of relation algebras using Routley-Meyer frames. In C. A. Anderson & M. Zelëny (Eds.), *Logic, meaning and computation. Essays in memory of Alonzo Church* (pp. 77–108). Dordrecht: Kluwer.
Dunn, J. M. (2008). Information in computer science. In P. Adriaans & J. van Benthem (Eds.), *Philosophy of information*, D. M. Gabbay, P. Thagard & J. Woods (Eds.), Handbook of the philosophy of science (Vol. 8, pp. 581–608). Amsterdam: Elsevier.
Hollings, C. (2014). *Mathematics across the iron curtain: A history of the algebraic theory of semigroups*. History of mathematics (Vol. 41). Providence: American Mathematical Society.
Maksimova, L. L. (1964). O sisteme aksiom ischisleniya strogoĭ implikatsii. *Algebra i logika, 3*(5), 59–68. [On the system of axioms of the calculus of rigorous implication].

Maksimova, L. L. (1966). Formal'nye vyvody v ischislenii strogoĭ implikatsii. *Algebra i logika*, *5*(6), 33–39. [Formal deductions in the calculus of rigorous implication].
Maksimova, L. L. (1967a). Nekotorye voprosy ischisleniya Akkermana. *Doklady AN SSSR*, *175*(6), 1222–1224. [Translation: Some problems of the Ackermann calculus, *Soviet Mathematics*, *8* (1967), pp. 997–999.].
Maksimova, L. L. (1967b). O modelyakh ischisleniya *E*. *Algebra i logika*, *6*(6), 5–20. [On models of the calculus *E*].
Maksimova, L. L. (1967c). Topologicheskie prostranstva i kvaziuporyadochennye mnozhestva. *Algebra i logika*, *6*(4), 51–59. [Topological spaces and quasi-ordered sets].
Maksimova, L. L. (1968a). *Logicheskie ischisleniya strogoĭ implikatsii*, Ph.D. thesis, SO AN SSSR, Novosibirsk. [Logical calculi of rigorous implication].
Maksimova, L. L. (1968b). Ob ischislenii strogoĭ implikatsii. *Algebra i logika*, *7*(2), 55–76. [On a calculus of rigorous implication].
Maksimova, L. L. (1969). Interpretatsiya sistem so strogoĭ implikatsieĭ, *10th All-Union Algebraic Colloquium (Abstracts)*, Novosibirsk, p. 113. [An interpretation of systems with rigorous implication].
Maksimova, L. L. (1970). O E-teoriyakh. *Algebra i logika*, *8*(5), 530–538. [On the theories of E].
Maksimova, L. L. (1971). Interpretatsiya i teoremy otdeleniya dlya ischisleniĭ *E* i *R*. *Algebra i logika*, *10*(4), 376–392. [Interpretations and separation theorems for the calculi *E* and *R*].
Maksimova, L. L. (1973a). A semantics for the system E of entailment. *Bulletin of the Section of Logic of the Polish Academy of Sciences*, *2*, 18–21.
Maksimova, L. L. (1973b). Struktury s implikatsieĭ. *Algebra i logika*, *12*(4), 445–467. [Structures with implication].
Maksimova, L. L. (1976). Printsip razdeleniya peremennykh v propozitsional'nykh logikakh. *Algebra i logika*, *15*(2), 168–184. [The principle of variable segregation in propositional logics].
Maksimova, L. L. (1989). Relevance principles and formal deducibility. In J. Norman & R. Sylvan (Eds.), *Directions in relevant logic* (Vol. 1, pp. 95–97). Reason and Argument. Dordrecht: Kluwer.
Meyer, R. K., & Dunn, J. M. (1969). E, R and γ. *Journal of Symbolic Logic*, *34*, 460–474. (Reprinted in Anderson, A. R., & Belnap, N. D. (1975). *Entailment: The logic of relevance and necessity* (Vol. 1, pp. 300–314). Princeton: Princeton University Press, Sect. 25.2.
Meyer, R. K., Dunn, J. M., & Leblanc, H. (1974). Completeness of relevant quantification theories. *Notre Dame Journal of Formal Logic*, *15*(1), 97–121.
Priestley, H. A. (1970). Representation of distributive lattices by means of ordered Stone spaces. *Bulletin of the London Mathematical Society*, *2*, 186–190.
Routley, R., & Meyer, R. K. (1972). The semantics of entailment - III. *Journal of Philosophical Logic*, *1*, 192–208.
Routley, R., & Routley, V. (1972). The semantics of first degree entailment. *Noûs*, *6*, 335–359.
Stone, M. H. (1936). The theory of representations for Boolean algebras. *Transactions of the American Mathematical Society*, *40*, 37–111.
Urquhart, A. (1984). The undecidability of entailment and relevant implication. *Journal of Symbolic Logic*, *49*, 1059–1073.
Urquhart, A. (1996). Duality for algebras of relevant logics. *Studia Logica*, *56*, 263–276.
Urquhart, A. (2007). Four variables suffice. *Australasian Journal of Logic*, *5*, 66–73.

Chapter 4
A Note on Strictly Positive Logics and Word Rewriting Systems

Lev Beklemishev

Abstract We establish a natural translation from word rewriting systems to strictly positive polymodal logics. Thereby, the latter can be considered as a generalization of the former. As a corollary we obtain examples of undecidable finitely axiomatizable strictly positive normal modal logics. The translation has its counterpart on the level of proofs: we formulate a natural deep inference proof system for strictly positive logics generalizing derivations in word rewriting systems. We also make some observations and formulate open questions related to the theory of modal companions of superintuitionistic logics that was initiated by L.L. Maksimova and V.V. Rybakov.

Introduction

In this note we study the fragment of polymodal logic consisting of implications of the form $A \to B$, where A and B are formulas built-up from $\top$ and propositional variables using just $\wedge$ and the diamond modalities. We call such formulas A and B *strictly positive* and will often omit the word 'strictly.'

The interest towards strictly positive logics independently emerged within two different disciplines: description logic and provability logic (see (Kurucz et al. 2010; Dashkov 2012; Beklemishev 2012), though the systems equivalent to strictly positive logics were considered in description logic much earlier). In both cases, it was observed that the strictly positive language combines simplicity and efficiency while retaining a substantial amount of expressive power of modal logic. Thus, strictly positive fragments of many standard modal and description logics are polytime decidable. The strictly positive fragment of the (Kripke incomplete) polymodal provability logic **GLP** is both polytime decidable and complete w.r.t. a natural class of finite Kripke frames (Dashkov 2012). The positive variable-free fragment of this logic

L. Beklemishev (✉)
Steklov Mathematical Institute, RAS, Moscow, Russia
e-mail: bekl@mi.ras.ru

L. Beklemishev
Moscow M.V. Lomonosov State University, Moscow, Russia

L. Beklemishev
National Research University Higher School of Economics, Moscow, Russia

S. Odintsov (ed.), *Larisa Maksimova on Implication, Interpolation, and Definability*, Outstanding Contributions to Logic 15, https://doi.org/10.1007/978-3-319-69917-2_4

gives rise to a natural ordinal notation system up to the ordinal ε_0 and provides for a proof-theoretic analysis of Peano arithmetic (Beklemishev 2012).

In the present paper we study some general questions related to strictly positive logics. In particular, we establish a link between proof systems for strictly positive logics and the standard word rewriting (semi-Thue) systems.

4.1 Strictly Positive Logics

Consider a modal language $\mathscr{L}_\Sigma$ with propositional variables $p, q, \ldots$, a constant $\top$, conjunction $\wedge$, and a possibly infinite set of symbols $\Sigma = \{a_i : i \in I\}$ understood as diamond modalities. The family Σ is called the *signature* of the language $\mathscr{L}_\Sigma$. Strictly positive formulas (or simply *formulas*) are built up by the grammar:

$$A ::= p \mid \top \mid (A \wedge A) \mid aA, \quad \text{where } a \in \Sigma.$$

Sequents are expressions of the form $A \vdash B$ where A, B are strictly positive formulas. We present two types of calculi for strictly positive logics: sequent-style and deep inference-style.

Sequent-style systems for several positive logics have been introduced and studied in (Beklemishev 2012, 2014). This was preceded by an equational logic characterizations of the same logics in (Dashkov 2012).

Basic sequent-style system, denoted $\mathbf{K}^+$, is given by the following axioms and rules:

1. $A \vdash A$; $A \vdash \top$; if $A \vdash B$ and $B \vdash C$ then $A \vdash C$ (syllogism);
2. $A \wedge B \vdash A$; $A \wedge B \vdash B$; if $A \vdash B$ and $A \vdash C$ then $A \vdash B \wedge C$;
3. if $A \vdash B$ then $aA \vdash aB$.

It has not been explicitly mentioned but easily follows from the techniques of (Dashkov 2012; Beklemishev 2014) that $\mathbf{K}^+$ axiomatizes the strictly positive fragment of basic polymodal logic $\mathbf{K}$, so we state this result without proof.

Theorem 1 *A sequent $A \vdash B$ is provable in $\mathbf{K}^+$ iff $\mathbf{K} \vdash A \to B$.*

If one wishes, one can adjoin some further axioms to $\mathbf{K}^+$, which correspond to some standard modal axioms.

(4) $aaA \vdash aA$;
(T) $A \vdash aA$;
(5) $aA \wedge aB \vdash a(A \wedge aB)$.

Let $\mathbf{K4}^+$ denote the logic axiomatized over $\mathbf{K}^+$ by Axiom (4); $\mathbf{S4}^+$ is axiomatized over $\mathbf{K}^+$ by (4) and (T); $\mathbf{S5}^+$ is $\mathbf{S4}^+$ together with (5).

If L is an extension of $\mathbf{K4}^+$, we write $A \vdash_L B$ for the statement that the sequent $A \vdash B$ is provable in L. Formulas A and B are called *L-equivalent* (written $A \sim_L B$) if $A \vdash_L B$ and $B \vdash_L A$.

The following theorem was obtained by Dashkov (2012) (the case $\mathbf{K4}^+$) and by Dashkov and Svyatlovsky (the cases $\mathbf{S4}^+$ and $\mathbf{S5}^+$), see (Svyatlovsky 2014). The latter paper also gives an infinite though explicit axiomatization of the strictly positive fragment of the logic **K4.3**.

Theorem 2 *Let L be any of the logics* **K4**, **S4**, **S5**. *Then $L \vdash A \to B$ iff $A \vdash_{L^+} B$.*

Let $C[A/p]$ denote the result of replacing in C all occurrences of a variable p by A. If a logic L contains $\mathbf{K}^+$ then $\vdash_L$ satisfies the following *positive replacement lemma*.

Lemma 1 *Suppose $A \vdash_L B$, then $C[A/p] \vdash_L C[B/p]$, for any C.*

Proof Induction on the build-up of C. □

A positive logic L is called *normal* if it contains $\mathbf{K}^+$ and is closed under the following *substitution rule*: if $A \vdash_L B$ then $A[C/p] \vdash_L B[C/p]$. It is clear that all the positive logics considered so far are normal.

4.2 Modal Companions of Strictly Positive Logics

The language of modal logic is obtained from $\mathscr{L}_\Sigma$ by adding the boolean connectives. Recall that a modal logic is called normal if it contains basic modal logic **K** and is closed under the rules *modus ponens*, necessitation and substitution.

There is a natural functor associating with each normal modal logic L its strictly positive fragment $\mathscr{P}(L)$ consisting of all sequents $A \vdash B$ with A, B strictly positive such that $L \vdash (A \to B)$. Vice versa, to each strictly positive normal logic P we can associate its modal counterpart $\mathscr{M}(P)$ axiomatized over **K** by all the implications $A \to B$ such that $A \vdash_P B$.

We note that both functors preserve inclusion, that is, are monotone. The following obvious lemma states that $\mathscr{M}$ and $\mathscr{P}$, in fact, form a *Galois connection*.

Lemma 2 *For any normal modal logic L and any strictly positive normal logic P,*

$$\mathscr{M}(P) \subseteq L \iff P \subseteq \mathscr{P}(L).$$

As a standard consequence we obtain that the composite operations $\mathscr{M}\mathscr{P}$ and $\mathscr{P}\mathscr{M}$ are monotone and idempotent on the corresponding classes of logics. Moreover,

(i) $\mathscr{M}(\mathscr{P}(L)) \subseteq L$;
(ii) $P \subseteq \mathscr{P}(\mathscr{M}(P))$.

The converse inclusions in (i) and (ii), generally, do not hold. For (i) we can refer to the results of Dashkov (Dashkov 2012). He has shown that for the standard modal logic **GL** of Gödel and Löb we have $\mathscr{P}(\mathbf{GL}) = \mathbf{K4}^+$. However, by Theorem 2, $\mathscr{M}(\mathbf{K4}^+) = \mathbf{K4} \neq \mathbf{GL}$.

For (ii) we refer to Kurucz et al. (2011). Let $\Sigma = \{\Diamond\}$ and consider the logic P obtained from $\mathbf{K}^+$ by adding the schema $\Diamond A \vdash A$. We claim that

$$\mathscr{M}(P) \vdash p \wedge \Diamond\top \to \Diamond p.$$

Indeed, substituting $\neg p$ for A we obtain $\mathscr{M}(P) \vdash p \to \Box p$. Furthermore, $\mathbf{K} \vdash \Diamond\top \wedge \Box p \to \Diamond p$, therefore $\mathscr{M}(P) \vdash p \wedge \Diamond\top \to \Diamond p$.

On the other hand, $p \wedge \Diamond\top \nvdash_P \Diamond p$. Consider a Kripke model (W, R, v) where $W = \{0, 1\}$ and the only R-related elements are $0R1$. We also let $v(p) = \{0\}$, and all the other variables are assumed to be false. For every positive formula A, the set of all the nodes of W where A is true is downward closed. Hence, it is easy to see that this model is sound for P. However $W, 0 \nvDash p \wedge \Diamond\top \to \Diamond p$.

It has to be noted that strictly positive logics not representable as strictly positive fragments of modal logics naturally occur in the study of reflection principles in arithmetic. For example, the system $\mathbf{RC}\omega$ axiomatizing the properties of uniform reflection principles over Peano arithmetic is of this kind (Beklemishev 2014).

A modal logic L such that $\mathscr{P}(L) = P$ is called a *modal companion* of a positive logic P. As we have seen, not every normal positive logic P has a companion. If it does, then $\mathscr{M}(P)$ is the least modal companion of P in the sense that $\mathscr{M}(P)$ is contained in any other companion of P. The set of modal companions of P, if it is not empty, also has maximal elements. This statement immediately follows from Zorn's lemma noting that the union of a chain of modal companions of P is also its modal companion.

The notion of modal companion of a strictly positive logic is parallel to the one of a superintuitionistic logic. The systematic study of maximal and minimal modal companions of superintuitionistic logics was initiated by Maksimova and Rybakov (1974) and followed by several important results including the Blok–Esakia theorem, see (Blok 1976; Esakia 1976) and also (Chagrov and Zakharyaschev 1992) for a survey. For normal strictly positive logics many natural questions regarding modal companions present themselves, however so far this interesting area has not been really explored. We mention some such questions here, all of which have well-known answers in the case of superintuitionistic logics.

Problem 1. Find useful criteria for a normal strictly positive logic P to have a modal companion. Equivalently, for which strictly positive logics P do we have $\mathscr{P}(\mathscr{M}(P)) = P$?

Problem 2. Are there normal strictly positive logics P, for which there is no greatest modal companion? Are $\mathbf{K}^+$ and $\mathbf{K4}^+$ such logics?

Problem 3. Is **GL** a maximal modal companion of $\mathbf{K4}^+$? In fact, except for the cases where maximal and minimal modal companions coincide, we do not know any specific examples of maximal modal companions.

Let us also note that modal logics L representable as the least modal companions of strictly positive logics are exactly those axiomatized over **K** by a set of strictly positive implications. Hence, if $L = \mathscr{M}(P)$, as a consequence of Lemma 2 we have

$$L = \mathscr{M}(P) = \mathscr{M}\mathscr{P}\mathscr{M}(P) = \mathscr{M}\mathscr{P}(L).$$

Strictly positive implications are Sahlqvist formulas, therefore such logics enjoy the nice properties ensured by Sahlqvist theorem, that is, their completeness with respect to an elementary class of frames and canonicity. Hence, we obtain the following theorem.

Theorem 3 *Let P be a strictly positive logic.*

(i) The logics $\mathscr{M}(P)$ and $\mathscr{P}\mathscr{M}(P)$ are complete w.r.t. an elementary class of frames. Moreover, they both are valid in the canonical frame for $\mathscr{M}(P)$.
(ii) P has a modal companion iff P is Kripke complete.

Proof We only prove Claim (ii). Clearly, if P is complete w.r.t. a class $\mathscr{C}$ of Kripke frames, then the modal logic of $\mathscr{C}$ is conservative over P, that is, P has a modal companion. If, on the other hand, $P = \mathscr{P}(L)$, for some normal modal logic L, then also $P = \mathscr{P}(\mathscr{M}(P))$. However, by Sahlqvist theorem $\mathscr{M}(P)$ is Kripke complete (w.r.t. an elementary class of frames). Hence, so is P. □

In particular, any normal modal logic has a Kripke complete strictly positive fragment. An example of this phenomenon is the Kripke incomplete polymodal provability logic **GLP**. Its strictly positive fragment **RC** according to Dashkov (2012) is complete w.r.t. a class of finite Kripke frames.

4.3 Strictly Positive Deep Inference Calculus

It is natural to axiomatize the consequence relation on $\mathscr{L}_\Sigma$ in such a way that the derived objects are positive formulas and $A \vdash B$ is understood as provability of B from hypothesis A.

We postulate the following conjunction introduction and elimination rules:

$$\frac{A}{A \wedge A} \qquad \frac{A \wedge B}{A} \qquad \frac{A \wedge B}{B}$$

The rule for $\top$ is just

$$\frac{A}{\top}$$

Notice that all the rules have one premiss. Rules in deep inference calculi are applied within a context. A *context* is a strictly positive formula $C(p)$ in which a variable p occurs only once. Let

$$\frac{A}{B}$$

be a rule instance. For any context $C(p)$, we say that $C(B)$ is obtained from $C(A)$ by a *rule application*. A *derivation* is a sequence of formulas in which every member, except for the first one, is obtained from the previous one by a rule application.

Let L be a normal positive logic given by a set S of sequents (schemata) over $\mathbf{K}^+$. We can naturally associate with L its deep inference version L_D where, in addition to the above mentioned rules for $\wedge$ and $\top$, for every axiom-sequent $A \vdash B$ from S a rule $\frac{A}{B}$ in L_D is postulated.

We note the following property of L_D.

Lemma 3 *If $A \vdash_{L_D} B$ then $C(A) \vdash_{L_D} C(B)$, for any context C.*

Proof Obvious induction on the length of the derivation $A \vdash_{L_D} B$ using the fact that if $C_1(p)$, $C_2(p)$ are contexts then so is $C_1(C_2(p))$. □

Theorem 4 *$A \vdash_L B$ iff B is provable from A in L_D.*

Proof Both implications are established by induction on the number of rule applications in the corresponding derivation.

($\Leftarrow$) Assume B is provable from A in L_D and consider the last rule application $\frac{B}{B'}$ in this derivation. By the induction hypothesis $A \vdash_L B'$. There is a context $C(p)$ such that $B = C(B'')$, $B' = C(A'')$ and $\frac{B''}{A''}$ is an instance of a postulated rule of L_D. Thus, we obtain $A'' \vdash_L B''$ and by positive replacement $B' = C(A'') \vdash_L C(B'') = B$. By the transitivity rule $A \vdash_L B$.

For the ($\Rightarrow$) we note that the syllogism rule corresponds to the composition of derivations. The conjunction elimination axioms match the corresponding rules.

To treat the conjunction introduction rule assume $C \vdash_L A$ and $C \vdash_L B$. By the IH we have L_D derivations of A from C and of B from C. Lemma 3 yields derivations of $A \wedge C$ from $C \wedge C$ and of $A \wedge B$ from $A \wedge C$. Hence, we can derive in L_D: C, $C \wedge C$, ..., $A \wedge C$, ..., $A \wedge B$, as required.

The modal rule is also interpreted by putting a deep inference proof within a context. If there is an L_D proof of B from A, then by Lemma 3 there is a proof of aB from aA. □

Notice that this yields deep inference systems for $\mathbf{K}^+$, $\mathbf{K4}^+$, $\mathbf{S4}^+$ and $\mathbf{S5}^+$.

4.4 Word Rewriting Systems

A *word rewriting system* over an alphabet Σ is given by a set of rules of the form $A \mapsto B$ where A, B are words in Σ. Such systems are also known as *semi–Thue* systems (see Chap. 7 in Davis et al. (1994)). A *rule application* is a substitution of an occurrence of A in any word by B:

$$XAY \to XBY.$$

A *derivation* in a system R is a sequence of words in which every member is obtained from the previous one by an application of one of the rules of R. We write $A \twoheadrightarrow_R B$ iff there is a derivation of B from A in R (the subscript R is omitted if understood from the context).

It is well-known that finite word rewriting systems (over a finite alphabet Σ) are a universal model of computation. In particular, there is a finite system R such that it is undecidable whether a given word B is derivable from a given word A.

To each word rewriting system R over Σ we associate a normal strictly positive logic L_R in $\mathscr{L}_\Sigma$. L_R is obtained from $\mathbf{K}^+$ by adding the axioms $Ap \vdash Bp$, for each of the rules $A \mapsto B$ from R. The words A and B are now understood as sequences of modalities.

Theorem 5 *$A \twoheadrightarrow_R B$ iff $Ap \vdash_{L_R} Bp$.*

Proof (only if) We argue by induction on the length of an R-derivation x of B from A. Basis is easy. Suppose x has the form:

$$A \twoheadrightarrow XUY \to XVY = B,$$

where $U \mapsto V$ is a rule from R. By IH we have $Ap \vdash_{L_R} XUYp$. By the L_R axiom we obtain $UYp \vdash_{L_R} VYp$, and then by positive replacement $XUYp \vdash_{L_R} XVYp$. Hence, $Ap \vdash_{L_R} Bp$.

The (if) part is based on the following two lemmas.

Lemma 4 *Assume $\top$ does not occur in A, B. If $A \vdash_{L_R} B$ then there is a derivation of B from A in $(L_R)_D$ in which the $\top$-rule is not applied.*

Proof Induction on the number of applications of the $\top$-rule. Consider any such application

$$\frac{C(A_1)}{C(\top)}$$

The part of the derivation after $C(\top)$ may contain some occurrences of $\top$ inherited from this one. Replacing them all by A_1 yields a derivation of B from $C(A_1)$ with the same number of the $\top$-rule applications. Then, the derivation $A, \dots, C(A_1), \dots, B$ has one less application of the $\top$-rule than the original derivation. □

Lemma 5 *Assume $A, B \in \Sigma^*$ and $Ap \vdash_{(L_R)_D} Bp$. Then there is a derivation of Bp from Ap in which no conjunction rule is applied.*

Proof By Lemma 4 we may assume that the $\top$-rule is not applied in the given derivation. We argue by induction on the number of conjunction introduction rule applications in the given derivation d. Since the $\top$-rule is not applied in d, every conjunction occurrence disappears as a result of conjunction elimination rule application either to itself, or to an external conjunction. Every formula containing at least one conjunction has the form $\gamma(C_1 \wedge C_2)$ where $\gamma \in \Sigma^*$ (and the displayed conjunction is the outermost one).

In all the formulas of the derivation consider the outermost conjunction. Notice that at least one outermost conjunction must be introduced in the derivation (e.g., such is the conjunction introduced first). We select the chronologically last introduced outermost conjunction. We notice that no conjunction is introduced outside this one before it is eliminated. Otherwise, the first such application would introduce an outermost conjunction later than the selected one. Hence, the selected conjunction has exactly one successor in each step of the derivation until it disappears as a result of conjunction elimination applied to itself:

$$\gamma C, \gamma(C \wedge C), \dots, \delta(C_1 \wedge C_2), \delta C_i.$$

Notice that the R-rules do not apply to conjunctions, and the conjunction rules can only be applied inside the selected conjunction. Therefore, there exist separate derivations of $\delta(q)$ from $\gamma(q)$, and of each C_j ($j \in \{1, 2\}$) from C, respectively. It follows that we can replace this subderivation by

$$\gamma C, \dots, \delta C, \dots, \delta C_i,$$

thus eliminating at least one application of conjunction introduction rule in the whole derivation. □

To complete the proof of Theorem 5 we notice that a deep inference format L_R-derivation of Bp from Ap in which no $\top$-rule and conjunction rules are applied is essentially an R-derivation of B from A. The only applicable rules are the R-rules whose effect is exactly that of R-substitutions. □

Corollary 1 *There is a finitely axiomatizable undecidable strictly positive logic.*

It has to be noted that the finitely axiomatized strictly positive logics that have naturally occurred so far all are polytime decidable, see (Dashkov 2012; Beklemishev 2014).

The results of the last section of this paper have a very close predecessor in the work of Valentin Shehtman and Alexander Chagrov, see (Shehtman 1982; Chagrov and Shehtman 1995). In particular, Chagrov and Shehtman exhibit undecidable propositional polymodal logics whose axioms are given by the implications of the form $A \to B$, where A and B are sequences of $\Box$-modalities. Clearly, such logics are the minimal modal companions of the positive logics we considered in this section. The authors, however, use semantical rather than syntactical arguments to establish a correspondence of their logics with the (semi-)Thue systems. Similar results, including a form of Corollary 1, occur in the work on description logics (Baader 2003,Theorem 18). In a sense, the correspondence between strictly positive logics of the considered kind and semi-Thue systems is even closer than for modal logics, for it extends to the level of derivations in the deep inference proof system.

Added in Proof

In the meantime new interesting results on strictly positive logics were announced in Kikot et al. (2016). The authors focus on the problem whether a finitely axiomatized strictly positive logic has a modal companion. In particular, their Theorem 1 shows that the problem whether a finitely axiomatized strictly positive logic has a modal companion (equivalently, is Kripke complete) is undecidable.

Acknowledgements We thank Daniyar Shamkanov and the anonymous referees for spotting some errors in the previous version of the paper, and Valentin Shehtman, Michael Zakharyaschev and Stas Kikot for pointing out connections with the previous work.

The investigation presented in this article was supported by Russian Foundation for Basic Research project No. 15–01–09218a and by the Presidential council for support of leading scientific schools.

References

Baader, F. (2003). Resricted role-value-maps in a description logic with existential restrictions and terminological cycles. In D. Calvanese, G. De Giacomo, E. Franconi (Ed.), *Proceeding of the 2003 International Workshop on Description Logics (DL 2003)*, Rome, Italy, September 5–7, 2003, Vol. 81, CEUR Workshop Proceedings, CEUR–WS.org.

Beklemishev, L. D. (2012). Calibrating provability logic: from modal logic to reflection calculus. In T. Bolander, T. Braüner, S. Ghilardi, & L. Moss (Eds.), *Advances in Modal Logic* (Vol. 9, pp. 89–94). London: College Publications.

Beklemishev, L. D. (2014). Positive provability logic for uniform reflection principles. *Annals of Pure and Applied Logic, 165*(1), 82–105.

Blok, W.J. (1976). Varieties of interior algebras, PhD thesis, University of Amsterdam.

Chagrov, A.V. & Shehtman, V.B. (1995). Algorithmic aspects of propositional tense logics, In *Lecture Notes in Computer Science* (Vol. 933, pp. 442–455).

Chagrov, A. V., & Zakharyaschev, M. (1992). Modal companions of intermediate propositional logics. *Studia Logica, 51*(1), 49–82.

Dashkov, E. V. (2012). O positivnom fragmente polimodalnoy logiki dokazuemosti GLP, Matematicheskie Zametki 91(3): 331–346. [Translation: "On the positive fragment of the polymodal provability logic GLP". Mathematical Notes, 91(3), 318–333.

Davis, M., Sigal, R., & Weyuker, E.J. (1994). *Computability, complexity, and languages: fundamentals of theoretical computer science* (2nd ed.). Academic Press.

Esakia, L. L. (1976). On modal companions of superintuitionistic logics. In *VII Soviet Symposium on Logic*. Kiev.

Kikot, S., Kurucz, A., Tanaka, Y., Wolter, F. & Zakharyaschev, M. (2016). On the completeness of EL-equiations: First results. In *11th International Conference on Advances in Modal Logic, Short Papers (Budapest, 30 August - 2 September, 2016)*, pp. 82–87.

Kurucz, A., Wolter, F. & Zakharyaschev, M. (2010). Islands of tractability for relational constraints: towards dichotomy results for the description logic **EL**. In *Advances in Modal Logic* Vol. 8 pp. 271–291. College Publications, London.

Kurucz, A., Tanaka, Y., Wolter, F., & Zakharyaschev, M. (2011). Conservativity of Boolean algebras with operators over semilattices with operators. *Proceedings of TACL*, 49–52.

Maksimova, L.L. & Rybakov, V.V. (1974). A lattice of normal modal logics. *Algebra i Logika*, 13(2): 188–216.

Shehtman, V.B. (1982). Undecidable propositional calculi. In*Problems of cybernetics. Non-classical logics and their applications*, Moscow, pp. 74–116. In Russian.

Svyatlovsky, M. (2014) *Positivnye fragmenty modalnykh logik*, Manuscript, [Positive fragments of modal logics]. http://www.mi.ras.ru/~bekl/Papers/work_2.pdf.

Wolter, F. & Zakharyaschev, M. (2014). On the Blok–Esakia theorem, *in* G. Bezhanishvili (ed.) *Leo Esakia on duality in modal and intuitionistic logics*, Vol. 4 of *Outstanding Contributions to Logic*, Springer, pp. 99–118.

Chapter 5
Characteristic Formulas Over Intermediate Logics

Alex Citkin

Abstract We expand the notion of characteristic formula to infinite finitely presented subdirectly irreducible algebras. We prove that there is a continuum of varieties of Heyting algebras containing infinite finitely presented subdirectly irreducible algebras. Moreover, we prove that there is a continuum of intermediate logics that can be axiomatized by characteristic formulas of countable algebras, while they are not axiomatizable by standard Yankov (Jankov) formulas. We also give the examples of intermediate logics that are not axiomatizable by characteristic formulas of infinite algebras. Further, using the Gödel–McKinsey–Tarski translation, we extend these results to the varieties of interior algebras and normal extensions of S4. For this, using Maksimova's Translation Lemma, we show that a finite presentation of a given Heyting algebra can be extended to its modal span. So, Maksimova's Lemma allows us to extend the properties established for the finitely presented Heyting algebras to interior algebras.

5.1 Introduction

This paper deals with intermediate propositional logics (called logics thereafter), that is, the logics intermediate between the intuitionistic propositional logic (IPL for short) and the classical propositional logic (CPL for short). The systematic study of this class of logics started in the mid 1950s, and it was soon established that the lattice of intermediate logics has infinite width and contains the chains ordered by type ω^ω (cf. Umezawa 1959). The traditional way of defining a given logic is by using a set of axioms, from which all those valid in this logic formulas (and only them) can be derived by inference rules, for instance, by modus ponens and (simultaneous) substitution. Often, while trying to survey the objects of a given infinite class, we decompose the objects into the simple ones: we decompose the natural numbers into products of prime numbers, or we decompose an algebra into subdirect product of subdirectly irreducible (s.i. for short) algebras, etc. So, we decompose an object

A. Citkin (✉)
Metropolitan Telecommunications, 30 Upper Warren Way, Warren, NJ 07059, USA
e-mail: acitkin@gmail.com

S. Odintsov (ed.), *Larisa Maksimova on Implication, Interpolation, and Definability*, Outstanding Contributions to Logic 15, https://doi.org/10.1007/978-3-319-69917-2_5

into “irreducible” objects, and then we study the properties of these irreducibles. In case of logics defined by axioms, it is natural to try to decompose the axioms into conjunctively irreducible (c.i. for short) formulas. Moreover, if a logic can be defined by a set of c.i. axioms, this logic can be defined by an independent set of axioms, i.e. such a set of axioms that none of the axioms is derivable from the rest. In (Yankov 1969) V.A. Yankov[1] described all c.i. formulas of IPL. More precisely, with each finite s.i. Heyting algebra **A**, he associated a formula $\chi(\mathbf{A})$, which he called a characteristic formula of **A**. Nowadays these formulas are known as Yankov formulas. Also, he had proven that the classes of c.i. and characteristic formulas coincide.

A very important property of Yankov formulas (proven in Yankov 1963) is the following:

$$\text{formula } \chi(\mathbf{A}) \text{ is valid in a logic } \mathcal{L} \text{ if and only if algebra } \mathbf{A} \text{ is not a model of } \mathcal{L}. \tag{CH}$$

Back then (prior to 1963), it was commonly accepted that every intermediate logic enjoys finite model property (fmp thereafter), that is, that every intermediate logic is defined by its finite models (Heyting algebras). And this had made Yankov formulas even more appealing.

Later, it was observed that not every logic can be defined by a set of c.i. formulas, i.e. not every formula is interderivable with a set of c.i. formulas. Indeed, it is not hard to see that if two distinct logics $\mathcal{L}_1$ and $\mathcal{L}_2$ have the same classes of finite models, their classes of valid characteristic formulas coincide. Hence, at least one of such logics cannot be axiomatized by Yankov formulas. For instance, if $\mathcal{L}_1$ is a logic lacking fmp (the first example of such a logic was introduced by Yankov in Yankov 1968) and $\mathcal{L}_2$ is a logic of all finite models of $\mathcal{L}_1$, then one of these logics (namely $\mathcal{L}_2$) cannot be defined by Yankov formulas. Moreover, it turned out that there is continuum many logics not axiomatizable by c.i. formulas (cf. Tomaszewski 2003, Corollary p.128).

Independently, and at about the same time, formulas enjoying the same properties as Yankov formulas have been introduced by K. Fine for the modal logics (Fine 1974) and by D. de Jongh for the intermediate logics (de Jongh 1968) (for this reason sometimes Yankov formulas are called Yankov-de Jongh formulas (e.g. Bezhanishvili 2006) or Yankov-Fine formulas in case of modal logics (e.d. Bezhanishvili 2006). K. Fine and D. de Jongh constructed their formulas based upon finite Kripke frames rather than algebras. Since frame formulas enjoy the same properties as Yankov formulas, the Yankov formula of a given algebra is interderivable with a frame formula of its frame. Later, the theory of the frame formulas was extended further to different types of subframes (for more details see e.g. Chagrov and Zakharyaschev 1997; Bezhanishvili 2006; Yang 2009). It also became apparent that the idea used by

[1]There are two transcriptions of the last name: ‘Yankov’, that is used by Zentralblatt, and ‘Jankov’ used by Mathematical Reviews. The latter transcription is used much more often even though the former is more accurate and, as V.A. Yankov mentioned to the author, the transcription “Yankov” is preferred by him.

Yankov for intermediate logics and Heyting algebras can be applied to modal logics and interior algebras (see e.g. Rautenberg 1979; Blok 1976).

In Zakharyashchev (1983, 1988, 1989), M. Zakharyaschev, using a modification of frame formula, introduced the canonical formulas, with which one can axiomatize any intermediate logic or any logic of transitive frames (cf. Zakharyaschev 1992, 1996, 1997). Canonical formulas proved to be very helpful in solving some problems in intermediate and modal logics (see e.g. Chagrov and Zakharyaschev 1997, Chap. 11). An algebraic account of the theory of canonical formulas for intermediate logics was offered in Tomaszewski (2003); Bezhanishvili N. and G. (2009) where the canonical formulas are regarded as modified Yankov formulas of finite s.i. algebras (the difference between these two approaches is outlined in Bezhanishvili N. and G. (2009)).

Since it turned out that not every intermediate logic is axiomatizable by Yankov formulas, one can try to modify the definition of Yankov formula and obtain a class of formulas that allows to axiomatize any intermediate logic. Let us note that already in his original papers (Yankov 1963, 1969), Yankov had observed that characteristic formula of an algebra **A** can be written using any set of elements generating **A**. So, not finiteness of an algebra, but rather the finiteness of a generating set of an algebra is, perhaps, important for constructing formulas enjoying (CH). If we want to associate a formula with not necessarily finite algebra and especially if we want to be able to effectively write down such a formula, the algebra must be defined in some constructive way. For finite algebra, Yankov was using its diagram. In case of infinite algebras, one can try to take the finitely presented algebras (for definition cf. Mal'cev 1973) and use the defining relations of the algebra. The idea of using the defining relations of algebra instead of its diagram was introduced in Citkin (1977) for quasi-characteristic rules, and for the varieties with equationally definable principal congruences (EDPC) the defining relations were used in Blok and Pigozzi (1982) and Wolter (1993), Definition 2.4.10. So, a notion of Yankov formula can be extended to finitely presented s.i. algebras. Even in cases of finite algebras, we can gain some benefits: using different defining relations of an s.i. algebra **A**, one can construct syntactically different formulas each of which is interderivable with Yankov formula of **A** and many of which are syntactically much simpler than Yankov formula.

So, the characteristic formulas of finitely presented s.i. Heyting algebras are good candidates for generalizing the notion of Yankov formula. But, as it was shown in Citkin (1979), every finitely presented Heyting algebra is finitely approximated and, therefore, every s.i. finitely presented algebra is finite.[2] Thus, all these formulas are simply interderivable with Yankov formulas.

On the other hand, finite presentability is relative to a given variety and an algebra can be finitely presented over some varieties while being not finitely presented over

---

[2]Independently the same observation was made in Butz (1998). In Blok and Pigozzi (1982) it was proven that in finitely approximated varieties with EDPC (and, as it is well known, the variety $\mathscr{H}$ of all Heyting algebras is finitely approximated and enjoys EDPC) every s.i. finitely presented algebra is finite.

others.[3] Thus, we can define characteristic formula that depends not only on a given algebra, but also on a given variety. And the first question that needs to be answered is whether there are varieties of Heyting algebras containing infinite s.i. finitely presented algebras.

Let us recall that not every variety of Heyting algebras is finitely approximated. In fact, in Yankov (1968), using characteristic formulas, Yankov proved that there is a continuum of not finitely approximated intermediate logics, thus, there is a continuum of not finitely approximated varieties of Heyting algebras. And we will see that infinitely many of varieties contain infinite finitely presented s.i. algebras. Therefore, one can construct the characteristic formulas relative to a particular variety and such relative characteristic formulas are not Yankov formulas.

In general, if we want to axiomatize a given intermediate logic L by Yankov formulas, we take all Yankov formulas not belonging to L and we add them to the axioms of intuitionistic logic; in such a way we obtain a logic $L' \subseteq L$. If $L' \neq L$, obviously, L cannot be axiomatized by Yankov formulas. But in all such cases logic L' is not finitely approximated, thus, there may be characteristic relative to L' formulas that are not Yankov formulas. So, we may try to add these characteristic formulas to L' and obtain with their help an axiomatization for L. And, as we will see, in continuum many cases we will succeed.

The goal of the paper is to show that

1. There exist characteristic formulas that are not (interderivable with) Yankov formulas;
2. There exist continuum many intermediate logics that can be axiomatized by characteristic formulas, while they are not axiomatizable by Yankov formulas;
3. There exist intermediate logics not axiomatizable by characteristic formulas.

In Sect. 5.3, we define characteristic formula of a given s.i. algebra **A** finitely presented in a given variety $\mathscr{V}$ of Heyting algebras. And we demonstrate that the introduced formula, if regarded as an axiom, defines a co-splitting variety.[4]

In order to demonstrate (1), in Sect. 5.4 we give an example of variety $\mathscr{V}$ and of infinite s.i. algebra finitely presented over $\mathscr{V}$. We also prove (see Sect. 5.6) that there is a continuum of varieties of Heyting algebras (or intermediate logics) that can be axiomatized by relative characteristic formulas, but cannot be axiomatized by Yankov formulas. Nevertheless, it turned out that not every intermediate logic can be axiomatized by relative characteristic formulas (see Sect. 5.6.2).

In Sect. 5.7, by use of Maksimova's Translation Lemma (cf. Maksimova 1979, Translation Lemma), we establish the relations between finite presentability of an algebra **A** in a variety of Heyting algebras and finite presentability of a modal span of **A** in the corresponding variety of Grzegorczyk algebras. This relation allows us to extend the results established for intermediate logics to normal extensions of **S4**.

[3] The Heyting algebras finitely presented over variety of all Heyting algebras are studied in Butz (1998).

[4] The general fact that every finitely presented in a variety with EDPC s.i. algebra defines a splitting was observed in Blok and Pigozzi (1982).

Let us also note that (CH) entails that every Yankov, or frame formula defines a splitting[5] variety. Instigated by the book Rautenberg (1979) and papers by W. Routenberg, the splittings in the lattices of normal extensions of logic **S4** and various modal logics were intensively studied by M. Kracht and F. Wolter (see, e.g. Kracht 1990, 1993; Wolter 1993). In (Kracht 1990) Kracht is using finitely presented algebras as a main tool while studying the splittings in non-transitive modal logics. We study the splittings in sublattices of the lattice of all intermediate logics. In this sense, the present paper complements the studies of splittings undertaken by M. Kraht and F. Wolter (the overview can be found in Kracht 1999).

5.2 Basic Definitions

We consider propositional formulas built in a regular way from the (propositional) variables and connectives $\wedge, \vee, \rightarrow \neg$. We denote the formulas by capital letters A, B, C, D, perhaps with indexes, and we use $\leftrightarrow$ as abbreviation: $A \leftrightarrow B \leftrightharpoons (A \rightarrow B) \wedge (B \rightarrow A)$.

Abbreviation IPC stands for Intuitionistic Propositional Calculus, and $\vdash_{IPC}$ denotes the derivability relation in IPC (with rule of substitution). Heyting algebras are (algebraic) models of IPC.

An algebra $\mathbf{A} = (\mathsf{A}; \wedge, \vee, \rightarrow, \neg)$, where $(\mathsf{A}; \wedge, \vee)$ is a distributive lattice and $\rightarrow$ and $\neg$ are relative pseudo-complement and pseudo-complement (see e.g. Rasiowa and Sikorski (1970) for definitions), is called a *Heyting algebra*. Each Heyting algebra $\mathbf{A}$ contains a top element that we denote by $\mathbf{1}_{\mathbf{A}}$, and the bottom element that we denote by $\mathbf{0}_{\mathbf{A}}$ (and we omit the reference to the algebra when no confusion arises).

By $\mathscr{H}$ we denote the variety of all Heyting algebras. We will often use the single-generated Heying algebras. Z_n denotes the n-element single-generated Heyting algebra, so Z_2 is the two-element Boolean algebra, Z_∞ is Rieger-Nishimura ladder that we also will denote by Z. Note that n-element single-generated Heyting algebra is unique modulo isomorphism. Figure 5.1 contains diagrams of single-generated Heyting algebras with the generator marked by $\odot$.

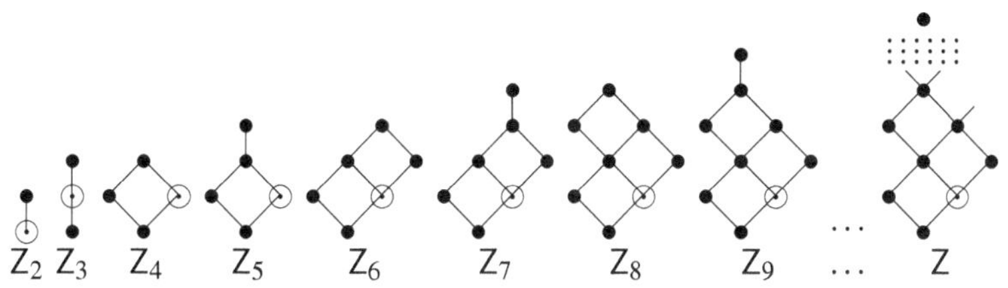

Fig. 5.1 Single-generated Heyting algebras

[5] For definition see McKenzie (1972).

Let $\mathbf{A}$ be an algebra and $\nabla \subseteq \mathbf{A}$. Then ∇ is said to be a *filter of algebra* $\mathbf{A}$ if for every $\mathsf{a}, \mathsf{b} \in \nabla$ we have $\mathsf{a} \wedge \mathsf{b} \in \nabla$, and for every $\mathsf{c} \in \mathbf{A}$ if $\mathsf{a} \leq \mathsf{c}$ then $\mathsf{c} \in \nabla$. Each filter defines a congruence: $\mathsf{a} \equiv \mathsf{b} \pmod{\theta(\nabla)}$ exactly when $\mathsf{a} \leftrightarrow \mathsf{b} \in \nabla$, thus, we can consider quotient-algebra $\mathbf{A}/\theta(\nabla)$, and we use the simplified notation $\mathbf{A}/\nabla$. If $\mathsf{a} \in \mathbf{A}$ we let $\nabla(\mathsf{a}) \leftrightharpoons \{\mathsf{b} \in \mathbf{A} \mid \mathsf{a} \leq \mathsf{b}\}$. Clearly $\nabla(\mathsf{a})$ is a filter of $\mathbf{A}$ and we say the $\nabla(\mathsf{a})$ *a filter generated by element* a.

If $\mathbf{A}, \mathbf{B}$ are algebras, by $\mathbf{A} + \mathbf{B}$ we denote a *concatenation*[6] of $\mathbf{A}$ and $\mathbf{B}$, that is $\mathbf{A} + \mathbf{B}$ is an algebra obtained by putting algebra $\mathbf{B}$ onto $\mathbf{A}$ and identifying the top element of $\mathbf{A}$ and the bottom element of $\mathbf{B}$. Let us observe the following rather simple property of concatenation.

Proposition 1 *If* $\mathbf{A}$ *and* $\mathbf{B}$ *are algebras and* $\nabla \subseteq \mathbf{B}$ *is a filter, then*

$$(\mathbf{A} + \mathbf{B})/\nabla \cong \mathbf{A} + \mathbf{B}/\nabla. \tag{5.1}$$

Given an algebra $\mathbf{A}$, by $\mathscr{V}(\mathbf{A})$ we denote the variety generated by algebra $\mathbf{A}$. If A and B are (propositional) formulas and p is a variable, by $A(B/p)$ we denote the result of the substitution of formula B for variable p in formula A. The strings of distinct variables are indicated by $\underline{p}, \underline{q}$ and if A contains variables only from the list $\underline{p} = p_1, \ldots, p_n$, we express this fact by the notation $A(\underline{p})$ or $A(p_1, \ldots, p_n)$. Accordingly, if $\mathsf{a}_1, \ldots, \mathsf{a}_n$ are elements of an algebra and $A(\underline{p})$ is a formula, we can write $A(\underline{\mathsf{a}})$ instead of $A(\mathsf{a}_1, \ldots, \mathsf{a}_n)$.

Element a is said to be *regular* if $\neg\neg\mathsf{a} = \mathsf{a}$, and element a is said to be *dense* if $\neg\neg\mathsf{a} = \mathbf{1}$ (see e.g. Rasiowa and Sikorski 1970). By $Rg(\mathbf{A})$ we denote the set of all regular elements of algebra $\mathbf{A}$, and by $Dn(\mathbf{A})$ we denote the set of all dense elements of algebra $\mathbf{A}$. Let us observe that $\mathbf{A}$ is a Boolean algebra if and only if $Dn(\mathbf{A}) = \{\mathbf{1}\}$. It is not hard to see that $Dn(\mathbf{A})$ is a filter of algebra $\mathbf{A}$ and $\mathbf{A}/Dn(\mathbf{A})$ is a Boolean algebra: the natural homomorphism sends all dense elements of $\mathbf{A}$ in $\mathbf{1}$. Moreover, $\mathbf{A}/Dn(\mathbf{A})$ isomorphic as a lattice to $R(\mathbf{A})$ (e.g. Rasiowa and Sikorski 1970).

By *valuation in algebra* $\mathbf{A}$, we understand a mapping $\nu : Var \to \mathbf{A}$, where Var is the set of all variables. Clearly, a valuation ν can be extended to a given formula A by letting $\nu(A(p_1, \ldots, p_n)) = A(\nu(p_1), \ldots \nu(p_n))$. We say that a formula A is valid in a given algebra $\mathbf{A}$ if for every valuation ν in $\mathbf{A}$, $\nu(A) = \mathbf{1}$, and we denote this by $\mathbf{A} \models A$.

If $\mathscr{V}$ is a variety and A is a formula then by $\vdash_{\mathscr{V}} A$ we denote the fact that formula A is valid in $\mathscr{V}$, that is $\mathbf{A} \models A$ for all $\mathbf{A} \in \mathscr{V}$. If $\mathscr{K}$ is a class of algebras, by $\mathsf{S}\mathscr{K}$ we denote a class of all (isomorphic copies) of subalgebras of all algebras from $\mathscr{K}$ and by $\mathsf{H}\mathscr{K}$ we denote a class of all homomorphic images of all algebras from $\mathscr{K}$. If class $\mathscr{K}$ consists of just one algebra $\mathbf{A}$, we will write $\mathsf{S}\mathbf{A}$ and $\mathsf{H}\mathbf{A}$.

Let us recall the following definition.

Definition 1 (cf. Mal'cev 1973) Let $\mathscr{V}$ be a variety, $A(\underline{p})$ be a formula and ν be a valuation in algebra $\mathbf{A}$. Then a pair $\langle A, \nu \rangle$ *defines algebra* $\mathbf{A}$ over $\mathscr{V}$ if

[6]The concatenations are often called ordered, linear or Troelstra sums. We are trying to avoid use of the term "sum" since it suggests some kind of commutativity which is not the case here.

1. Elements $a_1 = \nu(p_1), \ldots, a_n = \nu(p_n)$ generate algebra **A**;
2. $A(\underline{a}) = \mathbf{1}$;
3. For any formula $B(\underline{p})$ if $B(\underline{a}) = \mathbf{1}$, then $\vdash_{\mathcal{V}} (A(\underline{p}) \to B(\underline{p}))$.

Algebra **A** is said to be *finitely presented over variety* $\mathcal{V}$ if there exists a pair that defines algebra **A** over variety $\mathcal{V}$. We also will say that *formula* $A(\underline{p})$ *defines algebra* **A** *over variety* $\mathcal{V}$ *in generators* $\underline{a}$ or that a pair $\langle A, \nu\rangle$ is a *presentation of algebra* **A** over $\mathcal{V}$ (sometimes we will omit reference to $\mathcal{V}$).

The following criterion is very useful and it often serves as a definition of finitely presented algebra.

Proposition 2 (Mal'cev (1973)). *Let* $\mathcal{V}$ *be a variety,* $A(\underline{p})$ *be a formula and* ν *be a valuation in algebra* **A**. *Then the pair* $\langle A, \nu\rangle$ *defines algebra* **A** *over* $\mathcal{V}$ *if and only if*

1. *Elements* $a_1 = \nu(p_1), \ldots, a_n = \nu(p_n)$ *generate algebra* **A**;
2. $A(\underline{a}) = \mathbf{1}$;
3. *If* $\mathbf{B} \in \mathcal{V}$ *is an algebra,* $b_1, \ldots, b_n \in \mathbf{B}$ *and* $A(b_1, \ldots, b_n) = \mathbf{1_B}$, *then the mapping* $a_i \mapsto b_i; i = 1, \ldots, n$ *can be extended to a homomorphism of* **A** *in* **B**.

Let us note that Proposition 2 holds for any variety, and we will use it later for varieties of interior algebras.

Remark 1 Since any variety is closed relative to homomorphisms, in the Proposition 2, it is sufficient to take into consideration only s.i. algebras $\mathbf{B} \in \mathcal{V}$.

If $\mathcal{V}$ is a variety, by $SI(\mathcal{V})$ we will denote the class of all s.i. algebras from $\mathcal{V}$, by $FP(\mathcal{V})$ - a class of all algebras finitely presented over $\mathcal{V}$, and by $FPSI(\mathcal{V})$ - a class of all finitely presented over $\mathcal{V}$ s.i. algebras.

Let us note also the following properties of finitely presented algebras.

Proposition 3 *If pairs* $\langle A(\underline{p}), \nu\rangle$ *and* $\langle B(\underline{p}), \nu\rangle$ *define over* $\mathcal{V}$ *the same algebra* **A**, *then* $\vdash_{\mathcal{V}} A \leftrightarrow B$.

Proof Straight from the Definition 1(3).

Proposition 4 (Mal'cev 1973, Theorem 5, Chap. V Sect. 11) *Assume that formulas* $A(\underline{p})$ *and* $B(\underline{p})$ *define over variety* $\mathcal{V}$ *algebras* **A** *and* **B**. *Then, if* $\vdash_{\mathcal{V}} B \to A$, *then* **B** *is a homomorphic image of* **A**. *In particular, if* $\vdash_{\mathcal{V}} A \leftrightarrow B$ *the algebras* **A** *and* **B** *are isomorphic.*

We will also use the following property of finitely presented algebras (Mal'cev 1973, Corollary 7, Chap. V sect. 11):

Proposition 5 (Tietze's Theorem) *If an algebra* **A** *is finitely presented over* $\mathcal{V}$, *then* **A** *is finitely presented in any set of its generators.*

5.3 Characteristic Formulas

In this section we generalize the notion of Yankov formula to finitely presented s.i. algebras and study the properties of such formulas.

5.3.1 Yankov Formulas

Let us recall from Yankov (1968) the definition of Yankov formula.

Definition 2 Assume $\mathbf{A}$ is a finite s.i. algebra and $\mathbf{A} = \{\mathsf{a}_1, \ldots, \mathsf{a}_n\}$. With every element $\mathsf{a}_i \in \mathbf{A}$, we associate a variable p_{a_i}; $i = 1, \ldots, n$. Let

$$\tilde{D}(\mathbf{A}) \leftrightharpoons \{(p_{\mathsf{a}} \circ p_{\mathsf{b}}) \leftrightarrow p_{\mathsf{a}\circ\mathsf{b}} \mid \mathsf{a}, \mathsf{b} \in \mathbf{A}, \circ \in \{\wedge, \vee, \to\}\} \cup \{\neg p_{\mathsf{a}} \leftrightarrow p_{\neg\mathsf{a}} \mid \mathsf{a} \in \mathbf{A}\}.$$

$$D(\mathbf{A}) \leftrightharpoons \bigwedge \tilde{D}(\mathbf{A}). \tag{5.1}$$

Formula $D(\mathbf{A})$ is a *diagram formula* of algebra $\mathbf{A}$. Let o be an *opremum* of algebra $\mathbf{A}$, that is, the greatest element among all distinct from $\mathbf{1}$ elements of $\mathbf{A}$. Then formula

$$\chi(\mathbf{A}) = D(\mathbf{A}) \to p_{\mathsf{o}} \tag{5.2}$$

is called *Yankov formula*.

The most frequently used properties of Yankov formulas are presented in the following Proposition.

Proposition 6 (Yankov (1969)) *Assume* $\mathbf{A}$ *is a finite s.i. algebra,* $\mathbf{B}$ *is an algebra and* B *is a formula. Then*

(a) if $\mathbf{B} \not\models \chi(\mathbf{A})$*, then* $\mathbf{A} \in \mathsf{SHB}$*;*
(b) if $\mathbf{A} \not\models B$*, then* $B \vdash_{IPC} \chi(\mathbf{A})$*.*

The property (b) of Proposition 6 means that $\chi(\mathbf{A})$ is the weakest formula among formulas refutable in $\mathbf{A}$. Let us note that if A_1 and A_2 are two weakest formulas refutable in $\mathbf{A}$, then formulas A_1 and A_2 are interderivable in IPC.

5.3.2 Characteristic Formulas: Definition

Now, let us extend the definition of Yankov formula to the finitely presented algebras.

Definition 3 Assume $\mathcal{V}$ is a variety and $\mathbf{A} \in \mathcal{V}$ is an s.i. algebra finitely presented over $\mathcal{V}$. Suppose $\langle A(\underline{p}), \nu \rangle$ is a presentation of $\mathbf{A}$ over $\mathcal{V}$. If $B(\underline{p})$ is such a formula that $\nu(B)$ is the opremum of $\mathbf{A}$, then the formula

$$\chi_{\mathcal{V}}(\mathbf{A}) = A(\underline{p}) \to B(\underline{p}) \tag{5.3}$$

is a *characteristic formula of algebra* $\mathbf{A}$ *over variety* $\mathcal{V}$.

First of all, let us note that $\mathbf{A}$ always has an opremum, for $\mathbf{A}$ is an s.i. algebra. And, since elements $\nu(p_1), \ldots, \nu(p_n)$ generate algebra $\mathbf{A}$, there always is such a formula $B(\underline{p})$ that $\nu(B)$ is an opremum. Thus, for any s.i. finitely presented over $\mathcal{V}$ algebra, a characteristic formula can be defined. Let us establish properties of characteristic formulas similar to those of Yankov formulas.

Theorem 7 *Assume* $\mathbf{A} \in FPSI(\mathcal{V})$, $\mathbf{B} \in \mathcal{V}$ *is an algebra and* B *is a formula. Then*

(a) if $\mathbf{B} \not\models \chi_{\mathcal{V}}(A)$*, then* $\mathbf{A} \in \mathsf{SH}\mathbf{B}$*;*
(b) if $\mathbf{A} \not\models B$*, then* $B \vdash_{\mathcal{V}} \chi_{\mathcal{V}}(A)$*.*

Proof (a) Suppose $\mathbf{B} \not\models \chi_{\mathcal{V}}(A)$. By the definition of characteristic formula, $\chi_{\mathcal{V}}(\mathbf{A}) = A(\underline{p}) \to B(\underline{p})$, where $\langle A(\underline{p}), \nu \rangle$ is a defining pair and $\nu(B(\underline{p})) = \mathsf{a} \in \mathbf{A}$ is the opremum of algebra $\mathbf{A}$. Thus, since $\mathbf{B} \not\models (A(\underline{p}) \to B(\underline{p}))$, for some homomorphic image $\mathbf{B}'$ of algebra $\mathbf{B}$ and some elements $\mathsf{b}_1, \ldots, \mathsf{b}_n \in \mathbf{B}'$ we have

$$A(\mathsf{b}_1, \ldots, \mathsf{b}_n) = \mathbf{1}_{\mathbf{B}'} \text{ and } B(\mathsf{b}_1, \ldots, \mathsf{b}_n) \neq \mathbf{1}_{\mathbf{B}'}. \tag{5.4}$$

By Proposition 2, the mapping $\phi : \nu(p_i) \mapsto \mathsf{b}_i;\ i = 1, \ldots, n$ can be extended to a homomorphism $\varphi : \mathbf{A} \to \mathbf{B}'$. Let us observe that

$$\phi(\nu(B)) = B(\mathsf{b}_1, \ldots, \mathsf{b}_n) \neq \mathbf{1}_{\mathbf{B}}. \tag{5.5}$$

Recall that the opremum of a Heyting algebra belongs to every proper filter, and therefore it belongs to a kernel of any proper homomorphism. Hence, from (5.5) it follows that ϕ is an isomorphism. Thus, algebra $\mathbf{A}$ is embedded in $\mathbf{B}'$, i.e. $\mathbf{A} \in \mathsf{SH}\mathbf{B}$.

(b) Assume the contrary: $\mathbf{A} \not\models B$ and $B \nvdash_{\mathcal{V}} \chi_{\mathcal{V}}(A)$. If $B \nvdash_{\mathcal{V}} \chi_{\mathcal{V}}(A)$, then for some algebra $\mathbf{B} \in \mathcal{V}$ we have

$$\mathbf{B} \models B \text{ and } \mathbf{B} \not\models \chi_{\mathcal{V}}(\mathbf{A}). \tag{5.6}$$

But we have just proven in (a) that, if $\mathbf{B} \not\models \chi_{\mathcal{V}}(\mathbf{A})$, then $\mathbf{A} \in \mathsf{SH}\mathbf{B}$. Thus, if $\mathbf{B} \models B$, then $\mathbf{A} \models B$ and this contradicts the assumption.

Corollary 1 *Let* $\mathcal{V}$ *be a variety and* $\mathbf{A} \in FPSI(\mathcal{V})$*. Then all characteristic formulas of algebra* $\mathbf{A}$ *over* $\mathcal{V}$ *(regardless of which presentation we used) are interderivable over* $\mathcal{V}$*.*

The Corollary 1 means that characteristic formula of a finitely presented over $\mathcal{V}$ s.i. algebra is defined uniquely modulo interderivability in $\mathcal{V}$.

5.3.3 Independent Formulas

Let $\mathscr{V}$ be a variety. A set of formulas is called *$\mathscr{V}$-independent* if no one formula of this set is derivable over $\mathscr{V}$ from the rest of formulas. An $\mathscr{H}$-independent set of formulas we will call *independent*.

On the set $\mathscr{H}$ of all Heyting algebras, we can define the following quasi-order: $\mathbf{A} \leq \mathbf{B}$ if $\mathbf{A} \in \mathsf{SHB}$. The reflexivity of $\leq$ is trivial, while transitivity follows from the fact that variety of Heyting algebras has a congruence extension property (see, for instance, Blok and Pigozzi (1982)). A class $\mathscr{K}$ of algebras is said to be an *antichain* if for any $\mathbf{A}, \mathbf{B} \in \mathscr{K}$ we have $\mathbf{A} \not\leq \mathbf{B}$ and $\mathbf{B} \not\leq \mathbf{A}$.

Corollary 2 *Let $\mathscr{V}$ be a variety and $\mathscr{K} \subseteq FPSI(\mathscr{V})$. $\mathscr{K}$ is an antichain (relative to $\leq$) if and only if the set $\{\chi_{\mathscr{V}}(\mathbf{A});\ \mathbf{A} \in \mathscr{K}\}$ is $\mathscr{V}$-independent.*

Proof Let $\mathscr{K} \subseteq FPSI(\mathscr{V})$ be an antichain. Then if $\mathbf{A} \in \mathscr{K}$, we have $\mathbf{A} \not\models \chi_{\mathscr{V}}(\mathbf{A})$, but $\mathbf{B} \models \chi_{\mathscr{V}}(\mathbf{A})$, because $\mathscr{K}$ is an antichain and, by the Theorem 7(a), $\mathbf{A} \notin \mathsf{SHB}$.

Conversely, assume the contrary: $\mathbf{B} \in \mathsf{SHA}$. Then $\mathbf{A} \not\models \chi_{\mathscr{V}}(\mathbf{B})$, for $\mathbf{B} \not\models \chi_{\mathscr{V}}(\mathbf{B})$. By virtue of the Theorem 7(b), $\chi_{\mathscr{V}}(\mathbf{B}) \vdash_{\mathscr{V}} \chi_{\mathscr{V}}(\mathbf{A})$. And the latter contradicts $\mathscr{V}$-independence.

Let us note that if $\mathscr{V}_1, \mathscr{V}_2$ are varieties and $\mathscr{V}_1 \subseteq \mathscr{V}_2$, then $\mathscr{V}_1$-independence yields $\mathscr{V}_2$-independence. Thus, given a variety $\mathscr{V}$, any $\mathscr{V}$-independent set of formulas is $\mathscr{H}$-independent, i.e. it is independent. Therefore the following holds.

Corollary 3 *Let $\mathscr{V}$ be a variety and $\mathscr{K} \subseteq FPSI(\mathscr{V})$ and $\mathscr{K}$ is an antichain. Then the set $\{\chi_{\mathscr{V}}(\mathbf{A}) \mid \mathbf{A} \in \mathscr{K}\}$ is independent.*

Remark 2 It is obvious that any finite s.i. algebra $\mathbf{A}$ is finitely presented (over $\mathscr{H}$). Let us observe that diagram formula D in the definition of Yankov formula (5.1) defines algebra $\mathbf{A}$ in the trivial set of generators: the set of all elements of algebra $\mathbf{A}$. On the other hand, if we take a set of all distinct from $\mathbf{1}$ $\vee$-irreducible elements as a set of generators and use the diagram relations in order to construct a defining formula, we obtain de Jongh formula de Jongh (1968) (or frame-based formula Bezhanishvili 2006) of algebra $\mathbf{A}$.

Remark 3 Let us recall from McKenzie (1972): an algebra $\mathbf{A}$ from a variety $\mathscr{V}$ is called a *splitting algebra in the variety $\mathscr{V}$* if there is the greatest subvariety of $\mathscr{V}$ not containing algebra $\mathbf{A}$. From Theorem 7 it follows immediately that any algebra from $FPSI(\mathscr{V})$ is a splitting algebra in the variety $\mathscr{V}$. The finitely presented s.i. modal algebras as splitting algebras are studied in Kracht (1990).

5.4 An Example of Infinite Finitely Presented s.i. Heyting Algebra

The goal of this section is to give an example of a variety $\mathscr{V}$ and an infinite finitely presented in this variety s.i. algebra Z°. Then, in the following section, based on this

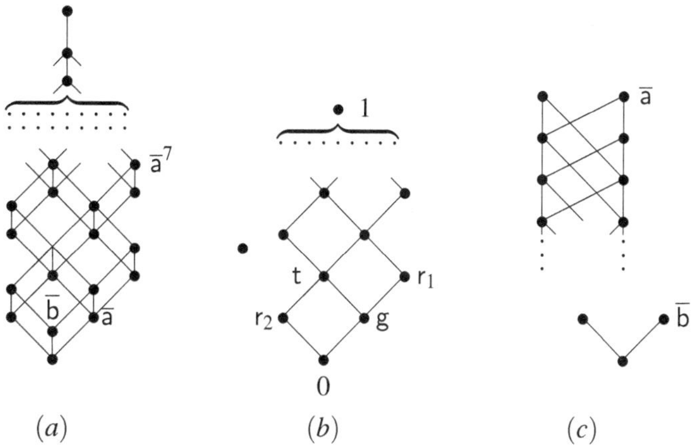

Fig. 5.2 An example of s.i. finitely presented algebra

example, we will construct an infinite set of such algebras. More precisely, we will show that algebra $\mathsf{Z}^\circ = \mathsf{Z} \times \mathsf{Z}_2 + \mathsf{Z}_2$ depicted in Fig. 5.2a (the corresponding frame is depicted in Fig. 5.2c) is finitely presented over every variety generated by algebra $\mathsf{Z} \times \mathsf{Z}_2 + \mathbf{A}$, where $\mathbf{A}$ is any non-degenerate algebra.[7]

The elements of algebra $\mathsf{Z} \times \mathsf{Z}_2$ will be regarded as pairs $\langle \mathsf{a}, \mathsf{b} \rangle$, where $\mathsf{a} \in \mathsf{Z}$ and $\mathsf{b} \in \{\mathbf{0}, \mathbf{1}\}$. So, $\langle \mathbf{1}, \mathbf{1} \rangle$ is an opremum of algebra Z° (see Fig. 5.1) and $\bar{\mathsf{b}} = \langle \mathbf{0}, \mathbf{1} \rangle$. Let us also note that elements $\bar{\mathsf{a}}, \bar{\mathsf{b}}$ generate algebra Z°.

Our goal is to prove the following theorem.

Theorem 8 *Let* $\mathbf{A}$ *be any non-degenerate Heyting algebra and* $\mathsf{Z}^* = \mathsf{Z} \times \mathsf{Z}_2 + \mathbf{A}$. *Then algebra* Z° *is finitely presented over variety* $\mathscr{V}(\mathsf{Z}^*)$.

We will prove that the formula

$$A = \neg(p \wedge q) \wedge (\neg\neg q \to q) \wedge (p^{10} \to (q \vee \neg q)), \tag{5.1}$$

where $p^{10} = (\neg\neg p \to p) \vee ((\neg\neg p \to p) \to (p \vee \neg p))$, and the valuation ν that sends p to $\bar{\mathsf{a}}$ and q to $\bar{\mathsf{b}}$, define algebra Z° over $\mathscr{V}$.

Let us observe that formula A is equivalent to the following formula:

$$\neg(p \wedge q) \wedge (\neg\neg q \to q) \wedge$$
$$((\neg\neg p \to p) \to (q \vee \neg q)) \wedge (((\neg\neg p \to p) \to (p \vee \neg p)) \to (q \vee \neg q)). \tag{5.2}$$

In order to prove the theorem, first, we need to establish some properties of formulas valid on elements $\bar{\mathsf{a}}, \bar{\mathsf{b}} \in \mathsf{Z}^\circ$.

[7] An infinite splitting modal algebra in an extension of **S4** was constructed in Kracht (1990), Sect. E. The Heyting algebra depicted at Fig. 5.2 is much simpler and it can be used for providing a simpler than in Kracht (1990) example of infinite splitting $S4$-algebra.

5.4.1 Auxiliary Lemmas

In this section, A always denotes the formula introduced by (5.1) in the proof of Theorem 8.

Lemma 1 *Suppose $B(p, q)$ is a formula and $B(\overline{\mathsf{a}}, \overline{\mathsf{b}}) = \mathbf{1}_{\mathsf{Z}^\circ}$. Then*

$$IPC \vdash B(p, q \wedge \neg q). \tag{5.3}$$

Proof Recall that $\overline{\mathsf{a}} = \langle \mathsf{g}, \mathbf{0}\rangle$ and $\overline{\mathsf{b}} = \langle \mathbf{0}, \mathbf{1}\rangle$. Thus, from $B(\overline{\mathsf{a}}, \overline{\mathsf{b}}) = \mathbf{1}$ we have $B(\langle \mathsf{g}, \mathbf{0}\rangle, \langle \mathbf{0}, \mathbf{1}\rangle) = \mathbf{1}$, hence $B(\mathsf{g}, \mathbf{0}) = \mathbf{1}_\mathsf{Z}$ and, obviously, $B(\mathsf{g}, \mathsf{g} \wedge \neg \mathsf{g}) = \mathbf{1}_\mathsf{Z}$. Let us also recall that Z an algebra freely generated by g (see Fig. 5.1), hence, $IPC \vdash B(p, p \wedge \neg p)$. Taking into consideration that $IPC \vdash (p \wedge \neg p) \leftrightarrow (q \wedge \neg q)$, we can conclude that $IPC \vdash B(p, q \wedge \neg q)$.

Lemma 2 *Suppose $B(p, q)$ is a formula and $B(\overline{\mathsf{a}}, \overline{\mathsf{b}}) = \mathbf{1}_{\mathsf{Z}^\circ}$. Then in the two-element Boolean algebra Z_2*

$$B(\mathbf{0}_{\mathsf{Z}_2}, \mathbf{0}_{\mathsf{Z}_2}) = B(\mathbf{0}_{\mathsf{Z}_2}, \mathbf{1}_{\mathsf{Z}_2}) = B(\mathbf{1}_{\mathsf{Z}_2}, \mathbf{0}_{\mathsf{Z}_2}) = \mathbf{1}_{\mathsf{Z}_2}.$$

Proof Let us consider the following three filters: $\nabla(\neg \overline{\mathsf{a}} \wedge \neg \overline{\mathsf{b}})$, $\nabla(\overline{\mathsf{b}})$, $\nabla(\overline{\mathsf{a}})$. And now let us observe that the corresponding homomorphisms send elements $\overline{\mathsf{a}}, \overline{\mathsf{b}}$ respectively in $\mathbf{0}_{\mathsf{Z}_2}, \mathbf{0}_{\mathsf{Z}_2}$, or $\mathbf{0}_{\mathsf{Z}_2}, \mathbf{1}_{\mathsf{Z}_2}$, or $\mathbf{1}_{\mathsf{Z}_2}, \mathbf{0}_{\mathsf{Z}_2}$. Since $B(\overline{\mathsf{a}}, \overline{\mathsf{b}}) = \mathbf{1}_{\mathsf{Z}^\circ}$ and any homomorphism preserves the top element, we can complete the proof.

Since in any Heyting algebra the set $\{\mathbf{0}, \mathbf{1}\}$ forms a subalgebra isomorphic to Z_2, the following holds.

Corollary 4 *If $B(p, q)$ is a formula and $B(\overline{\mathsf{a}}, \overline{\mathsf{b}}) = \mathbf{1}_{\mathsf{Z}^\circ}$, then in any Heyting algebra*

$$B(\mathbf{0}, \mathbf{0}) = B(\mathbf{0}, \mathbf{1}) = B(\mathbf{1}, \mathbf{0}) = \mathbf{1}.$$

Lemma 3 *Suppose $B(p, q)$ is a formula, $B(\overline{\mathsf{a}}, \overline{\mathsf{b}}) = \mathbf{1}_{\mathsf{Z}^\circ}$ and $A(\mathsf{c}, \mathbf{1}_\mathbf{A}) = \mathbf{1}_\mathbf{A}$ for some element c of an arbitrary algebra $\mathbf{A}$. Then $B(\mathsf{c}, \mathbf{1}_\mathbf{A}) = \mathbf{1}_\mathbf{A}$.*

Proof Since $A(\mathsf{c}, \mathbf{1}_\mathbf{A}) = \mathbf{1}_\mathbf{A}$, we have $\neg(\mathsf{c} \wedge \mathbf{1}_\mathbf{A}) = \neg \mathsf{c} = \mathbf{1}_\mathbf{A}$, that is, $\mathsf{c} = \mathbf{0}_\mathbf{A}$. Application of Corollary 4 completes the proof.

Corollary 5 *Suppose $B(p, q)$ is a formula, $B(\overline{\mathsf{a}}, \overline{\mathsf{b}}) = \mathbf{1}_{\mathsf{Z}^\circ}$ and $A(\mathsf{c}, \mathsf{d}) = \mathbf{1}_\mathbf{A}$ where c, d are some elements of an arbitrary s.i. algebra $\mathbf{A}$. Then*

(a) If $\mathsf{d} \vee \neg \mathsf{d} = \mathbf{1}_\mathbf{A}$ then $B(\mathsf{c}, \mathsf{d}) = \mathbf{1}_\mathbf{A}$;
(b) If $\mathsf{d} = \neg \mathsf{c}$ then $B(\mathsf{c}, \mathsf{d}) = \mathbf{1}_\mathbf{A}$;
(c) If $\neg\neg \mathsf{d} = \neg \mathsf{c}$ then $B(\mathsf{c}, \mathsf{d}) = \mathbf{1}_\mathbf{A}$.

Proof (a) Indeed, due to algebra $\mathbf{A}$ is s.i., $\mathsf{d} \vee \neg \mathsf{d} = \mathbf{1}_\mathbf{A}$ yields $\mathsf{d} = \mathbf{1}_\mathbf{A}$ or $\mathsf{d} = \mathbf{0}_\mathbf{A}$. Applications of Lemmas 1 and 3 complete the proof.

(b) Since $A(\mathsf{c}, \mathsf{d}) = \mathbf{1_A}$ we have

$$\begin{aligned} \mathbf{1_A} = ((\neg\neg\mathsf{c} \to \mathsf{c}) \to (\mathsf{d} \vee \neg\mathsf{d})) = \\ ((\neg\neg\mathsf{c} \to \mathsf{c}) \to (\neg\mathsf{c} \vee \neg\neg\mathsf{c})) = ((\neg\neg\mathsf{c} \to \mathsf{c}) \to (\mathsf{c} \vee \neg\mathsf{c})). \end{aligned} \tag{5.4}$$

From $A(\mathsf{c}, \mathsf{d}) = \mathbf{1_A}$ it also follows that

$$((\neg\neg\mathsf{c} \to \mathsf{c}) \to (\mathsf{c} \vee \neg\mathsf{c})) \to (\mathsf{d} \vee \neg\mathsf{d}) = \mathbf{1_A}. \tag{5.5}$$

From (5.4) and (5.5) it trivially follows that $\mathsf{d} \vee \neg\mathsf{d} = \mathbf{1_A}$ and application of (a) completes the proof of case (b).

(c) Immediately from $A(\mathsf{c}, \mathsf{d}) = \mathbf{1_A}$ it follows that $\neg\neg\mathsf{d} \to \mathsf{d} = \mathbf{1_A}$, that is, $\neg\neg\mathsf{d} = \mathsf{d}$. Thus, $\neg\neg\mathsf{d} = \neg\mathsf{c}$ yields $\mathsf{d} = \neg\mathsf{c}$ and we can apply (b).

5.4.2 The Proof of the Theorem

Proof In order to prove the theorem, we will demonstrate that formula $A(p, q)$ and the valuation ν such that $\nu(p) = \bar{\mathsf{a}}$ and $\nu(q) = \bar{\mathsf{b}}$ define algebra Z°. It is clear that elements $\bar{\mathsf{a}}, \bar{\mathsf{b}}$ generate algebra Z° and that $A(\bar{\mathsf{a}}, \bar{\mathsf{b}}) = \mathbf{1}$, thus, the conditions (1) and (2) of the Definition 1 are satisfied. So, we need to prove that for any formula $B(p, q)$ such that

$$B(\bar{\mathsf{a}}, \bar{\mathsf{b}}) = \mathbf{1} \tag{5.6}$$

we have

$$\vdash_{\mathcal{V}(\mathsf{Z}^*)} A \to B, \tag{5.7}$$

i.e. we have

$$\mathsf{Z}^* \models A \to B. \tag{5.8}$$

In order to prove (5.8) it is enough to show that for any s.i. homomorphic image $\mathbf{A}$ of algebra Z^* and any two elements $\mathsf{c}, \mathsf{d} \in \mathbf{A}$ if

$$A(\mathsf{c}, \mathsf{d}) = \mathbf{1_A}, \tag{5.9}$$

then

$$B(\mathsf{c}, \mathsf{d}) = \mathbf{1_A}. \tag{5.10}$$

We will consider the following cases:

	c	d
(1)	$c \in Dn(\mathbf{A})$	any
(2)	$c \in Rg(\mathbf{A})$	any
(3)	$c \notin Rg(\mathbf{A})$ and $c \notin Dn(\mathbf{A})$	any

Case (1). If $A(c, d) = \mathbf{1_A}$, then $\neg(a \wedge c) = \mathbf{1_A}$. Hence $c \wedge d = \mathbf{0_A}$. Since $c \in Dn(\mathbf{A})$, we have $d = \mathbf{0}$ and application of Lemma 1 completes the proof.

Case (2). If $A(c, d) = \mathbf{1_A}$, then $(\neg\neg c \to c) \to (d \vee \neg d) = \mathbf{1_A}$. Since $c \in Rg(\mathbf{A})$, that is, $(\neg\neg c \to c) = \mathbf{1_A}$, we can conclude that $d \vee \neg d = \mathbf{1_A}$ and apply the Corollary 5.

Case (3). Let c be neither regular, nor dense. Algebra **A** is an s.i. homomorphic image of algebra Z^* and let ∇ be a kernel of this homomorphism. Let us consider two cases:

(a) $\langle \mathbf{1}, \mathbf{1} \rangle \in \nabla$;
(b) $\langle \mathbf{1}, \mathbf{1} \rangle \notin \nabla$.

(a) Let us recall that **A** is an s.i. algebra and, therefore, $\langle \mathbf{0}, \mathbf{1} \rangle \vee \langle \mathbf{1}, \mathbf{0} \rangle = \langle \mathbf{1}, \mathbf{1} \rangle \in \nabla$ entails $\langle \mathbf{0}, \mathbf{1} \rangle \in \nabla$ or $\langle \mathbf{1}, \mathbf{0} \rangle \in \nabla$.

If $\langle \mathbf{0}, \mathbf{1} \rangle \in \nabla$, i.e. $\bar{b} \in \nabla$ (see Fig. 5.2), then $\mathbf{A} \cong Z^*/\nabla = Z^\circ/\nabla'$, where $\nabla' = Z^\circ \cap \nabla$. Note that **A** is a two-element Boolean algebra and we can apply Corollary 4 (because $c \wedge d = \mathbf{0_A}$ and, therefore, $c = \mathbf{0_A}$ or $d = \mathbf{0_A}$).

If $\langle \mathbf{1}, \mathbf{0} \rangle \in \nabla$, then $\mathbf{A} \cong Z^*/\nabla = Z^\circ/\nabla'$, where $\nabla' = Z^\circ \cap \nabla$ is a single-generated algebra. There is the only element of single-generated algebra which is not dense and regular, namely, its generator g. Let us observe that, since $A(g, d) = \mathbf{1_A}$, we have $g \wedge d = \mathbf{0_A}$ and there are just two possibilities for d: either $d = \mathbf{0}$, or $d = r_2$ (see Fig. 5.2). In the first case we can apply Lemma 1. In the second case we can apply Corollary 5(b), because $r_2 = \neg g$.

(b) Since element $c \in Z \times Z_2$ is neither dense, nor regular, there are exactly three sub-cases:

i. $c = \langle f, \mathbf{0} \rangle$, where $f \in Dn(Z)$;
ii. $c = \langle g, \mathbf{1} \rangle$;
iii. $c = \langle g, \mathbf{0} \rangle$.

Let us consider these sub-cases.

i. If $c = \langle f, \mathbf{0} \rangle$, then

$$(\neg\neg c \to c) \to (c \vee \neg c) = (\neg\neg\langle f, \mathbf{0} \rangle \to \langle f, \mathbf{0} \rangle) \to (\langle f, \mathbf{0} \rangle \vee \neg\langle f, \mathbf{0} \rangle) =$$
$$(\langle \mathbf{1}, \mathbf{0} \rangle \to \langle f, \mathbf{0} \rangle) \to (\langle f, \mathbf{0} \rangle \vee \langle \mathbf{0}, \mathbf{1} \rangle) = \langle f, \mathbf{1} \rangle \to \langle f, \mathbf{1} \rangle = \mathbf{1}. \tag{5.11}$$

By assumption, $A(c, d) = \mathbf{1}$, hence, $((\neg\neg c \to c) \to (c \vee \neg c)) \to (d \vee \neg d) = \mathbf{1}$. Therefore from (5.11) we have $d \vee \neg d = \mathbf{1}$ and we can apply Corollary 5(a).

ii. Let $c = \langle g, \mathbf{1} \rangle$. From $c \wedge d = \mathbf{0}$ it follows (see Fig. 5.2) that in this case $d = \mathbf{0}$, or $d = \langle r_2, \mathbf{0} \rangle$. In the first case we can apply Lemma 1. In the second case we can apply Corollary 5(b), because $\langle r_2, \mathbf{0} \rangle = \neg\langle g, \mathbf{1} \rangle = c$.

iii. Let $\mathsf{c} = \langle \mathsf{g}, \mathbf{0}\rangle = \bar{\mathsf{a}}$. From $\mathsf{c} \wedge \mathsf{d} = \mathbf{0}$ it follows that there are just four possibilities for d (see Fig. 5.2): $\mathsf{d} \in \{\mathbf{0}, \langle \mathbf{0}, \mathbf{1}\rangle, \langle \mathsf{r}_2, \mathbf{0}\rangle, \langle \mathsf{r}_2, \mathbf{1}\rangle\}$. If $\mathsf{d} = \mathbf{0}$ we can apply Lemma 1. If $\mathsf{d} = \langle \mathbf{0}, \mathbf{1}\rangle = \bar{\mathsf{d}}$, the statement trivially follows from the assumption (5.6).

Let us observe the following (see Fig. 5.2):

$$(\neg\neg c \to \mathsf{c}) \to (\mathsf{c} \vee \neg \mathsf{c}) = (\neg\neg\bar{\mathsf{a}} \to \bar{\mathsf{a}}) \to (\bar{\mathsf{a}} \vee \neg\bar{\mathsf{a}}) = \bar{\mathsf{a}}^7. \tag{5.12}$$

Since $A(\mathsf{c}, \mathsf{d}) = \mathbf{1}$, we have

$$((\neg\neg c \to \mathsf{c}) \to (\mathsf{c} \vee \neg \mathsf{c})) \to (\mathsf{d} \vee \neg \mathsf{d}) = \mathbf{1}, \tag{5.13}$$

hence,

$$(\neg\neg c \to \mathsf{c}) \to (\mathsf{c} \vee \neg \mathsf{c}) \leq (\mathsf{d} \vee \neg \mathsf{d}). \tag{5.14}$$

But

$$\begin{aligned} \mathsf{d} \vee \neg \mathsf{d} &= \langle \mathsf{r}_2, \mathbf{0}\rangle \vee \neg\langle \mathsf{r}_2, \mathbf{0}\rangle = \langle \mathsf{r}_2, \mathbf{0}\rangle \vee \langle \mathsf{r}_1, \mathbf{1}\rangle = \langle \mathsf{r}_2 \vee \mathsf{r}_1, \mathbf{1}\rangle;\\ \mathsf{d} \vee \neg \mathsf{d} &= \langle \mathsf{r}_2, \mathbf{1}\rangle \vee \neg\langle \mathsf{r}_2, \mathbf{1}\rangle = \langle \mathsf{r}_2, \mathbf{1}\rangle \vee \langle \mathsf{r}_1, \mathbf{0}\rangle = \langle \mathsf{r}_2 \vee \mathsf{r}_1, \mathbf{1}\rangle. \end{aligned} \tag{5.15}$$

The observation that $\langle \mathsf{r}_2 \vee \mathsf{r}_1, \mathbf{1}\rangle < \bar{\mathsf{a}}^7$ completes the proof.

5.5 Finite Presentability and Concatenations

In this section we construct an infinite set of infinite finitely presented algebras. For this we will use concatenations of finitely presented algebras.

Theorem 9 *Let $\mathcal{V}$ be a variety of Heyting algebras and $\mathbf{A} = \mathbf{A}' + \mathsf{Z}_2 \in \mathcal{V}$ and $\mathbf{B} = \mathsf{Z}_2 + \mathbf{B}' \in \mathcal{V}$. Suppose algebras $\mathbf{A}$ and $\mathbf{B}$ are finitely presented over $\mathcal{V}$ and $\mathbf{A}' + \mathbf{B}' \in \mathcal{V}$. Then algebra $\mathbf{A}' + \mathbf{B}'$ is finitely presented over $\mathcal{V}$.*

Proof Let pairs $\langle A(p_1, \dots, p_n); \nu\rangle$ and $\langle B(q_1, \dots, q_m); \mu\rangle$ define respectively algebras $\mathbf{A}$ and $\mathbf{B}$ and $\{p_1, \dots, p_n\} \cap \{q_1, \dots, q_m\} = \emptyset$. We will regard $\mathbf{A}$ and $\mathbf{B}$ as subalgebras of $\mathbf{A}' + \mathbf{B}'$, so ν and μ are the mappings into $\mathbf{A}' + \mathbf{B}'$. Assume that $A'(p_1, \dots, p_n)$ and $B'(q_1, \dots, q_m)$ are such formulas that $\nu(A') = \mathsf{a}$, where a is a coatom (opremum) of $\mathbf{A}$, and $\mu(B') = \mathsf{b}$, where b is an atom of $\mathbf{B}$. Let us note that $\nu(A') = \mu(B')$. Then the pair $\langle C; \phi\rangle$, where

$$\begin{aligned} &C(p_1, \dots, p_n, q_1, \dots, q_m) =\\ &A(p_1, \dots, p_n) \wedge B(q_1, \dots, q_m) \wedge (A'(p_1, \dots, p_n) \leftrightarrow B'(q_1, \dots, q_m))\\ &\phi(p_i) = \nu(p_i);\ i = 1, \dots, n \text{ and } \phi(q_j) = \mu(q_j);\ j = 1, \dots, m, \end{aligned} \tag{5.1}$$

defines algebra $\mathbf{A}' + \mathbf{B}'$ over $\mathcal{V}$.

Assume $\nu(p_1) = \mathsf{a}_i$; $i = 1, \dots, n$ and $\mu(q_j) = \mathsf{b}_j$; $j = 1, \dots, m$. It is easy to see that elements $G = \{\mathsf{a}_1, \dots, \mathsf{a}_n, \mathsf{b}_1, \dots, \mathsf{b}_m\}$ generate algebra $\mathbf{A}' + \mathbf{B}'$. From

Proposition 2 it follows that in order to prove our claim it is enough to demonstrate that for any algebra $\mathbf{C} \in \mathscr{V}$, any mapping $\psi : G \mapsto \mathbf{C}$ such that

$$C(\psi(\mathsf{a}_1), \ldots, \psi(\mathsf{a}_n), \psi(\mathsf{b}_1), \ldots, \psi(\mathsf{b}_m)) = \mathbf{1}_\mathbf{C} \tag{5.2}$$

can be extended to a homomorphism $\overline{\psi} : \mathbf{A}' + \mathbf{B}' \to \mathbf{C}$.

Let us consider the following reducts of ϕ:

$$\phi_1 : p_i \mapsto \mathsf{a}_i; i = 1, \ldots, n \text{ and } \phi_2 : q_i \mapsto \mathsf{b}_i; i = 1, \ldots, m. \tag{5.3}$$

Let us recall now that algebras **A** and **B** are finitely presented over $\mathscr{V}$. Hence, mappings ϕ_1 and ϕ_2 can be extended to homomorphisms $\overline{\phi_1} : \mathbf{A} \to \mathbf{C}$ and $\overline{\phi_2} : \mathbf{B} \to \mathbf{C}$. From (5.1), (5.2) and (5.3) it follows that

$$\overline{\phi_1}(\mathsf{a}) = \overline{\phi_2}(\mathsf{b}). \tag{5.4}$$

Moreover,

$$\overline{\phi_1}(\mathsf{a}') \leq \overline{\phi_1}(\mathsf{a}) \text{ for all } \mathsf{a}' \in \mathbf{A}' \text{ and } \overline{\phi_2}(\mathsf{b}) \leq \overline{\phi_2}(\mathsf{b}') \text{ for all } \mathsf{b}' \in \mathbf{B}'. \tag{5.5}$$

Thus, we can construct a homomorphism $\overline{\psi}$ in the following way:

$$\overline{\psi}(\mathsf{c}) = \begin{cases} \overline{\phi_1}(\mathsf{c}), \text{ when } \mathsf{c} \in \mathbf{A}'; \\ \overline{\phi_2}(\mathsf{c}), \text{ when } \mathsf{c} \in \mathbf{B}'. \end{cases} \tag{5.6}$$

$\overline{\psi}$ is a homomorphism because $\overline{\phi_1}$ and $\overline{\phi_2}$ are homomorphisms and for any $\mathsf{a}' \in \mathbf{A}'$ and $\mathsf{b}' \in \mathbf{B}'$

$$\begin{aligned} \mathsf{a}' \wedge \mathsf{b}' &= \mathsf{a}'; \\ \mathsf{a}' \vee \mathsf{b}' &= \mathsf{b}'; \\ \mathsf{a}' \to \mathsf{b}' &= \mathbf{1}; \\ \mathsf{b}' \to \mathsf{a}' &= \mathsf{a}'. \end{aligned} \tag{5.7}$$

Corollary 6 *Let algebra $\mathbf{A} = \mathbf{A}' + \mathsf{Z}_2 \in \mathscr{V}$ be finitely presented over $\mathscr{V}$. If* **B** *is a finite algebra and $\mathbf{A}' + \mathbf{B} \in \mathscr{V}$, then algebra $\mathbf{A}' + \mathbf{B}$ is finitely presented over $\mathscr{V}$.*

Proof If $\mathbf{A}' + \mathbf{B} \in \mathscr{V}$, then $\mathsf{Z}_2 + \mathbf{B} \in \mathscr{V}$. Clearly, algebra $\mathsf{Z}_2 + \mathbf{B}$ is finite, so, it is finitely presented and we can apply the theorem.

Corollary 7 *Variety $\mathscr{V} = \mathscr{V}(\mathsf{Z} \times \mathsf{Z}_2 + \mathsf{Z})$ contains infinitely many infinite finitely presented s.i. algebras.*

Proof From Theorem 8 it follows that algebra Z° is finitely presented over $\mathscr{V}$. On the other hand, for all $n = 1, 2, \ldots$ we have $\mathsf{Z}_2 + \mathsf{Z}_{2n+1} \in \mathscr{V}$. By virtue of Corollary 6 all algebras $\mathsf{Z}^\circ + \mathsf{Z}_{2n+1};\ n = 1, 2, \ldots$ are finitely presented over $\mathscr{V}$.

Remark 4 Theorem 30 from Kracht (1990) states that there is a variety of $S4$-algebras that contains infinitely many infinite splitting algebras.

5.6 Axiomatization by Characteristic Formulas

It is well-known that not every variety (or every intermediate logic for that matter) can be axiomatized by Yankov formulas. In fact, there is a continuum of varieties that cannot be axiomatized by Yankov formulas (Tomaszewski 2003, p. 128). In this section we will show that there is continuum many varieties that cannot be axiomatized by Yankov formulas but, nevertheless, can be axiomatized by characteristic formulas. In order to do so we will construct an infinite independent set of characteristic formulas and then demonstrate that any subset of this set defines a variety that cannot be axiomatized by Yankov formulas.

First, let us observe the following simple criterion (the proof in terms of frames can be found, for instance, in Bezhanishvili (2006), Corollary 3.4.14(1)).

Proposition 10 *A variety $\mathscr{V}$ can be axiomatized by Yankov formulas over a given variety $\mathscr{V}'$ extending $\mathscr{V}$ if and only if for every algebra $\mathbf{A} \in \mathscr{V}' \setminus \mathscr{V}$ there is such a finite algebra $\mathbf{B} \in \mathsf{SH}\mathbf{A}$ that $\mathbf{B} \in \mathscr{V}' \setminus \mathscr{V}$.*

In other words, a subvariety $\mathscr{V}_0$ can be axiomatized by Yankov formulas over its extension $\mathscr{V}$ if and only if every variety $\mathscr{V}' \subseteq \mathscr{V}$ strongly extending $\mathscr{V}_0$ can be separated from $\mathscr{V}_0$ by some finite algebra. That is, if $\mathscr{V}' \setminus \mathscr{V}_0$ contains a finite algebra for every $\mathscr{V}_0 \subset \mathscr{V}' \subseteq \mathscr{V}$.

Let us note the following simple corollary.

Corollary 8 *Suppose $\mathscr{V}$ is a variety and $\mathscr{V}_0$ is a subvariety of $\mathscr{V}$. If $\mathscr{V}_0$ cannot be axiomatized by Yankov formulas over $\mathscr{V}$, then $\mathscr{V}_0$ cannot be axiomatized by Yankov formulas over any extension of $\mathscr{V}$, in particular, $\mathscr{V}_0$ cannot be axiomatized by Yankov formulas over $\mathscr{H}$.*

Now we will use Proposition 10 and prove that every locally finite variety can be axiomatized by Yankov formulas.

Let us recall from Mal'cev (1973) that a variety $\mathscr{V}$ is said to be *locally finite* if for every number n there is a number m such that every n-generated $\mathscr{V}$-algebra contains at most m elements.

Corollary 9 (Bezhanishvili (2006); Tomaszewski (2003)). *Every locally finite variety of Heyting algebras can be axiomatized by Yankov formulas.*

Proof Let $\mathscr{V}$ be a locally finite variety. It suffices to demonstrate that every n-generated algebra $\mathbf{A} \notin \mathscr{V}$ can be separated from $\mathscr{V}$ by some Yankov formula. For finite algebras, the statement is trivial, so we can assume that $\mathbf{A}$ is an infinite algebra. Let m be an upper bound of cardinalities of n-generated algebras of $\mathscr{V}$. By virtue of theorem from Kuznetsov (1973), algebra $\mathbf{A}$ contains chain subalgebras of any finite length. Thus, it contains a finite subalgebra of cardinality greater than m and, therefore, the cardinality of $\mathbf{A}$ is greater than m. Hence, this subalgebra is not in $\mathcal{V}$ and we can apply Proposition 10.

Corollary 10 *If $\mathcal{V}$ is a variety and $\mathbf{A} \in FPSI(\mathcal{V})$ is an infinite algebra, then formula $\chi_{\mathcal{V}}(\mathbf{A})$ defines a variety that cannot be axiomatized over $\mathcal{V}$ by Yankov formulas.*

Proof Let $\mathcal{V}' = \{\mathbf{B} \mid \mathbf{B} \models \chi(\mathbf{A}), \mathbf{B} \in \mathcal{V}\}$ be a variety defined in $\mathcal{V}$ by formula $\chi_{\mathcal{V}}(\mathbf{A})$. Due to Proposition 10, in order to demonstrate that $\mathcal{V}'$ is not axiomatizable by Yankov formulas it suffices to show that all finite members of SHA belong to $\mathcal{V}'$. For contradiction: assume that $\mathbf{B} \in \mathsf{SHA}$ is a finite algebra and $\mathbf{B} \notin \mathcal{V}'$. Then $\mathbf{B} \not\models \chi_{\mathcal{V}}(\mathbf{A})$. Since $\mathbf{B} \in \mathsf{SHA} \subseteq \mathcal{V}$, we can apply Theorem 7(a) and conclude that $\mathbf{A} \in \mathsf{SH}\mathbf{B}$. The latter is impossible, for $\mathbf{A}$ is an infinite algebra, while all algebras from $\mathsf{SH}\mathbf{B}$ are finite, because $\mathbf{B}$ is a finite algebra.

Moreover, in a similar way one can prove the following.

Corollary 11 *Let $\mathcal{V}$ be a variety and $\mathcal{K}$ be a set of infinite algebras from $FPSI(\mathcal{V})$. Then the set of formulas $\{\chi_{\mathcal{V}}(\mathbf{A}) \mid \mathbf{A} \in \mathcal{K}\}$ defines in $\mathcal{V}$ a variety that cannot be axiomatized over $\mathcal{V}$ (and, hence, over $\mathcal{H}$) by Yankov formulas.*

It is important to note that in Corollary 11 all algebras from $\mathcal{K}$ are finitely presented over the same variety $\mathcal{V}$.

5.6.1 Varieties Not Axiomatizable by Yankov Formulas

The goal of this section is to prove that there is continuum many varieties of Heyting algebras that can be defined by characteristic formulas of infinite algebras, but cannot be defined by Ynakov formulas.

Lemma 4 *There is a variety $\mathcal{V}$ containing an infinite antichain of infinite finitely presented over $\mathcal{V}$ s.i. algebras.*

Proof Let

$$\mathbf{A} = \mathsf{Z} \times \mathsf{Z}_2 + \prod_{n=3}^{\infty} (\mathsf{Z}_{2n} + \mathsf{Z}_2 + \mathsf{Z}_2) \tag{5.1}$$

and let $\mathcal{V}$ be a variety generated by $\mathbf{A}$. Let us consider algebras

$$\mathbf{A}_n = \mathsf{Z} \times \mathsf{Z}_2 + \mathsf{Z}_{2n} + \mathsf{Z}_2 + \mathsf{Z}_2\ ;\ n = 3, 4, \ldots. \tag{5.2}$$

From Proposition 1 and the observation that algebra $\mathsf{Z}_{2n} + \mathsf{Z}_2 + \mathsf{Z}_2$ is a homomorphic image of algebra $\prod_{n=3}^{\infty}(\mathsf{Z}_{2n} + \mathsf{Z}_2 + \mathsf{Z}_2)$, it immediately follows that algebra $\mathbf{A}_k$ is a homomorphic image of $\mathbf{A}$ and, hence, $\mathbf{A}_k \in \mathcal{V}$ for all $k \geq 3$. We want to demonstrate that set $\{\mathbf{A}_k \mid k \geq 3\}$ forms an antichain. For this, we will demonstrate

(a) For every $k \geq 3$ algebra $\mathbf{A}_k$ is finitely presented over $\mathcal{V}(\mathbf{A})$;
(b) For any $m \neq k$ algebra $\mathbf{A}_m \notin \mathsf{SH}\mathbf{A}_k$, that is, the set $\{\mathbf{A}_k \mid k \geq 3\}$ forms an antichain.

(a) First, let us observe that for any k algebra $\mathsf{Z}_{2k} + \mathsf{Z}_2 + \mathsf{Z}_2$ is a homomorphic image of the direct product

$$\mathbf{B} = \prod_{n=3}^{\infty} (\mathsf{Z}_{2n} + \mathsf{Z}_2 + \mathsf{Z}_2).$$

Thus, for each k there is such a filter $\nabla_k \subseteq \mathbf{B}$ that $\mathbf{B}/\nabla_k \cong \mathbf{A}_k$. By virtue of Proposition 1, we have

$$A_k = \mathsf{Z} \times \mathsf{Z}_2 + (\mathsf{Z}_{2k} + \mathsf{Z}_2 + \mathsf{Z}_2) \cong \mathsf{Z} \times \mathsf{Z}_2 + (\mathbf{B}/\nabla_k) \cong (\mathsf{Z} \times \mathsf{Z}_2 + \mathbf{B})/\nabla_k) = \mathbf{A}/\nabla_k.$$

So, $\mathbf{A}_k \in \mathsf{HA}$, hence, $\mathbf{A}_k \in \mathcal{V}(\mathbf{A}) = \mathcal{V}$. Let us also observe that algebras $\mathsf{Z} \times \mathsf{Z}_2 + \mathsf{Z}_2$ and $\mathsf{Z}_2 + \mathsf{Z}_{2k} + \mathsf{Z}_2 + \mathsf{Z}_2$ are subalgebras of algebra $\mathbf{A}_k$. Therefore $\mathsf{Z} \times \mathsf{Z}_2 + \mathsf{Z}_2, \mathsf{Z}_2 + \mathsf{Z}_{2k} + \mathsf{Z}_2 + \mathsf{Z}_2 \in \mathcal{V}$. From the Theorem 8 it follows that algebra $\mathsf{Z} \times \mathsf{Z}_2 + \mathsf{Z}_2$ is finitely presented over $\mathcal{V}$. On the other hand, algebra $\mathsf{Z}_2 + \mathsf{Z}_{2k} + \mathsf{Z}_2 + \mathsf{Z}_2$ is finite and, therefore, is finitely presented over $\mathcal{V}$ too. Now we can apply Theorem 9 and conclude that algebra $\mathbf{A}_k$ is finitely presented over $\mathcal{V}$.

(b) Let $k \neq m$. We need to demonstrate that $\mathbf{A}_k \notin \mathsf{SH}\mathbf{A}_m$. Let $\nabla \subseteq \mathbf{A}_m$ be a filter, and we want to verify that $\mathbf{A}_k$ is not (isomorphic to) a subalgebra of $\mathbf{A}_m/\nabla$. Let us consider two cases:

1. ∇ contains the top element of algebra $\mathsf{Z} \times \mathsf{Z}_2$;
2. ∇ does not contain the top element of algebra $\mathsf{Z} \times \mathsf{Z}_2$.

Case 1. If ∇ contains the top element of algebra $\mathsf{Z} \times \mathsf{Z}_2$, then $\mathbf{A}_k/\nabla \cong (\mathsf{Z} \times \mathsf{Z}_2)/\nabla'$, where $\nabla' = \nabla \cap (\mathsf{Z} \times \mathsf{Z}_2)$. It is not hard to see that algebra $\mathsf{Z} \times \mathsf{Z}_2$ has just 2 infinite homomorphic images, namely, itself and algebra Z. Let us observe that algebra $\mathsf{Z} \times \mathsf{Z}_2 + \mathsf{Z}_2$ is not a subalgebra of $\mathsf{Z} \times \mathsf{Z}_2$ or of algebra Z, while algebra $\mathsf{Z} \times \mathsf{Z}_2 + \mathsf{Z}_2$ is a subalgebra of $\mathbf{A}_k$ Hence, algebra $\mathbf{A}_k$ is not embedded in any homomorphic image of $\mathbf{A}_m$ as long as its kernel contains the top element of algebra $\mathsf{Z} \times \mathsf{Z}_2$.

Case 2. The important point to note here is that in algebra $\mathbf{A}_k$ the top element of algebra $\mathsf{Z} \times \mathsf{Z}_2$ is at the same time the bottom element of algebra Z_{2k}. Thus, by virtue of Proposition 1, all considerations can be reduced to algebras $\mathsf{Z}_{2k} + \mathsf{Z}_2 + \mathsf{Z}_2$ and $\mathsf{Z}_{2m} + \mathsf{Z}_2 + \mathsf{Z}_2$. But it is well known (see, e.g. Gerčiu and Kuznecov 1970; Wroński 1974) that if $k \neq m$, then $\mathsf{Z}_{2k} + \mathsf{Z}_2 + \mathsf{Z}_2 \notin \mathsf{SH}(\mathsf{Z}_{2m} + \mathsf{Z}_2 + \mathsf{Z}_2)$.

Corollary 12 *There is a variety $\mathcal{V}$ containing continuum many subvarieties axiomatizable over $\mathcal{V}$ by characteristic formulas, but not axiomatizable over $\mathcal{V}$ by Yankov formulas.*

We can extend the above statement to the whole variety of Heyting algebras.

Theorem 11 *There is continuum of varieties that are axiomatized over $\mathscr{H}$ by characteristic formulas, but cannot be axiomatized by Yankov formulas.*

Proof Indeed, consider a variety $\mathscr{V}$ in which $\mathscr{K} \subseteq FPSI(\mathscr{V})$ is an antichain of infinite algebras, which exists according to Lemma 4. Then, by virtue of Corollary 3, the set of formulas $CH = \{\chi_{\mathscr{V}}(\mathbf{A}) \mid \mathbf{A} \in \mathscr{K}\}$ is independent. Thus, all the varieties defined by distinct subsets $CH' \subseteq CH$ are pairwise different. Hence we have continuum many distinct subvarieties of $\mathscr{V}$ defined in $\mathscr{V}$ by characteristic formulas. Due to Corollary 11, no variety defined by any formulas from CH can be axiomatized over $\mathscr{V}$ by Yankov formula. Now, we can apply Corollary 8 and conclude that all these varieties are not axiomatizable by Yankov formulas over $\mathscr{H}$ too. Thus, there is continuum many varieties axiomatizable over $\mathscr{H}$ by characteristic formulas that are not axiomatizable over $\mathscr{H}$ by Yankov formulas.

5.6.2 *Varieties Not Axiomatizable by Characteristic Formulas*

As we saw in the previous section, there is a continuum of intermediate logics that cannot be axiomatizable by Yankov formulas, but can be axiomatized by characteristic formulas. Naturally, the question arises whether any intermediate logic can be axiomatized by characteristic formulas. In this section, we give a negative answer to this question.

Theorem 12 *There is an intermediate logic that is not axiomatizable by characteristic formulas.*

Proof Let us consider the intermediate logic defined by the following axiom (the logic **KG** from Bezhanishvili et al. (2008)):

$$(p \to q) \lor (q \to r) \lor ((q \to r) \to r) \lor (r \to (p \lor q)). \tag{5.3}$$

And let $\mathscr{V}$ be a variety defined by the above formula.

From Kuznetsov and Gerčiu (1970) Lemma 4, it follows that any finitely generated s.i. algebra from $\mathscr{V}$ is a concatenation of finite number of some single-generated algebras. Thus, every infinite finitely generated s.i. algebra from $\mathscr{V}$ is a concatenation of single-generated algebras at least one of which is infinite, i.e. at least one of which is Z. From Citkin (2012) Theorem 2.14, it immediately follows that $\mathscr{V}$ does not contain infinite finitely presented s.i. algebras. Therefore, if a subvariety $\mathscr{V}_1 \subset \mathscr{V}$ can be axiomatized over $\mathscr{V}$ by some characteristic formulas, $\mathscr{V}_1$ can be axiomatized over $\mathscr{V}$ by characteristic formulas of finite algebras, that is, by Yankov formulas. So, if we demonstrate that variety $\mathscr{V}$ contains a subvariety $\mathscr{V}_1$ that is not axiomatizable by Yankov formula, we will complete the proof.

Let us recall from Kuznetsov and Gerčiu (1970) Theorem 4, that there are formulas A, B, C (formulas (3), (4) and (5) from Kuznetsov and Gerčiu 1970) such that formula A is invalid in the logic $\mathcal{L}$ defined by formulas (5.3), B and C, but A

cannot be separated from $\mathcal{L}$ by any finite algebra. In algebraic terms it means that varietiy $\mathscr{V}_1$ defined by formulas (5.3), A, B, C and variety $\mathscr{V}_2$ defined by formulas (5.3), B, C have the same finite algebras and $\mathscr{V}_1 \subset \mathscr{V}_2$. Since formula (5.3) is valid in $\mathscr{V}_2$, we have $\mathscr{V}_1 \subset \mathscr{V}_2 \subseteq \mathscr{V}$. Let us observe that formula C (that is formula (5) from Kuznetsov and Gerčiu (1970)) is invalid in algebra $\mathsf{Z}_9 + \mathsf{Z}_2$, while formula (5.3) is valid in this algebra. Thus, we have

$$\mathscr{V}_1 \subset \mathscr{V}_2 \subset \mathscr{V}.$$

Suppose **A** is a finite s.i. algebra. Then, from the properties of Yankov formulas, it follows that Yankov formula $\chi(\mathbf{A})$ is valid in $\mathscr{V}_1$ if and only if it is valid in $\mathscr{V}_2$, for $\mathscr{V}_1$ and $\mathscr{V}_2$ have the same finite algebras. Therefore, all Yankov formulas of finite s.i. algebras from $\mathscr{V} \setminus \mathscr{V}_1$ define in $\mathscr{V}$ a subvariety $\mathscr{V}'$ containing $\mathscr{V}_2$ and, hence, $\mathscr{V}_1 \subset \mathscr{V}_2 \subseteq \mathscr{V}'$, so $\mathscr{V}_1 \neq \mathscr{V}'$. It means that subvariety $\mathscr{V}_1$ cannot be defined in $\mathscr{V}$ by Yankov formulas.

5.7 Characteristic Formulas of Interior Algebras

In this section, using connections between varieties of Heyting and interior algebras, we will prove the analogous results for varieties of interior algebras.

5.7.1 *Basic Definition*

We will be using some facts regarding connections between varieties of Heyting and the interior algebras. All necessary information can be found in Maksimova and Rybakov (1974), Blok and Dwinger (1975), Jónsson and Tarski (1952).

We consider interior algebras in the signature $\wedge, \vee. \to, \neg, \Box$. By $\mathscr{H}$ we denote a variety of all Heyting algebras and by $\mathscr{I}$ we denote a variety of all interior algebras. Formulas without occurrences of $\Box$ are called *assertoric* as opposed to *modal* formulas in the extended signature. However, we will often omit "modal" if no confusion arises. If A is an assertoric formula, by $T(A)$ we denote the Gödel–McKinsey–Tarski translation of A. If A is a formula, $Var(A)$ denotes the set of all variables occurring in A.

We will use the following notation and statements from (Maksimova and Rybakov (1974)): if $\mathscr{M} \subseteq \mathscr{I}$ is a variety, then $\rho(\mathscr{M}) \subseteq \mathscr{H}$ is a variety defined by all assertoric formulas A whenever $\vdash_{\mathscr{M}} T(A)$. If **B** is an interior algebra, then $H(\mathbf{B})$ is the Heyting algebra of open elements of the algebra **B** that we will call *Heyting carcass of* **B**. Then, $\rho(\mathscr{M}) = \{H(\mathbf{B}) \mid \mathbf{B} \in \mathscr{M}\}$ and ρ is a homomorphism (see Maksimova and Rybakov (1974)) of the complete lattice of subvarieties of $\mathscr{I}$ onto complete lattice of subvarieties of $\mathscr{H}$.

We recall from Maksimova and Rybakov (1974) Lemma 3.1, that for any assertoric formula A and any interior algebra $\mathbf{B}$

$$H(\mathbf{B}) \models A \text{ if and only if } \mathbf{B} \models T(A); \tag{5.1}$$

If $\mathbf{A}$ is a Heyting algebra, then *a modal span of* $\mathbf{A}$ (span for short) is the smallest relative to embeddings interior algebra $s(\mathbf{A})$, the Heyting carcass of which is isomorphic to $\mathbf{A}$. The span of algebra $\mathbf{A}$ can be constructed by taking the free Boolean extension $B(\mathbf{A})$ of $\mathbf{A}$, and for each $\mathsf{a} \in B(\mathbf{A})$ letting $\Box\mathsf{a} = \bigwedge_{i=1}^{n}(\mathsf{a}_i \to \mathsf{a}'_i)$, where $\mathsf{a} = \bigwedge_{i=1}^{n}(\neg\mathsf{a}_i \vee \mathsf{a}'_i)$ (see Blok and Dwinger 1975, p. 191 or Rasiowa and Sikorski 1970, pp. 128–130). Then $(B(\mathbf{A}), \Box)$ is an interior algebra.

The spans and translations are related in the following way: for any assertoric formula A and any Heyting algebra $\mathbf{A}$

$$\mathbf{A} \models A \text{ if and only if } s(\mathbf{A}) \models T(A). \tag{5.2}$$

If $\mathbf{B}$ is an interior algebra, then by $\mathbf{B}^o$ we denote a subalgebra of $\mathbf{B}$ generated by its open elements, that is, by elements of $H(\mathbf{B})$. In fact, (Maksimova and Rybakov 1974, Lemma 3.4), $\mathbf{B}^o$ is a modal span of $H(\mathbf{B})$. More precisely, the following holds.

Proposition 13 *Let* $\mathbf{A}$ *be a Heyting algebra and* $\mathbf{B}$ *be an interior algebra. Then*

$$\mathbf{A} \cong H(\mathbf{B}^o) \textit{ if and only if } s(\mathbf{A}) \cong \mathbf{B}^o. \tag{5.3}$$

5.7.2 Finitely Presented Interior Algebras

For interior algebras, finite presentability can be defined in the following way.

Definition 4 (cf. Mal'cev 1973) Let $\mathscr{M} \subseteq \mathscr{I}$ be a variety of interior algebras, $A(\underline{p})$ be a formula and ν be a valuation in algebra $\mathbf{A}$. Then a pair $\langle A, \nu\rangle$ *defines algebra* $\mathbf{A}$ over $\mathscr{M}$ if

1. Elements $\mathsf{a}_1 = \nu(p_1), \ldots, \mathsf{a}_n = \nu(p_n)$ generate algebra $\mathbf{A}$;
2. $A(\underline{a}) = \mathbf{1}$;
3. For any formula $B(\underline{p})$, if $B(\underline{a}) = \mathbf{1}$, then $\vdash_{\mathscr{M}} (A(\underline{p}) \Rightarrow B(\underline{p}))$.

Proposition 2 gives us another way of defining finite representability of interior algebras.

Next, we prove that the Gödel–McKinsey–Tarski translation preserves finite presentability. For this we will use Maksimova's Translation Lemma.

Recall that a modal formula is said to be *special* (see Maksimova (1979)) if each of its variables is preceded by the necessity symbol $\Box$. For instance, $T(A)$ is a special formula for any assertoric formula A.

Lemma 5 *(Maksimova 1979, Translation Lemma) For any special modal formula B, there exists an assertoric formula A such that $S4 \vdash (\Box B \leftrightarrow T(A))$ (i.e. $\vdash_{\mathcal{I}} (B \Leftrightarrow T(A))$).*

The following corollary gives an algebraic meaning of Maksimova's lemma.

Corollary 13 *If elements $\mathsf{b}_1, \ldots, \mathsf{b}_n \in H(\mathbf{A})$ generate an algebra $\mathbf{A}$ (as an interior algebra), then these elements generate algebra $H(\mathbf{A})$ (as a Heyting algebra).*

Proof If elements $\mathsf{b}_1, \ldots, \mathsf{b}_n \in H(\mathbf{A})$ generate algebra $\mathbf{A}$, then for each element $\mathsf{b} \in H(\mathbf{A})$ there is a modal formula B such that $\mathsf{b} = B(\mathsf{b}_1, \ldots, \mathsf{b}_n)$. Since $\mathsf{b}_1, \ldots, \mathsf{b}_n \in H(\mathsf{b})$, we can assume that B is a special formula. Hence, by virtue of Maksimova's Lemma, there is an assertoric formula A such that $\vdash_{\mathcal{I}} B \Leftrightarrow T(A)$. It is not hard to see that in Heyting algebra $H(\mathbf{A})$ we have $\mathsf{b} = A(\mathsf{b}_1, \ldots, \mathsf{b}_n)$.

We will need the following a little bit stronger statement.

Lemma 6 *Algebra $\mathbf{B}^o$ is finitely generated as an interior algebra if and only if algebra $H(\mathbf{B}^o)$ is finitely generated as Heyting algebra.*

Proof It is clear that the generators of $H(\mathbf{B}^o)$ will generate the whole algebra $\mathbf{B}^o$, because, by the definition, algebra $\mathbf{B}^o$ is generated by elements from $H(\mathbf{B}^o)$.

Conversely, assume that $\mathbf{B}^o$ is generated by elements $\mathsf{b}_1, \ldots, \mathsf{b}_k$. Since $\mathbf{B}^o$ is also generated by some elements from $H(\mathbf{B})$, there exists a finite set of formulas $B_1, \ldots, B_k$ and elements $\mathsf{g}_1, \ldots, \mathsf{g}_n \in H(\mathbf{B}^o)$ such that $\mathsf{b}_i = B_i(\mathsf{g}_1, \ldots, \mathsf{g}_n)$, where $i = 1, \ldots, k$, that is, elements $\mathsf{g}_1, \ldots, \mathsf{g}_n$ generate algebra $\mathbf{B}^o$. Thus, for any element $\mathsf{b} \in H(\mathbf{B})$, there is such a formula $B(p_1, \ldots, p_n)$ that $B(\mathsf{g}_1, \ldots, \mathsf{g}_n) = \mathsf{b}$. Recall that $\mathsf{g}_i \in H(\mathbf{B})$ for all $i = 1, \ldots, n$. Therefore $\mathsf{g}_i = \Box \mathsf{g}_i$. Hence, we can assume that B is a special formula and, by Translation Lemma, there is such an assertoric formula A that $\vdash_{\mathcal{I}} (B \Leftrightarrow T(A))$. From the properties of translation, it follows that Heyting algebra $H(\mathbf{B}^o)$ is generated (as a Heyting algebra) by elements $\mathsf{g}_1, \ldots, \mathsf{g}_n$.

We will also need the following properties of special formulas.

Proposition 14 *Let B_1 and B_2 be special formulas containing variables only from the list $p_1, \ldots, p_n$. Then*

$$\vdash (B_1 \Rightarrow B_2) \tag{5.4}$$

if and only if

$$\vdash (\bigwedge_{i=1}^{n}(p_i \leftrightarrow \Box p_i) \wedge B_1) \Rightarrow B_2. \tag{5.5}$$

Proof It is clear that (5.5) follows from (5.4).

Conversely, suppose that in some algebra $\mathbf{B}$ we have

$$B_1(\mathsf{b}_1, \ldots, \mathsf{b}_n) \Rightarrow B_1(\mathsf{b}_1, \ldots, \mathsf{b}_n) \neq \mathbf{1}.$$

Since formulas B_1 and B_2 are special, we have

$$B_1(\Box \mathsf{b}_1, \ldots, \Box \mathsf{b}_n) \Rightarrow B_1(\Box \mathsf{b}_1, \ldots, \Box \mathsf{b}_n) \neq \mathbf{1}.$$

Thus, formula (5.5) is invalid on elements $\Box \mathbf{B}_1, \ldots, \Box \mathbf{B}_n$.

If $B(p_1, \ldots, p_n)$ is a formula, by $B^{\Box}$ we denote the special formula

$$B(\Box p_1, \ldots, \Box p_n)$$

and by B^* we denote the formula

$$\bigwedge_{i=1}^{n} (p_i \leftrightarrow \Box p_i) \wedge B^{\Box}.$$

Proposition 15 *Assume $\mathscr{M} \subseteq \mathscr{I}$ is a variety and algebra $\mathbf{B}$ is defined over $\mathscr{M}$ by a pair $\langle B(p_1, \ldots, p_n), \nu \rangle$ and $\nu(p_i) \in H(\mathbf{B})$ for all $i = 1, \ldots, n$. Then $\vdash_{\mathscr{M}} B^* \Leftrightarrow B$.*

Proof First, let us note that $\Box \nu(p_i) = \nu(p_i)$ for all $i = 1, \ldots, n$ and $\nu(B) = \mathbf{1}$. Hence, $\nu(B^*) = \mathbf{1}$ and, by the definition of finitely presented algebra, $\vdash_{\mathscr{M}} B \Rightarrow B^*$.

Conversely, if $\mathbf{C} \in \mathscr{M}$ is an algebra and $B^*(\mathsf{c}_1, \ldots, \mathsf{c}_n) = \mathbf{1}_{\mathbf{C}}$ for some elements $\mathsf{c}_1, \ldots, \mathsf{c}_n$ of algebra $\mathbf{C}$, then $\mathsf{c}_i = \Box \mathsf{c}_i$ and

$$B(\mathsf{c}_1, \ldots, \mathsf{c}_n) = B(\Box \mathsf{c}_1, \ldots, \Box \mathsf{c}_n) = B^{\Box}(\mathsf{c}_1, \ldots, \mathsf{c}_n) = \mathbf{1}_{\mathsf{C}},$$

i.e. $\vdash_{\mathscr{M}} B^* \Rightarrow B$.

Theorem 16 *Let $\mathscr{M}$ be a variety of interior algebras and $\mathbf{A}$ be a Heyting algebra. Algebra $s(\mathbf{A})$ is finitely presented over $\mathscr{M}$ if and only if algebra $\mathbf{A}$ is finitely presented over $\rho(\mathscr{M})$.*

Proof Let $\mathbf{B} = s(\mathbf{A})$ be finitely presentable over $\mathscr{M}$. By (5.3) algebras $H(\mathbf{B})$ and $s(\mathbf{A})$ are isomorphic. Let $\phi : H(\mathbf{B}) \to s(\mathbf{A})$ be an isomorphism. Since $\mathbf{B}$ is finitely generated, by Lemma 6 there are elements $\mathsf{b}_1, \ldots, \mathsf{b}_n \in H(\mathbf{B})$ generating $\mathbf{B}$. Using Proposition 5 – Titze's Theorem, we can conclude that there is a formula $B(p_1, \ldots, p_n)$ defining algebra $\mathbf{B}$ in generators $\mathsf{b}_1, \ldots, \mathsf{b}_n$. Let us consider formula $B^{\Box}$. By Maksimova's Lemma, there is an assertoric formula A which translation is equivalent to $B^{\Box}$. We will demonstrate that formula A defines algebra $\mathbf{A}$.

Let us consider elements $\mathsf{a}_i = \phi(\mathsf{b}_i), i = 1, \ldots, n$. It is clear that these elements generate algebra $\mathbf{A}$ and $A(\mathsf{a}_1, \ldots, \mathsf{a}_n) = \mathbf{1}$. All that remains to be checked is that for any given formula $A'(p_1, \ldots, p_n)$, from $A'(\mathsf{a}_1, \ldots, \mathsf{a}_n) = \mathbf{1}_{\mathbf{A}}$ it follows that $\vdash_{\rho(\mathscr{M})} A \to A'$. Or, by properties of translation, it suffices to check that $\vdash_{\mathscr{M}} T(A) \Rightarrow T(A')$.

Indeed, from $A'(\mathsf{a}_1, \ldots, \mathsf{a}_n) = \mathbf{1}_{\mathbf{A}}$ it follows that $T(A')(\mathsf{b}_1, \ldots, \mathsf{b}_n) = \mathbf{1}_{\mathbf{B}}$. Recall that B is a defining formula, hence, $\vdash_{\mathscr{M}} B \Rightarrow T(A')$. Using Proposition 15, we obtain $\vdash_{\mathscr{M}} B^* \Rightarrow T(A')$. Now, we can apply Proposition 14 and conclude that

$$\vdash_{\mathscr{M}} B^{\Box} \Rightarrow T(A'). \tag{5.6}$$

Let us also recall that we have selected formula A in such a way that

$$\vdash_{\mathscr{I}} B^{\Box} \Leftrightarrow T(A), \tag{5.7}$$

hence

$$\vdash_{\mathscr{M}} B^{\Box} \Leftrightarrow T(A). \tag{5.8}$$

Thus, from (5.6) and (5.8) we have

$$\vdash_{\mathscr{M}} T(A) \Rightarrow T(A').$$

Conversely, assume that a pair $\langle A(p_1, \dots, p_n), \mu \rangle$ defines algebra $\mathbf{A}$ over variety $\mathscr{H}' = \rho(\mathscr{M})$. We will prove that algebra $s(\mathbf{A})$ is finitely presentable, namely, it is defined by formula

$$B = T(A) \wedge (p_1 \leftrightarrow \Box p_1) \wedge \cdots \wedge (p_n \leftrightarrow \Box p_n). \tag{5.9}$$

More precisely, we will prove that algebra $\mathbf{B}$, that is defined over $\mathscr{M}$ by pair $\langle B, \nu \rangle$, where ν is a valuation in $\mathbf{B}$, is isomorphic to $s(\mathbf{A})$ by proving the following:

(a) $\mathbf{B} = \mathbf{B}^o$,
(b) $\mathbf{A} \cong H(\mathbf{B})$

and applying (5.3).

(a) Let $\mathsf{b}_i = \nu(p_i)$ for $i = 1, \dots, n$. By Definition 4, the elements $\mathsf{b}_1, \dots, \mathsf{b}_b$ generate algebra $\mathbf{B}$ and $\nu(B) = \mathbf{1_B}$. Hence, $\nu(p_i \leftrightarrow \Box p_i) = \mathbf{1_B}$ for all $i = 1, \dots, n$, and we can see that $\mathsf{b}_i \in H(\mathbf{B})$, that is, algebra $\mathbf{B}$ is generated by its open elements. Thus $\mathbf{B} = \mathbf{B}^o$.

(b) For $\nu(B) = \mathbf{1_B}$, we have $T(A)(\mathsf{b}_1, \dots, \mathsf{b}_n) = \mathbf{1_B}$ and, hence by Proposition 1, $A(\mathsf{b}_1, \dots, \mathsf{b}_n) = \mathbf{1}_{H(\mathbf{B})}$. Thus, by Proposition 2 the mapping

$$\mu(p_i) \mapsto \nu(p_i), \; i = 1, \dots, n \tag{5.10}$$

can be extended to a homomorphism $\phi : \mathbf{A} \to H(\mathbf{B})$. So algebra $H(\mathbf{B})$ is a homomorphic image of algebra $\mathbf{A}$. By virtue of Corollary 13, elements $\mathsf{b}_1, \dots \mathsf{b}_n$ are generators of $H(\mathbf{B})$. Therefore homomorphism ϕ is a homomorphism of $\mathbf{A}$ onto $H(\mathbf{B})$. What is left is to show that the kernel of homomorphism ϕ contains the only element, namely $\mathbf{1_A}$.

Indeed, let $\mathsf{c} \in \mathbf{A}$ and $\mathsf{c} \neq \mathbf{1_A}$. Then, there exists an assertoric formula C such that $C(\mu(p_1), \dots, \mu(p_n)) = \mathsf{c} < \mathbf{1_A}$. By definition of finitely presentable algebra, $\mu(A) = \mathbf{1_A}$. Thus, μ is a refuting valuation of the formula $A \to C$, i.e. $\mathbf{A} \not\models A \to C$. Hence, by (5.2), $s(\mathbf{A}) \not\models T(A) \Rightarrow T(C)$. Since $T(A)$ is obviously a special formula, using that formula B is defined by is (5.9), we can conclude that

$$s(\mathbf{A}) \nvDash B \Rightarrow T(C). \tag{5.11}$$

Our next goal is to prove that $\phi(\mathsf{c}) \neq \mathbf{1}_{H(\mathbf{B})}$. Let us assume the contrary:

$$\phi(c) = \phi(C(\mu(p_1), \ldots, p_n)) = C(\nu(p_1), \ldots, \nu(p_n)) = \mathbf{1}_{H(\mathbf{B})}. \tag{5.12}$$

Then

$$T(C)(\nu(p_1), \ldots, \nu(p_n)) = \mathbf{1}_{\mathbf{B}}. \tag{5.13}$$

From the definition of finitely presented algebra, due to B is a defining formula, it follows that

$$\vdash_{\mathscr{M}} B \Rightarrow T(C) \tag{5.14}$$

and, since $s(\mathbf{A}) \in \mathscr{M}$, we have

$$s(\mathbf{A}) \vDash B \Rightarrow T(C). \tag{5.15}$$

The latter contradicts (5.11). Thus, ϕ is an isomorphism and $\mathbf{A} \cong H(\mathbf{B})$. Let us recall that $\mathbf{B} = \mathbf{B}^o$, so $\mathbf{A} \cong H(\mathbf{B}^o)$ and by (5.3) $s(\mathbf{A}) \cong \mathbf{B}^o = \mathbf{B}$. Therefore, algebra $s(\mathbf{A})$ is finitely presentable.

The following theorem is a straight consequence of Theorems 8 and 16.

Theorem 17 *Let* $\mathbf{A}$ *be any non-degenerate Heyting algebra and* $\mathsf{Z}^* = \mathsf{Z} \times \mathsf{Z}_2 + \mathbf{A}$. *Then algebra* $s(\mathsf{Z}^\circ)$, *where* $\mathsf{Z}^\circ = \mathsf{Z} \times \mathsf{Z}_2 + \mathsf{Z}_2$, *is finitely presented over variety* $\mathscr{V}(s(\mathsf{Z}^*))$.

Example 1 Let $\mathbf{A} = s(\mathsf{Z} \times \mathsf{Z}_2 + \mathsf{Z}_2))$ (see Fig. 5.2). Obviously $\mathbf{A}$ is subdirectly irreducible and, according to the Theorem 17, algebra $\mathbf{A}$ is finitely presented over $\mathscr{V}(\mathbf{A})$.

Remark 5 M. Kracht shows (see Kracht 1990, Sect. E) the way how to construct an infinite set of infinite algebras that are splitting the variety corresponding to the logic of **S4**-frames of width 3.

Now we can define a characteristic formula for interior algebra similarly to how we did it for Heyting algebra.

Definition 5 Assume $\mathscr{M}$ is a variety of interior algebras and $\mathbf{A} \in \mathscr{M}$ is an s.i. algebra finitely presented over $\mathscr{M}$. Suppose $\langle A(\underline{p}), \nu \rangle$ is a presentation of $\mathbf{A}$ over $\mathscr{M}$. If $B(\underline{p})$ is such a formula that $\nu(B)$ is an opremum of $H(\mathbf{A})$, then the formula

$$\chi_{\mathscr{M}}(\mathbf{A}) = A(\underline{p}) \Rightarrow B(\underline{p}) \tag{5.16}$$

is a *characteristic formula of algebra* $\mathbf{A}$ *over variety* $\mathscr{M}$.

It is not hard to see that Theorem 7 holds true for interior algebras.

Theorem 18 *Assume* $\mathscr{M} \subseteq \mathscr{I}$, $\mathbf{A} \in FPSI(\mathscr{M})$, $\mathbf{B} \in \mathscr{M}$ *and* B *is a formula. Then*

(a) if $\mathbf{B} \not\models \chi_{\mathcal{M}}(A)$, *then* $\mathbf{A} \in \mathsf{SHB}$;
(b) if $\mathbf{A} \not\models B$, *then* $B \vdash_{\mathcal{M}} \chi_{\mathcal{V}}(A)$.

Using Theorem 18 and Lemma 4 one can prove the following theorem.

Theorem 19 *There is a continuum of varieties of interior algebras that are defined by characteristic formulas, but cannot be axiomatized by Yankov (or Fine) formulas.*

Acknowledgements Many thanks to A. Muravitsky, G. Bezhanishvili and F. Wolter for their fruitful discussions. The author is also indebted to L.L. Maksimova for pointing out that the Translation Lemma can be used in the proof of the Theorem 16.

References

Bezhanishvili, G., & Bezhanishvili, N. (2009). An algebraic approach to canonical formulas: Intuitionistic case. *Review of Symbolic Logic*, *2*(3), 517–549.

Bezhanishvili, G., Bezhanishvili, N., & de Jongh, D. (2008). The Kuznetsov-Gerčiu and Rieger-Nishimura logics. *The boundaries of the finite model property, Logic and Logical Philosophy*, *17*(1–2), 73–110.

Bezhanishvili, N. (2006). *Lattices of intermediate and cylindric modal logics*, PhD thesis, Institute for Logic, Language and Computation, University of Amsterdam.

Blok, W. (1976). *Varieties of interior algebras*, PhD thesis, University of Amsterdam.

Blok, W. J., & Dwinger, P. (1975). Equational classes of closure algebras. *I, Indagationes Mathematicae*, *37*, 189–198.

Blok, W. J., & Pigozzi, D. (1982). On the structure of varieties with equationally definable principal congruences. *I, Algebra Universalis*, *15*(2), 195–227.

Butz, C. (1998). *Finitely presented heyting algebras*, BRICS Reports, University of Aarhus.

Chagrov, A., & Zakharyaschev, M. (1997). *Modal logic*. *Oxford Logic Guides* (Vol. 35). New York: The Clarendon Press.

Citkin, A. (1977). On admissible rules of intuitionistic propositional logic. *Mathematics of the USSR-Sbornik 31*, 279–288. (A. Tsitkin).

Citkin, A. (1979) *Pravila vyvoda dopustimye v superintuicionistskikh logikakh*, PhD thesis, Uzhgorod State University, Ukraine (1979). [Inference rules addmissible in superintuitionistic logics].

Citkin, A. (2012). Not every splitting Heyting or interior algebra is finitely presentable. *Studia Logica*, *100*(1–2), 115–135.

de Jongh, D. (1968). *Investigations on Intuitionistic Propositional Calculus*, PhD thesis, University of Wisconsin.

Fine, K. (1974). An ascending chain of S4 logics. *Theoria*, *40*(2), 110–116.

Gerčiu, V. Y., & Kuznecov, A. V. (1970). Konechno aksiomatiziruemye superintuicionistskie logiki. *Doklady Akademii Nauk SSSR*, *195*, 1263–1266. [The finitely axiomatizable superintuitionistic logics].

Jónsson, B., & Tarski, A. (1952). Boolean algebras with operators. *II, American Journal of Mathematics*, *74*, 127–162.

Kracht, M. (1990). An almost general splitting theorem for modal logic. *Studia Logica*, *49*(4), 455–470.

Kracht, M. (1993). Splittings and the finite model property. *Journal of Symbolic Logic*, *58*(1), 139–157.

Kracht, M. (1999). *Tools and techniques in modal logic*. Studies in Logic and the Foundations of Mathematics (Vol. 142). Amsterdam: North-Holland Publishing Co.

Kuznetsov, A. V. (1973). O konechno porozhdennykh psevdo-bulevykh algebrakh i finitno appoksimiruemykh mnogoobraziyakh. In *Proceedings of the 12th USSR Algebraic Colloquium*, Sverdlovsk, p. 281. [On finitely generated pseudo-Boolean algebras and finitely approximable varieties].
Kuznetsov, A. V., & Gerčiu, V. Y. (1970). Superintuizionistskiye logiki i finitnaya approksimiruemost. *Doklady Akademii Nauk SSSR*, *195*, 1029–1032. [The superintuitionistic logics and finite approximability].
Maksimova, L. L. (1979). Interpolation properties of superintuitionistic logics. *Studia Logica*, *38*(4), 419–428.
Maksimova, L. L., & Rybakov, V. V. (1974). Reshetka normalnykh modalnykh logik. *Algebra i Logika*, *13*, 188–216. [The lattice of normal modal logics].
Mal'cev, A. (1973). *Algebraic systems*. Berlin: Springer.
McKenzie, R. (1972). Equational bases and nonmodular lattice varieties. *Transactiones of the American Mathematical Society*, *174*, 1–43.
Rasiowa, H., & Sikorski, R. (1970). *The mathematics of metamathematics* (3rd ed.). Warsaw: PWN.
Rautenberg, W. (1979). *Klassische und nichtklassische Aussagenlogik*. Braunschweig: Friedr. Vieweg & Sohn.
Tomaszewski, E. (2003). *On sufficiently rich sets of formulas*, PhD thesis, Institute for Philosophy, Jagellonian University, Krakov.
Umezawa, T. (1959). On intermediate propositional logics. *Journal of Symbolic Logic*, *24*, 20–36.
Wolter, F. (1993). *Lattices of modal logics*, PhD thesis, Freien Universität Berlin.
Wroński, A. (1974). On cardinality of matrices strongly adequate for the intuitionistic propositional logic. *Bulletin of the Section of Logic*, *3*(1), 34–40.
Yang, F. (2009). *Intuitionistic subframe formulas, NNIL-Formulas and n-universal models*, PhD thesis, Univerity of Amsterdam, Amsterdam, ILLC Dissertation Series MoL-2008-12.
Yankov, V. A. (1963). O svyazi mezhdu vyvodimost'yu v intuicionistskom propozicionalnom ischislenii i konechnymi implikativnymi strukturami. *Doklady Akademii Nauk SSSR*, *151*, 1293–1294. [Translation: "On the relation between deducibility in intuitionistic propositional calculus and finite implicative structures", *Soviet Mathematics, Doklady 4*, pp. 1203–1204.].
Yankov, V. A. (1968). Postroenie posledovatelnosti silno nezavisimykh superintuizionistskikh propozicionalnykh ischislenij. *Doklady Akademii Nauk SSSR*, *181*, 33–34. [Translation: "The construction of a sequence of strongly independent superintuitionistic propositional calculi.", *Soviet Mathematics, Doklady 9*, pp. 806–807.].
Yankov, V. A. (1969). Konjunktivno nerazreshimye formuly v propozicionalnom ischislenii. *Izvestiya Akademii Nauk SSSR, Ser. Mat.*, *33*, 18–38. [Translation: "Conjunctively irresolvable formulae in propositional calculi", *Mathematics of the USSR-Izvestiya 3*, 17–35.].
Zakharyaschev, M. (1992). Canonical formulas for K4. I. Basic results. *Journal of Symbolic Logic*, *57*(4), 1377–1402.
Zakharyaschev, M. (1996). Canonical formulas for K4. II. Cofinal subframe logics. *Journal of Symbolic Logic*, *61*(2), 421–449.
Zakharyaschev, M. (1997). Canonical formulas for K4. III. The finite model property. *Journal of Symbolic Logic*, *62*(3), 950–975.
Zakharyashchev, M. V. (1983). O promezhutochnykh logikakh. *Doklady Akademii Nauk SSSR*, *269*(1), 18–22. [On intermediate logics].
Zakharyashchev, M. V. (1988). Sintaksis i semantika modalnykh logik, soderzhaschikh S4. *Algebra i Logika*, *27*(6), 659–689. [Syntax and semantics of modal logics that contain S4].
Zakharyashchev, M. V. (1989). Sintaksis i semantika superintuicionistskikh logik. *Algebra i Logika*, *28*(4), 486–487 (1989). [Syntax and semantics of superintuitionistic logics].

Chapter 6
A Generalization of Maksimova's Criterion for the Disjunction Property

Alex Citkin

Dedicated to L.L. Maksimova who has always been for me a role model, as a scientist as well as an individual.

Abstract We prove that Maksimova's Criterion of disjunctive property for intermediate logics can be extended to the varieties with equationally definable principal meets in which disjunction is introduced via principal intersection terms. We also give a necessary condition for semiconstructvity, and show how it can be applied to admissibility of multiple-conclusion rules expressing semiconstructivity.

6.1 Introduction

In 1932 Gödel remarked about Heyting's intuitionistic propositional calculus H: "Besides, the following holds with full generality: a formula of form $A \vee B$ can only be provable in H if either A or B is provable in H" (see Gödel 1986, p. 225). Nowadays, this property is known as the Disjunction Property (DP for short). This observation was important, because the DP is directly linked to intuitionistic/constructivist critique of the classical logic, and, in particular, to the critique of classically understood disjunction and the law of excluded middle. This circumstance led Łukasiewicz to conjecture (see Łukasiewicz 1952) that intuitionistic (propositional) logic **Int** has no proper consistent extensions enjoying the DP (and he had formulated the DP in the form of a refutation rule: if formulas α and β are refuted, refute $\alpha \vee \beta$). In other words, he conjectured that the DP uniquely characterizes **Int** among intermediate logics.

But soon, Kreisel and Putnam discovered (see Kreisel and Putnam 1957) a proper consistent extension of **Int** enjoying the DP: the Kreisel–Putnam logic. And in Wroński (1973) Wronski proved that indeed there is continuum of proper consistent extensions of **Int** with the DP. At the same time Gabbay and de Jongh had

A. Citkin (✉)
Metropolitan Telecommunications, 55 Water Str., New York, NY 10041, USA
e-mail: acitkin@gmail.com

S. Odintsov (ed.), *Larisa Maksimova on Implication, Interpolation, and Definability*, Outstanding Contributions to Logic 15, https://doi.org/10.1007/978-3-319-69917-2_6

constructed an infinite sequence of finitely axiomatizable decidable extensions of **Int** with the DP (see Gabbay and De Jongh 1974). So, it became important to give a method to describe the intermediate logic enjoying the DP, i.e. having, in a way, a constructively acceptable disjunction.

In order to construct the logics with the DP, Wronski, as well as Gabbay and de Jongh, were considering the logics defined by the classes of Heyting algebras[1] $\mathscr{M}$ closed under following operation: if $\mathbf{M}_1, \mathbf{M}_2 \in \mathscr{M}$, then $(\mathbf{M}_1 \times \mathbf{M})^* \in \mathscr{M}$, where $*$ is the operation of adjoining a new top element. Let us observe that a Heyting algebra obtained by adjoining a new top element is always subdirectly irreducible. Maksimova had captured this property in the following criterion, which in the sequel we call Maksimova's Criterion.

Theorem 6.1 (Maksimova 1986) *Let L be a consistent extension of* ***Int*** *complete with respect to a class* $\mathscr{K}$ *of Heyting algebras. Then the following are equivalent:*

(i) L has the disjunction property,
(ii) for any Heyting algebras $\mathbf{A}_1, \mathbf{A}_2 \in \mathscr{K}$ *there exists a finitely subdirectly irreducible Heyting algebra* **B** *that is a model of L and such that there is a homomorphism from* **B** *onto* $\mathbf{A}_1 \times \mathbf{A}_2$.

Maksimova had established this criterion for intermediate logics, but it was clear that this criterion can be extended to a broad class of logics. And this is exactly what had happened (see e.g. Chagrov and Zakharyaschev 1993 or Galatos et al. 2007, Theorem 5.21 or Horčík and Terui 2011, Theorem 3.4).

Gödel did not provide a proof of the DP. The proof was given in Gentzen (1935) as a consequence of the Cut Elimination Theorem, and in McKinsey and Tarski (1948) by different means. Let us note that in McKinsey and Tarski (1948) this theorem was proven not only for **Int**, but also for Lewis system **S4**: If $\sim\Diamond\sim\alpha \vee \sim\Diamond\sim\beta$ is provable in **S4**, then either α or β is provable in the **S4**. It is clear that in **S4**, as well as in the classical logic, formula $\alpha \vee \beta$ can be derived even though neither disjunct is derived (take, for instance, the law of excluded middle). The example of **S4** shows that even though the DP does not hold for the disjunction declared in the language, the DP nevertheless holds for a disjunction introduced via formula(s), in this case, for the strong disjunction.

Thus, the question arises whether, given a logic $\mathcal{L}$, there is a way of introducing in a natural way a disjunction (that possesses regular properties) and for which the DP holds. There are different ways of introducing disjunction (see e.g. Czelakowski 1984; Font and Jansana 1996; Cintula and Noguera 2013; the reader can find the extensive information on different types of disjunction in Humberstone 2011, Chap. 6). We use the algebraic approach suggested in Blok and Pigozzi (1986), and in Sect. 6.3.2 we prove Maksimova's criterion for a broad class of logics admitting such a disjunction.

Let us note that the DP itself can be generalized. Ferrari and Miglioli (see Ferrari and Miglioli 1993) had introduced the following version of the generalized DP: given two logics $\mathcal{L}_1$ and $\mathcal{L}_2$, logic $\mathcal{L}_1$ is semicontructive in $\mathcal{L}_2$, if $\alpha \vee \beta \in \mathcal{L}_1$ entails

[1]In Gabbay and De Jongh (1974) the Kripke models are used.

that $\alpha \in \mathcal{L}_2$ or $\beta \in \mathcal{L}_2$. In Sect. 6.4 we study the generalizations of the DP and give a simple sufficient condition for semiconstructivity that is based on Maksimova's criterion.

The definitions of different variations of the DP and their connections to admissibility of multiple-conclusion inference rules (as well as criteria of the DP based on Kripke models) can be found in Goudsmit (2015), Sect. 3.5.1. We discuss these generalizations in Sect. 6.4.

The goal of this paper is to prove Maksimova's criterion for a very broad class of algebraizable logics, that covers all known to the author classes of logics for which Maksimova's criterion was established, and in such a way to provide a unified approach to studying the DP. We use the algebraic semantic, and we prove the criterion in terms of varieties of algebras but all results can be naturally reformulated in terms of logics.

6.2 Preliminaries

6.2.1 *Algebras and Congruences*

We consider algebras of a given type $\mathscr{F}$. We also assume that X is a given set of variables and $T(X)$ is the set of all terms of type $\mathscr{F}$ over X (cf. Burris and Sankappanavar 1981). Given an algebra $\mathbf{A}$, a mapping $\nu : X \longrightarrow \mathbf{A}$ is called an *assignment in* $\mathbf{A}$. A given assignment ν in $\mathbf{A}$ assigns the value to each term $t(x_1, \dots, x_n)$ by letting $\nu(t(x_1, \dots, x_n)) = t(\nu(x_1), \dots, \nu(x_n))$ and we denote this value by $\nu(t)$. And for a given identity $t \approx r$ and an assignment ν, if $\nu(t) = \nu(r)$, we say that ν *validates the identity* in $\mathbf{A}$, or, if $\nu(t) \neq \nu(r)$ in $\mathbf{A}$, we say that ν *refutes the identity* in $\mathbf{A}$. An identity ι is said to be *refutable in an algebra* $\mathbf{A}$ (in symbols $\mathbf{A} \not\models \iota$), if there is an assignment in $\mathbf{A}$ refuting ι. If ι is not refutable in $\mathbf{A}$ we say that ι *is valid in* $\mathbf{A}$ and we denote this by $\mathbf{A} \models \iota$. If I is a set of identities, $\mathbf{A} \models \mathsf{I}$ means that every identity from I is valid in $\mathbf{A}$. Also, if $\mathscr{K}$ is a class of algebras, $\mathscr{K} \models \mathsf{I}$ means that every identity from I is valid in any algebra from $\mathscr{K}$.

If $\mathscr{V}$ is a variety, by $\mathbf{F}_{\mathscr{V}}(\kappa)$ we denote the free algebra of $\mathscr{V}$ of rank κ. We will use the following property of free algebra $\mathbf{F}_{\mathscr{V}}(\omega)$ (comp. Grätzer 2008, Theorem 1 Sect. 26).

Proposition 6.2 *Let $\mathscr{V}$ be a variety, $t \approx r$ be an identity, and $\mu : X \longrightarrow \mathsf{G}$ be a 1-1-mapping of the set of variables in the set of free generators of $\mathbf{F}_{\mathscr{V}}(\omega)$ that will be regarded as an assignment in $\mathbf{F}_{\mathscr{V}}(\omega)$. Then*

$$\mathscr{V} \models t \approx r \text{ if and only if } \mu(t) = \mu(r). \tag{6.1}$$

The proof easily follows from the fact that any mapping of G in any algebra $\mathbf{A}$ from $\mathscr{V}$ can be extended to a homomorphism of $\mathbf{F}_{\mathscr{V}}(\omega)$ in $\mathbf{A}$.

Given an algebra $\mathbf{A}$, by $\mathsf{Con}(\mathbf{A})$ we denote the lattice of all congruences (see Burris and Sankappanavar (1981)) of $\mathbf{A}$: if $\theta_1, \theta_2 \in \mathsf{Con}(\mathbf{A})$, $\theta_1 \wedge \theta_2 = \theta_1 \cap \theta_2$ and $\theta_1 \vee \theta_2$ is the smallest congruence containing θ_1 and θ_2. Any two elements $\mathsf{a}, \mathsf{b} \in \mathbf{A}$ define a congruence $\theta_{\mathbf{A}}(\mathsf{a}, \mathsf{b})$: the least congruence of $\mathbf{A}$ containing (a, b) (and we will omit the reference to algebra when no confusion arises). A congruence that can be defined by a pair of elements is said to be *principal*. By $\varepsilon_{\mathbf{A}}$ we denote the identity congruence, that is $(\mathsf{a}, \mathsf{b}) \in \varepsilon_{\mathbf{A}}$ if only if $\mathsf{a} = \mathsf{b}$. Clearly, $\theta(\mathsf{a}, \mathsf{b})_{\mathbf{A}} = \varepsilon_{\mathbf{A}}$ if and only if $\mathsf{a} = \mathsf{b}$. The congruence $\varepsilon_{\mathbf{A}}$ also is called *trivial congruence of* $\mathbf{A}$.

Let $\mathbf{A}$ be an algebra. $\mathbf{A}$ is *subdirectly irreducible* (s.i. for short) if the meet of any system of non-trivial congruences of $\mathbf{A}$ is non-trivial congruence. And $\mathbf{A}$ is *finitely subdirectly irreducible* (f.s.i. for short) if the meet of any finite system of non-trivial congruences of $\mathbf{A}$ is non-trivial congruence. It is clear that every s.i. algebra is f.s.i., and that every finite f.s.i. algebra is s.i. Since every non-trivial congruence contains a non-trivial principal congruence, the following holds.

Proposition 6.3 *An algebra* $\mathbf{A}$ *is f.s.i. if and only if for any* $\mathsf{a}_1, \mathsf{b}_1, \mathsf{a}_2, \mathsf{b}_2 \in \mathbf{A}$ *such that* $\mathsf{a}_1 \neq \mathsf{b}_1$ *and* $\mathsf{a}_2 \neq \mathsf{b}_2$*, we have*

$$\theta_{\mathbf{A}}(\mathsf{a}_1, \mathsf{b}_1) \cap \theta_{\mathbf{A}}(\mathsf{a}_2, \mathsf{b}_2) \neq \varepsilon_{\mathbf{A}}.$$

That is, $\mathbf{A}$ *is f.s.i. if the meet of any two non-trivial principal congruences is non-trivial.*

If $\mathscr{K}$ is a class of algebras, by $\mathscr{K}_{FSI}$ we denote the class of all f.s.i. members of $\mathscr{K}$.

If $\mathsf{Con}(\mathbf{A})$ is distributive, algebra $\mathbf{A}$ is said to be *congruence distributive*. If every algebra of a given variety $\mathscr{V}$ is congruence distributive, the variety $\mathscr{V}$ is called *congruence distributive*. From Jónsson (1967) we know that congruence distributivity plays an important role in studying different varieties.

6.2.2 Joint Refutability

We say that an assignment ν in $\mathbf{A}$ *jointly refutes* a pair of identities $t_1 \approx r_1$ and $t_2 \approx r_2$ (in symbols $\mathbf{A} \not\models \langle t_1 \approx r_1, t_2 \approx r_2 \rangle$) if ν refutes each of these identities, that is, $\nu(t_1) \neq \nu(r_1)$ and $\nu(t_2) \neq \nu(r_2)$. A pair of identities is said to be *jointly refutable in* $\mathbf{A}$ if there is an assignment in $\mathbf{A}$ that jointly refutes these identities.

Let us observe the following simple properties of joint refutability.

Proposition 6.4 *For any algebras* $\mathbf{A}_1, \mathbf{A}_2$ *and any identities* ι_1, ι_2 *the following hold*

(a) If $\mathbf{A}_1 \not\models \iota_1$ *and* $\mathbf{A}_2 \not\models \iota_2$*, then* $\mathbf{A}_1 \times \mathbf{A}_2 \not\models \langle \iota_1, \iota_2 \rangle$*;*
(b) If $\mathbf{A}_1$ *is a homomorphic image of* $\mathbf{A}_2$*, then* $\mathbf{A}_2 \not\models \langle \iota_1, \iota_2 \rangle$ *as long as* $\mathbf{A}_1 \not\models \langle \iota_1, \iota_2 \rangle$*;*

(c) *If* $\mathbf{A_1} \not\models \langle \iota_1, \iota_2 \rangle$, *then* $\mathbf{F}_{\mathscr{V}}(\omega) \not\models \langle \iota_1, \iota_2 \rangle$. *Moreover, there is an assignment* ν *in* $\mathbf{F}_{\mathscr{V}}(\omega)$ *jointly refuting* ι_1 *and* ι_2 *and sending every variable in a free generator of* $\mathbf{F}_{\mathscr{V}}(\omega)$.

Proof (a) If ν_i is a refuting assignment in $\mathbf{A}_i$, $i = 1, 2$, then $\nu : x \mapsto (\nu_1(x), \nu_2(x))$ is an assignment in $\mathbf{A}_1 \times \mathbf{A}_2$ jointly refuting ι_1 and ι_2.

(b) Let ν be an assignment jointly refuting ι_1 and ι_2 in $\mathbf{A}_1$, and let $\varphi : \mathbf{A}_2 \longrightarrow \mathbf{A}_1$ be a homomorphism of $\mathbf{A}_2$ onto $\mathbf{A}_1$. Then any assignment $\nu' : x \mapsto \mathsf{a}'$, where a' is any element from $\varphi^{-1}(\nu(x))$, jointly refutes ι_1 and ι_2 in $\mathbf{A}_2$.

(c) If $\mathbf{A_1} \not\models \langle \iota_1, \iota_2 \rangle$, there is an at most countable subalgebra $\mathbf{A}'_1$ of $\mathbf{A_1}$ such that $\mathbf{A}'_1 \not\models \langle \iota_1, \iota_2 \rangle$. Algebra $\mathbf{A}'_1$ is a homomorphic image of $\mathbf{F}_{\mathscr{V}}(\omega)$, thus, we can apply (b). And we always can take a 1-1-mapping $\varphi : \mathsf{G} \longrightarrow \mathbf{A}'$, where G is a set of free generators and extend φ to a homomorphism $\overline{\varphi} : \mathbf{F}_{\mathscr{V}}(\omega) \longrightarrow \mathbf{A}'_1$. It is clear that an assignment constructed in (b) sends every variable in a free generator.

Let us note that refutability of disjunction is directly related to joint refutability: if an assignment refutes disjunction of two formulas, it refutes each of these formulas. The converse is not true: as we already know, any two non-valid in a variety identities (or non-valid in a logic formulas) are always jointly refutable in some algebra. It is important that identities (or formulas) are jointly refutable in some f.s.i. algebra. For instance, in the classical logic the formulas p and $\sim p$ are jointly refutable in the four-element Boolean algebra, but they are not jointly refutable in any f.s.i. Boolean algebra (so, $p \vee \sim p$ is valid in classical logic); while in the case of **Int**, the formulas p and $\sim p$ can be jointly refuted in the three-element Heyting algebra which is s.i.

6.2.3 Equationally Definable Principal Meets

The interest in studying varieties with equationally definable principal meets was instigated by the Baker's paper Baker (1977), because in such varieties every finite algebra generates a finitely axiomatizable subvariety. We will use the equationally definable principal meets in order to introduce disjunction.

Definition 6.1 (Blok and Pigozzi (1986)) A variety $\mathscr{V}$ has the *equationally definable principal meets (EDPM)* if there is a finite system

$$\nabla := \{\langle p_i(x, y, w, z), q_i(x, y, w, z)\rangle \mid i = 1, \ldots, n\}$$

of pairs of terms, such that for every algebra $\mathbf{A} \in \mathscr{V}$ and any elements $\mathsf{a}, \mathsf{b}, \mathsf{c}, \mathsf{d}$

$$\theta(\mathsf{a}, \mathsf{b}) \cap \theta(\mathsf{c}, \mathsf{d}) = \bigvee_{i=1}^{n} \theta(p_i(\mathsf{a}, \mathsf{b}, \mathsf{c}, \mathsf{d}), q_i(\mathsf{a}, \mathsf{b}, \mathsf{c}, \mathsf{d})). \tag{6.2}$$

∇ is called a *system of principal intersection terms*.

Let us observe the following straightforward property that immediately follows from the definition of EDPM.

Proposition 6.5 *Let $\mathscr{V}$ be a variety with EDPM and*

$$\nabla := \{\langle p_i(x, y, w, z), q_i(x, y, w, z)\rangle \mid i = 1, \dots, n\}$$

be its system of principal intersection terms. Then for any algebra $\mathbf{A} \in \mathscr{V}$ and any elements $\mathsf{a}_1, \mathsf{b}_1, \mathsf{a}_2, \mathsf{b}_2 \in \mathbf{A}$

$$\begin{aligned} &\theta_{\mathbf{A}}(\mathsf{a}_1, \mathsf{b}_1) \cap \theta_{\mathbf{A}}(\mathsf{a}_2, \mathsf{b}_2) = \varepsilon_{\mathbf{A}} \text{ if and only if} \\ &p_i(\mathsf{a}_1, \mathsf{b}_1, \mathsf{a}_2, \mathsf{b}_2) = q_i(\mathsf{a}_1, \mathsf{b}_1, \mathsf{a}_2, \mathsf{b}_2) \text{ for all } i = 1, \dots, n. \end{aligned} \tag{6.3}$$

It is worth noting that (6.3) entails that a system of principal intersection terms is unique modulo equivalence in the following sense. If $\mathscr{V}$ is a variety and Γ, Δ are the sets of identities, we say that Γ and Δ are *$\mathscr{V}$-equivalent* (in written, $\Gamma \dashv\vdash_{\mathscr{V}} \Delta$), if every assignment ν in any algebras from $\mathscr{V}$ makes valid all identities from Γ exactly when ν makes valid all identities form Δ. From (6.3) it follows that for any variety $\mathscr{V}$ with EDPM, any two of its systems of principal intersection terms ∇ and ∇' are $\mathscr{V}$-equivalent, that is,

$$\nabla \dashv\vdash_{\mathscr{V}} \nabla'. \tag{6.4}$$

Example 6.1 In many cases, a system of principal intersection terms consists of a single pair. For instance, if congruences are uniquely defined by filter, we can take $q(x, y, w, z) = \mathbf{1}$. Using abbreviations: $x \leftrightarrow y \leftrightharpoons (x \to y) \wedge (y \to x)$ and $[0]x := x$, $[n+1]x := \Box x \wedge [n]x$, we have

class of logics	principal intersection terms
extensions of **Int** :	$p(x, y, w, z) = (x \leftrightarrow y) \vee (w \leftrightarrow z)$
normal extensions of **S4** :	$p(x, y, w, z) = \Box(x \leftrightarrow y) \vee \Box(w \leftrightarrow z)$
$n-$ transitive logics :	$p(x, y, w, z) = [n](x \leftrightarrow y) \vee [n](w \leftrightarrow z)$,

where n-transitive logics are the logics extending **K** in which $[n]p \to [n+1]p$ holds (see Chagrov and Zakharyaschev 1997, Sect. 3.5).

Varieties (and quasivarieties) with EDPM have rather nice properties that were observed in Blok and Pigozzi (1986); Czelakowski and Dziobiak (1990); Agliano and Baker (1999) and summarized in Czelakowski (2001).

Proposition 6.6 (comp. Czelakowski (2001), Theorem Q.8.6) *For any variety $\mathscr{V}$ the following conditions are equivalent:*

1. *$\mathscr{V}$ has EDPM;*
2. *$\mathscr{V}$ is congruence distributive and the class $\mathscr{V}_{FSI}$ is closed under the formation of subalgebras;*
3. *For every algebra $\mathbf{A} \in \mathscr{V}$, the lattice $\mathsf{Con}_{\mathscr{V}}(\mathbf{A})$ is distributive and the set of compact elements of $\mathsf{Con}_{\mathscr{V}}(\mathbf{A})$ forms a sublattice of $\mathsf{Con}_{\mathscr{V}}(\mathbf{A})$;*

4. *The lattice* $\mathsf{Con}_{\mathscr{V}}(\mathbf{F}_{\mathscr{V}})$ *is distributive and the set of compact elements of* $\mathbf{F}_{\mathscr{V}}$ *forms a sublattice of* $\mathbf{F}_{\mathscr{V}}$;
5. *There exists a set* $\nabla(x, y, z, w)$ *of identities in four variables* x, y, z, w *such that*

$$\mathscr{V}_{FSI} \models (\forall x, y, z, w)(\bigwedge \nabla(x, y, z, w) \leftrightarrow (x \approx y \vee z \approx w)).$$

Let us note that from the above Proposition it immediately follows that for any variety $\mathscr{V}$ with EDPM, any subalgebra of any algebra from $\mathscr{V}_{FSI}$ is f.s.i.

Proposition 6.7 *Suppose* $\mathscr{V}$ *is a variety with EDPM and* $\mathbf{F}_{\mathscr{V}}(\omega)$ *is f.s.i. Then all free algebras of* $\mathscr{V}$ *are f.s.i.*

Proof Suppose $\mathscr{V}$ is a variety with EDPM and $\mathbf{F}_{\mathscr{V}}(\omega)$ is f.s.i. Then, every free algebra $\mathbf{F}_{\mathscr{V}}(\kappa)$ of $\mathscr{V}$ of rank $\kappa < \omega$ is (isomorphic to) a subalgebra of $\mathbf{F}_{\mathscr{V}}(\omega)$ and, hence, is f.s.i.

Let $\mathbf{F}_{\mathscr{V}}(\kappa)$ be a free algebra of $\mathscr{V}$ of rank $\kappa > \omega$. For contradiction: assume that $\mathbf{F}_{\mathscr{V}}(\kappa)$ is not f.s.i. Then for some elements $\mathsf{a}, \mathsf{b}, \mathsf{c}, \mathsf{d} \in \mathbf{F}_{\mathscr{V}}(\kappa)$ we have $\mathsf{a} \neq \mathsf{b}$ and $\mathsf{c} \neq \mathsf{d}$, while $\theta_{\mathbf{F}_{\mathscr{V}}(\kappa)}(\mathsf{a}, \mathsf{b}) \cap \theta_{\mathbf{F}_{\mathscr{V}}(\kappa)}(\mathsf{c}, \mathsf{d}) = \varepsilon$. Clearly, one needs only a finite set of free generators of $\mathbf{F}_{\mathscr{V}}(\kappa)$ to express all elements $\mathsf{a}, \mathsf{b}, \mathsf{c}, \mathsf{d}$. Let G be a finite subset of free generators such that elements $\mathsf{a}, \mathsf{b}, \mathsf{c}, \mathsf{d}$ belong to the subalgebra $\mathbf{F}$ of $\mathbf{F}_{\mathscr{V}}(\kappa)$ generated by G. Obviously, $\mathbf{F}$ is a subalgebra of $\mathbf{F}_{\mathscr{V}}(\omega)$ and $\mathsf{a}, \mathsf{b}, \mathsf{c}, \mathsf{d} \in \mathbf{F}$. Hence congruences $\theta_{\mathbf{F}}(\mathsf{a}, \mathsf{b})$ and $\theta_{\mathbf{F}}(\mathsf{c}, \mathsf{d})$ (the restrictions of congruences $\theta_{\mathbf{F}_{\mathscr{V}}(\kappa)}(\mathsf{a}, \mathsf{b})$ and $\theta_{\mathbf{F}_{\mathscr{V}}(\kappa)}(\mathsf{c}, \mathsf{d})$ to subalgebra $\mathbf{F}$) are distinct from $\varepsilon_{\mathbf{F}}$ and $\theta_{\mathbf{F}}(\mathsf{a}, \mathsf{b}) \cap \theta_{\mathbf{F}}(\mathsf{c}, \mathsf{d}) = \varepsilon_{\mathbf{F}}$. Thus $\mathbf{F}$, and, therefore, $\mathbf{F}_{\mathscr{V}}(\omega)$ are not f.s.i., and this contradicts the assumption of the proposition.

6.2.4 Disjunction in Equational Logic

We start with the following quotation from Blok and Pigozzi (1986), p. 2 regarding varieties with EDPM: "Many of these varieties arise from logic, and in these cases the principal intersection terms are closely related to the existence of a disjunction connective in the underlying logic. The existence of disjunction in a broad class of logics has been studied in the literature (Dzik and Suszko (1977) and Czelakowski (1983)). In this general setting a connective is taken to be a disjunction if it defines the intersection of two principal theories of the logic in essentially the same way principal intersection terms define the intersection of principal congruences in a variety with EDPM." In what follows, we study the disjunction property relative to disjunction defined via principal intersection terms.

Definition 6.2 Let $\mathscr{V}$ be a variety with EDPM and ∇ be its system of principal intersection terms, and let $t_1 \approx r_1$ and $t_2 \approx r_2$ be identities. Then

$$t_1 \approx r_1 \nabla\, t_2 \approx r_2 \leftrightharpoons \{p(t_1, r_1, t_2, r_2) \approx q(t_1, r_1, t_2, r_2) \mid \langle p, q \rangle \in \nabla\}. \quad (6.5)$$

That is, $t_1 \approx r_1 \nabla\, t_2 \approx r_2$ is a set of identities obtained from all pairs of principal intersection terms by replacing variables x, y, w, z respectively with t_1, r_1, t_2, r_2.

Let us note that in spite of (6.4), disjunction (6.5) is unique up to $\mathscr{V}$-equivalence. Moreover, it is not hard to verify that introduced above disjunction if commutative and associative (modulo $\dashv\vdash_{\mathscr{V}}$), and $\mathscr{V} \models t \approx r$ entails $\mathscr{V} \models t \approx r \nabla t' \approx r'$ for any terms t' and r'.

Example 6.2 Using the intersection terms from Example 6.1, we have

class of logics	disjunction defined by principal intersection terms
extensions of **Int** :	$x \vee y$
normal extensions of **S4** :	$\Box x \vee \Box y$
$n -$ transitive logics :	$[n]x \vee [n]y$.

Now, after we have introduced disjunction in varieties with EDPM, we can study when such a disjunction enjoys the DP.

6.3 Disjunction Property in Equational Logic

In this section we establish the criterion for the disjunction introduced via principal intersection terms to enjoy the DP.

Definition 6.3 Let $\mathscr{V}$ be a variety with EDPM and ∇ be its system of principal intersection terms, and let $\mathscr{T}$ be a corresponding equational theory. Then $\mathscr{V}$ is *disjunctive* or theory $\mathscr{T}$ enjoys the *Disjunction Property* (DP) if for any identities ι_1 and ι_2

$$\mathscr{V} \models \iota_1 \nabla\, \iota_2 \text{ entails } \mathscr{V} \models \iota_1 \text{ or } \mathscr{V} \models \iota_2, \tag{6.1}$$

or, in terms of equational logic,

$$\mathscr{T} \vdash \iota_1 \nabla\, \iota_2 \text{ entails } \mathscr{T} \vdash \iota_1 \text{ or } \mathscr{T} \vdash \iota_2. \tag{6.2}$$

Since free algebras can be viewed as term algebras (cf. Burris and Sankappanavar (1981)), the DP can be linked to properties of free algebras. Namely, the following holds.

Proposition 6.8 *A variety $\mathscr{V}$ with EDPM is disjunctive if and only if $\mathbf{F}_{\mathscr{V}}(\omega)$ is f.s.i.*

Proof Suppose $\mathscr{V}$ is disjunctive. We need to prove that $\mathbf{F}_{\mathscr{V}}(\omega)$ is f.s.i., that is, we need to prove that for any $\mathsf{a}_1, \mathsf{b}_1, \mathsf{a}_2, \mathsf{b}_2 \in \mathbf{F}_{\mathscr{V}}(\omega)$ if $\mathsf{a}_i \neq \mathsf{b}_j$, $j = 1, 2$, then $\theta_{\mathbf{F}_{\mathscr{V}}(\omega)}(\mathsf{a}_1, \mathsf{b}_1) \cap \theta_{\mathbf{F}_{\mathscr{V}}(\omega)}(\mathsf{a}_2, \mathsf{b}_2) \neq \varepsilon_{\mathbf{F}_{\mathscr{V}}(\omega)}$. Suppose $\mathsf{a}_1, \mathsf{b}_1, \mathsf{a}_2, \mathsf{b}_2 \in \mathbf{F}_{\mathscr{V}}(\omega)$ and $\mathsf{a}_j \neq \mathsf{b}_j$, $j = 1, 2$. Then there are terms t_1, r_1, t_2, r_2 using which we can express elements $\mathsf{a}_1, \mathsf{b}_1, \mathsf{a}_2, \mathsf{b}_2$ via free generators of $\mathbf{F}_{\mathscr{V}}(\omega)$, that is, there is an assignment $\nu : X \longrightarrow \mathsf{G}$, where G is the set of free generators of $\mathbf{F}_{\mathscr{V}}(\omega)$, such that

$$\mathsf{a}_j = \nu(t_j), \mathsf{b}_j = \nu(r_j), \ j = 1, 2.$$

Clearly, for $j = 1, 2$ we have $\mathbf{F}_{\mathcal{V}}(\omega) \not\models t_j \approx r_j$ and, hence,

$$\mathcal{V} \not\models t_j \approx r_j. \tag{6.3}$$

Assume for contradiction that $\theta_{\mathbf{F}_{\mathcal{V}}(\omega)}(\mathsf{a}_1, \mathsf{b}_1) \cap \theta_{\mathbf{F}_{\mathcal{V}}(\omega)}(\mathsf{a}_2, \mathsf{b}_2) = \varepsilon_{\mathbf{F}_{\mathcal{V}}(\omega)}$. Then, by (6.3), for all $i = 1, \ldots, n$

$$p_i(\mathsf{a}_1, \mathsf{b}_1, \mathsf{a}_2, \mathsf{b}_2) = q_i(\mathsf{a}_1, \mathsf{b}_1, \mathsf{a}_2, \mathsf{b}_2)$$

and, thus,

$$p_i(\nu(t_1), \nu(r_1), \nu(t_2), \nu(r_2)) = q_i(\nu(t_1), \nu(r_1), \nu(t_2), \nu(r_2)). \tag{6.4}$$

Let us recall that for any identity $t \approx q$ and any assignment $\nu : X \longrightarrow \mathsf{G}$ sending variables in free generators of $\mathbf{F}_{\mathcal{V}}(\omega)$, if $\nu(t) = \nu(q)$, then $\mathcal{V} \models t \approx q$. So, from (6.4) we have that for all i = 1, …, n,

$$\mathcal{V} \models p_i(t_1, r_1, t_2, r_2) \approx q_i(t_1, r_1, t_2, r_2).$$

And the latter means that

$$\mathcal{V} \models t_1 \approx r_1 \nabla\, t_2 \approx r_2,$$

and this, together with (6.3), contradicts that $\mathcal{V}$ is disjunctive.

Conversely, suppose that $\mathbf{F}_{\mathcal{V}}(\omega)$ is f.s.i. We need to prove that $\mathcal{V}$ is disjunctive. Let $t_i \approx r_i, i = 1, 2$ be identities and $\mathcal{V} \not\models t_i \approx r_i, i = 1, 2$. Then, by Proposition 6.4(a), there is an algebra $\mathbf{A} \in \mathcal{V}$ such that $\mathbf{A} \not\models \langle t_1 \approx r_1, t_2 \approx r_2\rangle$. By Proposition 6.4(c), there is a valuation $\nu : X \longrightarrow \mathsf{G}$, where G is a set of free generators of $\mathbf{F}_{\mathcal{V}}(\omega)$, jointly refuting $t_1 \approx r_1$ and $t_2 \approx r_2$, that is,

$$\nu(t_1) \neq \nu(r_1) \text{ and } \nu(t_2) \neq \nu(r_2).$$

Hence, for each $i = 1, 2$, $\theta(\nu(t_i), \nu(r_i)) \neq \varepsilon$. Let us recall that $\mathbf{F}_{\mathcal{V}}(\omega)$ is f.s.i. and therefore

$$\bigvee_{i=1}^{n} \theta(p_i(\nu(t_1), \nu(r_1), \nu(t)_2, \nu(r_2)), q_i(\nu(t_1), \nu(r_1), \nu(t)_2, \nu(r_2))) =$$
$$\theta(\nu(t_1), \nu(r_1)) \cap \theta(\nu(t_2), \nu(r_2)) \neq \varepsilon.$$

Thus, ν refutes $t_1 \approx r_1 \nabla\, t_2 \approx r_2$ and we have $\mathcal{V} \not\models t_1 \approx r_1 \nabla\, t_2 \approx r_2$.

A variety $\mathcal{V}$ is said to be *locally finite* (see e.g. Burris and Sankappanavar 1981) if any finitely generated algebra from $\mathcal{V}$ is finite.

Corollary 6.1 *Let $\mathscr{V}$ be a nontrivial disjunctive locally finite variety. Then for any given number k there is a finite s.i. algebra $\mathbf{A} \in \mathscr{V}$ having cardinality exceeding k.*

Proof Indeed, by Proposition 6.8, algebra $\mathbf{F}_{\mathscr{V}}(\omega)$ is f.s.i. and, hence, by Proposition 6.7, all free algebras of $\mathscr{V}$ are f.s.i. Thus, $\mathbf{F}_{\mathscr{V}}(k+1)$ has a cardinality greater than k, due to local finiteness of $\mathscr{V}$, $\mathbf{F}_{\mathscr{V}}(k+1)$ is finite, and, hence, it is s.i.

Let us recall from Jónsson (1967), Corollary 3.4 that any s.i. algebras of a congruence-distributive variety generated by a finite algebra $\mathbf{A}$ has cardinality not exceeding cardinality of $\mathbf{A}$. Thus, the following holds.

Corollary 6.2 [2] *Nontrivial disjunctive variety cannot be generated by a finite algebra.*

6.3.1 Disjunction and Joint Refutability

The following theorem establishes the ties between the disjunction introduced via principal intersection terms and joint refutability in f.s.i. algebras.

Theorem 6.9 *Let $\mathscr{V}$ be a variety with EDPM and a system of principal intersection terms $\nabla := \{\langle p_i(x,y,w,z), q_i(x,y,w,z)\rangle \mid i = 1,\dots,n\}$. And let $\iota_1 := t_1 \approx r_1$ and $\iota_2 := t_2 \approx r_2$ be identities. Then $\mathscr{V} \models \iota_1 \nabla \iota_2$ if and only if ι_1 and ι_2 are not jointly refutable in any f.s.i. algebra $\mathbf{A} \in \mathscr{V}$.*

Proof First, we prove that if $\mathscr{V} \not\models \iota_1 \nabla \iota_2$, then ι_1 and ι_2 are jointly refutable in an f.s.i. algebra $\mathbf{A} \in \mathscr{V}$.

Suppose $\mathscr{V} \not\models \iota_1 \nabla \iota_2$. Then $\iota_1 \nabla \iota_2$ is refutable in some f.s.i (and even s.i.) algebra $\mathbf{A} \in \mathscr{V}$. Let ν be a refuting assignment, that is, ν refutes one of the identities $p_i(t_1,r_1,t_2,r_2) \approx q_i(t_1,r_1,t_2,r_2)$ from $\iota_1 \nabla \iota_2$. Suppose $\nu(t_j) = \mathsf{a}_j$ and $\nu(r_j) = \mathsf{b}_j$, $j = 1,2$. Then we have $p_i(\mathsf{a}_1,\mathsf{b}_1,\mathsf{a}_2,\mathsf{b}_2) \neq q_i(\mathsf{a}_1,\mathsf{b}_1,\mathsf{a}_2,\mathsf{b}_2)$. Hence,

$$\theta_{\mathbf{A}}(p_i(\mathsf{a}_1,\mathsf{b}_1,\mathsf{a}_2,\mathsf{b}_2), r_i(\mathsf{a}_1,\mathsf{b}_1,\mathsf{a}_2,\mathsf{b}_2)) \neq \varepsilon$$

and, therefore,

$$\bigvee_{i=i}^{n} \theta_{\mathbf{A}}(p_i(\mathsf{a}_1,\mathsf{b}_1,\mathsf{a}_2,\mathsf{b}_2), r_i(\mathsf{a}_1,\mathsf{b}_1,\mathsf{a}_2,\mathsf{b}_2)) \neq \varepsilon.$$

By the definition of EDPM,

$$\theta_{\mathbf{A}}(\mathsf{a}_1,\mathsf{b}_1) \cap \theta_{\mathbf{A}}(\mathsf{a}_2,\mathsf{b}_2) \neq \varepsilon.$$

[2] In Horčík and Terui (2011) it is proven that nontrivial substructural logics with the DP are PSPACE-hard.

So,

$$\mathsf{a}_1 \neq \mathsf{b}_1 \text{ and } \mathsf{a}_2 \neq \mathsf{b}_2, \text{ that is, } \nu(t_1) \neq \nu(r_1) \text{ and } \nu(t_2) \neq \nu(r_2).$$

And the latter means that ν jointly refutes ι_1 and ι_2 in the f.s.i. algebra $\mathbf{A}$.

Conversely, suppose identities ι_1 and ι_2 are jointly refuted in an f.s.i. algebra $\mathbf{A} \in \mathcal{V}$. We need to demonstrate that $\mathcal{V} \not\models \iota_1 \nabla \iota_2$.

Let ν be an assignment in $\mathbf{A}$ jointly refuting ι_1 and ι_2. Suppose $\nu(t_i) = \mathsf{a}_i$ and $\nu(r_i) = \mathsf{b}_i$, $i = 1, 2$. That is,

$$\mathsf{a}_1 \neq \mathsf{b}_1 \text{ and } \mathsf{a}_2 \neq \mathsf{b}_2.$$

Thus,

$$\theta_{\mathbf{A}}(\mathsf{a}_1, \mathsf{a}_2) \neq \varepsilon \text{ and } \theta_{\mathbf{A}}(\mathsf{a}_2, \mathsf{b}_2) \neq \varepsilon.$$

Let us recall that $\mathbf{A}$ is f.s.i. and, therefore,

$$\theta_{\mathbf{A}}(\mathsf{a}_1, \mathsf{b}_1) \cap \theta_{\mathbf{A}}(\mathsf{a}_2, \mathsf{b}_2) \neq \varepsilon.$$

By the definition of principal intersection terms

$$\bigvee_{i=1}^{n} \theta_{\mathbf{A}}(p_i(\mathsf{a}_1, \mathsf{b}_1, \mathsf{a}_2, \mathsf{b}_2), q_i(\mathsf{a}_1, \mathsf{b}_1, \mathsf{a}_2, \mathsf{b}_2)) = \theta_{\mathbf{A}}(\mathsf{a}_1, \mathsf{a}_2) \cap \theta_{\mathbf{A}}(\mathsf{a}_2, \mathsf{b}_2) \neq \varepsilon.$$

Hence, for some $1 \leq i \leq n$

$$p_i(\mathsf{a}_1, \mathsf{b}_1, \mathsf{a}_2, \mathsf{b}_2) \neq q_i(\mathsf{a}_1, \mathsf{b}_1, \mathsf{a}_2, \mathsf{b}_2),$$

that is,

$$p_i(\nu(t_1), \nu(r_1), \nu(t_2), \nu(r_2)) \neq q_i(\nu(t_1), \nu(r_1), \nu(t_2), \nu(r_2)).$$

Thus, assignment ν refutes identity $p_i(t_1, r_1, t_2, r_2) \approx q_i(t_1, r_1, t_2, r_2)$, which is one of the identities from $\iota_1 \nabla \iota_2$. Hence, $\mathbf{A} \not\models \iota_1 \nabla \iota_2$, i.e. $\mathcal{V} \not\models \iota_1 \nabla \iota_2$

Remark 6.1 Let us note that in the variety of modal algebras corresponding to logic **S4** the formulas p and $\sim p$ are jointly refutable in the four-element modal algebra with exactly two open elements: the top and the bottom elements. This algebra is s.i., nevertheless formula $p \vee \sim p$ is a theorem of **S4**. It is because $\vee$ is not the disjunction defined by principal intersection terms. The principal intersection terms define the strong disjunction: $p \nabla q := \sim\Diamond\sim p \vee \sim\Diamond\sim q$ – the disjunction which was used in McKinsey and Tarski (1948).

6.3.2 Maksimova's Criterion

Now, we will extend Maksimova's Criterion to the varieties with EDPM.

Theorem 6.10 *Let $\mathscr{V}$ be a variety with EDPM and ∇ be its system of principal intersection terms. Then the following is equivalent*

(a) *$\mathscr{V}$ is disjunctive;*
(b) *Any algebra $\mathbf{A} \in \mathscr{V}$ is a homomorphic image of an f.s.i. algebra $\mathbf{B} \in \mathscr{V}$;*
(c) *The direct product $\mathbf{A}_1 \times \mathbf{A}_2$ of any two algebras from $\mathscr{V}$, is a homomorphic image of an f.s.i. algebra $\mathbf{B} \in \mathscr{V}$;*
(d) *The direct product $\mathbf{F}_{\mathscr{V}}(\omega) \times \mathbf{F}_{\mathscr{V}}(\omega)$ is a homomorphic image of some f.s.i. algebra $\mathbf{B} \in \mathscr{V}$;*
(e) *There is a set $\mathscr{K}$ generating $\mathscr{V}$, such that the direct product $\mathbf{A}_1 \times \mathbf{A}_2$ of any two algebras $\mathbf{A}_1, \mathbf{A}_2 \in \mathscr{K}$ is a homomorphic image of an f.s.i. algebra from $\mathscr{V}$.*

Proof (a) $\Rightarrow$ (b). Suppose $\mathscr{V}$ is disjunctive. Then, by Proposition 6.8, $\mathbf{F}_{\mathscr{V}}$ is f.s.i., and, hence, by Proposition 6.7, every free algebra of $\mathscr{V}$ is f.s.i. And any algebra from $\mathscr{V}$ is a homomorphic image of some free algebra of $\mathscr{V}$.

(b) $\Rightarrow$ (c) $\Rightarrow$ (d) is trivial. (d) $\Rightarrow$ (e) because $\mathscr{K} = \{\mathbf{F}_{\mathscr{V}}\}$ generates $\mathscr{V}$.

(e) $\Rightarrow$ (a). Let $\mathscr{K}$ be a set of algebras generating $\mathscr{V}$ and for any two algebras $\mathbf{A}_1, \mathbf{A}_2 \in \mathscr{K}$ there is an f.s.i. algebra $\mathbf{B} \in \mathscr{V}$ such that $\mathbf{A}_1 \times \mathbf{A}_2$ is a homomorphic image of $\mathbf{B}$. We need to prove that $\mathscr{V} \not\models \iota_1 \nabla \iota_2$ for any two identities ι_1, ι_2 such that $\mathscr{V} \not\models \iota_1$ and $\mathscr{V} \not\models \iota_2$.

Suppose that $\mathscr{V} \not\models \iota_1$ and $\mathscr{V} \not\models \iota_2$. Then, due to the fact that $\mathscr{K}$ generates $\mathscr{V}$, there are $\mathbf{A}_1, \mathbf{A}_2 \in \mathscr{K}$ such that $\mathbf{A}_1 \not\models \iota_1$ and $\mathbf{A}_2 \not\models \iota_2$. Then, by Proposition 6.4(a), $\mathbf{A}_1 \times \mathbf{A}_2 \not\models \langle \iota_1, \iota_2 \rangle$. By the assumption of the case (e), $\mathbf{A}_1 \times \mathbf{A}_2$ is a homomorphic image of an f.s.i. algebra $\mathbf{B} \in \mathscr{V}$. Hence, by Proposition 6.4(b), $\mathbf{B} \not\models \langle \iota_1, \iota_2 \rangle$. Now, we can apply Theorem 6.9 and conclude that $\mathscr{V} \not\models \iota_1 \nabla \iota_2$.

6.3.3 Some Examples

As we mentioned at the beginning of the paper, adjoining a new top element to a Heyting algebra $\mathbf{A}$ gives an s.i. algebra $\mathbf{A}^*$. Clearly, $\mathbf{A}$ is a homomorphic image of $\mathbf{A}^*$. Hence, the variety of Heyting algebras is disjunctive, i.e. **Int**enjoys the DP.

The same argument can be applied to the variety of Heyting algebras enriched by compatibly operations, that is, by operations which preserve Heyting congruences. For instance, in the variety of KM-algebras (also known as the Heyting algebras with successor, e.g. Castiglioni and Martín (2011)) adjoining a new top element gives an s.i. KM-algebra and therefore the variety of all KM-algebras is disjunctive (see Muravitsky (2014), Sect. 7.4.3). KM-algebras are the algebraic models of the proof-intuitionistic propositional calculus (for more information see Muravitsky (2014)).

Using Theorem 6.10 and the known relations between Heyting and **S4**-algebras, one can prove that the greatest modal companion of an intermediate logic with the DP also enjoys the DP (see Gudovschikov and Rybakov (1982)). Let us stress out that in the normal extensions of **S4** the system of principal intersection terms consists of a single term $\Box x \vee \Box y \approx \mathbf{1}$, hence, for such a logic L the DP means that $\Box\alpha \vee \Box\beta \in \mathsf{L}$ yields $\Box\alpha \in \mathsf{L}$ or $\Box\beta \in \mathsf{L}$ for any formulas α, β.

The examples of applications of Maksimova's Criterion for proving the DP in substractural logics can be found in Galatos et al. (2007), Sect. 5.3.1.

6.4 Final Remarks

6.4.1 Generalizations of the DP

In Ferrari and Miglioli (1993) the DP has been relativized in the following way: given two logics $\mathcal{L} \subseteq \mathcal{L}'$, $\mathcal{L}$ is 2-*semiconstructive*[3] in $\mathcal{L}'$ if whenever $\alpha \vee \beta \in \mathcal{L}$ then either $\alpha \in \mathcal{L}'$, or $B \in \mathcal{L}'$. For instance, if $\mathcal{L}'$ is classical logic (denoted thereafter by **Cl**), then (due to Glivenko's Theorem), an intermediate logic $\mathcal{L}$ is 2-semiconstructive in **Cl** if $\alpha \vee \beta \in \mathcal{L}$ yields $\neg\neg\alpha \in \mathcal{L}$ or $\neg\neg\beta \in \mathcal{L}$ (in Chagrov and Zakharyaschev (1993) the latter property is denoted by DP^*).

Let us observe that, in contrast to the DP, logic $\mathcal{L}$ may be 2-semiconstructive in $\mathcal{L}'$, but for some formulas α, β, γ we may have $\alpha \vee \beta \vee \gamma \in \mathcal{L}$, while neither of these formulas is in $\mathcal{L}'$. Hence, we can say that logic $\mathcal{L}$ is n-*semiconstructive* in $\mathcal{L}'$, if $\bigvee_{i=1}^{n} \alpha_i \in \mathcal{L}$ yields that $\alpha_i \in \mathcal{L}'$ at least for one $i \in \{1, \dots, n\}$. And we call a logic $\mathcal{L}$ *semiconstructive in* $\mathcal{L}'$ if $\mathcal{L}$ is n-semiconstructive in $\mathcal{L}'$ for every $n \geq 2$. In the case when $\mathcal{L}'$ is **Cl**, we will omit a reference to $\mathcal{L}'$.

If $\mathcal{V}$ is a variety with EDPM and ∇ is its system of principal intersection terms, we let

$$\nabla_{i=1}^{2}\iota_i := \iota_1 \nabla \iota_2 \text{ and } \nabla_{i=1}^{n+1}\iota_i := \nabla_{i=1}^{n}\iota_i \nabla \iota_{n+1}.$$

Remark 6.2 The distribution of parentheses is irrelevant, due to ∇ is associative (up to $\dashv\vdash_{\mathcal{V}}$).

In terms of equational logics, we say that a variety $\mathcal{V}$ with EDPM and a system of intersection terms ∇ is *n-disjunctive* ($n \geqslant 2$) relative to a subvariety $\mathcal{V}' \subseteq \mathcal{V}$ if for any identities $\iota_1, \dots, \iota_n$

$$\mathcal{V} \models \nabla_{i=1}^{n}\iota_i \text{ yields } \mathcal{V}' \models \iota_i, \text{ at least for one } i \in \{1, \dots, n\}, \tag{6.1}$$

and $\mathcal{V}$ is *disjunctive relative to* $\mathcal{V}'$, if $\mathcal{V}$ is n-disjunctive relative to $\mathcal{V}'$ for every $n \geqslant 2$.

Naturally the question arises whether Maksimova's criterion can be extended to n-disjunctive varieties.

[3]Ferrari and Miglioli use term "semiconstructive" that we will use in stronger sense.

6.4.2 n-Disjunctive Varieties

First, let us extend the notion of joint refutability: identities $t_i \approx r_i, i = 1, \dots, n$ are *jointly refutable in an algebra* **A**, if there is a valuation ν in **A** such that $\nu(t_i) \neq \nu(r_i)$ for all $i = 1, \dots, n$. And let us note that, by simple induction and using Theorem 6.9, one can prove the following.

Proposition 6.11 *Let $\mathcal{V}$ be a variety with EDPM and a system of principal intersection terms ∇. And let $\iota_i, i = 1, \dots, n$ be identities. Then $\mathcal{V} \models \nabla_{i=1}^n \iota_i$ if and only if identities $\iota_i, i = 1, \dots, n$ are not jointly refutable in any f.s.i. algebra $\mathbf{A} \in \mathcal{V}$.*

If identities $\iota_i, i = 1, \dots, n$ are jointly refutable in an algebra **A**, we denote this by $\mathbf{A} \not\models \langle \iota_i, i \leq n \rangle$

Next, one can easily extend Proposition 6.4 to joint refutability of more than two identities.

Proposition 6.12 *For any algebras $\mathbf{A}_i, i = 1, \dots, n$ and any identities $\iota_i, i = 1, \dots, n$ the following hold*

(a) If $\mathbf{A}_i \not\models \iota_i$ for every $i = 1, \dots, n$, then $\prod_{i=1}^n \mathbf{A}_i \not\models \langle \iota_i, i \leqslant n \rangle$;
(b) If $\mathbf{A}_1$ is a homomorphic image of $\mathbf{A}_2$, then $\mathbf{A}_2 \not\models \langle \iota_i, i \leqslant n \rangle$ as long as $\mathbf{A}_1 \not\models \langle \iota_i, i \leqslant n \rangle$;
(c) If $\mathbf{A}_1 \not\models \langle \iota_i, i \leqslant n \rangle$, then $\mathbf{F}_{\mathcal{V}}(\omega) \not\models \langle \iota_i, i \leqslant n \rangle$. Moreover, there is an assignment ν in $\mathbf{F}_{\mathcal{V}}(\omega)$ jointly refuting $\iota_i, i = 1, \dots, n$ and sending every variable in a free generator of $\mathbf{F}_{\mathcal{V}}(\omega)$.

Now, we are in a position to give a sufficient condition for relative n-disjunctivity.

Theorem 6.13 *Let $\mathcal{V}$ be a variety with EDPM and ∇ be its system of principal intersection terms, $\mathcal{V}' \subseteq \mathcal{V}$ be a subvariety, and $n \geqslant 2$ be an integer. If there is a set $\mathcal{K}$ generating $\mathcal{V}'$ such that the direct product $\prod_{i=1}^n \mathbf{A}_i$ of any n algebras $\mathbf{A}_i \in \mathcal{K}, i = 1, \dots, n$ is a homomorphic image of an f.s.i. algebra from variety $\mathcal{V}$, then $\mathcal{V}$ is n-disjunctive relative to $\mathcal{V}'$.*

Proof Let $\mathcal{K}$ be set $\mathcal{K}$ generating $\mathcal{V}'$ and satisfying the condition of the theorem. We need to show that for any n identities $\iota_i, i = 1, \dots, n$,

$$\text{if } \mathcal{V}' \not\models \iota_i \text{ for every } i = 1, \dots, n, \text{ then } \mathcal{V} \not\models \nabla_{i=1}^n \iota_i.$$

Indeed, suppose $\mathcal{V}' \not\models \iota_i$ for every $i = 1, \dots, n$. Then, there are algebras $\mathbf{A_i} \in \mathcal{V}'$ such that $\mathbf{A}_i \not\models \iota_i, i = 1, \dots, n$. Hence, by Proposition 6.12(a),

$$\prod_{i=1}^n \mathbf{A}_i \not\models \langle \iota_i, i \leqslant n \rangle.$$

By the assumption of the theorem, $\prod_{i=1}^n \mathbf{A}_i$ is a homomorphic image of an f.s.i. algebra $\mathbf{B} \in \mathcal{V}$, hence, we can apply Propositions 6.12(b) and conclude that

$$\mathbf{B} \not\models \langle \iota_i, i \leqslant n \rangle.$$

And, by virtue of Proposition 6.11, we have

$$\mathscr{V} \not\models \nabla_{i=1}^{n} \iota_i.$$

Corollary 6.3 *Let $\mathscr{V}$ be a variety with EDPM and ∇ be its system of principal intersection terms, $\mathscr{V}' \subseteq \mathscr{V}$ be a subvariety. If there is a set $\mathscr{K}$ generating $\mathscr{V}'$ such that the direct product $\prod_{i=1}^{n} \mathbf{A}_i$ of any n algebras $\mathbf{A}_i \in \mathscr{K}$ is a homomorphic image of an f.s.i. algebra from variety $\mathscr{V}$, then $\mathscr{V}$ is disjunctive relative to $\mathscr{V}'$.*

6.4.2.1 Some Examples

Let $\mathscr{B}$ be a variety of Boolean algebras, and let us describe the varieties of Heyting algebras that are 2-disjunctive relative to $\mathscr{B}$. As it is well-known, variety $\mathscr{B}$ is generated by a single two-element Boolean algebra that we denote $\mathbf{B}_2$. Hence, by Theorem 6.13, any variety $\mathscr{V}$ of Heyting algebras is 2-disjunctive relative to $\mathscr{B}$, as long as $(\mathbf{B}_2^2)^* \in \mathscr{V}$.

Let us observe that a variety of Heyting algebras does not contain algebra $(\mathbf{B}^2)^*$ if and only if the identity $\neg p \vee \neg\neg p \approx \mathbf{1}$ holds in it. In other words, the following holds (comp. Chagrov and Zakharyaschev (1993)).

Proposition 6.14 *A variety $\mathscr{V}$ of Heyting algebras is 2-disjunctive relative to $\mathscr{B}$ if and only if $(\mathbf{B}^2)^* \in \mathscr{V}$.*

Proposition 6.14 states, that an intermediate logic $\mathcal{L}$ is 2-semiconstructive if and only if $\mathcal{L}$ extends the logic of algebra $(\mathbf{B}^2)^*$ (the logic LV from Chagrov and Zakharyaschev (1993)). That is, the logic of $(\mathbf{B}^2)^*$ is the greatest 2-semisonstructive intermediate logic.

Also, Corollary 6.3 entails that a variety $\mathscr{V}$ of Heyting algebras containing all algebras $(\mathbf{B}_2^n)^*, n \geqslant 2$ is disjunctive relative to $\mathscr{B}$. In terms of logics, any intermediate logic extending the logic of algebras $(\mathbf{B}_2^n)^*, n \geqslant 2$ is semiconstructive, that is, any intermediate logic extending $\mathbf{BD_2}$ (see definition in Chagrov and Zakharyaschev (1997), Table 4.1) is semiconstructive.

Naturally, the following problem arises.

Problem 6.1 Is the condition of Corollary 6.3 necessary for relative disjunctivity?

6.4.3 Multiply-Conclusion Rules and the DP

Let us recall (e.g. Goudsmit (2015)) that an ordered pair of finite sets Γ, Δ (written as Γ/Δ) is called a *multiple-conclusion rule*. Given a logic $\mathcal{L}$, a rule Γ/Δ is said to be *admissible for* $\mathcal{L}$ if any substitution that makes all formulas of Γ valid in $\mathcal{L}$, makes at least one formula from Δ valid in $\mathcal{L}$. For instance, a rule $p \vee q/p, q$ is admissible for **Int**, and, as it is easy to see, the admissibility of this rule expresses the DP. So, in the context of admissibility, Maksimova's criterion is, at the same time, a criterion for admissibility of multiple-conclusion rule $p\nabla q/p, q$.

Similarly, the n-semiconstructivity of an intermediate logic $\mathcal{L}$ is equivalent to admissibility in $\mathcal{L}$ of multiple-conclusion rule

$$dn_n := p_1 \vee \cdots \vee p_n/\neg\neg p_1, \ldots, \neg\neg p_n.$$

Let us note that the rule

$$\widetilde{dn_n} := \neg p_1 \vee \cdots \vee \neg p_n/\neg p_1, \ldots, \neg p_n$$

is a consequence of the rule dn_n.

Now, we can rephrase the results of Sect. 6.4.2.1 in terms of admissibility of multiple-conclusion rules.

Proposition 6.15 *An intermediate logic $\mathcal{L}$ admits the rule dn_2 (or $\widetilde{dn_2}$ for this matter) if and only if algebra $(\mathbf{B}^2)^*$ is a model of this $\mathcal{L}$.*

Proposition 6.16 *If an intermediate logic $\mathcal{L}$ has logic $\mathbf{BD_2}$ among its extensions, then $\mathcal{L}$ admits all rules $dn_n, n \geqslant 2$ and, hence, logic $\mathcal{L}$ admits all rules $\widetilde{dn_n}, n \geqslant 2$.*

Thus, as we see, Maksimova's criterion can also be useful for establishing admissibility of the multiple-conclusion rules.

References

Agliano, P., & Baker, K. A. (1999). Congruence intersection properties for varieties of algebras. *Journal of the Australian Mathematical Society, Ser. A*, *67*(1), 104–121.

Baker, K. A. (1977). Finite equational bases for finite algebras in a congruence-distributive equational class. *Advances in Mathematics*, *24*(3), 207–243.

Blok, W. J., & Pigozzi, D. (1986). A finite basis theorem for quasivarieties. *Algebra Universalis*, *22*(1), 1–13.

Burris, S., & Sankappanavar, H. P. (1981). *A course in universal algebra* (Vol. 78)., Graduate texts in mathematics New York: Springer.

Castiglioni, J. L., & Martín, H. J. (2011). Compatible operations on residuated lattices. *Studia Logica*, *98*(1–2), 203–222.

Chagrov, A., & Zakharyaschev, M. (1993). The undecidability of the disjunction property of propositional logics and other related problems. *Journal of Symbolic Logic*, *22*, 967–1002.

Chagrov, A., & Zakharyaschev, M. (1997). *Modal logic*. New York: Oxford Science Publications.

Cintula, P., & Noguera, C. (2013). The proof by cases property and its variants in structural consequence relations. *Studia Logica*, *101*(4), 713–747.

Czelakowski, J. (1983). Matrices, primitive satisfaction and finitely based logics. *Studia Logica*, *42*(1), 89–104.

Czelakowski, J. (1984). Remarks on finitely based logics. *Models and sets (Aachen, 1983)* (Vol. 1103, pp. 147–168)., Lecture notes in mathematics Berlin: Springer.

Czelakowski, J. (2001). *Protoalgebraic logics* (Vol. 10)., Trends in logic Dordrecht: Kluwer Academic Publishers.
Czelakowski, J., & Dziobiak, W. (1990). Congruence distributive quasivarieties whose finitely subdirectly irreducible members form a universal class. *Algebra Universalis*, *27*(1), 128–149.
Dzik, W., & Suszko, R. (1977). On distributivity of closure systems. *Bulletin of the Section of Logic*, *6*(2), 64–66.
Ferrari, M., & Miglioli, P. (1993). Counting the maximal intermediate constructive logics. *Journal of Symbolic Logic*, *58*(4), 1365–1401.
Font, J. M., & Jansana, R. (1996). *A general algebraic semantics for sentential logics* (Vol. 7)., Lecture notes in logic Berlin: Springer.
Gabbay, D. M., & De Jongh, D. H. J. (1974). A sequence of decidable finitely axiomatizable intermediate logics with the disjunction property. *Journal of Symbolic Logic*, *39*, 67–78.
Galatos, N., Jipsen, P., Kowalski, T., & Ono, H. (2007). *Residuated lattices: an algebraic glimpse at substructural logics* (Vol. 151)., Studies in logic and the foundations of mathematics Amsterdam: Elsevier B. V.
Gentzen, G. (1935). Untersuchungen über das logische Schließen. *II, Mathematische Zeitschrift*, *39*(1), 405–431.
Gödel, K. (1986). *Collected works. Publications 1929–1936, Edited and with a preface by Solomon Feferman (1986)* (Vol. I). New York: The Clarendon Press, Oxford University Press.
Goudsmit, J. (2015). *Intuitionistic Rules Admissible Rules of Intermediate Logics*, Ph.D. thesis, Utrecht University.
Grätzer, G. (2008). *Universal algebra* (2nd ed.). New York: Springer.
Gudovschikov, V., & Rybakov, V. (1982). Dizyunktivnoe svoystvo v modalnykh logikakh. In *Proceedings of 8th USSR Conference "Logic and Methodology of Science"* (pp. 35–36). Vilnus.
Horčík, R., & Terui, K. (2011). Disjunction property and complexity of substructural logics. *Theoretical Computer Science*, *412*(31), 3992–4006.
Humberstone, L. (2011). *The connectives*. Cambridge: MIT Press.
Jónsson, B. (1967). Algebras whose congruence lattices are distributive. *Mathematica Scandinavica*, *21*, 110–121.
Kreisel, G., & Putnam, H. (1957). Eine Unableitbarkeitsbeweismethode für den intuitionistischen Aussagenkalkul. *Archiv für Mathematische Logik und Grundlagenforschung*, *3*, 74–78.
Łukasiewicz, J. (1952). On the intuitionistic theory of deduction. *Indagationes Mathematicae*, *14*, 202–212.
Maksimova, L. L. (1986). On maximal intermediate logics with the disjunction property. *Studia Logica*, *45*(1), 69–75.
McKinsey, J. C. C., & Tarski, A. (1948). Some theorems about the sentential calculi of Lewis and Heyting. *Journal of Symbolic Logic*, *13*, 1–15.
Muravitsky, A. (2014). Logic KM: a biography. In G. Bezhanishvili (Ed.), *Leo Esakia on duality in modal and intuitionistic logics* (Vol. 4, p. 155). Springer: Outstanding Contributions to Logic Dordrecht.
Wroński, A. (1973). Intermediate logics and the disjunction property. *Reports on Mathematical Logic*, *1*, 39–51.

Chapter 7
Rasiowa–Sikorski Sets and Forcing

Janusz Czelakowski

To Larisa

Abstract The paper is concerned with the problem of building models for first-order languages from the perspective of the classic paper of Rasiowa and Sikorski (1950). The central idea, due to Rasiowa and Sikorski and developed in this paper, is constructing first-order models from individual variables. The notion of a Rasiowa–Sikorski set of formulas of an arbitrary language L is introduced. Investigations are confined to countable languages. Each Rasiowa–Sikorski set defines a countable model for L. Conversely, each countable model for L is determined, up to isomorphism, by some Rasiowa–Sikorski set. Consequences of these facts are investigated.

On the Paper

The central idea, due to Rasiowa and Sikorski and developed in this paper, is constructing first-order models from individual variables. The key notion is that of a Rasiowa–Sikorski set. These sets enable one to build a substitutional semantics for first-order logic. This is due to the fact that the satisfaction relation in the model $\boldsymbol{A}_\Delta$ corresponding to a Rasiowa–Sikorski set Δ is expressed in a straightforward way in terms of a "double" substitutions of variables in the formulas of L. Since each consistent set of formulas Σ is the intersection of a family of Rasiowa–Sikorski sets, the Extended Completeness Theorem for first-order logic in terms of the above substitutional-semantics immediately follows. This fact shows that the class of Rasiowa–Sikorski sets (and even a narrower family of model sets) suffices for establishing an adequate substitutional semantics for first-order logic.

On may argue that the substitutional semantics is a tool that reduces the scope of formal semantics founded on set theory to a relatively modest fragment of the universe of sets, viz. to simple set-theoretic constructions performed on subsets of the sets of symbols, terms, and formulas. E.g. transfinite iterations of power sets above ω in the

---

J. Czelakowski (✉)
Insitute of Mathematics and Informatics, Opole University, Opole, Poland
e-mail: jczel@math.uni.opole.pl

S. Odintsov (ed.), *Larisa Maksimova on Implication, Interpolation, and Definability*, Outstanding Contributions to Logic 15, https://doi.org/10.1007/978-3-319-69917-2_7

universe of sets are not needed. The largest set in the sense of its cardinality which is needed in this approach is the uncountable family of all Lindenbaum sets, i.e., maximal, consistent sets of formulas of L. But the full strength of ZF together with the principle DC of dependent choices will be used as a set-theoretic tool.

These basic observations serve as a departure point for a deeper analysis of the title class. Rasiowa–Sikorski sets are strictly linked with the properties of the Lindenbaum–Tarski algebra $\boldsymbol{B}(L)$ of the formula algebra of classical logic. (The discussion of the Lindenbaum–Tarski algebras of elementary theories is omitted.) As is well known, $\boldsymbol{B}(L)$ is obtained from the formula-algebra for L by factoring the set $For(L)$ of formulas by the relation of logical equivalence with respect to the set of theses $\boldsymbol{CL}$ of classical logic. More specifically, two formulas φ and ψ are being identified if their equivalence $\varphi \leftrightarrow \psi$ is a thesis of classical logic. The universe of the Boolean algebra $\boldsymbol{B}(L)$ is formed by the equivalence classes of the above equivalence relation between formulas. Let us mark by $[\sigma]$ the equivalence class of the formula σ. We then define the notion of a forcing defined in terms of refined families of non-empty sets composed of equivalence classes $[\sigma]$ of *atomic* formulas σ. The forcing technique is a tool enabling one to define Rasiowa–Sikorski sets with various properties. But the latter topic is not analysed here in detail—we merely confine ourselves to presenting some general observations, leaving a more detailed examination of the problem under further scrutiny.

Since purely syntactic considerations play an important role in the narrative structure of the paper, the first three paragraphs provide a brief and systematic account (often without proofs) of these inferential aspects of classical first-order logic that are needed in the theory of Rasiowa–Sikorski sets. An emphasis is put on various aspects of substitution operations. Section 7.4 recalls basic semantic facts. Section 7.5 is central. The family of model sets as well as the larger family of Rasiowa–Sikorski sets of formulas in an arbitrary language L is defined there and their relationship with countable models for L is shown. Section 7.6 is concerned with Boolean valuations for L, forcing and their interrelations with Rasiowa–Sikorski sets.

7.1 Introduction

Helena Rasiowa and Roman Sikorski presented in (Rasiowa and Sikorski 1950) a purely algebraic proof of the completeness theorem for first-order logic. Their proof was formulated in the language of Boolean algebras and employed a result which is nowadays called *Rasiowa–Sikorski Lemma* in the literature:

Theorem 7.1.1 (Rasiowa–Sikorski Lemma) *Let* $\boldsymbol{B} = (B, +, \cdot, ^*, \mathbf{0}, \mathbf{1})$ *be a non-trivial Boolean algebra. Suppose that for each natural number n,* X_n *is a non-empty subset of B such that* $\sup(X_n)$ *exists in* $\boldsymbol{B}$. *Let* a_0 *be a non-zero element of* $\boldsymbol{B}$. *Then there exists an ultrafilter U in B such that* $a_0 \in U$ *and U preserves the suprema* $\sup(X_n)$, $n \in \mathbb{N}$, *that is, for each n, it is the case that*

$$\sup(X_n) \in U \quad \Leftrightarrow \quad \textit{there is } x_n \in X_n \textit{ such that } x_n \in U. \quad \square \qquad (*)$$

(In $(*)$ only the implication $(\Rightarrow)$ matters.) The original proof of the lemma was based on Stone's Theorem and employed Baire's Theorem. A simpler, algebraic proof of the lemma is due to Alfred Tarski (see Feferman 1952).

Various conditions pertinent or equivalent to Rasiowa–Sikorski Lemma are discussed by Goldblatt (1985).

Rasiowa–Sikorski Lemma has a direct reference to the Lindenbaum–Tarski algebra $\boldsymbol{B}(L)$. (This issue will be discussed later.) But the scope of the lemma is much wider. Its modified versions turned out to be useful in algebraic logic. But, more importantly, it plays a significant role in the Boolean-valued set theory and in the theory of forcing. As Dana Scott writes in *Foreword* to (Bell 2005): "Cohen's achievement lies in being able to *expand* models (countable, standard models) by adding new sets in a very economical fashion: they more or less have only the properties they are *forced* to have by the axioms (or by the truth of the given model)." Rasiowa–Sikorski Lemma is a tool that serves defining such *generic* expansions of models.

In this paper we argue that some forcing relations as well as appropriate Boolean valuations that validate classical logic can be also *directly* defined for *arbitrary* countable first-order languages. But, in contrast to set theory, the notion of forcing introduced in the article is restricted to refined families of sets of atomic formulas of a first-order language L or refined families formed by sets of equivalence classes of atomic formulas in appropriate Lindenbaum–Tarski algebras for L.

In the case of purely relational languages L (no function and constant symbols), each forcing relation $\Vdash$ for L is determined by a refined family $\boldsymbol{P} = (P, \subseteq)$ of subsets of the set $AFor(L)$ of *atomic* formulas of L ordered by inclusion. Elements of P are called *conditions*. (Since L is countable, each such a family P has cardinality equal or less than the continuum.) $\Vdash$ is a recursively defined relation holding between elements of P and first-order formulas of L. If $p \Vdash \sigma$ holds, the condition p is said to *force* the formula σ. Each refined family $\boldsymbol{P}$ defines the Boolean algebra of regular open sets $\boldsymbol{RO}(\boldsymbol{P})$ as well as a (canonical) Boolean valuation $[\sigma]^{\boldsymbol{P}}$, $\sigma \in For(L)$, of the set of formulas in $\boldsymbol{RO}(\boldsymbol{P})$. The problem is whether $[\,\cdot\,]^{\boldsymbol{P}}$ validates classical logic. Quite surprisingly, only the axioms for equality create a problem. But, after imposing a weak constraint on $\boldsymbol{P}$, $[\,\cdot\,]^{\boldsymbol{P}}$ validates the classical first-order logic as a whole. The main tool we employ is of course Rasiowa–Sikorski Lemma. This lemma enables one to determine Rasiowa–Sikorski sets associated with $\boldsymbol{P}$. Each such a set defines a model for L.

Instead of working with refined sets of atomic formulas of L we shall work with the Lindenbaum–Tarski algebra $\boldsymbol{B}(L)$ and consider refined families $\boldsymbol{P}$ of subsets of $\boldsymbol{B}(L)$. Each condition of $\boldsymbol{P}$ is then a set of the equivalence classes $[\sigma]$ of *atomic* formulas (in the sense of $\boldsymbol{B}(L)$). In this case the situation is logically much simpler—the forcing relation defined by each such a refined set $\boldsymbol{P}$ automatically validates the rules and axioms of classical logic defined in the set of first-order formulas of an *arbitrary* language L (also with operation symbols and constants). (No constraints on $\boldsymbol{P}$ are imposed.). One may even go further and work in the Lindenbaum–Tarski algebra $\boldsymbol{B}_T(L)$ of arbitrary *consistent* elementary theory T by suitably modifying the above definitions. In this approach, instead of working with generic extensions of a given model of T, one directly constructs models of T via forcing relations and Rasiowa–Sikorski Lemma. The assumption of consistency of T is crucial, because it guarantees that the algebra $\boldsymbol{B}_T(L)$ is non-trivial and the above forcing-related constructions do not trivialize as well.

7.2 Syntax

This and the following sections put in nutshell all indispensable definitions and results that will be utilised later. Most of them are well-known. The emphasis is put rather on notational issues.

A *language* L is a set of symbols. L encompasses three categories of symbols: *relational symbols* (predicates), *function symbols* (alias operation symbols), *constant symbols*, each of appropriate arity. 0 is the arity of constant symbols; they are therefore *nullary* symbols. (Constant symbols may be therefore treated as nullary function symbols.) Constant symbols denote some elements in the models for L.

A *model* for L is any ordered pair

$$\boldsymbol{A} = (A; I), \tag{7.1}$$

where A is a non-empty set, called the *universe* (or the *carrier*) of the model, and I is a function called an *interpretation* of the language L. It is assumed that the domain of I is equal to L. Moreover:

if P is an n-ary relational symbol, then $I(P)$ is an n-argument relation on A, $I(P) \subseteq A^n$;
if F is an m-ary function symbol, then $I(F)$ is an m-argument function (operation) on A, $I(F) : A^m \to A$;
if c is a constant symbol, then $I(c) \in A$.

First-Order Formalizations of Languages

We need two more sets: the set of logical symbols and the set of auxiliary symbols.

The set of *auxiliary symbols* encompasses:

parentheses:	), (	(the right and the left parenthesis)
individual variables:	$v_0, v_1, \ldots, v_n, \ldots$	

The set of variables is countably infinite and is denoted by Var. The above well-ordering of variables according to which variables are enumerated is called the *alphabetic order* of the set Var.

The set of (primitive) *logical symbols* encompasses:

logical connectives:	$\wedge$ (conjunction), $\neg$(negation)
existential quantifier:	$\exists$ (for some)
equality symbol:	$\approx$.

Any language is assumed to be disjoint with the sets of auxiliary symbols and logical symbols.

Let L be a language. $Te(L)$ is the set of terms of L. It is defined in the well-known recursive way. $Te(L)$ is the underlying set of the absolutely free algebra called the *term algebra* and denoted by $\boldsymbol{Te}(L)$. The set Var is the set of absolutely free generators of $\boldsymbol{Te}(L)$.

$AFor(L)$ is the set of atomic formulas of L. Thus (i) if s and t are terms, then the sequence $(s \approx t)$ is an atomic formula, (ii) if P is an n-ary predicate and $t_1, \ldots, t_n$ is a sequence of terms, then the sequence $P(t_1, \ldots, t_n)$ is an atomic formula.

$For(L)$ is the set of (first-order) formulas of L. Thus (i) every atomic formula is a formula, (ii) if φ and ψ are formulas, then the sequence $(\varphi \wedge \psi)$ is a formula, (iii) if ψ is a formula then the sequence $(\neg\psi)$ is a formula, (iv) if x is an individual variable and φ is a formula, then the sequence $((\exists x)\varphi)$ is a formula.

The other Boolean connectives as $\vee$, $\rightarrow$, $\leftrightarrow$ are defined in the standard way as useful abbreviations. The universal quantifier is also definitionally introduced. We also adopt the standard conventions concerning parentheses. E.g. outer parentheses will be omitted when it does not lead to confusion.

Other Useful Notions

Let L be a language.

$Var(t)$ is the set of individual variables *occurring in a term* t. The sets $Var(t)$, $t \in Te(L)$ are recursively defined in the well-known manner.

$Subfor(\sigma)$ is the set of subformulas of a formula σ. (Since the connectives of disjunction, implication and equivalence as well the universal quantifier are introduced definitionally, some care is needed while operating with subformulas.)

$FreeVar(\sigma)$ and $BoundVar(\sigma)$ are the sets of *free* and *bound* individual variables occurring in a formula σ, respectively. We only note that if σ is $(\exists x)\varphi$, then $FreeVar(\sigma) = FreeVar(\varphi) \setminus \{x\}$ and $BoundVar(\sigma) = BoundVar(\varphi) \cup \{x\}$. The sets $FreeVar(\sigma)$ and $BoundVar(\sigma)$ need not be disjoint. If x does not occur in φ at all, then $FreeVar((\exists x)\varphi) = FreeVar(\varphi)$.

σ is a sentence if $FreeVar(\sigma) = \varnothing$.

Notation

If t is a term and $\{x_1, \ldots, x_n\}$ is a finite set of individual variables, then $t(x_1, \ldots, x_n)$ means that $Var(t) \subseteq \{x_1, \ldots, x_n\}$. Similarly, if φ is a formula, then $\varphi(x_1, \ldots, x_n)$ means that $FreeVar(\varphi) \subseteq \{x_1, \ldots, x_n\}$.

Given a formula φ, a term t, and an individual variable x, we define

$$\varphi(x/t)$$

to be the formula obtained from φ by means of the uniform substitution of t for x in every *free* occurrence of x in φ. If x is not in $FreeVar(\varphi)$, then the above substitution does not change φ.

Generally, if $x_1, \ldots, x_n$ are different variables and $\langle t_1, \ldots, t_n \rangle$ is a sequence of terms, then

$$\varphi(x_1/t_1, \ldots, x_n/t_n)$$

is the formula obtained from φ by means of the uniform and simultaneous substitution of t_i for x_i in every free occurrence x_i in φ for $i = 1, \ldots, n$.

The substitution of a term t for x in φ is *free* if none of the variables y occurring in t becomes a bound variable in $\varphi(x/t)$ in the place where t is substituted.

Example Let φ be the formula $(\exists y)(x \approx y + 1)$. x is a free variable in φ. Let t be the variable y. $\varphi(x/t)$ is the sentence $(\exists y)(y \approx y + 1)$. Here the substitution of y for x in φ is not free, because the variable y is bound by $\exists$ in $\varphi(x/t)$.

If t is a term $x + z$, where z is different from x and y, then $\varphi(x/t)$ is the formula $(\exists y)(x + z \approx y + 1)$. The substitution of $x + z$ for x in φ is free, because neither x nor z is bound by $\exists$ in $\varphi(x/t)$. □

Substitutions are operations performed on terms and formulas. If $x_1, \ldots, x_n$ are different variables in the alphabetic order and $\langle t_1, \ldots, t_n \rangle$ a sequence of terms, then

$$\langle x_1/t_1, \ldots, x_n/t_n \rangle \qquad (*)$$

denotes the substitution which assigns the term t_i to the variable x_i for $i = 1, \ldots, n$. The value of the substitution $\langle x_1/t_1, \ldots, x_n/t_n \rangle$ for an arbitrary term s is marked as $s(x_1/t_1, \ldots, x_n/t_n)$. $(*)$ 'acts' only on those variables among $x_1, \ldots, x_n$, which actually occur in s and do not 'move' other variables. Hence when writing $s(x_1/t_1, \ldots, x_n/t_n)$ we shall generally assume that $x_1, \ldots, x_n$ are precisely the variables that occur in s unless otherwise stated. Formally, every substitution is a function defined on terms and whose values are terms as well. The substitutions of the form $(*)$ are called n-ary due to the number of variables on which each such a substitution acts.

To simplify notation, $(*)$ is marked simply as $x_1/t_1, \ldots, x_n/t_n$, dropping the external angle brackets. In particular, this notation will be used in the case of unary substitutions, which are uniformly marked as x/t.

Definition 1 A substitution $\langle x_1/t_1, \ldots, x_n/t_n \rangle$ is *free in a formula* $\varphi(x_1, \ldots, x_n)$ if all unary substitutions $x_1/t_1, x_2/t_2, \ldots, x_n/t_n$ are free in φ. □

Note $\langle x_1/t_1, \ldots, x_n/t_n \rangle$ is *not* the composition of the unary substitutions x_1/t_1, $x_2/t_2, \ldots, x_n/t_n$ in the following meaning of the word: first one makes the substitution x_1/t_1 in $\varphi(x_1, \ldots, x_n)$, then one makes x_2/t_2 in the resulting formula etc., because one arrives at the formula which differs from $\varphi(x_1/t_1, \ldots, x_n/t_n)$! □

7.3 Classical Logic

Let L be an arbitrary but fixed language. After (Adamowicz and Zbierski 1997), the following axiom system of classical logic in $For(L)$ is adopted. The axioms are divided into 3 groups:

1. *Propositional axioms*

Every formula of $For(L)$ obtained from a tautology of classical propositional logic $\alpha(p_1, \dots, p_n)$ by means of a simultaneous and uniform substitution of an arbitrary formula φ_i of $For(L)$ for each propositional variable p_i, for $i = 1, \dots, n$, is an axiom of logic.

The resulting axiom is written as $\alpha(p_1/\varphi_1, \dots, p_n/\varphi_n)$, or $\alpha(\varphi_1, \dots, \varphi_n)$ for short. (In the sequence $\langle \varphi_1, \dots, \varphi_n \rangle$ some formulas may be repeated.)

2. *Quantifier axioms*

(i) If φ and ψ are formulas of L and x is an individual variable, then the formula

$$(\forall x)(\varphi \to \psi) \to ((\forall x)\varphi \to (\forall x)\psi)$$

is an axiom of logic.

(ii) If φ is a formula of L, x is an individual variable, t is a term and the substitution x/t is free in φ, then

$$(\forall x)\varphi \to \varphi(x/t)$$

is an axiom of logic.

(i) are the axioms of *distributivity of the universal quantifier for implication*.
(ii) are the *substitution axioms*.

3. *Equality axioms*

Let $t(x_1, \dots, x_n)$ be a term and $\varphi(x_1, \dots, x_n)$—an atomic formula. Then

$$x \approx x,$$
$$x \approx y \to t(x_1, \dots, x_{i-1}, x, x_{i+1}, \dots, x_n) \approx t(x_1, \dots, x_{i-1}, y, x_{i+1}, \dots, x_n)$$

and

$$x \approx y \to (\varphi(x_1, \dots, x_{i-1}, x, x_{i+1}, \dots, x_n) \to \varphi(x_1, \dots, x_{i-1}, y, x_{i+1}, \dots, x_n))$$

for each i $(1 \leqslant i \leqslant n)$ are axioms of logic.

Note The formulas $x \approx y \to y \approx x$, $x \approx y \wedge y \approx z \to x \approx z$ are consequences of the above axioms. □

There are two primitive rules of inference.

4. *The detachment rule* (alias: ***Modus Ponens***): From φ and $\varphi \to \psi$ infer ψ.

5. *The rule of generalization*: From φ infer $(\forall x)\varphi$,
where x is an arbitrary individual variable.

We then define theses (alias: theorems) of logic. According to the standard definition, a formula φ is a *thesis of logic* if and only if there exists a finite sequence of formulas

$$\varphi_1, \ldots, \varphi_n \qquad (*)$$

such that the formula φ_n is identical with φ and, for each i ($i = 1, \ldots, n$), φ_i is an axiom of logic or φ_i is a conclusion obtained from 'earlier' formulas by an application of one of the above rules.

The sequence $(*)$ is called a *proof of* φ *on the grounds of axioms of logic with the use of the detachment and generalization rules.*

CL is the set of all theses of logic. (The acronym '***CL***' is formed from the initial letters of *classical logic*.)

By *classical (first-order) logic* we shall understand the above set ***CL*** of theses of logic. (This is one of the meanings of classical logic.)

Definition 2 If Σ is a set of formulas and σ a formula, then the notation

$$\Sigma \vdash \sigma$$

means that there exists a proof of σ from Σ, that is, there exists a finite sequence of formulas

$$\langle \varphi_1, \ldots, \varphi_n \rangle \qquad (**)$$

such that the formula φ_n is identical with σ and, for each i ($i = 1, \ldots, n$) either φ_i is a thesis of logic (i.e., $\varphi_i \in \boldsymbol{CL}$), or φ_i belongs to Σ, or φ_i is a conclusion from earlier formulas from the sequence $(*)$ obtained by a direct application of the detachment rule.

The relation $\vdash$, holding between set of formulas and individual formulas is called the relation of *inferential entailment* (alias the *deducibility relation*) of classical logic. □

Note In the above definition of $(**)$ the rule of generalization *is not* applied. This rule is needed only in the definition of the set ***CL*** of theses.

The above formalization implies (see Theorem 7.3.2 below) that for any two formulas $\varphi(\underline{x})$, $\psi(\underline{x})$:

$$\varphi(\underline{x}) \vdash \psi(\underline{x}) \quad \Leftrightarrow \quad \vdash (\forall \underline{x})(\varphi(\underline{x}) \rightarrow \psi(\underline{x})), \qquad (*)$$

where $\underline{x} = FreeVar(\varphi) \cup FreeVar(\psi)$.

If we let $\vdash_{GR}$ denote the consequence operation of the first-order logic in which the Rule of Generalization is a primitive rule of inference (and thus applied in arbitrary deductions), then instead of $(*)$ one obtains the following equivalence:

$$\varphi(\underline{x}) \vdash_{GR} \psi(\underline{x}) \quad \Leftrightarrow \quad \vdash_{GR} (\forall \underline{x})\varphi(\underline{x}) \rightarrow (\forall \underline{x})\psi(\underline{x}), \qquad (**)$$

(The consequences $\vdash$and $\vdash_{GR}$ coincide on the set ***CL***.)

The fact that the deducibility does not include the generalization rule is of basic importance, because it makes Lindenbaum sets non-closed under changing variables. The above fact is central in the paper because it enables one to build the semantics for first-order logic based on Rasiowa–Sikorski sets. □

If Σ is finite set of formulas, $\Sigma = \{\sigma_1, \dots, \sigma_n\}$, then we also write

$$\sigma_1, \dots, \sigma_n \vdash \sigma$$

instead of $\Sigma \vdash \sigma$.

If $\Sigma \vdash \sigma$ holds, then we say that *there is a proof of* σ *on the grounds of* Σ. We also say that σ is *deducible* (or *provable*) from the set Σ.

If $\Sigma = \varnothing$, then $\varnothing \vdash \sigma$ means that σ is a thesis of logic, that is, $\sigma \in$ ***CL***. We shall write $\vdash \sigma$ when $\varnothing \vdash \sigma$.

In the standard way one defines the notion of a secondary rule of inference of classical logic.

A set of formulas Σ is *inconsistent* when all formulas are deducible from Σ; otherwise Σ is *consistent*. A single formula σ is *inconsistent* (*consistent*), if the set $\{\sigma\}$ is inconsistent (consistent, respectively).

Theorem 7.3.1 *Let* Σ *be a set of formulas of* L. *The following conditions are equivalent:*

(1) Σ *is inconsistent.*
(2) There is a formula φ *such that* $\Sigma \vdash \varphi \wedge \neg \varphi$. □

We write

$$C(\Sigma) := \{\sigma \in For(L) : \Sigma \vdash \sigma\}.$$

$C(\Sigma)$ is the *set of consequences* of Σ. The mapping C itself, defined as above on sets of formulas, is called the *consequence operation of classical logic*. According to the second meaning of the term "classical logic", classical logic is identified not with the set ***CL***, but with the consequence operation C (or, equivalently, with the entailment relation $\vdash$).

We have that ***CL*** $= C(\varnothing)$.

A set of formulas Σ is *logically closed*, if $\Sigma = C(\Sigma)$. ***CL*** is the least logically closed set. Every logically closed set contains ***CL*** as a subset and it is closed on Modus Ponens (and any other rule of classical propositional logic.)

The following facts are well-known.

Theorem 7.3.2 *Let* Σ *be a set of formulas and* σ, τ *arbitrary formulas.*

(i) Σ *is consistent if and only if every finite subset of* Σ *is consistent.*
(ii) (Deduction Theorem) $\Sigma \cup \{\sigma\} \vdash \tau \quad \Leftrightarrow \quad \Sigma \vdash \sigma \to \tau$.
(iii) $\Sigma \cup \{\sigma\}$ *is inconsistent if and only if* $\Sigma \vdash \neg\sigma$. *Hence* $\Sigma \cup \{\sigma\}$ *consistent if and only if* $\neg\sigma$ *is not provable from* Σ.

(iv) If x is a variable not occurring as free variable in the formulas of Σ and $\Sigma \vdash \sigma$, then $\Sigma \vdash (\forall x)\sigma$. □

Notes 1. (iv) implies that if Σ is a set of *sentences* and $\Sigma \vdash \sigma$, then $\Sigma \vdash (\forall x)\sigma$.
2. Formulas φ and ψ are *deductively equivalent* (or: *logically equivalent*), when $C(\varphi) = C(\psi)$; equivalently, the formula $\varphi \leftrightarrow \psi$ is a thesis of logic. □

Theorem 7.3.3 *Let σ be a formula.*

1. *If a variable y does not occur in σ (neither as free nor bound variable), then*

$$\vdash (\forall x)\sigma \leftrightarrow (\forall y)\sigma(x/y) \quad \textit{and} \quad \vdash (\exists x)\sigma \leftrightarrow (\exists y)\sigma(x/y).$$

2. *Assume that $x \notin FreeVar(\sigma)$. Then*

$$\vdash \sigma \leftrightarrow (\forall x)\sigma \quad \textit{and} \quad \vdash \sigma \leftrightarrow (\exists x)\sigma.$$

3. *Let φ, ψ be any formulas and x, y arbitrary variables.*
   a. $\vdash \varphi \leftrightarrow \psi$ *implies* $\vdash (\forall x)\varphi \leftrightarrow (\forall x)\psi$ *and* $\vdash (\exists x)\varphi \leftrightarrow (\exists x)\psi$.
   b. $\vdash \varphi \leftrightarrow \psi$ *implies* $\vdash (\forall x)(\forall y)\varphi \leftrightarrow (\forall y)(\forall x)\psi$ *and* $\vdash (\exists x)(\exists y)\varphi \leftrightarrow (\exists y)(\exists x)\psi$.
4. *Let t be a term of L and x a variable such that $x \notin Var(t)$. Then $\vdash (\exists x)(t \approx x)$.*
5. *Let x and y be different variables. The formulas $x \approx y$ and $\neg(x \approx y)$ are not theses of logic.*
6. *If φ is a formula of L and x is a variable not occurring as a free variable in φ, that is, $x \notin FreeVar(\varphi)$, then $\vdash \varphi \to (\forall x)\varphi$ (the law of adjoining the superfluous quantifier.)* □

Note Substitutions need not preserve deductive equivalence of formulas, that is, if $\varphi(x_1, \ldots, x_n)$ and $\psi(x_1, \ldots, x_n)$ are deductively equivalent formulas in the same free variables and $x_1/y_1, \ldots, x_n/y_n$ is a substitution, then $\varphi(x_1/y_1, \ldots, x_n/y_n)$ and $\psi(x_1/y_1, \ldots, x_n/y_n)$ need not be deductively equivalent. $\varphi(x_1/y_1, \ldots, x_n/y_n)$ and $\psi(x_1/y_1, \ldots, x_n/y_n)$ are deductively equivalent when the substitution $\langle x_1/y_1, \ldots, x_n/y_n\rangle$ is free in φ and ψ.

Let $\varphi(x)$ and $\psi(x)$ be the formulas $(\exists y)(y < x)$, $(\exists z)(z < x)$, respectively, where the variables y, z are different. They are deductively equivalent. But the formulas $\varphi(x/y)$ and $\psi(x/y)$, that is, $(\exists y)(y < y)$, $(\exists z)(z < y)$ are not deductively equivalent. □

Lindenbaum Sets

Definition 3 A set Δ of formulas is a *Lindenbaum set* if it is maximal consistent, that is, Δ is consistent and for any set of formulas Σ, if $\Delta \subset \Sigma$, then Σ is inconsistent. ($\subset$ is the symbol of proper inclusion.) □

Every Lindenbaum set Δ is logically closed. In particular, Δ contains the set ***CL*** of logical theses.

Theorem 7.3.4 (Lindenbaum) *Every consistent set of formulas of L is contained in a Lindenbaum set.* □

The following fact is well known, its proof is straightforward:

Theorem 7.3.5 *Let Δ be a set of formulas. The following conditions are equivalent:*

(A) Δ is a Lindenbaum set.
(B) Δ is consistent and satisfies the conditions:

(1) For any $\sigma, \tau \in For(L)$, $\sigma \wedge \tau \in \Delta \;\Leftrightarrow\; \sigma \in \Delta$ and $\tau \in \Delta$.
(2) For any $\sigma \in For(L)$, $\neg\sigma \in \Delta \;\Leftrightarrow\; \sigma \notin \Delta$. □

7.4 Semantics

This and the following sections contain the material which is also well-known. We present it here because we shall introduce a certain notation which will be utilised later.

Assignments

Let $\boldsymbol{A}$ be a model for L. By an *assignment* in $\boldsymbol{A}$ (we shall also use the term *valuation*) we shall mean any function h whose domain $Dom(h)$ is a set of individual variables and the values of h belong to the universe of $\boldsymbol{A}$.

If $\{x_i : i \in I\}$ is the domain of h and $a_i = h(x_i)$ for all $i \in I$, h is marked as

$$[x_i/a_i : i \in I].$$

An assignment h is *global* if its domain coincides with the set Var of all individual variables.

An assignment h is *finite* if its domain is a finite set of variables.

Let $\{x_1, \ldots, x_n\}$ be a finite set of variables (this set may be empty). If h is a (finite) assignment with the domain $\{x_1, \ldots, x_n\}$ and $a_1 = h(x_1), \ldots, a_n = h(x_n)$, then h is also written as

$$[x_1/a_1, \ldots, x_n/a_n],$$

or

$$[a_1, \ldots, a_n]$$

for short, when the domain of h is known from context.

We will have mainly finite valuations within our eyeshot. Infinite valuations will be pertinent in situations of (simultaneous) satisfiability of infinite sets of formulas.

Let $h = [x_1/a_1, \ldots, x_n/a_n]$ be a finite assignment with the domain $Dom(h) = \{x_1, \ldots, x_n\}$. If x is an arbitrary variable and $a \in A$, then we define:

$$h(x/a) := (h \setminus \{(x, h(x))\}) \cup \{(x, a)\}.$$

(Every function is a set of ordered pairs.) Thus, if $x \in \{x_1, \ldots, x_n\}$ and $x = x_i$ for some i ($1 \leqslant i \leqslant n$), then $Dom(h(x/a)) = \{x_1, \ldots, x_n\}$ and $h(x/a) = [x_1/a_1, \ldots, x_{i-1}/a_{i-1}, x_i/a, x_{i+1}/a_{i+1}, \ldots, x_n/a_n]$. If $x \notin \{x_1, \ldots, x_n\}$, then $Dom(h(x/a)) = \{x_1, \ldots, x_n\} \cup \{x\}$ and $h(x/a) = [x_1/a_1, \ldots, x_{i-1}/a_{i-1}, x_i/a, x_{i+1}/a_{i+1}, \ldots, x_n/a_n]$.

Let $h = [x_1/a_1, \ldots, x_n/a_n]$ be a finite valuation. To each term $t(x_1, \ldots, x_n)$ (in the variables contained in $\{x_1, \ldots, x_n\}$) an element of the model $\boldsymbol{A}$ is assigned that is called the *value of the term t for the sequence* $a_1, \ldots, a_n$. This element is denoted by

$$t^{\boldsymbol{A}}[h] \quad \text{(or, in a more developed form, by } t^{\boldsymbol{A}}[a_1, \ldots, a_n]). \qquad (*)$$

$t^{\boldsymbol{A}}[h]$ is recursively defined in the standard way.

Notes 1. A given term t may contain no individual variables. Then the value $t^{\boldsymbol{A}}[\epsilon]$ makes sense, where ϵ is the empty assignment. (The length of the sequence ϵ equals 0.) E.g. if t is a constant c, then $t^{\boldsymbol{A}}[\epsilon] := c^{\boldsymbol{A}}$. In a similar way one defines $t^{\boldsymbol{A}}[\epsilon]$ for compound terms which do not involve individual variables.

2. Generally, the value $t^{\boldsymbol{A}}[h]$ of $t(x_1, \ldots, x_n)$ makes sense for any valuation h, even infinite, whose domain $Dom(h)$ includes $\{x_1, \ldots, x_n\}$. $\square$

Satisfaction

Let n be a natural number and $h = [x_1/a_1, \ldots, x_n/a_n]$ a finite valuation in a model $\boldsymbol{A}$. The relation:

A formula $\sigma(x_1, \ldots, x_n)$ *is satisfied by h in* $\boldsymbol{A}$

is denoted in the standard way as

$$\boldsymbol{A} \models \sigma(x_1, \ldots, x_n)[h],$$

or, in a more developed form, as

$$\boldsymbol{A} \models \sigma(x_1, \ldots, x_n)[x_1/a_1, \ldots, x_n/a_n],$$

or $\boldsymbol{A} \models \sigma(x_1, \ldots, x_n)[a_1, \ldots, a_n]$ too. We also say that

A formula $\sigma(x_1, \ldots, x_n)$ *is satisfied by the sequence* $a_1, \ldots, a_n$ *in the model* $\boldsymbol{A}$,

when the valuation $[x_1/a_1, \ldots, x_n/a_n]$ is clear from context.

The satisfaction relation (between formulas $\sigma(x_1, \ldots, x_n)$ in free variables $x_1, \ldots, x_n$ and valuations h with the domain $\{x_1, \ldots, x_n\}$) is recursively defined in the well-known way. We shall note here the case of existential formulas:

If $\sigma(x_1, \ldots, x_n)$ is of the form $(\exists x)\varphi$, then

$$\boldsymbol{A} \models \sigma(x_1, \ldots, x_n)[h] \quad \Leftrightarrow \quad \boldsymbol{A} \models \varphi[h(x/a)] \text{ for some } a \in A. \qquad \square$$

Let σ be a sentence. σ is *true in the model* $\boldsymbol{A}$ if $\boldsymbol{A} \models \sigma[\epsilon]$, that is, σ is satisfied in $\boldsymbol{A}$ by the empty valuation ϵ.

In the light of the above remarks:
For any sentence σ *the following conditions are equivalent*:

$$A \models \sigma[\epsilon],$$
$$A \models \sigma[h] \text{ for every valuation } h,$$
$$A \models \sigma[h] \text{ for some valuation } h.$$

Logical Entailment

Definition 4 Let Σ be a set of formulas and σ a formula. The symbol

$$\Sigma \models \sigma,$$

marks the fact that σ *logically follows* from Σ, which means that for every model A and every global valuation h in A, if $A \models \varphi[h]$ for all $\varphi \in \Sigma$, then $A \models \sigma[h]$. □

(One should distinguish here the meaning of the big turnstyle symbol $\models$ from the meaning of $\models$!)

If Σ is empty, we write $\models \sigma$ instead of $\varnothing \models \sigma$. $\models \sigma$ means that σ is a tautology in L. □

The above definition yields that:

$$\Sigma \cup \{\varphi\} \models \psi \quad \Leftrightarrow \quad \Sigma \models \varphi \rightarrow \psi,$$

for any formulas φ, ψ. Moreover, if x is an individual variable such that $x \notin FreeVar(\Sigma)$, then

$$\Sigma \models \varphi \quad \Leftrightarrow \quad \Sigma \models (\forall x)\varphi.$$

Theorem 7.4.1 (The Extended Completeness Theorem) *Let* Σ *be an arbitrary set of formulas of* L *and* σ *a formula. Then*

$$\Sigma \models \sigma \quad \Leftrightarrow \quad \Sigma \vdash \sigma,$$

i.e., the relation $\models$ *of logical entailment coincides with the relation* $\vdash$ *of inferential entailment.* □

We shall return to the proof of the above theorem.

7.5 Models Constructed from Individual Variables

We shall prove the following fact concerning countable languages.

Theorem 7.5.1 *Let* L *be a countable language. Let* ψ_n, $n \in \mathbb{N}$, *be an enumeration of all formulas of* $For(L)$ *beginning with the existential quantifier. Mark* ψ_n *as*

$(\exists x_n)\varphi_n$, for all n. Moreover, let Σ be a consistent set of formulas such that the set of variables $Var \setminus FreeVar(\Sigma)$ is infinite. (In particular, let Σ be a set of sentences.) There exists a sequence of individual variables $\langle y_n : n \in \mathbb{N}\rangle$ such that the set

$$\Sigma \cup \{(\exists x_n)\varphi_n \to \varphi_n(x_n/y_n) : n \in \mathbb{N}\}$$

is consistent. Moreover, for every n, $y_{n+1} \notin FreeVar(\Sigma)$ and y_{n+1} does not occur in the formulas of $(\exists x_i)\varphi_i \to \varphi_i(x_i/y_i)$, $(1 \leqslant i \leqslant n)$.

Note 1 As the substitution x_n/y_n is free in φ_n for all n, the reverse implications $\varphi_n(x_n/y_n) \to (\exists x_n)\varphi_n$, $n \in \mathbb{N}$, are theses of logic. The theorem can be therefore formulated in the form:

The set

$$\Sigma \cup \{(\exists x_n)\varphi_n \leftrightarrow \varphi_n(x_n/y_n) : n \in \mathbb{N}\}$$

is consistent and the substitution x_n/y_n is free in φ_n for all n.

The above theorem thus states that under the above assumption the set Σ can be extended to a consistent set by means of the elimination of the existential quantifier in the formulas of L. □

Proof We first prove:

Lemma 7.5.2 *Let Σ be a consistent set of formulas of L. Let x be an individual variable and φ a formula. For every variable y not belonging to $FreeVar(\Sigma)$ and not occurring in $(\exists x)\varphi$, the set*

$$\Sigma \cup \{(\exists x)\varphi \leftrightarrow \varphi(x/y)\}$$

is consistent.

Proof (*of the lemma*) Fix x and φ. It suffices to show that every variable y not belonging to $FreeVar(\Sigma)$ and not occurring in $(\exists x)\varphi$, the set

$$\Sigma \cup \{(\exists x)\varphi \to \varphi(x/y)\}$$

is consistent. (The reverse implication $\varphi(x/y) \to (\exists x)\varphi$ is a thesis of logic.) Suppose otherwise. Hence for *some* variable y not belonging to $FreeVar(\Sigma)$ and not occurring in $(\exists x)\varphi$, the set $\Sigma \cup \{(\exists x)\varphi \to \varphi(x/y)\}$ is inconsistent.

We mark by $\alpha(y)$ the formula $(\exists x)\varphi \to \varphi(x/y)$ (with y with the above property). $\alpha(y)$ may contain other free variables; they are all different from x. As $\Sigma \cup \{\alpha(y)\}$ is inconsistent, we have that $\Sigma \cup \{\alpha(y)\} \vdash \neg\alpha(y)$. This gives that $\Sigma \vdash \neg\alpha(y)$, by the Deduction Theorem. In virtue of Theorem 7.3.2.(iv) we get that

$$\Sigma \vdash (\forall y)\neg\alpha(y), \tag{1}$$

(because y is not free in the formulas of Σ). Since Σ is consistent, (1) implies the formula $(\forall y)\neg\alpha(y)$ itself is consistent, i.e., the formula

$$(\forall y)((\exists x)\varphi \wedge \neg\varphi(x/y)) \tag{2}$$

is consistent.

As y is not free in φ (because it does not occur in φ), formula (2) is deductively equivalent to the formula

$$(\exists x)\varphi \wedge (\forall y)\neg\varphi(x/y). \tag{3}$$

Applying once more the assumption that y does not occur in φ, we obtain that $(\forall y)\neg\varphi(x/y)$ is deductively equivalent to $(\forall x)\neg\varphi(x)$ (Theorem 7.3.3.(1) is applied here.) It follows that (3) is deductively equivalent to $(\exists x)\varphi \wedge (\forall x)\neg\varphi(x)$, that is, (3) is equivalent to

$$\neg(\forall x)\neg\varphi \wedge (\forall x)\neg\varphi(x).$$

Concluding, the consistent formula (2) is equivalent to the inconsistent formula $\neg(\forall x)\neg\varphi \wedge (\forall x)\neg\varphi(x)$, which is excluded.

This proves the lemma. □

We pass to the proof of the theorem.

The sequence $\langle y_n : n \in \mathbb{N}\rangle$ is recursively defined. The case $n = 1$ is a direct corollary from the above lemma. Fix a positive $n \in \mathbb{N}$ and assume that a sequence of variables $\langle y_1, \ldots, y_n\rangle$ has been defined so that the set

$$\Sigma_n := \Sigma \cup \{(\exists x_1)\varphi_1 \rightarrow \varphi_1(x_1/y_1), \ldots, (\exists x_n)\varphi_n \rightarrow \varphi_n(x_n/y_n)\}$$

is consistent. The set $Var \setminus FreeVar(\Sigma_n)$ is infinite. We then apply the above lemma to the set Σ_n, the variable x_{n+1} and the formula φ_{n+1}. Let y be the first variable (with the lowest index in the alphabetic ordering of variables) not belonging $FreeVar(\Sigma_n)$ and not occurring in $(\exists x_{n+1})\varphi_{n+1}$. In virtue of the above lemma the set

$$\Sigma_n \cup \{(\exists x_{n+1})\varphi_{n+1} \rightarrow \varphi_{n+1}(x_{n+1}/y)$$

is consistent. We then put $y_{n+1} := y$.

This concludes the proof of the theorem. □

Note 2 The above theorem is provable in a weak arithmetic. (The full strength of ZF is not needed.) The assumption that L and the set Var of individual variables are countable sets implies that the set of finite sequences of elements of $L \cup Var \cup \{\,(,\,)\}$ is well-ordered (by a countable ordinal). Consequently, the set of terms and the set of formulas $For(L)$ are well-ordered too. Every set of formulas is well-ordered; so is in particular the set of formulas beginning with the existential quantifier.

In the above proof of the theorem, for each n one pick out y_{n+1} from $Var \setminus FreeVar(\Sigma_n)$ as the variable with the *least* index in the alphabetic ordering among the variables and not occurring φ_n. Thus the proof is carried out in accordance with the principle of definability by the arithmetic recursion. Neither the principle DC of dependent choices nor any other weak form of the axiom of choice is needed here. Similarly, the proof of Lindenbaum's Theorem for countable languages merely requires the above arithmetic principle.

The above theorem can be extended onto uncountable languages. The proof of the following theorem is similar to the proof of the previous theorem but it requires stronger set-theoretic assumptions (see e.g. Dzik 1981). We omit the details.

Theorem 7.5.3 *Assume* AC. *Let L be a language of cardinality μ. We assume that the set Var of individual variables is of cardinality μ as well. Let $(\exists x_\alpha)\varphi_\alpha$, $\alpha < \mu$, be an enumeration of all formulas of $For(L)$ beginning with the existential quantifier. Let Σ be a consistent set of formulas such that the set $Var \setminus FreeVar(\Sigma)$ is of cardinality μ. (In particular, let Σ be a set of sentences.) Then there exists a transfinite sequence of individual variables $\langle y_\alpha : \alpha < \mu\rangle$ such that the set*

$$\Sigma \cup \{(\exists x_\alpha)\varphi_\alpha \rightarrow \varphi_\alpha(x_\alpha/y_\alpha) : \alpha < \mu\}$$

is consistent. Moreover, the substitution x_α/y_α is free in φ_α, for all $\alpha < \mu$. □

Model Sets

Definition 5 A set Δ of formulas L is called a *model set* (for L) if and only if Δ is a Lindenbaum set with the additional property that for every variable x and every formula φ of L:

$$(1)(\exists x)\varphi \in \Delta \quad \Leftrightarrow \quad \varphi(x/y) \in \Delta \text{ for some variable } y \text{ such that the substitution } x/y \text{ is free in } \varphi. \quad \square$$

Every model set is therefore maximal consistent and logically closed.

The following fact immediately follows from Theorem 7.3.5 and the above definition:

Corollary 1 *A subset $\Delta \subseteq For(L)$ is a model set if and only if it is consistent and satisfies the following conditions:*

(1) For any $\sigma, \tau \in For(L)$,

$$\sigma \wedge \tau \quad \Leftrightarrow \quad \sigma \in \Delta \text{ and } \tau \in \Delta.$$

(2) For any $\sigma \in For(L)$,

$$\neg\sigma \in \Delta \quad \Leftrightarrow \quad \sigma \notin \Delta,$$

(3) For any individual variable x and any formula φ,

$$(\exists x)\varphi \in \Delta \quad \Leftrightarrow \quad \varphi(x/y) \in \Delta \text{ for some variable } y \text{ such that the substitution } x/y \text{ is free in} \varphi. \quad \square$$

Note 3 The consistency of Δ is an essential assumption in the above corollary; this condition does not follow from (1)–(3). An appropriate example is given in Sect. 7.6. $\square$

Corollary 2 *Let Δ be a model set for L. Then for any variable x and any formula φ,*

$$(4) (\forall x)\varphi \in \Delta \quad \Leftrightarrow \quad \varphi(x/y) \in \Delta \text{ for every variable } y \text{ such that the substitution } x/y \text{ is free in } \varphi.$$

Proof Straightforward. $\square$

Model sets exist:

Theorem 7.5.4 *Let Σ be an arbitrary consistent set of formulas of L such that the set $Var \setminus FreeVar(\Sigma)$ is infinite. Σ is contained in a model set for L.*

Proof Σ is first extended to a consistent set Γ which possesses the following property: for every variable x and every formula φ there is a variable y such that y does not occur in $(\exists x)\varphi$ and the formula $(\exists x)\varphi \leftrightarrow \varphi(x/y)$ belongs to Γ. In virtue of Theorem 7.5.1, such an extension Γ exists. We then apply Lindenbaum's Theorem (Theorem 7.3.5) and extend Γ to a maximal consistent set Δ. It is then easy to check that Δ is a model set. $\square$

One may say more:

Corollary 3 *Let Σ be any consistent set of formulas of L such that the set $Var \setminus FreeVar(\Sigma)$ is infinite. (In particular, let Σ be a set of sentences.) The set $C(\Sigma)$ of all consequences of Σ coincides with the intersection of all model sets Δ such that $\Sigma \subseteq \Delta$.*

Proof Suppose that $\sigma \notin C(\Sigma)$. In light of Theorem 7.3.2.(iii), the set $\Sigma \cup \{\neg\sigma\}$ is consistent. The set $Var \setminus FreeVar(\Sigma \cup \{\neg\sigma\})$ is infinite. According to Theorem 7.5.4, there exists a model set Δ for L such that $\Sigma \cup \{\neg\sigma\} \subseteq \Delta$. The last inclusion and the maximality of Δ yield that $\sigma \notin \Delta$. $\square$

Rasiowa–Sikorski Sets

Let φ be a formula and x a free variable in φ. Let y be an arbitrary variable. $\varphi(x /\!\!/ y)$ is the formula obtained from φ in the following way: (1) in every subformula of φ beginning with the quantifier $(\exists y)$ and in which x appears as a free variable (if there is such a formula), one uniformly replaces every occurrence of y by the variable not occurring in φ with the least index (in the alphabetic order); one obtains a formula φ', being a variant of φ, in which the substitution x/y is free, (2) one uniformly substitutes y for x in φ' where x occurs as a *free* variable. Thus $\varphi(x /\!\!/ y)$ is $\varphi'(x/y)$.

Note 4 Let φ and ψ be different formulas. Form the implication $\varphi \to \psi$. The formulas $(\varphi \to \psi)(x /\!/ y)$ and $\varphi(x /\!/ y) \to \psi(x /\!/ y)$ need not be identical, even if x occurs as a free variable in φ and ψ. But the above formulas are logically equivalent, because the antecedent of the implication $(\varphi \to \psi)(x /\!/ y)$ is logically equivalent to $\varphi(x /\!/ y)$, and the succedent of $(\varphi \to \psi)(x /\!/ y)$ is equivalent to $\psi(x /\!/ y)$. Analogous interdependencies hold if φ and ψ are conjoined by other binary connectives. Thus the double substitution distributes with respect to all Boolean connectives up to deductive equivalence. □

Definition 6 Let L be a language. A *Rasiowa–Sikorski* set for L is an arbitrary Lindenbaum set for L with the additional property that for any variable x and any formula φ:

$$(\exists x)\varphi \in \Delta \quad \Leftrightarrow \quad \varphi(x /\!/ y) \in \Delta \text{ for some variable } y.$$

Every Rasiowa–Sikorski set is logically closed (as being a maximal consistent set).

Note 5 The term 'Rasiowa–Sikorski set' originates from the algebraic proof of the completeness theorem for first-order logic given by Rasiowa and Sikorski in 1950. Their proof is based on Theorem 1.1 (*Rasiowa–Sikorski Lemma*). This lemma is utilised by them in the proof of the existence of so called Q-ultrafilters in Lindenbaum–Tarski algebras of elementary theories. Each Q-ultrafilter defines a Rasiowa–Sikorski set for L. □

Theorem 7.5.5 *$\Delta \subseteq For(L)$ is a Rasiowa–Sikorski set if and only if Δ is consistent and satisfies:*

(1) For any $\sigma, \tau \in For(L)$,

$$\sigma \wedge \tau \quad \Leftrightarrow \quad \sigma \in \Delta \text{ and } \tau \in \Delta.$$

(2) For any $\sigma \in For(L)$,

$$\neg\sigma \in \Delta \quad \Leftrightarrow \quad \sigma \notin \Delta.$$

(3) For any variable x and any formula φ

$$(\exists x)\varphi \in \Delta \quad \Leftrightarrow \quad \varphi(x /\!/ y) \in \Delta \text{ for some variable } y.$$

Proof Straightforward. □

Example Let L be a language and $\boldsymbol{A}$ a model for L such that there is a surjection h from Var onto the universe of $\boldsymbol{A}$. (This condition holds if $\boldsymbol{A}$ is countable.) h is thus a global valuation of Var in $\boldsymbol{A}$. Define:

$$\Delta(h) := \{\sigma \in For(L) : \boldsymbol{A} \models \sigma[h]\}.$$

Lemma 7.5.6 *$\Delta(h)$ is a Rasiowa–Sikorski set.*

Proof It is clear that $\Delta(h)$ consistent, logically closed and satisfies the above conditions (1)–(2), that is, Δ is a Lindenbaum set. We shall check that (3) holds.

We assume $(\exists x)\varphi \in \Delta(h)$, that is, $\boldsymbol{A} \models (\exists x)\varphi[h]$. Hence $\boldsymbol{A} \models \varphi[h(x/a)]$ for some $a \in A$. We select a with this property. Since h is surjective, there is a variable y such that $a = h(y)$. Hence $\boldsymbol{A} \models \varphi[h(x/h(y))]$.

The substitution x/y need not be free in φ. Form φ' as in the definition of $\varphi(x /\!\!/ y)$. Since $\varphi' \leftrightarrow \varphi$ is a thesis of logic, we obtain that $\boldsymbol{A} \models \varphi'[h(x/h(y))]$. But the substitution x/y is free in φ'. Hence $\boldsymbol{A} \models \varphi'[h(x/h(y))]$ implies that $\boldsymbol{A} \models \varphi'(x/y)[h]$. Since $\varphi'(x/y)$ is identical with $\varphi(x /\!\!/ y)$, we obtain that $\boldsymbol{A} \models \varphi(x /\!\!/ y)[h]$, which means that $\varphi(x /\!\!/ y) \in \Delta(h)$.

The fact that $\varphi(x /\!\!/ y) \in \Delta(h)$ entails $(\exists x)\varphi \in \Delta(h)$ is straightforward. □

Theorem 7.5.7 *Every model set for L is a Rasiowa–Sikorski set.*

Proof Straightforward. □

The above theorem cannot be reversed:

Theorem 7.5.8 *There exist Rasiowa–Sikorski sets that are not model sets.*

Proof Let $L = \{\leqslant\}$, where $\leqslant$ is a binary predicate. We consider the model $\boldsymbol{N} = (N, \leqslant_N)$ for L, where N is the set of natural numbers (with zero), and $\leqslant_N$ is the standard order in N. Let $h : Var \to N$ be the surjection such that $h(v_0) = 1$, $h(v_1) = 0$ and $h(v_i) = i$ for any natural number $i \geqslant 2$. According to Lemma 7.5.6, $\Delta(h) := \{\sigma \in For(L) : \boldsymbol{N} \models \sigma[h]\}$ is a Rasiowa–Sikorski set. But $\Delta(h)$ is not a model set. For let

$$\varphi(v_0) := (\forall v_1)(v_0 \leqslant v_1).$$

The sentence $(\exists v_0)\varphi$ is true in $\boldsymbol{N}$, and hence $(\exists v_0)\varphi \in \Delta(h)$. $\varphi(v_0 /\!\!/ v_1)$ is the formula $(\forall v_2)(v_1 \leqslant v_2)$ (and obviously $\varphi(v_0 /\!\!/ v_1) \in \Delta(h)$). But, on the other hand, we also have:

Claim *If y is whatever variable such that the substitution v_0/y is free in φ, then $\varphi(v_0/y) \notin \Delta(h)$.*

Proof (of the claim) A substitution v_0/y is free in φ if and only if $y = v_0$ or $y = v_i$, where $i = 2, 3, \ldots$. For $y = v_0$, the formula $\varphi(v_0/y)$ is identical with φ. The formula φ is not satisfied by h. For $y = v_i$, where $i = 2, 3, \ldots$, the formula $\varphi(v_0/y)$ is identical with $(\forall v_1)(v_i \leqslant v_1)$. Each such a formula is satisfied by h neither. Thus in any case, $\varphi(v_0/y)$ is not satisfied by h, which proves the claim. □

Since $(\exists v_0)\varphi \in \Delta(h)$, the above claim shows that $\Delta(h)$ is not a model set. □

According to the above theorems, the notion of a Rasiowa–Sikorski set is a weakening of the notion of a model set.

Lemma 7.5.9 *Each Rasiowa–Sikorski set for L is unambiguously determined by the atomic formulas belonging to it, that is, for any Rasiowa–Sikorski sets Δ, Δ':*

$$\Delta = \Delta' \quad \Leftrightarrow \quad \Delta \cap AFor(L) = \Delta' \cap AFor(L).$$

Proof Implication ($\Rightarrow$) is obvious.

($\Leftarrow$). Assume $\Delta \cap AFor(L) = \Delta' \cap AFor(L)$. By induction on the length of formulas σ one proves the equivalence: $\sigma \in \Delta \;\Leftrightarrow\; \sigma \in \Delta'$. □

Models Determined by Rasiowa–Sikorski Sets

Every Rasiowa–Sikorski set Δ defines a model for L. We shall present here the pertinent definitions.

Definition 7 Let Δ be a Rasiowa–Sikorski set for L. $\equiv_\Delta$ is the binary relation on the set $Var = \{v_n : n \in \mathbb{N}\}$ of individual variables defined by:

$$x \equiv_\Delta y \quad \Leftrightarrow_{df} \quad x \approx y \in \Delta$$

for any $x, y \in Var$.

$\equiv_\Delta$ is an equivalence relation. Let

$$A_\Delta := Var/ \equiv_\Delta$$

be the quotient set, i.e., A_Δ is the set of equivalence classes of $\equiv_\Delta$. Thus $A_\Delta = \{[v_n] : n \in \mathbb{N}\}$.

The interpretation I_Δ of L on A_Δ is defined as follows.

If P is an m-ary relational symbol in L and $[x_1], \ldots, [x_m] \in A_\Delta$, then:

$$I_\Delta(P)([x_1], \ldots, [x_m]) \quad \Leftrightarrow_{df} \quad P(x_1, \ldots, x_m) \in \Delta.$$

If F is an n-ary function symbol in L and $[x_1], \ldots, [x_n] \in A_\Delta$, then:

$$I_\Delta(F)([x_1], \ldots, [x_n]) := \text{ the unique class } [y] \text{ such that } F(x_1, \ldots, x_n) \approx y \in \Delta.$$

As the formula $(\exists x)(F(x_1, \ldots, x_n) \approx x)$ is a thesis of logic (where x is different from $x_1, \ldots, x_n$), we obtain that $(\exists x)(F(x_1, \ldots, x_n) \approx x) \in \Delta$. Since Δ is a Rasiowa–Sikorski set, there exists a variable y such that $F(x_1, \ldots, x_n) \approx y \in \Delta$.

In the light of the equality axioms and properties of Rasiowa–Sikorski sets, both the relation $I_\Delta(P)$ and the operation $I_\Delta(F)$ are well-defined.

If c is a constant symbol and x a variable, then $(\exists x)(x \approx c)$ is a thesis of logic and therefore $(\exists x)(x \approx c) \in \Delta$. It follows that $c \approx y \in \Delta$ for some variable y. Then

$$I_\Delta(c) := \text{ the unique class } [y] \text{ such that } c \approx y \in \Delta.$$

The resulting model for L is marked as

$$\mathbf{A}_\Delta := (A_\Delta; I_\Delta),$$

and called the *model determined by* Δ. □

For each term $t(x_1, \ldots, x_n)$, t^{Δ} is its interpretation in $\boldsymbol{A}_{\Delta}$ defined in the standard way.

Lemma 7.5.10 *For any term* $t(x_1, \ldots, x_n)$ *and any valuation* $[[y_1], \ldots, [y_n]]$ *in* $\boldsymbol{A}_{\Delta}$ *(with the domain* $\{x_1, \ldots, x_n\}$*):*

$$t^{\Delta}([y_1], \ldots, [y_n]) = \textit{the unique class } [y] \textit{ such that } t(x_1/y_1, \ldots, x_n/y_n) \approx y \in \Delta.$$

Proof Induction on the length of terms t in the variables $x_1, \ldots, x_n$. □

We shall consider mappings that assign formulas to formulas. These functions in a special way replace variables in formulas. In the case of open formulas these mappings act as substitutions. If a formula contains quantifiers, the value of such a function is evaluated in a more complex way.

Let n be a positive integer. Let $\{x_1, \ldots, x_n\}$ be an n-element set of variables and $\langle y_1, \ldots, y_n\rangle$ a sequence of variables. The variables of $\langle y_1, \ldots, y_n\rangle$ are ordered by the alphabetic order, though they may repeat.

Let $\sigma(x_1, \ldots, x_n)$ be a formula whose free variables belong to $\{x_1, \ldots, x_n\}$.

$$\sigma(x_1 /\!\!/ y_1, \ldots, x_n /\!\!/ y_n) \qquad (/\!\!/)$$

denotes the formula obtained from σ in the following way. In the first step, for each i $(1 \leqslant i \leqslant n)$ and for each subformula of σ beginning with $(\exists y_i)$ and in which x_i occurs as a free variable, we uniformly replace every occurrence of y_i by the variable z_i with the lowest index (in the alphabetic order) and not occurring in σ, paying heed to the order of $\langle z_1, \ldots, z_n\rangle$. (The order of the variables occurring in $\langle z_1, \ldots, z_n\rangle$ should agree with the order of $\langle y_1, \ldots, y_n\rangle$.) As a result, we get the formula $\sigma'(x_1, \ldots, x_n)$ with the property that the substitution $\langle x_1/y_1, \ldots, x_n/y_n\rangle$ is free in it. In the second step the substitution $\langle x_1/y_1, \ldots, x_n/y_n\rangle$ is applied to σ'. Formula $(/\!\!/)$ is the result of this substitution. Thus $\sigma(x_1 /\!\!/ y_1, \ldots, x_n /\!\!/ y_n)$ is identical with $\sigma'(x_1/y_1, \ldots, x_n/y_n)$.

The implication $\sigma(x_1 /\!\!/ y_1, \ldots, x_n /\!\!/ y_n) \to (\exists x_1) \ldots (\exists x_n)\sigma(x_1, \ldots, x_n)$ is a thesis of logic. This follows from the fact that the formulas $\sigma'(x_1/y_1, \ldots, x_n/y_n) \to (\exists x_1) \ldots (\exists x_n)\, \sigma'(x_1, \ldots, x_n)$ and $(\exists x_1) \ldots .(\exists x_n)\, \sigma'(x_1, \ldots ., x_n) \leftrightarrow (\exists x_1) \ldots .(\exists x_n)\, \sigma(x_1, \ldots, x_n)$ are theses of logic.

The following theorem, which is an extension of a result proved in Rasiowa and Sikorski (1950), ties together the satisfaction relation in the model $\boldsymbol{A}_{\Delta}$ with the above 'double' substitution:

Theorem 7.5.11 *Let* Δ *be a Rasiowa–Sikorski set. For any* n*, for any formula* $\sigma(x_1, \ldots, x_n) \in For(L)$ *and any sequence of individual variables* $\langle y_1, \ldots, y_n\rangle$

$$\boldsymbol{A}_{\Delta} \models \sigma(x_1, \ldots, x_n)[[y_1], \ldots, [y_n]] \;\Leftrightarrow\; \sigma(x_1 /\!\!/ y_1, \ldots, x_n /\!\!/ y_n) \in \Delta. \qquad (*)$$

In particular, if σ *is a sentence, then*

$$\boldsymbol{A}_{\Delta} \models \sigma \;\Leftrightarrow\; \sigma \in \Delta.$$

Note 6 Theorem 7.5.4 is an extension of one of the main results proved in Rasiowa–Sikorski (1950). They proved a special case of (∗), viz.,

$$\boldsymbol{A}_\Delta \models \sigma(x_1, \ldots, x_n)[[x_1], \ldots, [x_n]] \quad \Leftrightarrow \quad \sigma(x_1, \ldots, x_n) \in \Delta,$$

for any formula $\sigma(x_1, \ldots, x_n)$. □

Proof For each mapping $e : Var \to Var$, we define h_e to be the global valuation in $\boldsymbol{A}_\Delta$ which to each variable x assigns the element $[ex]$.

It suffices to show that:

For every mapping $e : Var \to Var$ and every formula $\sigma(x_1, \ldots, x_n)$:

$$\boldsymbol{A}_\Delta \models \sigma(x_1, \ldots, x_n)[h_e] \quad \Leftrightarrow \quad \sigma(x_1 /\!\!/ ex_1, \ldots, x_n /\!\!/ ex_n) \in \Delta. \qquad (**)$$

The proof is by induction on the length of formulas.

If $\sigma(x_1, \ldots, x_n)$ is an atomic formula, then for every $e : Var \to Var$ the formula $\sigma(x_1/ex_1, \ldots, x_n/ex_n)$ is identical with $\sigma(x_1 /\!\!/ ex_1, \ldots, x_n /\!\!/ ex_n)$.

If $\sigma(x_1, \ldots, x_n)$ is of the form $s \approx t$, where $s(x_1, \ldots, x_n)$ and $t(x_1, \ldots, x_n)$ are terms, then for every $e : Var \to Var$

$$\begin{aligned}
&\boldsymbol{A}_\Delta \models \sigma(x_1, \ldots, x_n)[h_e] \quad \Leftrightarrow \\
&s^\Delta[h_e] = t^\Delta[h_e] \quad \Leftrightarrow \text{(Lemma 7.5.10)} \\
&s(x_1/ex_1, \ldots, x_n/ex_n) \approx t(x_1/ex_1, \ldots, x_n/ex_n) \in \Delta \quad \Leftrightarrow \\
&\sigma(x_1/ex_1, \ldots, x_n/ex_n) \in \Delta \quad \Leftrightarrow \\
&\sigma(x_1 /\!\!/ ex_1, \ldots, x_n /\!\!/ ex_n) \in \Delta.
\end{aligned}$$

If $\sigma(x_1, \ldots, x_n)$ is of the form $P(t_1, \ldots, t_m)$, where P is an m-ary predicate and $t_1(x_1, \ldots, x_n), \ldots, t_m(x_1, \ldots, x_n)$ are terms, then

$$\begin{aligned}
&\boldsymbol{A}_\Delta \models \sigma(x_1, \ldots, x_n)[h_e] \quad \Leftrightarrow \\
&P^\Delta(t_1[h_e], \ldots, t_m[h_e]) \text{ in } \boldsymbol{A}_\Delta \quad \Leftrightarrow \\
&\sigma(x_1/ex_1, \ldots, x_n/ex_n) \in \Delta \quad \Leftrightarrow \\
&\sigma(x_1 /\!\!/ ex_1, \ldots, x_n /\!\!/ ex_n) \in \Delta.
\end{aligned}$$

If $\sigma(x_1, \ldots, x_n)$ is of the form $\varphi \wedge \psi$, then

$$\begin{aligned}
&\boldsymbol{A}_\Delta \models \sigma(x_1, \ldots, x_n)[h_e] \quad \Leftrightarrow \\
&\boldsymbol{A}_\Delta \models \varphi(x_1, \ldots, x_n)[h_e] \text{ and } \boldsymbol{A}_\Delta \models \psi(x_1, \ldots, x_n)[h_e] \quad \Leftrightarrow \\
&\varphi(x_1 /\!\!/ ex_1, \ldots, x_n /\!\!/ ex_n) \in \Delta \text{ and } \psi(x_1 /\!\!/ ex_1, \ldots, x_n /\!\!/ ex_n) \in \Delta \quad \Leftrightarrow \\
&\sigma(x_1 /\!\!/ ex_1, \ldots, x_n /\!\!/ ex_n) \in \Delta.
\end{aligned}$$

If $\sigma(x_1, \ldots, x_n)$ is of the shape $\neg\varphi$, then

$$\begin{aligned}&\boldsymbol{A}_\Delta \models \sigma(x_1, \ldots, x_n)[h_e] \quad \Leftrightarrow\\ &\text{It is not the case that } \boldsymbol{A}_\Delta \models \varphi(x_1, \ldots, x_n)[h_e] \quad \Leftrightarrow\\ &\text{It is not the case that } \varphi(x_1 /\!\!/ ex_1, \ldots, x_n /\!\!/ ex_n) \in \Delta \quad \Leftrightarrow\\ &\neg\varphi(x_1 /\!\!/ ex_1, \ldots, x_n /\!\!/ ex_n) \in \Delta \quad \Leftrightarrow\\ &\sigma(x_1 /\!\!/ ex_1, \ldots, x_n /\!\!/ ex_n) \in \Delta.\end{aligned}$$

(We make use here of the maximality of Δ.)

The only non-trivial case is when $\sigma(x_1, \ldots, x_n)$ is of the form $(\exists x)\varphi$, where the variable x is different from $x_1, \ldots, x_n$. We must show that for every function $e : Var \to Var$

$$\boldsymbol{A}_\Delta \models ((\exists x)\varphi)(x_1, \ldots, x_n)[h_e] \quad \Leftrightarrow \quad ((\exists x)\varphi)(x_1 /\!\!/ ex_1, \ldots, x_n /\!\!/ ex_n) \in \Delta. \qquad (***)$$

($\Rightarrow$). We assume that $\boldsymbol{A}_\Delta \models ((\exists x)\varphi)(x_1, \ldots, x_n)[h_e]$. Then:

$$\begin{aligned}&\text{There is } [y] \in A_\Delta \text{such that } \boldsymbol{A}_\Delta \models \varphi[h_e(x/[y])] \quad \Leftrightarrow\\ &\text{There is a variable } y \text{ such that } \boldsymbol{A}_\Delta \models \varphi[h_e(x/[y])] \quad \Leftrightarrow\\ &\text{There is a variable } y \text{ such that } \boldsymbol{A}_\Delta \models \varphi[x/[y], x_1/[ex_1], \ldots, x_n/[ex_n]].\end{aligned}$$

Let us then assume that y is a variable such that

$$\boldsymbol{A}_\Delta \models \varphi(x, x_1, \ldots, x_n)[x/[y], x_1/[ex_1], \ldots, x_n/[ex_n]]. \qquad (1)$$

We define the function $e' : Var \to Var$ as follows: $e'z := ez$ for any variable $z \neq x$ and $e'x := y$.

(1) gives that

$$\boldsymbol{A}_\Delta \models \varphi(x, x_1, \ldots, x_n)[x/[e'x], x_1/[e'x_1], \ldots, x_n/[e'x_n]]. \qquad (2)$$

As φ is shorter than $(\exists x)\varphi$, the induction hypothesis and (1) give that $\varphi(x /\!\!/ e'x, x_1 /\!\!/ e'x_1, \ldots, x_n /\!\!/ e'x_n) \in \Delta$. This, in view of the definition of $\varphi(x /\!\!/ e'x, x_1 /\!\!/ e'x_1, \ldots, x_n /\!\!/ e'x_n)$, means that

$$\varphi'(x/e'x, x_1/e'x_1, \ldots, x_n/e'x_n) \in \Delta. \qquad (3)$$

(see ($/\!\!/$) above and the remarks following it). Since the substitution $x/e'x$, $x_1/e'x_1, \ldots, x_n/e'x_n$ is free for φ', the formula $\varphi'(x/e'x, x_1/e'x_1, \ldots, x_n/e'x_n) \to ((\exists x)\varphi')(x_1/e'x_1, \ldots x_n/e'x_n)$ is a logical thesis. This observation together with (3) give that

$$((\exists x)\varphi')(x_1/e'x_1, \ldots, x_n/e'x_n) \in \Delta.$$

i.e.,

$$((\exists x)\varphi')(x_1/ex_1, \dots, x_n/ex_n) \in \Delta.$$

As $((\exists x)\varphi)(x_1 /\!/ ex_1, \dots, x_n /\!/ ex_n)$ is identical with $((\exists x)\varphi')(x_1/ex_1, \dots, x_n/ex_n)$, we get that

$$((\exists x)\varphi)(x_1 /\!/ ex_1, \dots, x_n /\!/ ex_n) \in \Delta.$$

($\Leftarrow$). We assume that $(\exists x)\varphi(x_1 /\!/ ex_1, \dots, x_n /\!/ ex_n) \in \Delta$. We mark the formula $(\exists x)\varphi(x_1 /\!/ ex_1, \dots, x_n /\!/ ex_n)$ as $\psi(y_1, \dots, y_m)$, where $y_1, \dots, y_m$ are all variables occurring in the sequence $\langle ex_1, \dots, ex_n\rangle$. (Here $m \leqslant n$, because some variables occurring in the sequence $\langle ex_1, \dots, ex_n\rangle$ may repeat.) Thus $((\exists x)\psi)(y_1, \dots, y_m) \in \Delta$. As Δ is a Rasiowa–Sikorski set, $\psi(x /\!/ y, y_1, \dots, y_m) \in \Delta$ for some variable y. (It may happen that y is among the variables $y_1, \dots, y_m$.) But $\psi(x /\!/ y, y_1, \dots, y_m)$ is logically equivalent with $\varphi(x /\!/ y, x_1 /\!/ ex_1, \dots, x_n /\!/ ex_n)$. Hence

$$\varphi(x /\!/ y, x_1 /\!/ ex_1, \dots, x_n /\!/ ex_n) \in \Delta. \tag{4}$$

Let $e' : Var \to Var$ be defined as above, that is, $e'z := ez$ for any variable $z \neq x$ and $e'x := y$. In view of (4) we have that

$$\varphi(x /\!/ e'x, x_1 /\!/ e'x_1, \dots, x_n /\!/ e'x_n) \in \Delta.$$

But the induction hypothesis gives that

$$\boldsymbol{A}_\Delta \models \varphi(x, x_1, \dots, x_n)[x/[e'x], x_1/[e'x_1], \dots, x_n/[e'x_n]].$$

Hence

$$\boldsymbol{A}_\Delta \models ((\exists x)\varphi)(x_1, \dots, x_n)[x_1/[e'x_1], \dots, x_n/[e'x_n]],$$

i.e.,

$$\boldsymbol{A}_\Delta \models ((\exists x)\varphi)(x_1, \dots, x_n)[x_1/[ex_1], \dots, x_n/[ex_n]].$$

So $(* * *)$ holds. This concludes the proof of $(**)$. The theorem has been proved. □

Let L be a countable language. *Every countable* model for L is determined by some Rasiowa–Sikorski set:

Theorem 7.5.12 *Let L be a countable language. Let $\boldsymbol{A}$ be any countable model for L. There exists a Rasiowa–Sikorski set $\Delta \subseteq For(L)$ such that $\boldsymbol{A}$ is isomorphic with the model $\boldsymbol{A}_\Delta$.*

Note 7 The above theorem carries over the languages in which the set Var of individual variables is uncountable and for models $\boldsymbol{A}$ such that there exists a surjection from Var onto A. □

Proof The set Var of individual variables is countably infinite. Let h be a surjection from Var onto the universe A of the model $\boldsymbol{A}$. h is a global valuation in $\boldsymbol{A}$. We define the set $\Delta(h)$ as in Lemma 7.5.6, that is,

$$\Delta(h) := \{\sigma \in For(L) : \boldsymbol{A} \models \sigma[h]\}.$$

$\Delta(h)$ is a Rasiowa–Sikorski set.

Let $\boldsymbol{A}_{\Delta(h)}$ be the model determined by $\Delta(h)$.

h recursively extends to a mapping from the set of terms $Te(L)$ onto A. (This extension is also marked by h.) Thus $h(t) = t^{\boldsymbol{A}}(h)$ for every term t. h is therefore an epimorphism from the term algebra $\boldsymbol{Te}(L)$ onto the algebra of $\boldsymbol{A}$.

We will show that the model $\boldsymbol{A}_{\Delta(h)}$ is isomorphic with $\boldsymbol{A}$.

We define the mapping $f : A_{\Delta(h)} \to A$ as follows:

$$f([x]) := h(x) \quad \text{for every variable } x.$$

f is a well-defined bijection. It remains to show that f preserves relations, operations, and constants.

Let F be an n-ary function symbol, let $\langle x_1, \ldots, x_n\rangle$ be an n-tuple of variables and let y be a variable such that $F(x_1, \ldots, x_n) \approx y \in \Delta(h)$. (In virtue of the fact that $\Delta(h)$ is a Rasiowa–Sikorski set, such a variable y exists.) It follows that $F^{\Delta(h)}([x_1], \ldots, [x_n]) = [y]$ in $\boldsymbol{A}_{\Delta(h)}$, and therefore $f(F^{\Delta(h)}([x_1], \ldots, [x_n])) = f([y]) = h(y)$. On the other hand, the definition of h gives that $F^{\boldsymbol{A}}(f([x_1]), \ldots, f([x_n])) = F^{\boldsymbol{A}}(h(x_1), \ldots, h(x_n)) = h(F(x_1, \ldots, x_n)) = h(y)$. Hence

$$f(F^{\Delta(h)}([x_1], \ldots, [x_n])) = F^{\boldsymbol{A}}(f([x_1]), \ldots, f([x_n])).$$

In a similar way one verifies that f preserves relations and constants.

This concludes the proof of the theorem. $\square$

Model Sets and the Extended Completeness Theorem

According to Theorem 7.5.12, the family of Rasiowa–Sikorski sets determine, up to isomorphism, *all* countable models for L. Although in the light of Theorems 7.5.7–7.5.8, model sets form a narrower family than Rasiowa–Sikorski sets, the class of models determined by model sets is adequate for first-order logic, that is, this class suffices for proving the Extended Completeness Theorem for any countable language L. (This issue is briefly discussed below.) Thus the family of model sets allows for the entrenchment of principal properties of various metalogical properties of elementary theories.

We shall briefly outline the proof of the Extended Completeness Theorem (Theorem 7.4.1) that employs model sets.

The relation $\vdash$ is included in the relation $\models$. To prove the opposite inclusion, it suffices to show that for any set of formulas Σ and any formula σ of L, if $\Sigma \nvdash \sigma$, then $\Sigma \not\models \sigma$. Let us then assume that $\Sigma \nvdash \sigma$. In view of Theorem 7.3.2.(iii), the set $\Sigma \cup \{\neg\sigma\}$ is consistent.

We first consider the case when the set $Var \setminus FreeVar(\Sigma)$ is infinite. In virtue of Theorem 7.5.4, $\Sigma \cup \{\neg\sigma\}$ is contained in a model set Δ. The set Δ, being a Rasiowa–Sikorski set, determines the model $\boldsymbol{A}_\Delta$. Let h be the global valuation in $\boldsymbol{A}_\Delta$ defined by the formula: $h(x) := [x]$ for every individual variable x. According to Theorem 7.5.11 we have that $\boldsymbol{A}_\Delta \models \tau[h]$ for all formulas $\tau \in \Delta$. In particular, $\boldsymbol{A}_\Delta \models \tau[h]$ for all formulas $\tau \in \Sigma$ and $\boldsymbol{A}_\Delta \models \neg\sigma[h]$. It follows that σ is not a logical consequence of Σ, that is, $\Sigma \not\models \sigma$.

If $Var \setminus FreeVar(\Sigma)$ is finite, we select an infinite subset $V \subseteq Var$ such that $Var \setminus V$ is also infinite. We define $For_V(L)$ to be the set of formulas of L containing only the variables of V as bound and free variables. Let e be a bijection from Var onto V. We consider the set of formulas $e(\Sigma) \cup \{e(\sigma)\}$, being the e-image of $\Sigma \cup \{\sigma\}$ and hence a subset of $For_V(L)$. The set of variables $Var \setminus FreeVar(e(\Sigma))$ is infinite. Arguing as in the first case, we obtain that $e(\Sigma) \not\models e(\sigma)$ (in the set of formulas $For_V(L)$). Applying the inverse isomorphism $e^{-1} : \boldsymbol{For}_V(L) \to \boldsymbol{For}(L)$, we obtain that $\Sigma \not\models \sigma$ (in $For(L)$). $\square$

The above observations give rise to open problems. Suppose we are given two model sets (two Rasiowa–Sikorski, respectively) Δ and Δ'. Find conditions imposed on Δ and Δ' equivalent to the fact that the models $\boldsymbol{A}_\Delta$ and $\boldsymbol{A}_{\Delta'}$ are isomorphic. Find an analogous characterization of the elementary equivalence of $\boldsymbol{A}_\Delta$ and $\boldsymbol{A}_{\Delta'}$.

7.6 Forcing and Rasiowa–Sikorski Sets

Lemma 7.5.9 states that every Rasiowa–Sikorski set Δ for L is determined by the set of atomic formulas that belong to Δ. For a given Rasiowa–Sikorski set Δ we define $\chi_{\Delta \cap AFor(L)} : AFor(L) \to \{0, 1\}$ to be the characteristic function of $\Delta \cap AFor(L)$:

$$\chi_{\Delta \cap AFor(L)}(\varphi) := \begin{cases} 1 & \text{if} \quad \varphi \in \Delta \cap AFor(L), \\ 0 & \text{if} \quad \varphi \notin \Delta \cap AFor(L). \end{cases}$$

for any atomic formula φ.

Each Rasiowa–Sikorski set Δ is therefore fully determined by a zero-one function whose domain is $AFor(L)$, viz., by $\chi_{\Delta \cap AFor(L)}$. But it is *not* true that every zero-one function defined on $AFor(L)$ corresponds to some Rasiowa–Sikorski set. For let x, y, z be three variables. If h_0 is a zero-one function on $AFor(L)$ such that $h_0(x \approx y) = h_0(y \approx z) = 1$ and $h_0(x \approx z) = 0$, then h_0 is not a characteristic function $\chi_{\Delta \cap AFor(L)}$ of a Rasiowa–Sikorski set Δ, because each Δ contains the transitivity law for $\approx$.

This section is concerned with the problem of whether a zero-one function whose domain is $AFor(L)$ corresponds to Rasiowa–Sikorski set. More exactly, we shall outline a general method, which, under some input assumptions, defines zero-one functions h_0 on $AFor(L)$ such that $h_0 = \chi_{\Delta \cap AFor(L)}$ for some Rasiowa–Sikorski set Δ. This is the well-known *forcing method*, invented by Paul Cohen, which has been successfully applied to set theory for years. Here the forcing method is presented from the more general perspective of arbitrary first-order languages, and not only of the language of set theory. On the other hand, the techniques of Boolean valuations we shall exhibit are not so advanced as compared with set theory (they are limited here to a narrower class of complete Boolean algebras) but nevertheless they show the essence of the forcing method. The emphasis is therefore put here on the scope of the method—it is applied to *arbitrary* first-order languages, and not only to the language of set theory.

Before entering the topic, let us consider the situation when L is a purely relational language—it does not contain function and constant symbols. A simple and general way of building countable models for L consists in assigning to each m-ary predicate P a set P_A of m-tuples of variables, $P_A \subseteq Var^m$. The set Var is the universe of the resulting model $\boldsymbol{A}$.

We define $h_0 : AFor(L) \to \{0, 1\}$ as follows. For each m-ary predicate and each m-tuple $\langle x_1, \ldots, x_m \rangle$ of individual variables we put:

$$h_0(P(x_1, \ldots, x_m)) := \begin{cases} 1 & \text{if} \quad \langle x_1, \ldots, x_m \rangle \in P_A, \\ 0 & \text{otherwise.} \end{cases}$$

Moreover, for any variables x, y:

$$h_0(x \approx y) := \begin{cases} 1 & \text{if} \quad x \text{ is identical with } y, \\ 0 & \text{otherwise.} \end{cases}$$

Then there exists a Rasiowa–Sikorski set Δ such that $h_0 = \chi_{AFor(L) \cap \Delta}$. Indeed, let $h : Var \to Var$ be the identity map. h is a therefore a global valuation in $\boldsymbol{A}$. We define:

$$\Delta(h) := \{\sigma \in For(L) : \boldsymbol{A} \models \sigma[h]\}.$$

$\Delta(h)$ is a Rasiowa–Sikorski set and the model $\boldsymbol{A}_{\Delta(h)}$ is isomorphic with $\boldsymbol{A}$. It is easy to see that h_0 coincides with $\chi_{AFor(L) \cap \Delta(h)}$. Thus h_0 corresponds to a Rasiowa–Sikorski set.

If $L = \{\in\}$ is the language of set theory, then every countable model for L is isomorphic with some model $\boldsymbol{A}$ defined as above (see the remarks following Theorem 7.5.12). But of course $\boldsymbol{A}$ need not be a model of ZF.

Regular Open Sets

Let $(X, \boldsymbol{T})$ be a topological space. $\mathrm{Int}(\,\cdot\,)$ and $\mathrm{Cl}(\,\cdot\,)$ stand for the interior and closure operations determined by $\boldsymbol{T}$. A set $A \subseteq X$ is called *regular open* if $A = \mathrm{Int}(\mathrm{Cl}(A))$. The sets X and $\varnothing$ are regular open. It is well-known that the family $RO(X)$ of regular open subsets of X endowed with the operations:

$$A + B := \mathrm{Int}(\mathrm{Cl}(A \cup B)),$$
$$A \cdot B := A \cap B,$$
$$A^* := \mathrm{Int}(X \setminus A),$$

where $A, B \in RO(X)$, and two constants $\mathbf{0} := \varnothing$ and $\mathbf{1} := X$, forms a Boolean algebra, denoted by

$$\boldsymbol{RO}(X) = (RO(X), +, \cdot, {}^*, \mathbf{0}, \mathbf{1}).$$

The algebra $\boldsymbol{RO}(X)$ is complete and

$$\sup\{A_i : i \in I\} = \mathrm{Int}\left(\mathrm{Cl}\left(\bigcup_{i \in I} A_i\right)\right)$$

$$\inf\{A_i : i \in I\} = \mathrm{Int}\left(\bigcap_{i \in I} A_i\right)$$

for any family $\{A_i : i \in I\}$ of regular open subsets of X.

Refined Posets

Let $\boldsymbol{P} = (P, \leqslant)$ be a poset. The family of sets $\downarrow p := \{q \in P : q \leqslant p\}$, $p \in P$, forms a base of a topology $\boldsymbol{T}$ on P. Thus the open sets of $\boldsymbol{T}$ are set-theoretic unions of some sets of the form $\downarrow p$. Int and Cl denote the interior and the closure operations corresponding to $\boldsymbol{T}$, respectively.

Lemma 7.6.1 *For any set $A \subseteq P$ and $p \in P$. Then:*

$$p \in \mathrm{Cl}(A) \quad \Leftrightarrow \quad \downarrow p \cap A \neq \varnothing, \tag{Cl}$$
$$p \in \mathrm{Int}(A) \quad \Leftrightarrow \quad \downarrow p \subseteq A, \tag{Int}$$
$$p \in \mathrm{Int}(\mathrm{Cl}(A)) \quad \Leftrightarrow \quad (\forall q)(q \leqslant p \Rightarrow A \cap \downarrow q \neq \varnothing). \tag{Int(Cl)}$$

Proof Use the fact that in any topological space X and for any $A \subseteq X, x \in \mathrm{Cl}(A) \Leftrightarrow A \cap G \neq \varnothing$ for every base set G such that $x \in G$. □

Definition 8 A poset $\boldsymbol{P} = (P, \leqslant)$ is called *refined* (Bell 2005, p. 55) or *polar*, if it obeys the condition: for every pair $p, q \in P$ such that $p \nleqslant q$ there exists $r \in P$ such that $r \leqslant p$ and there does not exist $s \in P$ being a lower bound of r and q. □

The reverse implication trivially holds, that is, if there exists $r \in P$ such that $r \leqslant p$ and there does not exist $s \in P$ being a lower bound of r and q, then $p \nleqslant q$. It follows that

$\boldsymbol{P} = (P, \leqslant)$ *is refined if and only if* (1)
$(\forall p, q \in P)(p \leqslant q \Leftrightarrow (\forall r \in P)(r \leqslant p \Rightarrow$ *a lower bound of* r *and* q *exists*$))$.

For example, the poset consisting of non-zero elements of a Boolean algebra (with the order inherited from the algebra) is refined.

The following lemma is well known. Its proof is straightforward.

Lemma 7.6.2 *A poset $\boldsymbol{P} = (P, \leqslant)$ is refined if and only if the sets $\downarrow p$, $p \in P$, are regular open in $\boldsymbol{T}$.* □

If $\boldsymbol{P}$ is a refined set, then the complete Boolean algebra of regular open sets of the topological space $(P, \boldsymbol{T})$ is denoted by

$$\boldsymbol{RO}(\boldsymbol{P}).$$

(The family of sets $\downarrow p$, $p \in P$, is a base of $\boldsymbol{T}$.) The universe $RO(\boldsymbol{P})$ of this algebra consists of all regular open sets. $RO(\boldsymbol{P})$ is endowed with the Boolean operations defined as above. The set P is the unit element $\mathbf{1}$ of $\boldsymbol{RO}(\boldsymbol{P})$ and the empty set is the zero element $\mathbf{0}$.

Boolean Methods and Rasiowa–Sikorski Sets

The question is: how to construct Rasiowa–Sikorski sets for L? The procedure presented in the previous paragraph (Theorem 7.5.4) consisting in recursive extending consistent sets to model sets, though useful, does not always yield models for L with required properties e.g., models of a definite elementary theory. We need other tools. Some techniques will be outlined here which seem to be more promising and which refer to well-known properties of global Boolean valuations of formulas. The heart of the matter consists in defining an appropriate interrelationship of elements of Boolean algebras with atomic formulas of L. The so defined Boolean valuations of atomic formulas are recursively extended onto arbitrary formulas. The next step requires an able application of Rasiowa–Sikorski Lemma to define an ultrafilter in a given Boolean algebra and then a Rasiowa–Sikorski set for L.

Boolean Valuations

Let $\boldsymbol{B} = (B, +, \cdot, {}^*, \mathbf{0}, \mathbf{1})$ be a complete Boolean algebra and L a countable language. Moreover, let $[\,\cdot\,]$ be a mapping defined on the set of atomic formulas $AFor(L)$ of L with values in $\boldsymbol{B}$. Thus, for every atomic σ, $[\sigma]$ is an element of $\boldsymbol{B}$. $[\,\cdot\,]$ is then extended onto $For(L)$ in the standard way, that is:

$$[\varphi \wedge \psi] := [\varphi] \cdot [\psi],$$
$$[\neg\varphi] := [\varphi]^* \quad (= \text{the complement of } [\varphi] \text{ in } \boldsymbol{B}),$$
$$[(\exists x)\varphi] := \sup\{[\varphi(x /\!\!/ y)] : y \in Var\}.$$

(This is a recursive definition with respect to the number of occurrences of the connectives $\wedge$ and $\neg$ and of the existential quantifier in formulas.)

The set $\{[\sigma] : \sigma \in For(L)\}$ forms a countable subalgebra of $\boldsymbol{B}$, denoted by $\boldsymbol{B}(L)$.

Any mapping $[\,\cdot\,]$ satisfying the above clauses is called a *Boolean valuation* of the set $For(L)$ in the algebra $\boldsymbol{B}$. $[\,\cdot\,]$ is at the same time a Boolean valuation in the algebra $\boldsymbol{B}(L)$. In virtue of Rasiowa–Sikorski Lemma, there exists an ultrafilter U

in $\boldsymbol{B}$ preserving the supremums $[(\exists x)\varphi] = \sup\{[\varphi(x /\!\!/ y)] : y \in Var\}$, $\varphi \in For(L)$. I.e., U satisfies:

$$[(\exists x)\varphi] \in U \quad \Leftrightarrow \quad [\varphi(x /\!\!/ y)] \in U \text{ for some variable } y \in Var,$$

for every formula of the form $(\exists x)\varphi$.

The set of formulas

$$\Delta_U := \{\sigma \in For(L) : [\sigma] \in U\}$$

satisfies conditions (1)–(3) of Theorem 7.5.5. But Δ_U need not be a Rasiowa–Sikorski set, because it may be inconsistent. Here is a simple example. Let L be the empty language. All atomic formulas take the form $x \approx y$, where x and y are variables. Let $\mathbf{2}$ be a two-element Boolean algebra with the universe $\{\mathbf{0}, \mathbf{1}\}$. Let $[\,\cdot\,]$ be a Boolean valuation in $\mathbf{2}$ such that $[\sigma] = \mathbf{0}$ for every atomic formula σ. In particular, $[x \approx x] = \mathbf{0}$ for every variable x. $[\,\cdot\,]$ is then defined on compound formulas in accordance with the above rules. $U = \{\mathbf{1}\}$ is the only ultrafilter of $\mathbf{2}$ and it preserves the suprema $[(\exists x)\varphi] = \sup\{[\varphi(x /\!\!/ y)] : y \in Var\}$, $\varphi \in For(L)$.

$$\Delta_U := \{\sigma \in For(L) : [\sigma] = \mathbf{1}\}$$

is a proper subset of $For(L)$ and it satisfies conditions (1)–(3) of Theorem 7.5.5; Δ_U is however inconsistent. Indeed, $\neg(x \approx x) \in \Delta_U$ for any x. On the other hand, every formula $x \approx x$ is provable on the grounds of Δ_U, because it is a logical axiom.

The above definition leaves the discretion of making a choice of $\boldsymbol{B}$ and defining the values of $[\,\cdot\,]$ on the set of atomic formulas. There is therefore the need of defining such an interrelationship of Boolean valuations in $\boldsymbol{B}$ with atomic formulas that preserves the logical structure adopted in the language: Boolean valuations must also validate equality axioms (they validate the adopted rules of inference and the remaining logical axioms). To this end we shall briefly outline the method of forcing for arbitrary first-order languages. This method enables us to produce sets of formulas that satisfy conditions (1)–(3) of Theorem 7.5.5. Moreover the resulting sets are consistent; they are therefore Rasiowa–Sikorski sets for L. Since the consistency of the defined sets is a crucial (and critical) property, the forcing constructions will be carried out for refined sets whose elements are certain non-empty sets of equivalence classes of atomic formulas in the Lindenbaum–Tarski algebra $\boldsymbol{B}(L)$ of L. Such an assumption facilitates the presentation of forcing constructions. In the case on consistent elementary theories T, one suitably modifies the presented definitions and accommodates them to refined sets derived from the Lindenbaum–Tarski algebra $\boldsymbol{B}_T(L)$ of the theory T. (Yet another option is to take refined posets consisting of sets of atomic formulas (ordered by inclusion) in languages without function and constant symbols and appropriately define the pertinent forcing relations. This option, though theoretically important, is logically more intricate and is not discussed in this paper.)

Given a formula of L we define: $[\sigma] := \{\sigma' \in For(L) : \ \vdash \sigma \leftrightarrow \sigma'\}$. The universe of the Lindenbaum–Tarski algebra $\boldsymbol{B}(L)$ of L consists of the abstraction classes $[\sigma]$ with σ ranging over $For(L)$.

Let $\boldsymbol{P} = (P, \subseteq)$ be a *non-empty* refined family (with respect to inclusion) consisting of *non-empty* subsets of the set $\{[\sigma] : \sigma \in AFor(L)\}$. The elements of $\boldsymbol{P}$ are called *conditions*. Every condition $p \in P$ is therefore a non-empty subset of the set $\{[\sigma] : \sigma \in AFor(L)\}$.

$\boldsymbol{T}$ is the topology on P whose basis is formed by the regular open sets $\downarrow p, p \in P$.[1] $\boldsymbol{RO}(\boldsymbol{P})$ is the Boolean algebra of regular open sets of the topological space $(P, \boldsymbol{T})$. The family P is the unit $\mathbf{1}^{\boldsymbol{P}}$ of $\boldsymbol{RO}(\boldsymbol{P})$ and the empty set $\varnothing$ is the zero $\mathbf{0}^{\boldsymbol{P}}$ of $\boldsymbol{RO}(\boldsymbol{P})$. Since $P \neq \varnothing$, the algebra $\boldsymbol{RO}(\boldsymbol{P})$ is non-degenerate, i.e., $\mathbf{0}^{\boldsymbol{P}} \neq \mathbf{1}^{\boldsymbol{P}}$.

Let $[\,\cdot\,]^{\boldsymbol{P}}$ be the function on $AFor(L)$ (with values in $\boldsymbol{RO}(\boldsymbol{P})$) defined as follows: for every *atomic* formula σ we put:

$$[\sigma]^{\boldsymbol{P}} := \sup\{\downarrow p : p \in P \text{ and } [\sigma] \in p\}$$

in the complete Boolean algebra $\boldsymbol{RO}(\boldsymbol{P})$. Since $\boldsymbol{P}$ is refined, $\downarrow p$ is in $RO(\boldsymbol{P})$ for all $p \in P$. Each $[\sigma]^{\boldsymbol{P}}$ is thus a regular open subset of P; more exactly,

$$[\sigma]^{\boldsymbol{P}} = \operatorname{Int}\left(\operatorname{Cl}\left(\bigcup\{\downarrow p : p \in P \text{ and } [\sigma] \in p\}\right)\right).$$

The function $[\,\cdot\,]^{\boldsymbol{P}}$ is extended onto $For(L)$ according to the above general definition of Boolean valuations. Thus

$$\begin{aligned}
&[\varphi \wedge \psi]^{\boldsymbol{P}} := [\varphi]^{\boldsymbol{P}} \cdot [\psi]^{\boldsymbol{P}},\\
&[\neg\varphi]^{\boldsymbol{P}} := ([\varphi]^{\boldsymbol{P}})^* \quad (= \text{the complement of } [\varphi]^{\boldsymbol{P}} in\ \boldsymbol{RO}(\boldsymbol{P})),\\
&[(\exists x)\varphi]^{\boldsymbol{P}} := \sup\{[\varphi(x /\!/ y)]^{\boldsymbol{P}} : y \in Var\}.
\end{aligned}$$

for any formulas φ, ψ and any individual variable x.

The set

$$B(\boldsymbol{P}) := \{[\sigma]^{\boldsymbol{P}} : \sigma \in For(L)\}$$

forms a countable subalgebra of $\boldsymbol{RO}(\boldsymbol{P})$, denoted by $\boldsymbol{B}(\boldsymbol{P})$.

Lemma 7.6.3 *The mapping $f : \boldsymbol{B}(L) \to \boldsymbol{RO}(\boldsymbol{P})$ defined by*

$$f([\varphi]) := [\varphi]^{\boldsymbol{P}}, \qquad \varphi \in For(L),$$

is a homomorphism from the Lindenbaum–Tarski algebra $\boldsymbol{B}(L)$ to $\boldsymbol{RO}(\boldsymbol{P})$. Moreover f preserves the suprema $[(\exists x)\varphi] = \sup\{[\varphi(x /\!/ y)] : y \in Var\}$ from the algebra $\boldsymbol{B}(L)$, that is,

[1] $\boldsymbol{T}$ is an example of an *Alexandroff space*. In Alexandroff spaces arbitrary intersections of open sets are open and this property characterizes the topologies obtained from posets—P.S. Alexandroff, *Diskrete Räume*, Matem. Sbornik **2** (1937), 501–518.

$$f([(\exists x)\varphi]) = \sup\{f([\varphi(x/\!\!/y)]) : y \in Var\}$$

for any formula φ *and any individual variable* x.

(f is also an epimorphism from $\boldsymbol{B}(L)$ onto $\boldsymbol{B}(\boldsymbol{P})$ preserving the above suprema.)

Proof Immediate. □

The fact that f is well-defined is noted down as a separate corollary:

Corollary 4 *If* $\vdash \varphi \leftrightarrow \psi$, *then* $[\varphi]^{\boldsymbol{P}} = [\psi]^{\boldsymbol{P}}$ *in the algebra* $\boldsymbol{RO}(\boldsymbol{P})$. □

We therefore obtain:

Corollary 5 *If* φ *is a thesis of logic, i.e.,* $\vdash \varphi$, *then* $[\varphi]^{\boldsymbol{P}} = \mathbf{1}^{\boldsymbol{P}}$ *in* $\boldsymbol{RO}(\boldsymbol{P})$.

Proof φ is a thesis of logic if and only if $\vdash \varphi \leftrightarrow \psi$, where ψ is the formula $(x \approx y) \vee \neg(x \approx y)$ and x and y are arbitrary variables. Since $[\psi]$ is the unit of $\boldsymbol{B}(L)$, it follows that $[\psi]^{\boldsymbol{P}} = \mathbf{1}^{\boldsymbol{P}}$. According to Corollary 4, $\vdash \varphi \leftrightarrow \psi$ implies that $[\varphi]^{\boldsymbol{P}} = [\psi]^{\boldsymbol{P}}$. Hence $[\varphi]^{\boldsymbol{P}} = \mathbf{1}^{\boldsymbol{P}}$. □

As the family P is non-empty, the algebra $\boldsymbol{RO}(\boldsymbol{P})$ has at least two elements. In virtue of Rasiowa–Sikorski Lemma, there exists an ultrafilter U in $\boldsymbol{RO}(\boldsymbol{P})$ preserving the suprema $[(\exists x)\varphi]^{\boldsymbol{P}} := \sup\{[\varphi(x/\!\!/y)]^{\boldsymbol{P}} : y \in Var\}$. U is called a *generic* ultrafilter. We then define:

$$\Delta_U := \{\sigma \in For(L) : [\sigma]^{\boldsymbol{P}} \in U\}.$$

Theorem 7.6.4 Δ_U *is a Rasiowa–Sikorski set.*

Proof It follows from the definition that Δ_U satisfies conditions (1)–(3) of Theorem 7.5.5. We will show that Δ_U is consistent.

In the light of Corollary 5 we have that $\boldsymbol{CL} \subseteq \Delta_U$. Moreover Δ_U is closed with respect to the detachment rule. These facts imply that Δ_U is a logically closed set of formulas. Since $\mathbf{0}^{\boldsymbol{P}}$ does not belong to U, it follows that the formulas of the form $\sigma \wedge \neg\sigma$ do not belong to Δ_U. Thus Δ_U is a logically closed set of formulas being a proper subset of $For(L)$, that is, Δ_U is consistent. □

Forcing

We return to the Lindenbaum–Tarski algebra $\boldsymbol{B}(L)$ of L. Let $\boldsymbol{P} = (P, \subseteq)$ be a refined family of subsets of $\{[\sigma] : \sigma \in AFor(L)\}$. As above, the elements of P are called *conditions*.

Let σ be an *atomic* formula and p a condition. We shall say that p *forces* σ, if for every condition q such that $q \subseteq p$ there exist conditions r and s such that $r \subseteq q$, $s \supseteq r$ and $[\sigma] \in s$. We then write $p \Vdash \sigma$. Thus

$$p \Vdash \sigma \quad \Leftrightarrow \quad (\forall q \subseteq p)(\exists r \subseteq q)(\exists s \supseteq r)([\sigma] \in s). \tag{1}$$

(The standard notation for quantifiers bounded by formulas is used here. Similar shorthands will appear below. This notation is applied in (Bell 2005, pp. 58–59).

It follows from (1) that for any atomic σ, if $[\sigma] \in p$, then $p \Vdash \sigma$; the converse need not hold—see Lemma 7.6.5).

The above definition of $\Vdash$ is recursively extended onto all formulas of L:

$$p \Vdash \neg\varphi \quad \Leftrightarrow \quad (\forall q)(q \subseteq p \Rightarrow \text{ it is not the case that } q \Vdash \varphi). \tag{2}$$
$$p \Vdash \varphi \wedge \psi \Leftrightarrow \quad p \Vdash \varphi \text{ and } p \Vdash \psi. \tag{3}$$
$$p \Vdash (\exists x)\varphi \Leftrightarrow \quad (\forall q \subseteq p)(\exists r \subseteq q)(\exists y \in Var)\, r \Vdash \varphi(x /\!/ y). \tag{4}$$

It follows from the above conditions and the assumption that $\boldsymbol{P}$ is refined that

$$p \Vdash \varphi \vee \psi \quad \Leftrightarrow \quad (\forall q \subseteq p)(\exists r \subseteq q)(r \Vdash \varphi \text{ or } r \Vdash \psi). \tag{5}$$
$$p \Vdash \varphi \rightarrow \psi \quad \Leftrightarrow \quad (\forall q \subseteq p)(q \Vdash \varphi \Rightarrow q \Vdash \psi). \tag{6}$$
$$p \Vdash \varphi \leftrightarrow \psi \quad \Leftrightarrow \quad (\forall q \subseteq p)(q \Vdash \varphi \Leftrightarrow q \Vdash \psi). \tag{7}$$
$$p \Vdash (\forall x)\varphi \quad \Leftrightarrow \quad (\forall y \in Var)\; p \Vdash \varphi(x /\!/ y). \tag{8}$$

Lemma 7.6.5 *For any formula φ and any $p, q \in P$,*

$$p \Vdash \varphi \;\textit{ and }\; q \subseteq p \;\textit{ imply }\; q \Vdash \varphi.$$

Proof Induction on the length of formulas. It follows from (1) that the lemma holds for atomic formulas. We shall check the case of negated formulas. We assume that $p \Vdash \neg\varphi$ and $q \subseteq p$. Equation (2) gives that for every $r \subseteq p$ it is not the case that $r \Vdash \varphi$. Consequently, for every $r \subseteq q$, it is not the case that $r \Vdash \varphi$. Hence $q \Vdash \neg\varphi$. The remaining cases are also easily verified. □

According to the standard interpretation of $\Vdash$, elements of P are viewed as states of information. For $p, q \in P$, $p \subseteq q$ means that the state p is a refinement of the state q. $p \Vdash \sigma$ is viewed as the assertion that in p one is in "the definite possession of the 'fact' σ, or have been 'forced' to accept σ as true" (Bell 2005, p. 59). The fact that p and q do not possess a lower bound is interpreted that p and q are inconsistent.

We pass to the algebra of regular open sets $\boldsymbol{RO}(\boldsymbol{P})$ and its subalgebra $\boldsymbol{B}(\boldsymbol{P})$ defined as in the previous section.

Theorem 7.6.6 below shows the relationship between the above Boolean valuation $[\,\cdot\,]^{\boldsymbol{P}}$ in $\boldsymbol{RO}(\boldsymbol{P})$ and the forcing relation $\Vdash$. (It seems that that this theorem and the results following it have not been considered in the literature on first-order logic.)

Theorem 7.6.6 *For any formula σ and any $p \in P : \downarrow p \subseteq [\sigma]^{\boldsymbol{P}} \;\Leftrightarrow\; p \Vdash \sigma$.*

Proof

Claim 1 *Let σ be an atomic formula. Then for any $p \in P$, $\downarrow p \subseteq [\sigma]^{\boldsymbol{P}} \;\Leftrightarrow\; p \Vdash \sigma$.*

Proof (of the claim)

$$\begin{aligned}
&\downarrow p \subseteq [\sigma]^{\boldsymbol{P}} \quad \Leftrightarrow \\
&\downarrow p \subseteq \mathrm{Int}\left(\mathrm{Cl}\left(\bigcup\{\downarrow u : u \in P \text{ and } [\sigma] \in u\}\right)\right) \quad \Leftrightarrow \\
&(\forall q \subseteq p)\; q \in \mathrm{Int}\left(\mathrm{Cl}\left(\bigcup\{\downarrow u : u \in P \text{ and } [\sigma] \in u\}\right)\right) \quad \Leftrightarrow \text{ (by Lemma 7.6.1)} \\
&(\forall q \subseteq p)(\forall r \subseteq q)\left(\downarrow r \cap \bigcup\{\downarrow u : u \in P \text{ and } [\sigma] \in u\} \neq \varnothing\right) \quad \Leftrightarrow \\
&(\forall q \subseteq p)\left(\downarrow q \cap \bigcup\{\downarrow u : u \in P \text{ and } [\sigma] \in u\} \neq \varnothing\right) \quad \Leftrightarrow \\
&(\forall q \subseteq p)(\exists r \subseteq q)\; r \in \bigcup\{\downarrow u : u \in P \text{ and } [\sigma] \in u\} \quad \Leftrightarrow \\
&(\forall q \subseteq p)(\exists r \subseteq q)(\exists u \supseteq r)([\sigma] \in u) \quad \Leftrightarrow \quad \text{(by (1))} \\
&p \Vdash \sigma.
\end{aligned}$$

Claim 2 *For any formulas φ, ψ and any p, $\downarrow p \subseteq [\varphi \wedge \psi]^{\boldsymbol{P}} \;\Leftrightarrow\; p \Vdash \varphi \wedge \psi$.*

Proof (of the claim)

$$\begin{aligned}
&\downarrow p \subseteq [\varphi \wedge \psi]^{\boldsymbol{P}} \quad \Leftrightarrow \\
&\downarrow p \subseteq [\varphi]^{\boldsymbol{P}} \cdot [\psi]^{\boldsymbol{P}} \quad \Leftrightarrow \\
&\downarrow p \subseteq [\varphi]^{\boldsymbol{P}} \text{ and } \downarrow p \subseteq [\psi]^{\boldsymbol{P}} \quad \Leftrightarrow \\
&p \Vdash \varphi \text{ and } p \Vdash \psi \quad \Leftrightarrow \\
&p \Vdash \varphi \wedge \psi. \quad \square
\end{aligned}$$

Claim 3 *For any formula φ and any p, $\downarrow p \subseteq [\neg\varphi]^{\boldsymbol{P}} \;\Leftrightarrow\; p \Vdash \neg\varphi$.*

Proof (of the claim) As $[\neg\varphi]^{\boldsymbol{P}} = ([\varphi]^{\boldsymbol{P}})^*$ by the definition of $[\,\cdot\,]^{\boldsymbol{P}}$, we have that

$$\downarrow p \subseteq [\neg\varphi]^{\boldsymbol{P}} \quad \Leftrightarrow \quad \downarrow p \subseteq ([\varphi]^{\boldsymbol{P}})^*.$$

Therefore it suffices to prove the equivalence

$$\downarrow p \subseteq ([\varphi]^{\boldsymbol{P}})^* \quad \Leftrightarrow \quad p \Vdash \neg\varphi, \tag{9}$$

for all p.

Fix p and φ. By the induction hypothesis we have that for all $q \in P$,

$$\downarrow q \subseteq [\varphi]^{\boldsymbol{P}} \quad \Leftrightarrow \quad q \Vdash \varphi. \tag{10}$$

To prove (9), assume $\downarrow p \subseteq ([\varphi]^{\boldsymbol{P}})^*$. Then $\downarrow p \cap [\varphi]^{\boldsymbol{P}} = \mathbf{0}$ $(= \varnothing)$ in $\boldsymbol{RO}(\boldsymbol{P})$. It follows that if $\downarrow q \subseteq \downarrow p$, then $\downarrow q \not\subseteq [\varphi]^{\boldsymbol{P}}$. But by (10), $\downarrow q \not\subseteq [\varphi]^{\boldsymbol{P}}$ is equivalent to $q \nVdash \varphi$. Thus $\downarrow p \subseteq ([\varphi]^{\boldsymbol{P}})^*$ implies that $(\forall q)(q \subseteq p \Rightarrow q \nVdash \varphi)$, that is, $\downarrow p \subseteq ([\varphi]^{\boldsymbol{P}})^*$ implies $p \Vdash \neg\varphi$. So the $(\Rightarrow)$-part of (9) holds.

To prove the ($\Leftarrow$)-part of (9), we argue by contraposition. We assume $\downarrow p \not\subseteq ([\varphi]^P)^*$. Then $\downarrow p \cap [\varphi]^P \neq \mathbf{0}$ in $\boldsymbol{RO}(\boldsymbol{P})$. As $\downarrow p \cap [\varphi]^P$ is a non-empty open set, there is $\downarrow q$ in the base of open sets such that $\downarrow q \subseteq \downarrow p \cap [\varphi]^P$. In other words, there is $q \subseteq p$ such that $\downarrow q \subseteq [\varphi]^P$. Hence, by (10), $q \Vdash \varphi$. Thus there $q \subseteq p$ such that $q \Vdash \varphi$. This shows that $p \nVdash \neg\varphi$. $\square$

Claim 4 $\downarrow p \subseteq [(\exists x)\varphi]^P \;\;\Leftrightarrow\;\; p \Vdash (\exists x)\varphi.$

Proof (of the claim)

$$
\begin{aligned}
&\downarrow p \subseteq [(\exists x)\varphi]^P \quad \Leftrightarrow\\
&\downarrow p \subseteq \sup\{[\varphi(x/\!\!/y)]^P : y \in Var\} \quad \Leftrightarrow\\
&\downarrow p \subseteq \mathrm{Int}(\mathrm{Cl}(\{[\varphi(x/\!\!/y)]^P : y \in Var\})) \quad \Leftrightarrow\\
&(\forall q \subseteq p)(q \in \mathrm{Int}(\mathrm{Cl}(\{[\varphi(x/\!\!/y)]^P : y \in Var\})) \quad \Leftrightarrow \text{ (by Lemma 7.6.1)}\\
&(\forall q \subseteq p)(\forall r \subseteq q)(\downarrow r \cap \{[\varphi(x/\!\!/y)]^P : y \in Var\} \neq \varnothing) \quad \Leftrightarrow\\
&(\forall q \subseteq p)(\downarrow q \cap \{[\varphi(x/\!\!/y)]^P : y \in Var\} \neq \varnothing) \quad \Leftrightarrow\\
&(\forall q \subseteq p)(\exists y \in Var)(\downarrow q \cap [\varphi(x/\!\!/y)]^P \neq \varnothing) \quad \Leftrightarrow \text{ (by the fact that}\\
&\qquad\qquad\qquad\qquad\qquad\qquad \text{the family}\{\downarrow r : r \in P\}\text{is a base)}\\
&(\forall q \subseteq p)(\exists r \subseteq q)(\exists y \in Var)\ \downarrow r \subseteq [\varphi(x/\!\!/y)]^P \quad \Leftrightarrow \text{ (by the induction}\\
&\qquad\qquad\qquad\qquad\qquad\qquad \text{hypothesis)}\\
&(\forall q \subseteq p)(\exists r \subseteq q)(\exists y \in Var)\ r \Vdash \varphi(x/\!\!/y) \quad \Leftrightarrow \text{ (by (4))}\\
&p \Vdash (\exists x)\varphi. \quad \square
\end{aligned}
$$

Claim 5 $\downarrow p \subseteq [(\forall x)\varphi]^P \;\;\Leftrightarrow\;\; p \Vdash (\forall x)\varphi.$

Proof (of the claim)

$$
\begin{aligned}
&\downarrow p \subseteq [(\forall x)\varphi]^P \quad \Leftrightarrow\\
&\downarrow p \subseteq \inf\{[\varphi(x/\!\!/y)]^P : y \in Var\} \quad \Leftrightarrow\\
&\downarrow p \subseteq [\varphi(x/\!\!/y)]^P, \quad \text{for all } y \in Var \quad \Leftrightarrow \text{ (by the induction hypothesis)}\\
&p \Vdash \varphi(x/\!\!/y), \quad \text{for all } y \in Var \quad \Leftrightarrow\\
&p \Vdash (\forall x)\varphi. \quad \square
\end{aligned}
$$

The theorem follows from the above claims. $\square$

Corollary 6 *For any formula* σ,

$$[\sigma]^P = \bigcup\{\downarrow p : p \in P \text{ and } p \Vdash \sigma\}.$$

The above corollary states that the regular open set $[\sigma]^P$ is the set-theoretic union of those down-sets $\downarrow p$ from the topological basis for which $p \Vdash \sigma$.

Proof Since the family $\{\downarrow p : p \in P\}$ forms a base of $\boldsymbol{T}$, we have:

$$[\sigma]^{\boldsymbol{P}} = \bigcup\{\downarrow p : p \in P \text{ and } \downarrow p \subseteq [\sigma]^{\boldsymbol{P}}\} = \text{ (by the above theorem)}$$
$$\bigcup\{\downarrow p : p \in P \text{ and } p \Vdash \sigma\}. \qquad \square$$

Corollary 7 *For any formula σ and any $p \in P$,*

$$p \Vdash \sigma \quad \Leftrightarrow \quad (\forall q \subseteq p)(\exists r \supseteq q)\, r \Vdash \sigma.$$

Proof

$$p \Vdash \sigma \quad \Leftrightarrow \quad \downarrow p \subseteq [\sigma]^{\boldsymbol{P}} \quad \Leftrightarrow$$
$$\downarrow p \subseteq \bigcup\{\downarrow u : \downarrow u \subseteq [\sigma]^{\boldsymbol{P}}\} \quad \Leftrightarrow \text{ (by the above theorem)}$$
$$\downarrow p \subseteq \bigcup\{\downarrow u : u \Vdash \sigma\} \quad \Leftrightarrow$$
$$(\forall q \subseteq p)(\exists r \supseteq q)\, r \Vdash \sigma. \qquad \square$$

Corollary 8 *For any formula σ,*

$$[\sigma]^{\boldsymbol{P}} = \mathbf{1}^{\boldsymbol{P}} \quad \Leftrightarrow \quad p \Vdash \sigma \text{ for all } p \in P.$$

Proof We have:

$$[\sigma]^{\boldsymbol{P}} = \mathbf{1}^{\boldsymbol{P}} \quad \Leftrightarrow \text{ (by Corollary 6)}$$
$$\bigcup\{\downarrow r : r \in P \text{ and } r \Vdash \sigma\} = \mathbf{1}^{\boldsymbol{P}} \quad \Leftrightarrow$$
$$\bigcup\{\downarrow r : r \in P \text{ and } r \Vdash \sigma\} = P \quad \Leftrightarrow$$
$$(\forall p \in P)(\exists r \supseteq p)\, r \Vdash \sigma \quad \Leftrightarrow$$
$$(\forall p \in P)(\forall q \subseteq p)(\exists r \supseteq q)\, r \Vdash \sigma \quad \Leftrightarrow \text{ (by Corollary 7)}$$
$$(\forall p \in P)\, p \Vdash \sigma. \qquad \square$$

Note Let T be an elementary theory in L, that is, T is a set of sentences of $For(L)$. $\boldsymbol{B}_T(L)$ is the Lindenbaum–Tarski algebra of T. (The universe of $\boldsymbol{B}_T(L)$ is constituted by the set of equivalence classes $[\sigma]_T$ of formulas with respect to the relation $\sim_T$, where $\varphi \sim_T \varphi$ means that $T \vdash \varphi \leftrightarrow \psi$, for any formulas φ, ψ.) The above definitions are modified and accommodated to refined subsets in the algebra $\boldsymbol{B}_T(L)$. Thus, in this case, a refined poset $\boldsymbol{P} = (P, \subseteq)$ is a family of non-empty subsets of $\{[\sigma]_T : \sigma \in AFor(L)\}$ of the set of abstraction classes of *atomic* formulas ordered by inclusion. The crucial assumption is that of consistency of T, which makes that the above notions do not trivialize. $\square$

The question arises: how to define refined families of subsets of $\{[\varphi]: \varphi \in AFor(L)\}$ (or subsets of $\{[\varphi]_T : \varphi \in AFor(L)\}$)? This seems to be one of central issues pertinent to model theory. Due to lack of space the discussion of this problem is postponed and not investigated in this work. Of course, there is a largest refined family consisting of all non-empty subsets of $\{[\varphi] : \varphi \in AFor(L)\}$. But we are mainly interested in 'small' refined posets consisting of 'narrow', specialized non-empty subsets of $\{[\varphi] : \varphi \in AFor(L)\}$. Natural candidates for refined sets are fields of subsets of $\{[\varphi] : \varphi \in AFor(L)\}$ with the empty set removed.

Acknowledgements The author is indebted to the anonymous referees for many insightful comments that have considerably improved the paper.

References

Adamowicz, Z., & Zbierski, P. (1997). *Logic of mathematics: A modern course of classical logic*. New York: Wiley.

Bell, J. B. (2005). *Set theory. Boolean-valued models and independence proofs* (3rd ed.). Clarendon Press: Oxford Science Publications.

Dzik, W. (1981). The existence of Lindenbaum's extensions is equivalent to the axiom of choice. *Reports on mathematical logic*, *13*, 29–31.

Feferman, S. (1952). Review of the paper: H. Rasiowa and R. Sikorski, A proof of the completeness theorem of Gödel. *Journal of Symbolic Logic*, *19*, 72.

Fitting, M. (1969). *Intuitionistic logic, model theory and forcing*. Amsterdam: North-Holland.

Goldblatt, R. (1985). On the role of the Baire category theorem and dependent choices in the foundations of logic. *Journal of Symbolic Logic*, *50*, 412–422.

Henkin, L. (1951). The completeness of the first-order functional calculus. *Journal of Symbolic Logic*, *14*, 159–166.

Hodges, W. (2008). *Model theory*. Cambridge: Cambridge University Press. The digitally printed version.

Rasiowa, H., & Sikorski, R. (1950). A proof of the completeness theorem of Gödel. *Fundamenta Mathematicae*, *37*, 193–200.

Rasiowa, H., & Sikorski, R. (1963). *The mathematics of metamathematics*. Warsaw: PWN.

Scott, D. (1967). *Boolean-Valued Models for Set Theory*. Mimeographed notes for the 1967 American Mathematical Society Symposium on axiomatic set theory.

Weaver, N. (2014). *Forcing for mathematicians*. Singapore: World Scientific.

Zygmunt, J. (1973). A survey of the methods of proof of the Gödel-Malcev's completeness theorem. In S. J. Surma (Ed.), *Studies in the history of mathematical logic* (pp. 165–238). Wrocław: Ossolineum.

Chapter 8
μ-Levels of Interpolation

Giovanna D'Agostino

Abstract In this paper we discuss the problem of interpolation in the alternation levels of the μ-Calculus. In particular, we consider interpolation and uniform interpolation for the alternation free fragment, and, more generally, for the level Δ_n of the alternation hierarchy of the μ-calculus.

8.1 Introduction

Interpolation is a desirable property for a logic. In very general terms it states that if a property is a consequence of another one, then only the common language between the two properties is important in order to have this consequence. In this way interpolation connects syntax with semantics, and, especially in its uniform version, it allows modularization. Larisa Maksimova has been a pioneer in this area, writing many papers and books on interpolation for classical and non classical logics. Here we consider the μ-Calculus, a very expressive and well known formalism extending Modal Logic with fixed points of monotone operators; in particular, we consider the stratification of μ-formulas defined by means of the alternation of least and greatest fixed point operators. This alternation is a major cause of complexity for μ-formulas, and many logics used in applications sit in the lower levels of the hierarchy. In particular, we consider the fragment of the μ-Calculus where formulas do not contain *real* alternations between least and greatest fixed point operators: the Alternation Free Fragment AF. Formulas in this fragment behave better from a computational point of view than formulas allowing alternation, and hence this fragment has been quite considered in the literature on the μ-Calculus (see Gutierrez et al. 2014 and Facchini et al. 2013 for recent results on AF).

G. D'Agostino (✉)
Department of Mathematics and Computer Science, Via delle Scienze 208,
33100 Udine, Italy
e-mail: giovanna.dagostino@uniud.it

S. Odintsov (ed.), *Larisa Maksimova on Implication, Interpolation, and Definability*,
Outstanding Contributions to Logic 15, https://doi.org/10.1007/978-3-319-69917-2_8

Another theme of this paper is automata which have been used with great success in studying the μ-Calculus, helping to prove completeness and complexity results and the like. As for interpolation, D'Agostino and Hollenberg (2000) shows how to construct the uniform interpolant of a formula starting from an automaton equivalent to the formula. Uniform interpolation via automata has been studied also in (Lutz et al. 2012) in the context of Descriptive Logics. In this paper we use automata to obtain the ordinary interpolant of a valid implication between alternation free formulas. This is the maximum we could hope, since we also prove that the alternation free fragment does not enjoy the uniform version of interpolation.

The paper is structured as follows: in Sect. 8.2 we recall the definitions of μ-formulas, alternating modal automata and levels of the Alternation Hierarchy. In Sect. 8.3 we summarize the results we already know about interpolation and levels, and finally in Sects. 8.4 and 8.5 we give the interpolation result for AF.

8.2 μ-Calculus and Automata

We consider the μ-formulas as constructed from a set of propositional constants $Prop$, their negations $\{\neg P : P \in Prop\}$, and a set of variables Var, using the operators: $\varphi_1 \vee \varphi_2, \varphi_1 \wedge \varphi_2, \Box(\varphi_1), \Diamond(\varphi), \mu X\varphi$, and $\nu X\varphi$, where $\varphi_1, \varphi_2, \varphi \in \mu$ and X in Var.

The language $\mathscr{L}(\varphi)$ of a formula φ is the set of propositional constants appearing in it.

The semantics of μ-formulas is defined as usual over pointed labelled graphs (e.g. Kripke models): given a graph $G = (V, R, L)$ (with $L : Prop \cup Var \to Pow(V)$), a μ-formula φ is interpreted as a subset $[\![\varphi]\!]_G$ of V, defined as follows:

$$
\begin{array}{lll}
[\![P]\!]_G & := L(P) & \text{for } P \in \mathscr{P};\\
[\![\neg P]\!]_G & := V \setminus L(P) & \text{for } P \in \mathscr{P};\\
[\![X]\!]_G & := L(X) & \text{for } X \in \mathscr{V}ar;\\
[\![\varphi \wedge \psi]\!]_G & := [\![\varphi]\!]_G \cap [\![\psi]\!]_G; & \\
[\![\Diamond\varphi]\!]_G & := \{s \in V \mid [\![\varphi]\!]_G \cap \{t : sRt\} \neq \varnothing\}; & \\
[\![\Box\varphi]\!]_G & := \{s \in V \mid [\![\varphi]\!]_G \supseteq \{t : sRt\}\}; & \\
[\![\mu X.\varphi]\!]_G & := \bigcap\{S \subseteq V \mid [\![\varphi]\!]_{G[X:=S]} \subseteq S\}; & \\
[\![\nu X.\varphi]\!]_G & := \bigcup\{S \subseteq V \mid [\![\varphi]\!]_{G[X:=S]} \supseteq S\}; &
\end{array}
$$

where $G,[X := S]$ is equal to G except that $L(X) = S$. Note that $[\![\mu X.\varphi]\!]_G$ is the least fixpoint of the monotone operator $S \mapsto [\![\varphi]\!]_{G[X:=S]}$, and $[\![\nu X.\varphi]\!]_G$ is the greatest fixpoint.

In the following, we denote $s \in [\![\varphi]\!]_G$ by $G, s \models \varphi$. If the graph G is a tree T, then by $T \models \varphi$ we mean $T, r \models \varphi$ where r is the root of T. Since μ formulas are bisimulation invariant, we may restrict the semantics to the class of trees.

8.2.1 Automata

In this section we define a class of automata running on trees that correspond to μ-formulas.

Definition 1 An alternating modal automaton over a finite alphabet A is a tuple

$$\mathscr{A} = \langle Q, q_0, \delta, \Omega \rangle,$$

such that:

- Q is a finite set of states;
- $q_0 \in Q$ is the initial state;
- Ω is a function from Q to the natural numbers;
- δ is a function which associates to every $q \in Q$ and $\sigma \in A$ a positive modal formula over Q.

If we want to stress the alphabet of the automaton, we will write it inside the tuple:

$$\mathscr{A} = \langle Q, A, q_0, \delta, \Omega \rangle.$$

An alternating automaton is said to be *non deterministic* if for all q, a there exists $X \subseteq Pow(Q)$ such that

$$\delta(q, a) = \bigvee_{J \in X} \left(\bigwedge_{q \in J} \Diamond(q) \wedge \Box \left(\bigvee_{q \in J} q \right) \right).$$

The notion of acceptance of a A-labeled tree $\mathscr{T} = (T, \lambda)$ by an automaton

$$\mathscr{A} = \langle Q, q_0, \delta, \Omega \rangle$$

is defined in terms of a two players game, as follows. Positions are of the form (t, F), where $F \in Mod^+(Q)$ (the positive modal formulas with variables in Q). The initial position is (t_0, q_0), where t_0 is the root of T. In a position (t, F) with $F \in Mod^+(Q)$ the play proceeds as follows:

1. if $F = \Diamond(G)$ $(F = \Box(G))$ then it is Player I's turn (Player II, respectively); he choses a son t' of t in T, and goes to position (t', G);
2. if $F = \bigvee_i G_i$ $(F = \bigwedge_i G_i)$ then it is Player I's turn (Player II, respectively); he chooses one disjunct G_i (conjunct, respectively) and goes to position (t, G);
3. if $F = q$, the game starts again from position $(t, \delta(q, \lambda(t)))$.

If a Player cannot move, it looses. Otherwise an infinite play is generated, which goes through infinite positions in $T \times Q$:

$$(t_0, q_0) \ldots (t_1, q_1) \ldots$$

The play is won by Player I if $limsup_i\ \Omega(q_i)$ is even, or, in other words, if the maximum among all $\Omega(q_i)$, for q_i appearing infinitely often along the play, is even.

A strategy for Player I is a function σ having as domain the set of partial plays π ending in a position for Player I, and such that $\sigma(\pi)$ is a legal move for Player I. The strategy σ is winning if Player I wins any play in which he follows σ. Similarly, we can define (winning) strategies for Player II. A tree is accepted by the automaton $\mathscr{A}$ if Player I has a winning strategy in the game of $\mathscr{A}$ over $\mathscr{T}$.

When dealing with non deterministic automata, we shall represent the transition function as a set of subsets of the set of states: $\delta(q, \lambda) = \{J_1, \ldots, J_k\}$ with $J_i \in Pow(Q)$; an acceptance game from a position (q, t) will then proceeds as follows: first Player I chooses a $J \in \delta(q, \lambda)$, then Player II chooses either a conjunct $\Diamond(q')$ with $q' \in J$ or $\Box(\bigvee_{q \in J} q)$; in the first case, Player I chooses a son t' of t and the play starts again from (q', t'); in the second case, Player II chooses a son t' of t, Player I must choose a $q' \in J$ and the play starts again from (q', t').

A well known result states:

Theorem 1 (Arnold and Niwinski 2001) *The class of alternating automata and the class of non deterministic automata accepts the same tree languages. Moreover, if the alphabet A is equal to $Pow(Prop)$ for a finite set of proposition $Prop$, then automata (alternating or non deterministic) have the same expressive power as the μ-sentences over $Prop$.*

In view of the previous result we shall confuse automata and formulas and write e.g. that a certain implication between automata is valid: $\models \mathscr{A} \rightarrow \mathscr{B}$ or has an interpolant (see Definition 4).

We shall use the following construction of the dual automaton.

Definition 2 The dual automaton $\tilde{\mathscr{A}}$ of the alternating automaton $\mathscr{A}$ is defined as follows.

1. the transition function of $\tilde{\mathscr{A}}$ is dualized: conjunctions in the $\mathscr{A}$-transitions become disjunctions in $\tilde{\mathscr{A}}$, disjunctions become conjunctions, diamonds become boxes and boxes become diamonds, while atomic propositions $q \in Q$ remain the same;
2. the set of states, the initial state, and the alphabet of the dual automaton are the same as the ones in $\mathscr{A}$;
3. the priority function of $\tilde{\mathscr{A}}$ is obtained from the one in $\mathscr{A}$ by adding 1.

Lemma 1 *The language recognized by the dual automaton $\tilde{\mathscr{A}}$ of an alternating automaton $\mathscr{A}$ is the complement of the language recognized by $\mathscr{A}$.*

8.2.2 Alternation Hierarchy

μ-formulas are divided into levels, depending on the alternation of fixed point operators inside the formula:

Definition 3 The *fixpoint alternation-depth hierarchy* of the μ-calculus is the sequence $\Pi_0 = \Sigma_0, \Pi_1, \Sigma_1, \ldots$ of sets of μ-formulas defined inductively as follows.

1. $\Pi_0 = \Sigma_0$ is defined as the set of all modal fixpoint free formulas.
2. Π_{k+1} is the closure of $\Pi_k \cup \Sigma_k$ under the operations described in (a), (b) below.
   (a) Positive Substitution: if $\varphi(P_1, \ldots P_n), \varphi_1, \ldots, \varphi_n$ are in Π_{k+1}, then $\varphi(\varphi_1 \ldots \varphi_n)$ is in Π_{k+1}, provided $P_1, \ldots, P_n$ are positive in φ and no occurrence of a variable which was free in one of the φ_i becomes bound in $\varphi(\varphi_1 \ldots \varphi_n)$.
   (b) If φ is in Π_{k+1}, then $\nu X.\varphi \in \Pi_{k+1}$.
3. Likewise, Σ_{k+1} is the closure of $\Pi_k \cup \Sigma_k$ under positive substitution and the μ-operator.

The previous hierarchy is defined at a syntactical level. In the following we will always consider the levels of the alternation hierarchy modulo semantics, i.e. we shall say that a formula φ belongs to the class Σ_n if it is semantically equivalent to a formula in this class. For each n we define the ambiguous class Δ_n as the set of μ formulas which are equivalent to both a formula in Σ_n and a formula in Π_n, while the class $Comp(\Sigma_n, \Pi_n)$ is defined as the least class closed under positive substitutions and containing Σ_n, Π_n.

We shall use the well known result that $\Delta_2 = Comp(\Sigma_1, \Pi_1)$ (the equality $\Delta_n = Comp(\Sigma_n, \Pi_n)$ is false for $n > 2$).

8.2.2.1 Levels and Automata

Levels of the Alternation Hierarchy are characterizable by automata:

Theorem 2 (Arnold and Niwinski 2001)

1. *A μ-formula φ is equivalent to a formula in the class Π_n if and only if there exists an alternating modal automaton which is equivalent to φ and an even $m \geq n-1$ such that the range of the priority function of the automaton is contained in $\{m-n+1, \ldots, m\}$.*
2. *If $n = 0, 1, 2$, then a μ-formula φ is equivalent to a formula in the class Π_n if and only if there exists a* non deterministic *modal automaton which is equivalent to φ and an even $m \geq n-1$ such that the range of the priority function of the automaton is contained in $\{m-n+1, \ldots, m\}$. The same result does not hold for $n > 2$.*
3. *A formula φ is equivalent to a formula in the class $Comp(\Sigma_n, \Pi_n)$ if and only if there exists an alternating modal automaton $\mathscr{A} = \langle Q, q_0, \delta, \Omega\rangle$ which is equivalent to φ and a preorder $\preceq$ of the set of states of the automaton such that, if $Q_0, \ldots, Q_k$ are the equivalence classes induced by the preorder, it holds:*
   a. *if $q \in Q_h$ then $\delta(q, a) \in Mod^+(\bigcup_{i\in\{h,\ldots,k\}} Q_i)$, for all $a \in A$;*

b. *for all* $i \in \{0, \ldots, k\}$ *there exists* $m \geq n-1$ *such that the range of the priority function restricted to an equivalence class* Q_i *is either contained in* $\{m-n+1, \ldots, m\}$ *or in* $\{m-n, \ldots, m-1\}$.

In view of the equivalence between automata and μ-formulas we shall be free to say, e.g., that an automaton is in the class Π_n (meaning that it is equivalent to a formula in this class).

8.3 Interpolation and Uniform Interpolation

Let us begin by stating what we mean by ordinary (Craig) interpolation and uniform interpolation.

Definition 4 Let φ and ψ be two μ-sentences such that $\models \varphi \rightarrow \psi$. Then θ is an *interpolant* of φ, ψ iff:

1. $\models \varphi \rightarrow \theta$ and $\models \theta \rightarrow \psi$;
2. $\mathscr{L}(\theta) \subseteq \mathscr{L}(\varphi) \cap \mathscr{L}(\psi)$. □

In words: if $\models \varphi \rightarrow \psi$, an interpolant of φ, ψ is a formula in the common language of φ and ψ which sits in between φ and ψ.

Definition 5 Given a μ-sentence φ and a language $\mathscr{L}' \subseteq \mathscr{L}(\varphi)$, the *uniform interpolant* of φ with respect to $\mathscr{L}'$ is a formula θ such that:

1. $\models \varphi \rightarrow \theta$;
2. Whenever $\models \varphi \rightarrow \psi$ and $\mathscr{L}(\varphi) \cap \mathscr{L}(\psi) \subseteq \mathscr{L}'$ then $\models \theta \rightarrow \psi$.
3. $\mathscr{L}(\theta) \subseteq \mathscr{L}'$. □

When we say that the μ-calculus has (uniform) interpolation we mean that we can always find a (uniform) interpolant when the appropriate conditions are satisfied. Clearly, if the μ-calculus has uniform interpolation, it also enjoys Craig interpolation. For if $\models \varphi \rightarrow \psi$, simply choose $\mathscr{L}' = \mathscr{L}(\varphi) \cap \mathscr{L}(\psi)$. The interpolant is then the uniform interpolant of φ relative to $\mathscr{L}'$. This explains why we call this formula a *uniform* interpolant: no information is needed about the formula ψ except which non-logical symbols it has in common with φ.

In (D'Agostino and Hollenberg 2000) the uniform interpolation property for the μ-Calculus was proved using non deterministic automata: given a μ-formula φ, a propositional constant $p \in \mathscr{L}(\varphi)$, and a non deterministic automaton $\mathscr{A} = (Q, q_0, \delta, \Omega)$ over the alphabet $Pow(\mathscr{L}(\varphi))$ equivalent to φ, the *Uniform Interpolant Automaton* $UI(\mathscr{A})$ in the alphabet $Pow(\mathscr{L}(\varphi)) \setminus \{p\})$ is defined as follows:

$$UI(\mathscr{A}) := (Q, q_0, \delta', \Omega)$$

with:

$$\delta'(q, \lambda) := \delta(q, \lambda) \cup \delta(q, \lambda \cup \{p\})$$

In (D'Agostino and Hollenberg 2000) it is proved that $UI(\mathscr{A})$ corresponds to a μ-sentence ψ in the language $\mathscr{L}(\varphi) \setminus \{p\}$ which is a uniform interpolant for φ w.r.t. the language $\mathscr{L}(\varphi) \setminus \{p\}$; by iterating this construction we may obtain all uniform interpolants of a formula.

Uniform interpolation also holds for the levels Π_0, Σ_1, Π_1, Π_2: here, to construct the uniform interpolant, we may use the same automaton $UI(\mathscr{A})$ as above, simply because these levels have the property that an automaton corresponding to a formula in the level is equivalent to a nondeterministic automaton in the same class.

However, uniform interpolation does not hold if we restrict to other levels of the Alternation Hierarchy: in (D'Agostino and Lenzi 2006) it is proved that, except for Π_0, Σ_1, Π_1, Π_2, the levels of the Alternation Hierarchy are not closed under uniform interpolation:

Lemma 2 (D'Agostino and Lenzi 2006) *The levels of the Alternation Hierarchy are not closed under uniform interpolation, except for Π_0, Σ_1, Π_1, Π_2. Among uniform interpolants of Σ_2-formulas we may find formulas of arbitrarily high alternations.*

Since $\Sigma_2 \subseteq \Delta_3$, the previous lemma implies that for $n \geq 3$ the level Δ_n does not have the uniform interpolation property. However, this result leaves open the question of whether Δ_2, that is, the Alternation Free Fragment AF, enjoys uniform Interpolation or Craig interpolation, and this is the content of the following section.

8.4 Alternation Free Fragment and Uniform/Craig Interpolation

In this section we investigate uniform/Craig interpolation for the Alternation Free Fragment AF of the μ-calculus, which is defined as $Comp(\Pi_1, \Sigma_1)$. A well known result about the alternation free fragment gives a characterization of AF as the intersection of levels Π_2, Σ_2. This fragment has many interesting properties, e.g. it allows less complex decision procedure for model checking than the full μ-calculus, still maintaining strong expressive power.

In this section we prove that AF does not enjoy the uniform interpolation property, while retaining the Craig Interpolation one.

8.4.1 Failure of Uniform Interpolation in AF

Lemma 3 *There exists a formula $\varphi(p) \in AF$ without a uniform interpolant in AF w.r.t. $\mathscr{L}(\varphi) \setminus \{p\}$.*

Proof Consider a formula $\psi \in \Delta_2$ such that $\nu X \psi \notin \Delta_2$. We claim that the uniform interpolant of the formula $\varphi(p) := p \wedge \Box^*(p \rightarrow \psi(p|X))$, w.r.t. the language

$\mathscr{L}(\varphi) \setminus \{p\}$, where $\Box^*(p) := \nu Y(p \wedge \Box(Y))$, is equivalent to $\nu X \psi$. To show this it suffices to prove:

1. $\models \varphi(p) \rightarrow \nu X \psi$;
2. if θ is a formula not containing p and $\models \varphi \rightarrow \theta$, then $\models \nu X \psi \rightarrow \theta$.

Suppose (G, w) is a model with $(G, w) \models \varphi(p)$; then, by definition of φ, the set of points $\{v \in V_G : (G, v) \models \varphi(p)\}$ is a post-fixed point of ψ and contains w. By Tarski–Knaster Theorem it follows that $(G, w) \models \nu X \psi$.

As for the second point, suppose θ does not contain p, $\models \varphi \rightarrow \theta$, and (G, w) is a model such that $(G, w) \models \nu X \psi$. Then, if $A := \{v \in V_G : (G, v) \models \nu X \psi\}$ we have $\psi(A) = A$, and $(G, p := A, w) \models p \wedge \Box^*(p \rightarrow \psi(p|X))$, that is, $(G, p := A, w) \models \varphi$. From $\models \varphi \rightarrow \theta$ it follows $(G, p := A, w) \models \theta$, and we obtain $(G, w) \models \theta$, since p is not contained in θ.

Finally, we notice that the formula $\varphi(p) := p \wedge \Box^*(p \rightarrow \psi(p|X))$ belongs to $Comp(\Pi_1, \Sigma_1) = \Delta_2$.

8.4.2 Craig Interpolation for AF

In this section we prove the Craig Interpolation property for the Alternation Free Fragment using automata. More precisely, we obtain the interpolation property for AF by adapting the proof of a separation result from (Santocanale and Arnold 2005). In this paper it is proved that given two Π_{n+1} non deterministic automata $\mathscr{A}, \mathscr{B}$ with no common models, there exists an automaton $\mathscr{C}$ in the class $Comp(\Pi_n, \Sigma_n)$ such that $\models \mathscr{A} \rightarrow \mathscr{C}$ and $\mathscr{C}, \mathscr{B}$ have no common models, where all automata $\mathscr{A}, \mathscr{B}, \mathscr{C}$ are over the same alphabet. We consider this result for $n = 1$, but we have to modify it in order to consider as input two non deterministic automata $\mathscr{A}, \mathscr{B}$ in Π_2 over the alphabets $A_1 = Pow(P_1)$, $A_2 = Pow(P_2)$, respectively, and obtain as output an automaton $\mathscr{C}$ in the class $Comp(\Pi_1, \Sigma_1)$ over the alphabet $A_1 \cap A_2$:

Lemma 4 (Santocanale and Arnold 2005) *If $\mathscr{A}, \mathscr{B}$ are non deterministic Π_2 automata in the alphabets A_1, A_2, respectively, such that $L(\mathscr{A}) \cap L(\mathscr{B}) = \varnothing$, there exists an automaton $\mathscr{C}$ in the class $Comp(\Pi_1, \Sigma_1)$ over the alphabet $A_1 \cap A_2$ with $\models \mathscr{A} \rightarrow \mathscr{C}$ and $\models \mathscr{C} \rightarrow \tilde{\mathscr{B}}$, for the dual automaton $\tilde{\mathscr{B}}$ of $\mathscr{B}$.*

We give a proof of this Lemma in the next section, but first we prove Craig Interpolation from it.

Theorem 3 *Craig Interpolation for AF.*
The Alternation Free Fragment of the modal μ-Calculus enjoys the Craig Interpolation Property.

Proof Suppose $\models \varphi \rightarrow \psi$ with $\varphi, \psi \in AF$. Since AF is closed under negation and any automaton in Π_2 is equivalent to a non deterministic automaton in the same class, both $\varphi, \neg\psi$ are equivalent to non deterministic automata $\mathscr{A}, \mathscr{B}$ in Π_2, for which the previous lemma applies. Hence, there exists an automaton $\mathscr{C}$ in Δ_2 over the alphabet $Pow(\mathscr{L}(\varphi) \cap \mathscr{L}(\psi))$, with $\models \mathscr{A} \rightarrow \mathscr{C}$ and $\models \mathscr{C} \rightarrow \tilde{\mathscr{B}}$, for the dual automaton $\tilde{\mathscr{B}}$ of $\mathscr{B}$. If θ is the formula which is equivalent to the automaton $\mathscr{C}$, then θ is an AF formula in $\mathscr{L}(\varphi) \cap \mathscr{L}(\psi)$ with $\models \varphi \rightarrow \theta$ and $\models \theta \rightarrow \psi$, that is, $\theta \in AF$ is the required interpolant between $\varphi, \psi \in AF$.

8.5 The Interpolation Automaton

Given two non deterministic automata $\mathscr{A}, \mathscr{B}$ in Π_2 (over the alphabets $A_1 = Pow(P_1)$, $A_2 = Pow(P_2)$, respectively) such that $L(\mathscr{A}) \cap L(\mathscr{B}) = \varnothing$, we will define the interpolation automaton $INT(\mathscr{A}, \tilde{\mathscr{B}})$ between $\mathscr{A}$ and the dual $\tilde{\mathscr{B}}$ of $\mathscr{B}$ using $\mathscr{A}, \mathscr{B}$ and a winning strategy for Player II in the satisfiability game defined below.

Definition 6 (Niwinski and Walukiewicz 1996; Santocanale and Arnold 2005) Let $\mathscr{A} = (Q, q_0, \delta_{\mathscr{A}}, \Omega_{\mathscr{A}})$ and $\mathscr{B} = (R, r_0, \delta_{\mathscr{B}}, \Omega_{\mathscr{B}})$ be non deterministic modal automata over the same alphabet A. The satisfiability game $\mathscr{G}(\mathscr{A}, \mathscr{B})$ is defined as follows:

1. Positions for Player I:

$$Pos_1 = (Q \times R) \cup \{(q, K) : q \in Q, K \in Pow(R)\} \cup \{(J, r) : r \in R, J \in Pow(Q)\};$$

2. Position for Player II:

$$Pos_2 = \{(J, K) : J \in Pow(Q), K \in Pow(R)\};$$

3. $Pos = Pos_1 \cup Pos_2$;
4. the initial position is $p_0 = (q_0, r_0)$.
5. from a position $(q, r) \in Q \times R$, Player I chooses $a \in A$ and move to $(J, K) \in \delta_{\mathscr{A}}(q, a) \times \delta_{\mathscr{B}}(r, a)$;
6. from a position $(J, K) \in Pow(Q) \times Pow(R)$, Player II can move either to (q, K) with $q \in J$ or to (J, r) with $r \in K$;
7. from a position (q, K) with $q \in Q$, $K \subseteq R$, Player I can move to (q, r) with $r \in K$; from a position (J, r) with $r \in R$, $J \subseteq Q$, Player I can move to (q, r) with $q \in J$.

If a Player cannot move, it looses. Player I wins a play $(q_0, r_0)(J_0, K_0)\ldots(q_1, r_1)\ldots$ if and only if both $limsup_i \Omega_{\mathscr{A}}(q_i)$ and $limsup_i \Omega_{\mathscr{B}}(r_i)$ are even.

Consider now two non deterministic modal automata $\mathscr{A}, \mathscr{B}$ such that $L(\mathscr{A}) \cap L(\mathscr{B}) = \varnothing$; by a result in (Niwinski and Walukiewicz 1996), Player II has a winning strategy in the satisfiability game $\mathscr{G}(\mathscr{A}, \mathscr{B})$. Notice that this kind of *biparity game* can be easily transformed into a Muller game, and since Muller games have finite memory winning strategies, it follows:

Lemma 5 *If $L(\mathscr{A}) \cap L(\mathscr{B}) = \varnothing$ then Player II has a finite memory winning strategy σ in the game $\mathscr{G}(\mathscr{A}, \mathscr{B})$*

$$\sigma = \begin{cases} nxt : Pos_2 \times M \to Pos; \\ updt : Pos \times M \to M; \\ init : Pos \to M, \end{cases}$$

(where M is a finite set, the memory), such that for all positions (q, r) reached by a play in which Player II follows σ it holds $L(\mathscr{A}, q) \cap L(\mathscr{B}, r) = \varnothing$.

The strategy of the above Lemma can be presented as a finite *strategy* graph $\mathscr{S}$ defined as follows. Consider the following binary relation on $Pos \times M$:

1. if $p \in Pos_1$, $p' \in Pos$ there is an edge from (p, m) to (p', m) iff $p \to p'$ is a possible move for Player I;
2. if $p \in Pos_2$ there is an edge from (p, m) to $(nxt(p, m), updt(p, m))$.

The finite strategy graph $\mathscr{S}$ is defined as the set of vertices and edges which are reachable from the vertex $(p_0, init(p_0))$, where p_0 is the initial position of the game $\mathscr{G}(\mathscr{A}, \mathscr{B})$.

In order to find the interpolant between AF formulas, we concentrate now on non deterministic modal automata over power set alphabets of the form $A = Pow(P)$, where P is a finite set of *propositions*. If $P \subseteq P'$, where P' is another set of propositions, a $Pow(P)$ automaton $\mathscr{A}$ will be also considered as an automaton over $A' = Pow(P')$, by extending the transition function as follows, for $a' \in A'$:

$$\delta_{\mathscr{A}}(q, a') = \delta_{\mathscr{A}}(q, a' \cap P)$$

If (T, λ) is an A'-labelled tree, then it holds:

$$(T, \lambda) \text{ is accepted by (the extension of)} \mathscr{A} \Leftrightarrow (T, \lambda \cap P) \text{ is acceped by } \mathscr{A},$$

where $\lambda \cap P(t) := \lambda(t) \cap P$. In the following, we shall use this kind of extension freely, every time an automaton originally defined over an alphabet $A = Pow(P)$ is considered as an automaton over a bigger alphabet $A' = Pow(P')$, with $P \subseteq P'$.

Definition 7 Let $\mathscr{A} = (Q, q_0, \delta_{\mathscr{A}}, \Omega_{\mathscr{A}})$, $\mathscr{B} = (R, r_0, \delta_{\mathscr{B}}, \Omega_{\mathscr{B}})$ be two non deterministic Π_2 modal automata over the alphabets $A_1 = Pow(P_1)$, $A_2 = Pow(P_2)$, respectively, with $\Omega_{\mathscr{A}}(Q), \Omega_{\mathscr{B}}(R) \subseteq \{1, 2\}$ and $L(\mathscr{A}) \cap L(\mathscr{B}) = \varnothing$ (considering both as automata over the alphabet $A = Pow(P_1 \cup P_2)$). Then if $\mathscr{S}$, σ are the strategy graph and the winning strategy for Player II in the game $\mathscr{G}(\mathscr{A}, \mathscr{B})$, as defined above, the automaton $INT(\mathscr{A}, \tilde{\mathscr{B}}) = (S, s_0, \delta, \Omega)$ over the alphabet $A_1 \cap A_2 = Pow(P_1 \cap P_2)$ is defined as follows:

1. S is the vertex set of the strategy graph $\mathscr{S}$.
2. $s_0 = (p_0, init(p_0))$, where p_0 is the initial position of the game $\mathscr{G}(\mathscr{A}, \mathscr{B})$;

3. if $a \in A_1 \cap A_2 = Pow(P_1 \cap P_2)$ then

$$\delta((q,r,m),a) = \bigvee_{\left\{\substack{a_1\in A_1,\ a_1\cap P_2=a \\ J\in\delta_{\mathscr{A}}(q,a_1)}\right\}} \ \bigwedge_{\left\{\substack{a_2\in A_2,\ a_2\cap P_1=a \\ K\in\delta_{\mathscr{B}}(r,a_2)}\right\}} F_{J,K,m},$$

where, if σ is defined as in Lemma 5, the formula $F_{J,K,m}$ is defined as

$$F_{J,K,m} = \begin{cases} \Diamond(\bigwedge_{r\in K}(q,r,updt(J,K,m))), & \text{if } nxt(J,K,m) = (q,K); \\ \Box(\bigvee_{q\in J}(q,r,updt(J,K,m))), & \text{if } nxt(J,K,m) = (J,r). \end{cases}$$

4. To define Ω, consider a strongly connected component $C(q,r,m)$ of the strategy graph $\mathscr{S}$. Then we consider the following cases:
   a. $|C(q,r,m)| = 1$;
   b. $|C(q,r,m)| > 1$ and $\forall(q',r',m') \in C(q,r,m)$ we have $\Omega_{\mathscr{A}}(q') = 1$;
   c. $|C(q,r,m)| > 1$ and $\exists(q',r',m') \in C(q,r,m)$ with $\Omega_{\mathscr{A}}(q') = 2$.

   Accordingly, we define:

$$\Omega(q,r,m) = \begin{cases} 1 \text{ if case (a) or (b) applies;} \\ 2 \text{ if case (c) applies.} \end{cases}$$

Consider the partial order $\preceq$ over the set of states S of the automaton $INT(\mathscr{A},\tilde{\mathscr{B}})$ defined by: $s \preceq s'$ iff there is a path from s to s' in the strategy graph S; then we easily see that the automaton $INT(\mathscr{A},\tilde{\mathscr{B}})$ defined above is in Δ_2 by using the preorder $\preceq$: if we are in a non trivial strongly connected component of the graph and there is a (q,r,m) in the component with $\Omega_{\mathscr{A}}(q) = 2$ then by definition the priority function is always 2 on the component, otherwise it is always 1.

We are now ready to prove:

Lemma 6 *Let $\mathscr{A}, \mathscr{B}$ be non deterministic Π_2 automata in the alphabets $A_1 = Pow(P_1)$, $A_2 = Pow(P_2)$, respectively, such that $L(\mathscr{A}) \cap L(\mathscr{B}) = \emptyset$. If $\mathscr{C} = INT(\mathscr{A},\tilde{\mathscr{B}})$ is the automaton over $A_1 \cap A_2$ defined above we have:*

$$\models \mathscr{A} \to \mathscr{C} \text{ and } \models \mathscr{C} \to \tilde{\mathscr{B}}$$

where $\tilde{\mathscr{B}}$ is the dual automaton of $\mathscr{B}$.

Proof We first prove that every A_1-labeled tree (T,λ) which is accepted by $\mathscr{A}$, is accepted by $\mathscr{C}$ (i.e. $\mathscr{C}$ accepts the tree $(T,\lambda \cap P_2)$ where $\lambda \cap P_2(t) = \lambda(t) \cap P_2$). If (T,λ) is accepted by $\mathscr{A}$, let τ be a winning strategy for Player I in the game of $\mathscr{A}$ over T. Using τ and a winning strategy σ for Player II in the game $\mathscr{G}(\mathscr{A},\mathscr{B})$ we shall define a strategy η for Player I in the game of $INT(\mathscr{A},\tilde{\mathscr{B}})$ over $(T,\lambda \cap P_2)$, and prove it is a winning one.

Claim We claim that a strategy η can be defined in such a way to preserve the following invariant: any η-play of $\mathscr{C}$ over $(T, \lambda \cap P_2)$,

$$(q_0, r_0, m_0, t_0), \ldots, (q_1, r_1, m_1, t_1), \ldots, (q_i, r_i, m_i, t_i) \ldots,$$

leaves a trace

$$(q_0, t_0), \ldots, (q_1, t_1), \ldots, (q_i, t_i) \ldots$$

which is a τ-play of $\mathscr{A}$ over (T, λ), and a trace

$$(q_0, r_0, m_0), \ldots, (q_1, r_1, m_1), \ldots, (q_i, r_i, m_i) \ldots,$$

which is a σ-play in the game $\mathscr{G}(\mathscr{A}, \mathscr{B})$.

The $\mathscr{C}$-play over $(T, \lambda \cap P_2)$ starts form the initial position (q_0, r_0, m_0, t_0), where t_0 is the root of T, while the play of $\mathscr{A}$ over (T, λ) starts from the initial position (q_0, t_0), and the play of the game $\mathscr{G}(\mathscr{A}, \mathscr{B})$ starts from (q_0, r_0, m_0). From the position (q_0, t_0) of the $\mathscr{A}$ play, the strategy τ suggests to Player I a move $J \in \delta_{\mathscr{A}}(q_0, \lambda(t_0))$; since $\lambda(t_0) \cap P_2 := \lambda \cap P_2(t_0)$, this J is also a possible move for Player I in the game of $\mathscr{C}$ over $(T, \lambda \cap P_2)$; hence we define $\eta(q_0, r_0, m_0, t_0) = J$. In the $\mathscr{C}$-play it is now Player II's turn, and he can choose a $K \in \delta_{\mathscr{B}}(r_0, a_2)$, with $a_2 \in A_2$ and $a_2 \cap P_1 = \lambda(t_0) \cap P_2$. At this point we shift to the game $\mathscr{G}(\mathscr{A}, \mathscr{B})$, starting from (q_0, r_0, m_0), where Player I can choose to move to position (J, K, m_0), which is a legal position for him. Since (J, K, m_0) is a position for Player II in the game $\mathscr{G}(\mathscr{A}, \mathscr{B})$, we may use the σ strategy. We consider two cases:

1. $nxt((J, K), m_0) = (q_1, K)$, with $q_1 \in J$. In this case

$$F_{J,K,m_0} = \Diamond \left(\bigwedge_{r \in K} (q_1, r, updt((J, K), m_0)) \right),$$

   and it is Player I's turn in the $\mathscr{C}$-play; to decide the value of the strategy η, we consider the $\mathscr{A}$ play, where now in position J it is Player II's turn; since $q_1 \in J$, the conjunct $(\Diamond(q_1), t_0)$ is a legal move for Player II in the $\mathscr{A}$ play; remaining in this play the winning strategy τ will suggest a son t_1 of t_0, and Player I will move to the position (q_1, t_1). We choose this t_1 also to be the suggestion that the strategy η gives to Player I in the $\mathscr{C}$-play over $(T, \lambda \cap P_2)$, and move to the position $(\bigwedge_{r \in K} (q_1, r, m_1), t_1)$, where $m_1 = updt((J, K), m_0))$. Now it is Player II's turn in the $\mathscr{C}$-play, say he chooses $r_1 \in K$ and moves to the position (q_1, r_1, m_1, t_1). Since $r_1 \in K$, (q_1, r_1, m_1) is also a possible move for Player I in the game $\mathscr{G}(\mathscr{A}, \mathscr{B})$, and the invariant is satisfied.

2. $nxt((J, K), m_0)) = (J, r_1)$, with $r_1 \in K$. In this case

$$F_{J,K,m_0} = \Box \left(\bigvee_{q \in J} (q, r_1, updt((J, K), m_0)) \right),$$

and it is Player II's turn in the $\mathscr{C}$-play: he will choose a son t_1 of t_0 in T and move to position $(\bigvee_{q\in J}(q, r_1, m_1), t_1)$, with $m_1 = updt((J, K), m_0)$. We consider again the $\mathscr{A}$ play, and let Player II choose the position $(\Box(\bigvee_{q\in J} q), t_0)$ in this play; from this position it is still Player II turn and he can choose position $(\bigvee_{q\in J} q, t_1)$, which is now a position for Player I. Using the strategy τ for Player I, we find $q_1 \in J$ so that (q_1, t_1) is a legal position in the τ-play over (T, λ); in the $\mathscr{C}$-play we choose this q_1 to be the suggestion that the strategy η gives to Player I in position $(\bigvee_{q\in J}(q, r_1, m_1), t_1)$ of the $\mathscr{C}$-play over $(T, \lambda \cap P_2)$, and move to (q_1, r_1, m_1, t_1); since $q_1 \in J$, (q_1, r_1, m_1) is also a legal move for Player I in the $\mathscr{G}(\mathscr{A}, \mathscr{B})$, and the invariant is satisfied.

Since the same pattern can be applied to a position (q_i, r_i, m_i, t_i) already reached by the η play, the claim is proved.

To conclude the proof of $\models \mathscr{A} \to \mathscr{C}$ we have to check that the strategy η described above is winning for Player I in the game of $\mathscr{C}$ over $(T, \lambda \cap P_2)$. By construction, any η-play

$$(q_0, r_0, m_0, t_0), \ldots, (q_1, r_1, m_1, t_1), \ldots, (q_i, r_i, m_i, t_i) \ldots,$$

leaves a trace

$$(q_0, t_0), \ldots, (q_1, t_1), \ldots, (q_i, t_i) \ldots$$

which is a τ-play of $\mathscr{A}$ over T, and a trace

$$(q_0, r_0, m_0), \ldots, (q_1, r_1, m_1), \ldots, (q_i, r_i, m_i) \ldots,$$

which is a σ-play in the game $\mathscr{G}(\mathscr{A}, \mathscr{B})$. Since τ is winning for I in the $\mathscr{A}$ play, $limsup_i(\Omega_{\mathscr{A}}(q_i)) = 2$. Moreover, the σ play

$$(q_0, r_0, m_0), \ldots, (q_1, r_1, m_1), \ldots, (q_i, r_i, m_i) \ldots,$$

will sooner or later get trapped in a non trivial strongly connected component $\mathscr{C}$ of the strategy graph $\mathscr{S}$. Hence, we must be in case (c) in the definition of Ω; hence, from a certain point on $\Omega(q_i, r_i, m_i) = 2$, and η is winning. This concludes the proof that $\models \mathscr{A} \to \mathscr{C}$.

Finally, to prove that $\models \mathscr{C} \to \tilde{\mathscr{B}}$, consider the pair of automata $(\mathscr{B}, \mathscr{A})$. This pair satisfies the hypothesis of Definition 7; by the previous part of the proof, we have $\models \mathscr{B} \to INT(\mathscr{B}, \tilde{\mathscr{A}})$, where $INT(\mathscr{B}, \tilde{\mathscr{A}}) = (S, (q_0, r_0, m_0), \delta', \Omega')$ and δ', Ω' are defined as follows:

1. if $a \in A_1 \cap A_2$ then

$$\delta'((q,r,m),a) = \bigvee_{\left\{\substack{a_2 \in A_2,\ a_2 \cap P_1 = a \\ K \in \delta_{\mathscr{B}}(r,a_2)}\right\}} \bigwedge_{\left\{\substack{a_1 \in A_1,\ a_1 \cap P_2 = a \\ J \in \delta_{\mathscr{A}}(q,a_1)}\right\}} G_{K,J,m},$$

where, if σ is as in Lemma 5, the formula $G_{J,K,m}$ is defined as

$$G_{J,K,m} = \begin{cases} \Diamond(\bigwedge_{q \in J}(q, r, updt((J,K),m))), & \text{if } \sigma(J,K,m) = (J, r, updt((J,K),m); \\ \Box(\bigvee_{r \in K}(q, r, updt((J,K),m))), & \text{if } \sigma(J,K,m) = (q, K, updt((J,K),m). \end{cases}$$

2.

$$\Omega'(q,r,m) = \begin{cases} 1 \text{ if case (a) or (b) applies;} \\ 2 \text{ if case or (c) applies,} \end{cases}$$

where

a. $|\mathscr{C}(q,r,m)| = 1$;
b. $|\mathscr{C}(q,r,m)| > 1$ and $\forall (q',r',m') \in \mathscr{C}(q,r,m)$ we have $\Omega_{\mathscr{B}}(r') = 1$;
c. $|\mathscr{C}(q,r,m)| > 1$ and $\exists (q',r',m') \in \mathscr{C}(q,r,m)$ with $\Omega_{\mathscr{B}}(r') = 2$.

From the first part of the proof we already know that $\models \mathscr{B} \to INT(\mathscr{B}, \tilde{\mathscr{A}})$ and hence $\models I\tilde{N}T(\mathscr{B}, \tilde{\mathscr{A}}) \to \tilde{\mathscr{B}}$, for the dual automaton $I\tilde{N}T(\mathscr{B}, \tilde{\mathscr{A}})$ of $INT(\mathscr{B}, \tilde{\mathscr{A}})$, to conclude the proof of the Lemma we only have to show that $\models INT(\mathscr{A}, \tilde{\mathscr{B}}) \to I\tilde{N}T(\mathscr{B}, \tilde{\mathscr{A}})$.

Notice that the transition function ε of $I\tilde{N}T(\mathscr{B}, \tilde{\mathscr{A}})$ is

$$\varepsilon((q,r,m),a) = \bigwedge_{\left\{\substack{a_2 \subseteq A_2,\ a_2 \cap A_1 = a \\ K \in \delta_{\mathscr{B}}(r,a_2)}\right\}} \bigvee_{\left\{\substack{a_1 \subseteq A_1,\ a_1 \cap A_2 = a \\ J \in \delta_{\mathscr{A}}(q,a_1)}\right\}} \tilde{G}_{K,J,m},$$

where

$$\tilde{G}_{J,K,m} = F_{J,K,m} = \begin{cases} \Box(\bigvee_{q \in J}(q, r, updt((J,K),m))), & \text{if } \sigma(J,K,m) = (J, r, updt((J,K),m); \\ \Diamond(\bigwedge_{r \in K}(q, r, updt((J,K),m))), & \text{if } \sigma(J,K,m) = (q, K, updt((J,K),m). \end{cases}$$

Calling δ the transition function of $INT(\mathscr{A}, \tilde{\mathscr{B}})$, for propositional reasoning one can easily check that for all $a \in A$ it holds:

$$\models \delta((q,r,m),a) \to \varepsilon((q,r,m),a). \tag{8.1}$$

Moreover, if Ω'' is the priority function of $I\tilde{N}T(\mathscr{B}, \tilde{\mathscr{A}})$ then

$$\Omega''(q,r,m) = \begin{cases} 2 \text{ if case (a) or (b) of } \Omega' \text{ definition applies;} \\ 3 \text{ if case or (c) of } \Omega' \text{ definition applies.} \end{cases}$$

Suppose that $INT(\mathscr{A}, \tilde{\mathscr{B}})$ accepts a tree, via a winning strategy η for Player I; then:

1. since $\models \delta((q, r, m), a) \rightarrow \varepsilon((q, r, m), a)$, the strategy η is also a possible strategy for Player I in the game of $I\tilde{N}T(\mathscr{B}, \tilde{\mathscr{A}})$ over the same tree;
2. sooner or later the η-play will get trapped in a strongly connected component $\mathscr{C}(q, r, m)$ of the strategy graph $\mathscr{S}$ such that both $|\mathscr{C}(q, r, m)| > 1$ and $\exists(q', r', m') \in \mathscr{C}(q, r, m)$ with $\Omega_{\mathscr{A}}(q') = 2$ hold. Since the strategy graph $\mathscr{S}$ is built using a winning strategy for Player II, we must have $\Omega_{\mathscr{B}}(r') = 1$, $\forall(q', r', m') \in C(q, r, m)$. Hence, from some point in the play the priority function Ω'' will always assume value 2.

This proves that η is also winning for Player I in the game of $I\tilde{N}T(\mathscr{B}, \tilde{\mathscr{A}})$ over the tree and hence $\models INT(\mathscr{A}, \mathscr{B}) \rightarrow I\tilde{N}T(\mathscr{B}, \tilde{\mathscr{A}})$ follows.

8.6 Conclusion

In this paper we considered the property of interpolation for levels of the μ-Calculus hierarchy. We proved that a uniform version of interpolation does not hold in levels Δ_n, for $n \geq 2$, and in particular uniform interpolation does not hold in the Alternation Free Fragment AF of the μ-Calculus. However, a Craig interpolant for a valid implication in AF can always be found in AF, and we provided a construction of the interpolant via automata. This construction uses non deterministic automata for AF-formulas, and it is not directly generalizable to the other ambiguous classes Δ_n with $n > 2$. Hence, the problem of the validity of ordinary interpolation in these levels is left open in this paper.

Acknowledgements This paper has been partially supported by the GNCS-INDAM Project 'Algoritmica per il model checking e la sintesi di sistemi safety-critica'.
The author wish to thank Michael T. Vanden Boom for pointing out various mistakes in a previous version of the paper, and the anonimous referees for being very patient and positive in spite of all these mistakes.

References

Arnold, A., & Niwinski, D. (2001). *Rudiments of the mu-Calculus*. Amsterdam: North Holland.

D'Agostino, G., & Hollenberg, M. (2000). Logical questions concerning the μ-calculus: Interpolation. Lyndon and Łoś-Tarski, Journal of Symbolic Logic, *65*(1), 310–332.

D'Agostino, G., & Lenzi, G. (2006). On modal mu-calculus with explicit interpolants. *Journal of Applied Logic*, *4*(3), 256–278.

Facchini, A., Venema, Y., & Zanasi, F. (2013). A characterization theorem for the alternation-free fragment of the modal μ-calculus. In *LICS '13 Proceedings of the 2013 28th Annual ACM/IEEE Symposium on Logic in Computer Science*, pp. 478–487.

Gutierrez, J., & Klaedtke, F. (2014). The μ-calculus alternation hierarchy collapses over structures with restricted connectivity. *Theoretical Computer Science*, *560*, 292–306.

Lutz, C., Seylan, I., & Wolter, F. (2012). An automata-theoretic approach to uniform interpolation and approximation in the description logic EL. *Principles of Knowledge Representation and Reasoning: Proceedings of the Thirteenth International Conference, KR*. Rome, Italy: AAAI Press.

Niwinski, D., & Walukiewicz, I. (1996). Games for the μ-calculus. *Theoretical Computer Science*, *163*(1–2), 99–116.

Santocanale, L., & Arnold, A. (2005). Ambiguous classes in μ-calculi hierarchies. *Theoretical Computer Science*, *333*(1–2), 265–296.

Chapter 9
Decidability of Some Interpolation Properties for Weakly Transitive Modal Logics

Anastasia Karpenko

Abstract This article surveys results on different versions of interpolation property in extensions of weakly transitive modal logic $wK4$ and of difference logic DL as well as the results on algebraic analogs of these properties for varieties of weakly transitive and DL-algebras. In particular, we describe all DL-extensions with *IPD* and prove that this property is decidable over DL. We also establish the decidability of WIP over $wK4$.

Introduction

This article surveys author's results on different versions of interpolation property in weakly transitive modal logics. Part of these results was obtained in coauthorship with L.L. Maksimova (see Karpenko and Maksimova 2010).

The first modal systems were introduced by C.I. Lewis (Lewis and Langford, 1932) in the early years of the 20th century. Many of the currently known modal systems originate from such areas of knowledge as philosophy, foundations of mathematics, computer science, cognitive science, and mathematical linguistics.

The increasing interest in modal logic is based on the connection between modal logic and the logical foundations of programming. This connection makes it possible to use the methods of modal logic and to employ the obtained results in the development of the programming theory.

The algebraic semantics as well as the relational semantics of modal logic introduced by S. Kripke (1963) create an opportunity for studying all of the modal systems as a whole. The works on this topic were written by W. Blok, L.L. Maksimova, K. Segerberg, E. Raseva, W. Rautenberg, V.V. Rybakov, S.K. Thomason, K. Fine, V. Shehtman, L.L. Esakia, etc.

A. Karpenko (✉)
Novosibirsk State University, Novosibirsk, Russia
e-mail: anastasia.v.karpenko@gmail.com

S. Odintsov (ed.), *Larisa Maksimova on Implication, Interpolation, and Definability*, Outstanding Contributions to Logic 15, https://doi.org/10.1007/978-3-319-69917-2_9

We will consider the extensions of some well-known modal systems such as $K4$, $S4$, and $S5$ as well as some less known systems like $wK4$ and DL. $S4$ and $S5$ were obtained by the characterization of various interpretations of the operator $\Box$ ("necessity") but K and $wK4$ were obtained from technical considerations.

The logic $wK4$ has been studied by several authors. L. Esakia (2001) studies $wK4$ by considering Kripke frames with a weakly transitive accessibility relation, which is a relation that satisfies the following condition: $(x \neq z \& xRy \& yRz \Rightarrow xRz)$. The logic $wK4$ is intermediate between the minimal normal modal logic K and the logic $K4$. L. Esakia has found the topological semantics for $wK4$ and defined the algebras with derivation that form the algebraic semantics for this modal system. We will refer to these algebras as weakly transitive.

The final section of Esakia (2001) is devoted to KS (Krister Segerberg's logic). This logic was obtained by K. Segerberg (1971) when he was building an axiomatization for a modal system characterized by frames $\langle W, R\rangle$, where the accessibility relation R is the inequality relation. Later KS has been studied by several authors, including V. Goranko (1990) and M. De Rijke (1992). According to De Rijke (1992), we will refer to KS as DL (Difference logic). Furthermore, it was noted (e.g. Kudinov 2006) that adding the modality of inequality to the other modal operators significantly increases the expressiveness of the language.

As a rule, the study of modal logics allocates different classes of logics that have certain properties. Such properties are completeness with respect to semantics, decidability, finite approximability, tabularity, disjunction property and interpolation property. A property P is said to be decidable over logic L if given a formula A there exists an algorithm that can determine whether the smallest extension of L containing A has this property. In this paper we will consider the last of the listed properties.

The interpolation theorem, proved by Craig (1957) for classical predicate logic, initiated the study of interpolation in various formal theories. In classical predicate logic the interpolation theorem of Craig has several equivalent statements that become inequivalent for modal logics. In this regard a few cases of interpolation properties were formulated: a Craig interpolation property (*CIP*), a deductive interpolation property (*IPD*), a restricted interpolation property (*IPR*) suggested by Maksimova (2003), and finally a weak interpolation property (*WIP*) (see Maksimova 2006a).

D.M. Gabbay (1972) has proved the interpolation theorems for predicate versions of $S4$ and K. George F. Shumm (1976) has proved the interpolation theorem for $S5$. These results were significantly expanded by L.L. Maksimova. She has proved that there exist precisely four logics among the extensions of $S5$, including $S5$ itself and the contradictory logic For, that have *CIP* (see Maksimova 1980). In Gabbay and Maksimova (2005) there is a list of 37 logics that contains all of the consistent extensions of $S4$ in which Craig's theorem is true. There is also a list of 49 logics that includes all of the consistent extensions of $S4$ with *IPD*. It is proved that *CIP* and *IPD* are decidable over the logic $S4$ but undecidable over $K4$. The latter has been proved by A.V. Chagrov (1990) (see also Chagrov and Zakharyaschev 1997). The family of the extensions of $K4$ that have *CIP* has the cardinality of the continuum Gabbay and Maksimova (2005).

We will consider the interpolation properties of the extensions of $K4$, $wK4$ and DL.

9.1 Preliminaries

In this section we will recall some notions and theorems from metamathematics of modal logics.

9.1.1 Frames

To begin with, we recall some basic definitions and axiomatizations of the logics that will be considered in the paper.

The formulae of a propositional modal logic are constructed by using the operations $\to$ and $\Box$ and the constant $\bot$. The remaining connectives (&, $\vee$ and $-$) are expressed in terms of the fundamental operations in the usual fashion. Put $\Diamond A := -\Box - A$, $\top := \bot \to \bot$.

Refer as a *normal modal logic* to a set of modal formulae that contains all classical tautologies and the axiom $\Box(A \to B) \to (\Box A \to \Box B)$ and is closed under substitutions and the two rules:

R1. $\dfrac{A, (A \to B)}{B}$ (modus ponens);

R2. $\dfrac{A}{\Box A}$ (necessitation).

In this paper we only consider normal modal logics and so we omit the word "normal". Denote the family of all modal logics that extend L by $NE(L)$. For a logic L and formula A we denote by $L + A$ the smallest extension of L containing A. With every L we associate the entailment relation $\vdash_L$ as follows: $\Gamma \vdash_L A$ means that A can be obtained from elements of $\Gamma \cup L$ with the help of R1 and R2.

The minimal modal logic is denoted by K. Let us mention some other known extensions of K:

$wK4 = K + ((A\&\Box A) \to \Box\Box A)$;
$DL = wK4 + (A \to \Box\Diamond A)$;
$K4 = K + (\Box A \to \Box\Box A)$;
$S4 = K4 + (\Box A \to A)$;
$S4.1 = S4 + (\Box\Diamond A \to \Diamond\Box A)$;
$S5 = S4 + (A \to \Box\Diamond A)$.

Given a list of propositional variables $\mathbf{p}$, $A(\mathbf{p})$ is a formula such that all its variables appear in $\mathbf{p}$. Let $\mathbf{p}$, $\mathbf{q}$, and $\mathbf{r}$ be the pairwise disjoint lists of variables.

A logic L has a *Craig interpolation property* (*CIP*) if for all formulae $A(\mathbf{p}, \mathbf{q})$ and $B(\mathbf{p}, \mathbf{r})$ the condition $\vdash_L A(\mathbf{p}, \mathbf{q}) \to B(\mathbf{p}, \mathbf{r})$ implies the existence of a formula $C(\mathbf{p})$ such that $\vdash_L A(\mathbf{p}, \mathbf{q}) \to C(\mathbf{p})$ and $\vdash_L C(\mathbf{p}) \to B(\mathbf{p}, \mathbf{r})$.

The logics K, $K4$, $S4$, $S4.1$, $S5$, as well as the inconsistent logic *For*, have *CIP* (see Gabbay and Maksimova 2005).

A logic L has a *deductive interpolation property* (*IPD*) if for all formulae $A(\mathbf{p}, \mathbf{q})$ and $B(\mathbf{p}, \mathbf{r})$ the condition $A(\mathbf{p}, \mathbf{q}) \vdash_L B(\mathbf{p}, \mathbf{r})$ implies the existence of a formula $C(\mathbf{p})$ such that $A(\mathbf{p}, \mathbf{q}) \vdash_L C(\mathbf{p})$ and $C(\mathbf{p}) \vdash_L B(\mathbf{p}, \mathbf{r})$.

A logic L has a *weak interpolation property* (WIP) if for all formulae $A(\mathbf{p}, \mathbf{q})$ and $B(\mathbf{p}, \mathbf{r})$ the condition $A(\mathbf{p}, \mathbf{q}),\ B(\mathbf{p}, \mathbf{r}) \vdash_L \bot$ implies the existence of a formula $C(\mathbf{p})$ such that $A(\mathbf{p}, \mathbf{q}) \vdash_L C(\mathbf{p})$ and $C(\mathbf{p}),\ B(\mathbf{p}, \mathbf{r}) \vdash_L \bot$.

The interpolation properties mentioned above relate as follows: $CIP \Rightarrow IPD \Rightarrow WIP$. In the class of modal logics, the reverse arrows fail. Notice that all these properties hold for classical logic.

For any logic $L \in NE(S5)$ the properties *CIP*, *IPD*, and WIP are equivalent (see Maksimova 2006a). Besides, any propositional logic over $S4.1$ has WIP. The family of $S4.1$-extensions is continual so there exists a continuum of $S4$-extensions with WIP. Thus WIP differs significantly from *CIP* and *IPD* because there only exists a finite number of $S4$-extensions with *IPD* and, therefore, with *CIP* (see Gabbay and Maksimova 2005).

Now let us describe the algebraic semantics of modal logic. It is known that there exists a dual isomorphism between the class of extensions of a logic and the class of subvarieties of the corresponding variety of algebras (see, e.g., Chagrov and Zakharyaschev 1997, Gabbay and Maksimova 2005). This allows us to explore the classes of logics using the methods of universal algebra.

Refer as a *modal algebra* to an algebra $\mathfrak{A} = (|\mathfrak{A}|, \to, 0, \Box)$ satisfying the identities of a Boolean algebra for the operations $\to$, 0 and the conditions $\Box 1 = 1$ and $\Box(x \to y) \leqslant \Box x \to \Box y$, where $1 = 0 \to 0$. A modal algebra $\mathfrak{A}$ is called *transitive* whenever $\Box x \leqslant \Box\Box x$ and *weakly transitive* whenever $\boxdot x = x \& \Box x \leqslant \Box\Box x$. A weakly transitive modal algebra $\mathfrak{A}$ is called a *DL-algebra* if $\mathfrak{A}$ satisfies the condition $x \leqslant \Box\Diamond x$, where $\Diamond A := -\Box - A$.

Let A be a formula of the modal language. We call A a *tautology* in $\mathfrak{A}$ (and write $\mathfrak{A} \models A$) whenever the identity $A = 1$ holds in $\mathfrak{A}$.

For a modal logic L, we write $\mathfrak{A} \models L$ instead of $(\forall A \in L)(\mathfrak{A} \models A)$. Put $V(L) = \{\mathfrak{A} \mid \mathfrak{A} \models L\}$. It is obvious that $V(L)$ is a variety for every normal modal logic L.

Moreover, it is known that $L = \{A \mid (\forall \mathfrak{A} \in V(L))(\mathfrak{A} \models A)\}$ (Gabbay and Maksimova 2005). For every class $\mathscr{K}$ of modal algebras the set $L(\mathscr{K}) = \{A \mid (\forall \mathfrak{A} \in \mathscr{K})(\mathfrak{A} \models A)\}$ is a modal logic. In particular, if $\mathscr{K}$ is a family of weakly transitive modal algebras then $L(\mathscr{K})$ is a logic extending $wK4$, and if $\mathscr{K}$ is a family of DL-algebras then $L(\mathscr{K})$ extends DL.

An algebra is called *simple* if it has precisely two congruences. A modal algebra is called *finitely indecomposable* if it does not decompose into a subdirect product of a finite number of proper quotient algebras.

A class $\mathscr{K}$ of algebras is called *amalgamable* whenever $\mathscr{K}$ has an amalgamation property (AP). That is, for any $\mathfrak{A}, \mathfrak{B}, \mathfrak{C} \in \mathscr{K}$ if $\beta : \mathfrak{A} \to \mathfrak{B}$ and $\gamma : \mathfrak{A} \to \mathfrak{C}$ are monomorphisms then there exist an algebra $\mathfrak{D} \in \mathscr{K}$ and monomorphisms $\delta : \mathfrak{B} \to \mathfrak{D}$, $\varepsilon : \mathfrak{C} \to \mathfrak{D}$ such that $\delta\beta = \varepsilon\gamma$ for all $x \in \mathfrak{A}$.

A class $\mathscr{K}$ of algebras is called *superamalgamable* whenever $\mathscr{K}$ is amalgamable and for the existing monomorphisms δ, ε the following stands:

$\delta(x) \leqslant \varepsilon(y) \Longleftrightarrow (\exists z \in \mathfrak{A})(x \leqslant \beta(z), \gamma(z) \leqslant y)$;

$\delta(x) \geqslant \varepsilon(y) \Longleftrightarrow (\exists z \in \mathfrak{A})(x \geqslant \beta(z), \gamma(z) \geqslant y)$.

There exists a weaker version of the amalgamation property.

A class $\mathscr{K}$ of algebras is called *weakly amalgamable* whenever $\mathscr{K}$ has a weak amalgamation property (WAP). That is, for any $\mathfrak{A}$, $\mathfrak{B}$, $\mathfrak{C} \in \mathscr{K}$, if $\beta : \mathfrak{A} \to \mathfrak{B}$, $\gamma : \mathfrak{A} \to \mathfrak{C}$ are monomorphisms then there exist $\mathfrak{D} \in \mathscr{K}$ and homomorphisms $\delta : \mathfrak{B} \to \mathfrak{D}$, $\varepsilon : \mathfrak{C} \to \mathfrak{D}$ such that $\delta\beta = \varepsilon\gamma$ for all $x \in \mathfrak{A}$ and the algebra $\mathfrak{D}$ is nondegenerate provided that so is $\mathfrak{A}$.

Recall that an algebra is called *nondegenerate* whenever it contains at least two elements.

For varieties of modal algebras the following stands: $SAP \Longrightarrow AP \Longrightarrow WAP$ (see Maksimova 1992).

Maksimova has studied various cases of the amalgamation property of varieties of modal algebras. In particular, she has shown Maksimova (1979) that precisely 50 varieties of topoboolean algebras are amalgamable, among which 38 have the strong amalgamation property. Maksimova (1992) has found the necessary and sufficient conditions for the amalgamation property, which enable us to reduce the question of the presence of the properties for this variety of modal algebras to the consideration of a subclass of finitely generated and finitely indecomposable algebras. *CIP* comes down to AP of the class of finitely generated simple algebras Maksimova (2008).

Here are the key theorems of L. L. Maksimova that will be used further.

Theorem 1 (Maksimova 1992) *For every modal logic L the following statements are equivalent:*

1. *L has CIP;*
2. *the variety $V(L)$ is superamalgamable;*
3. *given finitely indecomposable and finitely generated algebras $\mathfrak{A}$, $\mathfrak{B}$, $\mathfrak{C} \in V(L)$, any monomorphisms $\beta : \mathfrak{A} \to \mathfrak{B}$, $\gamma : \mathfrak{A} \to \mathfrak{C}$, and elements $b_0 \in \mathfrak{B}$, $c_0 \in \mathfrak{C}$, such that there is no $a \in \mathfrak{A}$ that meets the following conditions: $b_0 \leqslant \beta(a)$ and $\gamma(a) \leqslant c_0$, there exist an algebra $\mathfrak{D} \in V(L)$ and homomorphisms $\delta : \mathfrak{B} \to \mathfrak{D}$, $\varepsilon : \mathfrak{C} \to \mathfrak{D}$, such that $\delta\beta = \varepsilon\gamma$ and $\delta(b_0) \not\leqslant \varepsilon(c_0)$.*

Theorem 2 (Maksimova 1992) *For every modal logic L the following statements are equivalent:*

1. *L has IPD;*
2. *the variety $V(L)$ is amalgamable;*
3. *given finitely indecomposable and finitely generated algebras $\mathfrak{A}$, $\mathfrak{B}$, $\mathfrak{C} \in V(L)$ and any monomorphisms $\beta : \mathfrak{A} \to \mathfrak{B}$, $\gamma : \mathfrak{A} \to \mathfrak{C}$ there exist an algebra $\mathfrak{D} \in V(L)$ and monomorphisms $\delta : \mathfrak{B} \to \mathfrak{D}$, $\varepsilon : \mathfrak{C} \to \mathfrak{D}$ such that $\delta\beta = \varepsilon\gamma$.*

Theorem 3 (Maksimova 2006, 2008) *For every modal logic L the following statements are equivalent:*

1. *L has WIP;*
2. *the class of simple algebras of the variety $V(L)$ is amalgamable;*
3. *the class of finitely generated simple algebras of the variety $V(L)$ is amalgamable.*

Besides, Maksimova (2006a) proved that having WIP is equivalent to the variety $V(L)$ being weakly amalgamable.

Thus, the study of weak amalgamation focuses on finitely generated simple algebras of the corresponding variety. In the joint paper Karpenko and Maksimova (2010) the following theorems have been proved:

Theorem 4 *A weakly transitive algebra $\mathfrak{A}$ is simple iff the following equality holds:*

$$\boxdot x = \begin{cases} 1 \text{ if } x = 1 \\ 0 \text{ otherwise.} \end{cases}$$

Theorem 5 *Every simple weakly transitive algebra is a DL-algebra.*

Element a_0 of a Boolean algebra is called an *atom* if $a_0 \neq 0$ and there is no element $a \leqslant a_0$ such that $0 \neq a \neq a_0$. Accordingly, a Boolean algebra is called *atomic* if for every element $a \neq 0$ there exists an atom $a_0 \leqslant a$. It is known that every finite Boolean algebra $\mathfrak{A}$ is atomic Rasiowa and Sikorski (1963).

In accordance with (Karpenko 2008; Karpenko and Maksimova 2010) by V_n^m we denote a finite modal algebra with $(n + m)$ atoms $a_1, \ldots, a_n, b_1, \ldots, b_m$ such that for every atom x the following equality holds:

$$\Diamond x = \begin{cases} 1 \;\; if \;\; x = a_i \text{ for some } 1 \leqslant i \leq n; \\ -x \; if \;\; x = b_j \text{ for some } 1 \leqslant j \leq m. \end{cases}$$

Since any element of algebra V_n^m is uniquely represented as a sum of atoms and $\Diamond(y \vee z) = \Diamond y \vee \Diamond z$, the above equality uniquely defines the operation $\Diamond$ for all elements of V_n^m.

Theorem 6 (Karpenko and Maksimova 2010) *Every finitely generated and finitely indecomposable DL-algebra is simple and isomorphic to V_n^m for suitable n, m, where $n + m > 0$.*

Therefore, the class of finitely generated simple DL-algebras up to isomorphism coincides with the class $\{V_n^m \mid m + n > 0\}$, which we denote by $\mathscr{K}(DL)$. According to Theorem 5, the class of finitely generated simple weakly transitive modal algebras also coincides with $\mathscr{K}(DL)$ up to isomorphism.

Now let us turn to Kripke semantics. This approach to the study of different logics is the most common. It uses "geometric" features of Kripke frames.

A *Kripke frame* is a pair $\mathscr{W} = \langle W, R\rangle$, where W is a set of possible worlds and R is a binary relation on W. A *Kripke model* is a triple $M = \langle W, R, \Vdash\rangle$, where $\langle W, R\rangle$ is a Kripke frame and $\Vdash$ is a relation between possible worlds and formulae such that for every $x \in W$ and formulae A and B (we write $x \nVdash A$ meaning that $x \Vdash A$ fails to hold) we have

(M0) $x \nVdash \bot$;
(M1) $x \Vdash A \to B \Leftrightarrow (x \nVdash A$ or $x \Vdash B)$;
(M2) $x \Vdash \Box A \Leftrightarrow \forall y\,(xRy \Rightarrow y \Vdash A)$.

It is known from Esakia (2001) that $wK4$ is Kripke complete and is characterized by frames with a weakly transitive relation R, and DL is complete with respect to the frames such that R is the inequality relation (De Rijke 1992). $K4$ is complete with respect to the frames with a transitive relation, $S4$ is complete with respect to the frames with a reflexive and transitive relation, and $S5$ is complete with respect to the frames with a reflexive, transitive, and symmetric relation (see e.g. Chagrov and Zakharyaschev 1997; Gabbay and Maksimova 2005).

Kripke semantics for all of the listed logics is definitely more demonstrative, but not all modal logics are Kripke complete (S.K. Thomason 1974; K. Fine 1974). In this regard, the approach that combines semantic and algebraic methods seems convenient. The method is based on a representation theorem proved by Jonsson and Tarski (1951) for Boolean algebras with operators.

Let $\mathscr{W} = \langle W, R\rangle$ be a Kripke frame. Let W^+ denote a set of subsets of W. Then

$$\mathscr{W}^+ = \langle W^+, \to, 0, \Box\rangle,$$

where $0 = \emptyset$, $X \to Y = (W - X) \cup Y$, $\Box X = \{x \in W | \forall y(xRy \Rightarrow y \in X)\}$ for $X, Y \subseteq W$, is a modal algebra.

An analog of representation theorem by Jonsson and Tarski is valid for modal algebras (see e.g. Gabbay and Maksimova 2005):

Theorem 7 *Every modal algebra $\mathfrak{A}$ is isomorphically embeddable in a suitable algebra $\mathscr{W}^+$. If $\mathfrak{A}$ is finite then it is isomorphic to $\mathscr{W}^+$.*

Refer as a *frame $T_{\mathfrak{A}}$ of algebra $\mathfrak{A}$* to the set of all atoms with the relation $R_{\mathfrak{A}}$ defined as $aR_{\mathfrak{A}}b \rightleftharpoons (a \leqslant \Diamond b)$ for all $a, b \in T_{\mathfrak{A}}$.

Consider the set of atoms of the algebra V_n^m:

$$At\,V_n^m = \{a_1, \dots, a_n, b_1, \dots, b_m\}$$

with a relation R: $a\,R\,b \Leftrightarrow a \leqslant \Diamond b$. From the definition of V_n^m we immediately have that $a_i R_{V_n^m} a_j$ for $1 \leqslant i, j \leqslant n$; $b_j R_{V_n^m} a_i$ for $1 \leqslant i \leq n$, $1 \leqslant j \leq m$; $a_i R_{V_n^m} b_j$ for $1 \leqslant i \leq n$, $1 \leqslant j \leq m$ and $b_i R_{V_n^m} b_j$ for $1 \leqslant i, j \leqslant m, i \neq j$.

Also, for $1 \leqslant i \leq m$ the condition $b_i R_{V_n^m} b_i$ is not met. Indeed, let us have that $b_i R_{V_n^m} b_i$, then by definition of the relation R we have $b_i \leqslant \Diamond b_i = \neg b_i$. Hence there is a contradiction.

So we have that in $At\,V_n^m$ a_i are reflexive atoms, and b_j are irreflexive atoms, and, in addition, all of the atoms are pairwise accessible.

For finite simple DL-algebras $\mathfrak{A}$ and $\mathfrak{B}$ we write $\mathfrak{A} \preccurlyeq \mathfrak{B}$ iff $\mathfrak{A}$ is isomorphically embeddable in $\mathfrak{B}$. Clearly, if $\mathfrak{A} \preccurlyeq \mathfrak{B}$ and $\mathfrak{B} \preccurlyeq \mathfrak{A}$ then $\mathfrak{A}$ and $\mathfrak{B}$ are isomorphic. A relation $\mathfrak{A} \prec \mathfrak{B}$ will mean that $\mathfrak{A}$ is isomorphically embeddable in $\mathfrak{B}$ but not isomorphic to $\mathfrak{B}$.

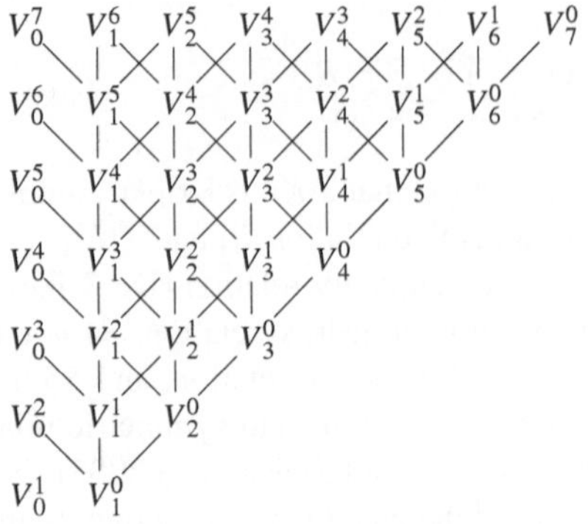

Fig. 9.1 The structure of embeddings

Karpenko and Maksimova (2010) proved that for $m \geqslant 0, n \geqslant 1$ we have

$$V_n^m \prec V_{n+1}^m, \quad V_n^m \prec V_n^{m+1}, \quad V_n^m \prec V_{n-1}^{m+2}. \tag{9.1}$$

Moreover, the following theorem takes place.

Theorem 8 *The relation $\preccurlyeq$ is the reflexive and transitive closure of (9.1).*

The diagram of $\preccurlyeq$ is depicted above (Fig. 9.1).

9.2 Characteristic Formulae for Algebras V_n^m

The Birkhoff Theorem (see, e.g., Maltsev 1970) implies the following important proposition:

Proposition 1 *Every variety is generated by the class of its finitely generated and finitely indecomposable algebras.*

Yankov (1963) defined the notion of a characteristic formula for Heyting algebras. A similar notion for modal algebras was defined in Rautenberg (1980). Karpenko (2010) defined characteristic formulae for subdirectly indecomposable weakly transitive finite modal algebras. However, we are interested in the varieties of DL-algebras for which the class of finitely indecomposable and finitely generated algebras coincides with the class $\mathscr{K}(DL)$. Thus, when studying this class it is sufficient to define the notion of a characteristic formula for algebras V_n^m $(n + m > 0)$.

Let V_n^m be defined as before. Associate a variable p_x to every x from $At\,V_n^m = \{a_1, \ldots, a_n, b_1, \ldots, b_m\}$. Formulas $\theta(V_n^m)$ that will be defined below have the properties similar to the properties of characteristic formulae:

$$\theta(V_n^m) = \Bigg(\Bigg(\Big(\mathop{\&}_{\substack{x,y\in At V_n^m\\ x\neq y}} \boxdot-(p_x\&p_y)\Big)\&\boxdot(\bigvee_{x\in At V_n^m} p_x)\&$$

$$\&\mathop{\&}_{x\in\{a_1,\dots,a_n\}}\boxdot\Diamond p_x\&\mathop{\&}_{x\in\{b_1,\dots,b_m\}}\boxdot(\Diamond p_x\leftrightarrow -p_x)\Big)\Bigg)\to 0\Bigg) \qquad (9.2)$$

Now let us mention some of the properties of characteristic formulae that have been proved in Karpenko (2012).

Lemma 1 *Let $\mathfrak{A}$ be a subdirectly indecomposable weakly transitive finite DL-algebra. Then the following conditions are equivalent:*

1. $\mathfrak{A} \nvDash \theta(\mathfrak{A})$;
2. every algebra $\mathfrak{B}$ satisfies $\mathfrak{B} \nvDash \theta(\mathfrak{A})$ iff there exists a monomorphism from $\mathfrak{A}$ into a suitable homomorphic image of $\mathfrak{B}$.

This lemma implies the following:

Lemma 2 *Let $\mathfrak{A}$ and $\mathfrak{B}$ be finitely indecomposable and finitely generated DL-algebras. $\mathfrak{B} \nvDash \theta(\mathfrak{A})$ iff $\mathfrak{A}$ is isomorphically embeddable in $\mathfrak{B}$.*

Lemma 3 *For every logic L that extends DL and a subdirectly indecomposable finite DL-algebra $\mathfrak{A}$ we have*

$$\mathfrak{A} \vDash L \Leftrightarrow \theta(\mathfrak{A}) \notin L.$$

A class $\mathcal{K}$ is called *downward closed* if for every algebras $\mathfrak{A} \in \mathcal{K}$ and $\mathfrak{B} \in \mathcal{K}(DL)$ we have $(\mathfrak{B} \preccurlyeq \mathfrak{A} \Rightarrow \mathfrak{B} \in \mathcal{K})$.

The above lemmas was used in Karpenko (2010) to prove the following:

Theorem 9 *There exists a bijective correspondence between the varieties of DL-algebras and the downward closed subclasses of the class $\mathcal{K}(DL)$.*

Note that every downward closed subclass of the class $\mathcal{K}(DL)$ can be characterized by minimal algebras from $\mathcal{K}(DL)$ that are not in the subclass. Thus, every variety of DL-algebras can also be characterized by minimal algebras from $\mathcal{K}(DL)$ that are not in the subclass.

9.3 Decidability of *IPD* and WIP for DL-logic and $wK4$

We have mentioned earlier that amalgamable varieties of DL-algebras have been studied thoroughly in Karpenko (2010). Particularly, the following theorems have been proved:

Theorem 10 (Karpenko (2010)) *There exist precisely 16 weakly amalgamable varieties of DL-algebras that correspond to the following subclasses of $\mathcal{K}(DL)$:*

1. *the empty subclass; in this case the variety contains only a one-element algebra and corresponds to the contradictory logic For;*
2. *the subclass generated by V_0^1;*
3. *the subclass generated by $V_n^0 (n > 0)$; in this case $V(L)$ corresponds to the logic S5;*
4. *the subclass generated by $V_n^0 (n > 0)$ and V_0^1;*
5. *the subclasses generated by V_j^0 for $(j = 1, 2)$;*
6. *the subclasses generated by V_j^0 and V_0^1, for $(j = 1, 2)$;*
7. *the subclass generated by V_1^1;*
8. *the subclass generated by V_1^1 and V_0^1;*
9. *the subclass generated by V_0^k for $(k = 2, 3, 4)$;*
10. *the subclass generated by V_0^k and V_0^1 for $(k = 2, 3, 4)$.*

Theorem 11 (Karpenko (2010)) *All varieties in Theorem 10, and only those, are amalgamable varieties of DL-algebras.*

So, for varieties of DL-algebras the amalgamation property and the weak amalgamation property are equivalent. However, L.L. Maksimova (1992) has proved that the variety corresponding to $\mathcal{K}(DL)$ that is the variety generated by the algebra V_1^1 does not have SAP. Consequently, for varieties of DL-algebras, the amalgamation property is not equivalent to the superamalgamation property.

As previously mentioned, there exists a one-to-one correspondence between the logics extending DL and the varieties of DL-algebras. In addition, Theorem 9 implies that there exists a one-to-one correspondence between the varieties of DL-algebras and the downward closed subclasses of $\mathcal{K}(DL)$. Thus, a class $\mathcal{K}(L) = \{V_n^m | V_n^m \models L, n + m > 0\}$ corresponds to every logic $L \in NE(DL)$, and, conversely, a logic extending DL corresponds to every subclass $\mathcal{K}(DL)$.

The following theorem has been proved in Karpenko (2010):

Theorem 12 *A logic L extending DL has IPD iff $\mathcal{K}(L) = \{V_n^m | V_n^m \in V(L), n + m > 0\}$ is one of the 16 classes from Theorem 10.*

This theorem implies that there exist precisely 16 extensions of DL with *IPD*. The axiomatization of these logic was obtained in Karpenko (2012).

Theorem 13 *Let $\mathcal{K}$ be one of the 16 amalgamable subclasses of $\mathcal{K}(DL)$, and let $V_{n_1}^{m_1}, \ldots, V_{n_k}^{m_k}$ be minimal algebras that are not in $\mathcal{K}$. Then $L = DL + \theta(V_{n_1}^{m_1}) + \cdots + \theta(V_{n_k}^{m_k})$ has IPD.*

Theorem 14 *The following extensions of DL, and only those, have IPD:*

1. $For = DL + \theta(V_1^0) + \theta(V_0^1)$;
2. $(For)' = DL + \theta(V_1^0)$;
3. $LV_1^0 = DL + \theta(V_0^2) + \theta(V_1^1) + \theta(V_2^0) + \theta(V_0^1)$;
4. $(LV_1^0)' = DL + \theta(V_0^2) + \theta(V_1^1) + \theta(V_2^0)$;
5. $LV_0^2 = DL + \theta(V_1^1) + \theta(V_2^0) + \theta(V_0^1)$;
6. $(LV_0^2)' = DL + \theta(V_1^1) + \theta(V_2^0)$;

7. $LV_1^1 = DL + \theta(V_2^0) + \theta(V_0^2) + \theta(V_0^3) + \theta(V_0^1)$;
8. $(LV_1^1)' = DL + \theta(V_2^0) + \theta(V_0^2) + \theta(V_0^3)$;
9. $LV_2^0 = DL + \theta(V_3^0) + \theta(V_1^1) + \theta(V_0^2) + \theta(V_0^1)$;
10. $(LV_2^0)' = DL + \theta(V_3^0) + \theta(V_1^1) + \theta(V_0^2)$,
11. $LV_0^3 = DL + \theta(V_0^2) + \theta(V_2^0) + \theta(V_0^1)$;
12. $(LV_0^3)' = DL + \theta(V_0^2) + \theta(V_2^0)$;
13. $LV_0^4 = DL + \theta(V_0^2) + \theta(V_0^3) + \theta(V_3^0) + \theta(V_2^1) + \theta(V_0^1)$;
14. $(LV_0^4)' = DL + \theta(V_0^2) + \theta(V_0^3) + \theta(V_3^0) + \theta(V_2^1)$;
15. $S5 = DL + \theta(V_0^2) + \theta(V_1^1) + \theta(V_0^1)$;
16. $(S5)' = DL + \theta(V_0^2) + \theta(V_1^1)$.

A convenient algorithm for verifying the validity in $S5$, which is characterized by algebras V_n^0, $n \geqslant 1$, has been built in Shreiner (2003).

Proposition 2 (Shreiner 2003) *Let a formula A contain n modalities and m variables, and* $k = \min(n+1, 2^m)$. *Then A is true in S5 iff A is valid in* V_k^0.

In addition, P.A. Shreiner created a computer program that given a finite list of axioms added to $S5$ checks whether a logic coincides with one of the four extensions that have *CIP*. In particular, it checks whether the conjunction of added axioms is valid in frames corresponding to the algebras of the form V_n^0, $n > 0$.

A problem of amalgamation of varieties of algebras is said to be decidable if given a finite basis of identities of the variety there exists an algorithm that can determine whether the variety is amalgamable.

Theorem 15 (Karpenko 2012)

1. *IPD is decidable over* DL.
2. AP *is decidable for varieties of* DL*-algebras.*

We have mentioned earlier that *IPD* and WIP are equivalent over DL. *IPD* and WIP are also equivalent to *CIP* over $S5$ (Maksimova 2006a). However, this is not true over the extensions of DL. The results of Maksimova (1992) imply that LV_1^1 does not have *CIP*, although it has *IPD*.

Theorem 16 (Karpenko 2010) *A variety V of weakly transitive modal algebras is weakly amalgamable iff* $V \cap V(DL)$ *is weakly amalgamable.*

Moreover, the following theorem is established.

Theorem 17 (Karpenko 2010) *A logic L extending* $wK4$ *has* WIP *iff the logic* $L + (A \to \Box\Diamond A)$ *has IPD.*

Theorem 16 directly implies the following.

Theorem 18 (Karpenko 2012)

1. WIP *is decidable over* $wK4$.
2. WAP *is decidable for varieties of weakly transitive modal algebras.*

Acknowledgements The author expresses her gratitude and deep appreciation to her scientific advisor L.L. Maksimova for stating interesting problems, and for the all-around support throughout the work.

References

Blok, W. J. (1980). The lattice of modal logics, an algebraic investigation. *Journal of Symbolic Logic*, *45*(6), 221–236.

Chagrov, A. V. (1990). Nerazreshimye svoĭstva rasshireniĭ logiki dokazuemosti. *Algebra i Logika*, *29*(3), 350–367. [Undecidable properties of extensions of provability logic].

Chagrov, A., & Zakharyaschev, M. (1997). *Modal logic*. Oxford: Clarendon Press.

Craig, W. (1957). Three uses of Herbrand-Gentzen theorem in relating model theory and proof theory. *Journal of Symbol Logic*, *22*, 269–285.

De Rijke, M. (1992). The modal logic of inequality. *Journal of Symbol Logic*, *57*(2), 566–584.

Esakia, L. L. (2001). Slabaya tranzitivnost' — restituciya. *Logical Investigations*, *8*, 244–255. [Weak Transitivity — A Restitution].

Fine, K. (1974). An incomplete logic containing S4. *Theoria*, *40*, 23–29.

Gabbay, D. M. (1972). Craig's interpolation theorem for modal logics, *in* W. Hodges (ed.) *Conference in Mathematical Logic London 70*. Vol. 255 of *Lecture Notes in Mathematics*, Springer, Berlin, Heidelberg.

Gabbay, D. M. and Maksimova, L. L. (2005). *Interpolation and definability: modal and intuitionistic logics*, Vol. 46 of *Oxford Logic Guides*, Clarendon Press, Oxford.

Goranko, V. (1990). Modal definability in enriched languages. *Notre Dame Journal of Formal Logic*, *31*(1), 81–105.

Jonsson, B., & Tarski, A. (1951). Boolean algebras with operators. *American Journal of Mathematics*, *73*, 891–939.

Karpenko, A. V. (2008). Weak interpolation in extensions of the logics S4 and K4. *Algebra and Logic*, *47*(6), 395–404.

Karpenko, A. V. (2010). Interpolation properties in the extensions of the logic of inequality. *Siberian Mathematical Journal*, *51*(3), 439–451.

Karpenko, A. V. (2012). Interpolation in weakly transitive modal logics. *Algebra and Logic*, *51*(2), 131–43.

Karpenko, A. V., & Maksimova, L. L. (2010). Simple weakly transitive modal algebras. *Algebra and Logic*, *49*(3), 233–245.

Kripke, S. (1963). Semantic analysis of modal logic I. *Normal modal propositional calculi, Zeitschrift für mathematische Logik und Grundlagen der Mathematik*, *9*, 67–96.

Kudinov A. (2006). Topological modal logics with difference modality, *in* Vol. 6 of *Advances in Modal Logic*, College Publications, London, pp. 319–332.

Lewis, C. I., & Langford, C. H. (1932). *Symbolic Logic*. New York: Appleton-Centyre-Croft.

Maksimova, L. L. (1979). Interpolacionnye teoremy v modalnykh logikakh i amalgamiruemye mnogoobraziya topobulevykh algebra. *Algebra i Logika*, *18*(5), 556–586. [Interpolation theorems in modal logics and amalgamable varieties in topoboolean algebras].

Maksimova, L. L. (1980). Interpolacionnye teoremy v modalnykh logikakh: dostatochnye usloviya. *Algebra i Logika*, *19*, 194–213. [Interpolation theorems in modal logics: Sufficient conditions.].

Maksimova, L. L. (1992). Modalnye logiki i mnogoobraziya modalnykh algebr: usloviya Betha, interpolaciya i amalgamiruemost'. *Algebra i Logika*, *31*(2), 145–166. [Modal logics and varieties of modal algebras: Beth's conditions, interpolation, and amalgamability].

Maksimova, L. L. (2008). A weak form of interpolation in equational logic. *Algebra and Logic*, *47*(1), 56–64.

Maksimova, L. (2003). Restricted interpolation in modal logics, *in* P. Balbiani, N.- Y. Suzuki, F. Wolter, and M. Zakharyashev (eds.), Vol. 4 of *Advances in Modal Logics*, Kings's College London Publications, London, pp. 297–312.

Maksimova, L. (2006). On a form of interpolation in modal logic. *Bulletin of Symbolic Logic*, *12*(2), 340.

Maksimova, L. (2006a). Definability and Interpolation in Non-Classical Logics. *Studia Logica*, *82*(2), 271–291.

Maltsev, A. I. (1970). *Algebraicheskie Systemy*. Moscow: Nauka. [Algebraic Systems].

Rasiowa, H., & Sikorski, R. (1963). *The mathematics of metamathematics*. Warszawa: Państwowe Wydawnistwo Naukowe.

Rautenberg, W. (1979). *Klassische und nicht-classische Aussagenlogik*. Braunschweig: Friedr. Vieweg & Sohn.

Rautenberg, W. (1980). Splitting lattice of logics. *Archiv für. mathematische Logik*, *20*, 155–159.

Segerberg, K. (1971). *An Essay in Classical Modal Logic*, Vol. 1–3 of *Filosofiska Föreningen och Filosofiska Institutionen vid Uppsala Universitet*, Filosofiska Studier, Uppsala.

Segerberg, K. (1976). "Somewhere else" and "some other time", *in Wright and Wrong: mini-essay in honor of G.H. von Wright*, Publication of the group in logic and methodology of Real Finland, pp. 61–64.

Segerberg, K. (1980). A note on the logic of elsewhere. *Theoria*, *46*(2–3), 183–187.

Shehtman, V. (1990). Everywhere "and" Here. *Journal of Applied Non-classical Logic*, *9*(2–3), 369–380.

Shreiner, P. A. (2003). Avtomaticheskoe raspozanvanie interpolacionnogo svojstva u nekotorykh superintuizionistskikh propozicionalnykh logik. *Vestnik NGU, Ser.: Matematika, Mekhanika i Informatika*, *3*, 85–92. [Automatic recognition of the interpolation property in some superintuitionistic propositional logics].

Shumm, G. F. (1976). Interpolation in S5 and related systems. *Reports on Mathematical Logic*, *6*, 107–110.

Thomason, S. K. (1974). An incompleteness theorem in modal logic. *Theoria*, *40*, 30–34.

Yankov, V. A. (1963). O svyazi mezhdu vyvodimost'yu v intuicionistskom propozicionalnom ischislenii i konechnymi implikativnymi strukturami. *Doklady Akademii Nauk SSSR*, *151*, 1293–1294. [On the relation between deducibility in intuitionistic propositional calculus and finite implicative structures].

Chapter 10
Lattice NExtS4 from the Embedding Theorem Viewpoint

Alexei Muravitsky

Dedicated to Larisa L. Maksimova, teacher and friend.

Abstract We follow along the line of research begun with the pioneering work Maksimova and Rybakov (1974) which has initiated comparative analysis of the lattices of normal extensions of S4 and of intermediate logics. Grounding on the Gödel–McKinsey–Tarski embedding, for any S4-logic, we define the notion of τ-decomposition and that of a modal component of the logic. Then, we investigate conditions, when a modal logic has its least modal component.

10.1 Preliminaries

Before starting our discussion (in the next section), we introduce some basic facts and notation.

We will be dealing with logics in the modal propositional language, L, containing countable set of propositional variables $p, q, r, \ldots$ (with or without subscripts) and the connectives: $\wedge$, $\vee$, $\rightarrow$, $\neg$, and $\Box$ (modality). The parentheses, (and), will be used, in a usual way, as punctuation marks. As usual, symbol $\Diamond$ abbreviates $\neg\Box\neg$. Unspecified formulas of this language, L-*formulas*, will be denoted by letters α, β, γ (maybe, with subscripts). Also, we will be using the *assertoric* fragment of L, which consists only of the $\Box$-free formulas of L. We denote this fragment by L_a and unspecified formulas of it, L_a-*formulas*, will be denoted by $A, B, C, \ldots$ (with or without subscripts). Treating the connectives as operations on the set of all L-formulas or L_a-formulas, we obtain the *formula algebra* for L or L_a, respectively.

We will be focusing only on those logics formulated in L which are consistent normal extensions of modal logic **S4** calling them **S4**-*logics* or *modal logics*. Given a set Γ of L-formulas, we obtain a normal extension $\mathbf{S4} + \Gamma$ by adding Γ to the

A. Muravitsky (✉)
Louisiana Scholars' College, Northwestern State University, Natchitoches, LA 71497, USA
e-mail: alexeim@nsula.edu

S. Odintsov (ed.), *Larisa Maksimova on Implication, Interpolation, and Definability*, Outstanding Contributions to Logic 15, https://doi.org/10.1007/978-3-319-69917-2_10

theorems of **S4** and, then, closing the union by (simultaneous) substitution, modus ponens and the necessitation rule $\alpha/\Box\alpha$. Occasionally, following Maksimova and Rybakov (1974), we use the following notation:

$$[\Gamma] = \mathbf{S4} + \Gamma,$$

provided that the right-hand side defines a consistent logic.

All logics formulated in L_a we will be dealing with are consistent extensions of intuitionistic propositional logic **Int** with substitution as a postulated inference rule. It is well known that these logics are the sets of L_a-formulas closed by substitution and modus ponens, which contain **Int** and are contained in classical propositional logic **Cl** — the reason according to which they are called *intermediate*. For definitions of **S4**, **Int**, and their (normal) extensions the reader is referred to books such as, e.g., Chagrov and Zakharyaschev (1997) and Hughes and Cresswell (1996).

It is well known that all **S4**-logics form a bounded distributive lattice, NExt**S4**, with the least element **S4** and the greatest $S_1 = \mathbf{S4} + p \to \Box p$. Also, the intermediate logics form a bounded distributive lattice, Ext**Int**, with the least element **Int** and the greatest **Cl**. Both NExt**S4** and Ext**Int** are Heyting algebras.

The fact that a formula α is valid in a logic $M \in$ NExt**S4** will be denoted either by $M \vdash \alpha$ or by $\alpha \in M$. The same notation applies to a formula A and a logic $L \in$ Ext**Int**.

We designate the following **S4**-logics which will play special role in our discussion.

$$\mathbf{S5} = \mathbf{S4} + p \to \Box\Diamond p = \mathbf{S4} + \Box\neg\Box\neg\Box p \to \Box p$$
$$\mathbf{S4.1} = \mathbf{S4} + \Box\Diamond p \to \Diamond\Box p \qquad (\textit{McKinsey logic})$$
$$\mathbf{Grz} = \mathbf{S4} + \Box(\Box(p \to \Box p) \to p) \to p \qquad (\textit{Grzegorczyk logic})$$
$$M_0 = \mathbf{Grz} \cap \mathbf{S5}.$$

Since we will be focusing on classes of (normal) extensions of modal and intermediate logics rather than on particular logics, we introduce the corresponding notation. Given $M \in$ NExt**S4**, NExt$M$ denotes the partially ordered set of all consistent modal logics containing M. In particular, in Sect. 10.2 we will discuss the structure of NExt**Grz**. Also, given modal logics M and M' with $M \subseteq M'$, by $[M, M']$ we denote the interval of NExt**S4** with the given logics as its endpoints. As to intermediate logics, only lattice Ext**Int** will be employed.

We assume that the reader is familiar with the algebraic semantics for logics in NExt**S4**, that is with **S4**-*algebras*, known also under the names of topological Boolean algebras in Rasiowa and Sikorski (1970), interior algebras in Blok (1976) and topoboolean algebras in Gabbay and Maksimova (2005); Chagrov and Zakharyaschev (1997) uses two terms — interior algebra and closure algebra. On the other hand, *Heyting algebras* discussed in Burris and Sankappanavar (1981) and Grätzer (1978), known also as pseudo-Boolean algebras in Rasiowa and Sikorski (1970) and Chagrov and Zakharyaschev (1997), will be employed here as the algebraic semantics for the intermediate logics. As is known, the collection of the open elements (x is open if $\Box x = x$) of an **S4**-algebra, constitutes a Heyting algebra, the *Heyting carcass* of the former. The Heyting carcass of an **S4**-algebra B is denoted by B°. Also, both **S4**-algebras and Heyting algebras are known to be bounded lattices. Given such an algebra, we denote its least element by **0** and its greatest element by **1**. The *logic of an algebra*, whether the latter is an **S4**- or Heyting algebra, is the set of the formulas that take **1** at each *valuation* on the algebra. (A valuation is understood as a homomorphism from the formula algebra into the given algebra.) If a formula α takes **1** at each valuation on an algebra B, it is called *valid* on B, symbolically $\mathrm{B} \models \alpha$. Otherwise α is *invalid* on B, symbolically $\mathrm{B} \not\models \alpha$.

Our discussion will go in the framework of the following notions, properties, and notations. Thus we expect that the reader is familiar with:

- algebraic class operators **H**, **S**, $\mathbf{P_u}$; consult, e.g., Grätzer (1978, 1979) or Burris and Sankappanavar (1981);
- the definitions of an intuitionistic relational model and frame (an **I**-*model* and **I**-*frame* for short) and those of relational model and frame for the modal logic **S4** (an **S4**-*model* and **S4**-*frame* for short); consult, e.g., Chagrov and Zakharyaschev (1997) or Hughes and Cresswell (1996);
- the fact that a Heyting algebra is *subdirectly irreducible* see definition in Burris and Sankappanavar (1981) or Grätzer (1978, 1979), if and only if it has a pre-top element and an **S4**-algebra is *subdirectly irreducible* if and only if the Heyting carcass of the latter is subdirectly irreducible; for Heyting algebras see Balbes and Dwinger (1974), Chapter IX, and for **S4**-algebras (alias interior algebras) see Blok (1976), Sect. 1.1; also, see Chagrov and Zakharyaschev (1997);
- that, given a finite subdirectly irreducible **S4**-algebra B, its *characteristic formula* χ_{B} is defined so that for any **S4**-algebra B′,

  (a) $\mathrm{B} \not\models \chi_{\mathrm{B}}$ (which also follows from (c) below), and
  (b) $\mathrm{B}' \not\models \chi_{\mathrm{B}} \Leftrightarrow \mathrm{B} \in \mathbf{HS}(\mathrm{B}')$, and
  (c) for any formula α, $\mathrm{B} \not\models \alpha \Leftrightarrow \mathbf{S4} + \alpha \vdash \chi_{\mathrm{B}}$;

  consult Fine (1974) or Rautenberg (1979), pp. 223–224, or Rautenberg (1980), Splitting Theorem, and Citkin (2013, 2014) for a more general setting; the properties (a) – (c) have originated with Jankov (1969);
- that for any Heyting algebra A, there is a unique (up to isomorphism) **S4**-algebra, $s(\mathrm{A})$, called the *Boolean span* of A, such that A is a sublattice of $s(\mathrm{A})$, $s(\mathrm{A})$ is generated with its Boolean operations by the elements of A and $s(\mathrm{A})^{\circ} = \mathrm{A}$; consult Maksimova (1975), Proposition 1, or Gabbay and Maksimova (2005), Theorem 3.8;
- that NExt**S4** is a distributive lattice relative to $\cap$ (set intersection) as meet and $\oplus$ (closure of set union) as join; moreover, given sets Γ and Σ of L-formulas, the following holds:

$$\left.\begin{array}{c}[\Gamma] \cap [\Sigma] = \left[\{\Box\alpha \vee' \Box\beta \mid \alpha \in \Gamma,\ \beta \in \Sigma\}\right], \\ [\Gamma] \oplus [\Sigma] = [\Gamma \cup \Sigma];\end{array}\right\} \tag{10.1}$$

where $\Box\alpha \vee' \Box\beta$ is the "disjoint" disjunction of $\Box\alpha$ and $\Box\beta$; see Maksimova and Rybakov (1974), Theorem 1;

- that, moreover, Ext**Int** and NExt**S4** are complete Heyting algebras; (we will be denoting infinite inf and sup by $\wedge$ and $\vee$, respectively);
- *Gödel–McKinsey–Tarski translation* see, e.g., Chagrov and Zakharyaschev (1997) or Gabbay and Maksimova (2005) $\mathbf{t} : A \mapsto A^{\mathbf{t}}$, where A is any L_a-formula, which is defined according to the sample: $(\neg\neg p \to p)^{\mathbf{t}} = \Box(\Box\neg\Box\neg\Box p \to \Box p)$; this translation is extended to any set Λ of L_a-formulas:

$$\Lambda^{\mathbf{t}} = \{A^{\mathbf{t}} \mid A \in \Lambda\};$$

- the map $\tau : \text{Ext}\mathbf{Int} \longrightarrow \text{NExt}\mathbf{S4}$ defined by the equality

$$\tau(\mathbf{Int} + \Lambda) = \mathbf{S4} + \Lambda^{\mathbf{t}},$$

  which is a lattice embedding; e.g., $\tau(\mathbf{Int}) = \mathbf{S4}$ and $\tau(\mathbf{Cl}) = \mathbf{S5}$; consult Maksimova and Rybakov (1974) or Chagrov and Zakharyaschev (1997) or Gabbay and Maksimova (2005);
- the map $\rho : \text{NExt}\mathbf{S4} \longrightarrow \text{Ext}\mathbf{Int}$ defined by the equality

$$\rho(M) = \mathbf{Int} + \{A \mid M \vdash A^{\mathbf{t}}\},$$

  which is a lattice epimorphism (or onto homomorphism); and as an easy consequence of the last definition, we observe:

$$\rho(M) = \{A \mid M \vdash A^{\mathbf{t}}\}; \tag{10.2}$$

  consult Maksimova and Rybakov (1974) or Chagrov and Zakharyaschev (1997) or Gabbay and Maksimova (2005);
- the map $\sigma : \text{Ext}\mathbf{Int} \longrightarrow \text{NExt}\mathbf{S4}$ defined by the equality

$$\sigma(L) = \mathbf{Grz} \oplus \tau(L), \tag{10.3}$$

  which is a lattice isomorphism of Ext**Int** onto NExt**Grz**; see Esakia (1976) or Blok (1976) or Chagrov and Zakharyaschev (1997) or Muravitsky (2006);
- that for any intermediate logic L and any modal logic M, the following hold:

$$\begin{array}{l}\text{(d) } \rho\tau(L) = L\\ \text{(e) } \rho\sigma(L) = L\\ \text{(f) } \tau\rho(M) \subseteq M \subseteq \sigma\rho(M);\end{array}$$

  consult Maksimova and Rybakov (1974), Sect. 3, or Chagrov and Zakharyaschev (1997) or Gabbay and Maksimova (2005);
- that for intermediate logics L_i and modal logics M_i, the following properties are true:

(g) $\tau(\vee\{L_i\}_{i<\omega}) = \vee\{\tau(L_i)\}_{i<\omega}$
(h) $\rho(\vee\{M_i\}_{i<\omega}) = \vee\{\rho(M_i)\}_{i<\omega}$;

about (g) and (h) see Maksimova and Rybakov (1974), Sects. 2 and 3, respectively, or Chagrov and Zakharyaschev (1997) or Gabbay and Maksimova (2005);

- the set $\mathscr{L} = \{\tau(L) \mid L \in \text{Ext}\mathbf{Int}\}$ which is a sublattice of NExt**S4** with the least element **S4** and the greatest **S5**; consult Maksimova and Rybakov (1974) or Chagrov and Zakharyaschev (1997) or Gabbay and Maksimova (2005).

An unspecified logic of $\mathscr{L}$ will be denoted by $\boldsymbol{\tau}$. In this notation, it is obvious (use, e.g., (d) above) that for any $\boldsymbol{\tau} \in \mathscr{L}$,

$$\tau\rho(\boldsymbol{\tau}) = \boldsymbol{\tau}.$$

Also, we note that for any $\boldsymbol{\tau} \in \mathscr{L}$,

$$\tau(\mathbf{Int}) = \mathbf{S4} \subseteq \boldsymbol{\tau} \subseteq \mathbf{S5} = \tau(\mathbf{Cl}). \tag{10.4}$$

Other notions used in this paper will be introduced as needed.

The remaining part of the paper is organized as follows. In the next section, *Introduction*, we show how the Embedding Theorem, the idea of which had been suggested by Gödel (1933)[1] and fifteen years later was materialized as a theorem proved in McKinsey and Tarski (1948), has led through the works Dummett and Lemmon (1959) and Maksimova and Rybakov (1974), as well as Grzegorczyk (1967), Blok (1976), and Esakia (1976), to the question of Problem 1.[2] Section 10.3 is of preliminary character. Its results will be applied in Sect. 10.4 to obtain Proposition 13, which will be a fundamental tool in Sect. 10.6. In Sect. 10.5 we will take another, abstract view on Problem 1 and reformulate it as Problem 2. In Sect. 10.6 we define a partial function $d(X, Y)$ and its derivative, $d^*(X) = d(X, \tau\rho(X))$, so that Problem 1 can be reformulated once again, now in terms of definability of $d^*(X)$ (Proposition 27). Further, we begin the process of reduction of the definability of $d^*(X)$ in NExt**S4** to definability of $d(X, Y)$ in [**S4**, **Grz**] (Propositions 28 and 29). Furthermore, using embeddings defined in Sect. 10.7, we reduce (Proposition 32) the question of the definability of $d^*(X)$ to its definability in the 0th S-slice defined in Sect. 10.3 and thus (through Proposition 29) to the definability of $d(X, Y)$ in the 0th E-slice defined in Sect. 10.4.

---

[1] See also (Gödel, 1986), pp. 296–302, where the translation into English and commentary by A. Troelstra are provided.

[2] Although the spirit of the Embedding Theorem is fundamental for our discussion, the Theorem will not appear here in its original form, but only as its generalization (10.5) due to Dummett and Lemmon (1959).

10.2 Introduction

What do we know about NExt**S4** as a lattice? What do we know about it as a Heyting algebra? The answer to the first question has been given by V. Rybakov in Maksimova and Rybakov (1974),[3] where he showed that NExt**S4** is a complete Heyting algebra with lattice operations defined according to (10.1). This might have been obtained in a manner similar to what T. Hosoi earlier claimed in Hosoi (1969), Corollary 1.7, about Ext**Int**, however, Rybakov gave in Maksimova and Rybakov (1974), Theorem 2, his original proof of this fact, which was grounded on (10.1) rather than on that how Hosoi proved that the lattice Ext**Int** is distributive. No less remarkable in Maksimova and Rybakov (1974) was that the authors pioneered in the systematic comparative investigation of Ext**Int** and NExt**S4** as algebraic structures. In the course of this investigation not only the maps τ, ρ, and σ were defined, but also the presence of Ext**Int** as a sublattice of NExt**S4**, namely the lattice $\mathscr{L}$, was discovered; see Maksimova and Rybakov (1974), Theorem 2. However, the definition of σ in Maksimova and Rybakov (1974) differs from definition (10.3). Also, it should be noted that the authors based their work on an earlier publication Dummett and Lemmon (1959), where a generalized version of the Gödel–McKinsey–Tarski embedding theorem was presented: For any set Λ of L_a-formulas and any L_a-formula A,

$$\mathbf{Int} + \Lambda \vdash A \Longleftrightarrow \tau(\mathbf{Int} + \Lambda) \vdash A^{\mathbf{t}}. \tag{10.5}$$

As we know today, the picture given in Dummett and Lemmon (1959) was incomplete. The reason for this was lack of the knowledge of what was discovered in Grzegorczyk (1967)[4]: For any L_a-formula A,

$$\mathbf{Int} \vdash A \Longleftrightarrow \mathbf{Grz} \vdash A^{\mathbf{t}}. \tag{10.6}$$

Relating the two last equivalences had to wait until in Esakia (1976) obtained the following: For any set Λ of L_a-formulas and any L_a-formula A,

$$\mathbf{Int} + \Lambda \vdash A \Longleftrightarrow \mathbf{Grz} + \Lambda^{\mathbf{t}} \vdash A^{\mathbf{t}}. \tag{10.7}$$

The equivalence (10.5) might suggest that any **S4**-logic could be axiomatized as $\mathbf{S4} + \Lambda^{\mathbf{t}}$. However, since $\tau(\mathbf{Int}) \subset \mathbf{Grz}$ (see Grzegorczyk 1967) and because of (10.6), it is not true. In turn, (10.7) suggests that any extension M of **Grz** could be defined as $M = \mathbf{Grz} + \Lambda^{\mathbf{t}}$, for some set Λ of L_a-formulas. This guess proved to be true, as we will see below.

Speaking of properties (e) and (f) of Sect. 10.1, we should say that they were proved by Maksimova not for σ defined by (10.3), but for the following map σ_M :

[3] As is indicated in the introductory part of Maksimova and Rybakov (1974), Sects. 1, 2, and 4 were written by V. Rybakov and Sect. 3 by L. Maksimova.

[4] We note that Grzegorczyk's original axiomatization of **Grz** was different from the one given above. The equivalence of the two axiomatizations was proved in Segerberg (1971), vol. 2.

Ext**Int** $\longrightarrow$ NExt**S4** defined as follows:

$$\sigma_M(L) \vdash \alpha \Longleftrightarrow \text{for any Heyting algebra A, A} \models L \Rightarrow s(\text{A}) \models \alpha. \qquad (10.8)$$

To compare the last definition with (10.3), we observe that

$$\sigma(L) \vdash \alpha \Longleftrightarrow \text{for any } \mathbf{Grz}\text{-algebra B, B}^\circ \models L \Rightarrow \text{B} \models \alpha. \qquad (10.9)$$

Proposition 1 *The right-hand conditions of* (10.8) and (10.9) are equivalent. Hence $\sigma_M = \sigma$.

Proof Let the right-hand condition of (10.8) be fulfilled and B be a **Grz**-algebra with $\text{B}^\circ \models L$. First, we notice that, by premise, $s(\text{B}^\circ) \models \alpha$. Second, by virtue of Lemma 7.6 of Blok (1976), Chapter III, $\text{B} \in \mathbf{SP_u}(s(\text{B}^\circ))$. Hence, $\text{B} \models \alpha$.

Now assume that the right-hand condition of (10.9) is satisfied. Let A be a Heyting algebra with $\text{A} \models L$. By virtue of Maksimova (1975), Proposition 4.1, $s(\text{A})$ is a **Grz**-algebra. Since $s(\text{A})^\circ$ is isomorphic to A, we conclude that $s(\text{A}) \models \alpha$.

Proposition 1 not only confirms the properties (e) – (f) of σ defined by (10.3), but also answers affirmatively the question about the extensions of **Grz**. Indeed, if $\mathbf{Grz} \subseteq M$, then, in virtue of (f), also $\mathbf{Grz} \oplus \tau\rho(M) \subseteq M \oplus \tau\rho(M) \subseteq \mathbf{Grz} \oplus \tau\rho(M)$. Since $\tau\rho(M) \subseteq M$, we obtain that $M = \mathbf{Grz} \oplus \tau\rho(M)$. The immediate consequence of this is what is known as the Blok–Esakia theorem, namely that σ establishes a lattice isomorphism between Ext**Int** and NExt**Grz**, which indicates another presence of Ext**Int** in NExt**S4**, different from $\mathscr{L}$. As we will see below (Proposition 16), there are infinitely many more pairwise disjoint sublattices of NExt**S4**, each isomorphic to Ext**Int**. Also, now one can easily derive L. Esakia's observation in Esakia (1979), Theorem 5.11, that for any $M \in$ NExt**S4**,

$$\rho(M) = \mathbf{Int} \Longleftrightarrow M \subseteq \mathbf{Grz}.$$

Indeed, the $\Longleftarrow$ implication follows from (10.6). Next, assume that $M \not\subseteq \mathbf{Grz}$. Then $\mathbf{Grz} \subset M' = M \oplus \mathbf{Grz}$. The latter implies that $M' = \mathbf{Grz} + \Lambda^{\mathbf{t}}$. In virtue of (10.6), there is $A \in \Lambda$ such that $\mathbf{Int} \nvdash A$. Therefore, $\rho(M') \neq \mathbf{Int}$ and hence $\rho(M) \neq \mathbf{Int}$.

Property (f) of Sect. 10.1 inspires another question which has been a stimulating engine of this research. Namely, (f) suggests that any **S4**-logic M consists of two components. One is $\tau\rho(M)$ and we call it the *assertoric component* of M. (It is certainly unique.) The other one is a logic that, along with the assertoric component, forms M. An easy consideration shows that $M = (M \cap \mathbf{Grz}) \oplus \tau\rho(M)$. This gives rise to the following definition.

Definition 10.2.1 (*modal component, τ -decomposition*) Given a modal logic M, a modal logic $M^* \in [\mathbf{S4}, \mathbf{Grz}]$ is called a modal component of M if $M = M^* \oplus \tau\rho(M)$. Given a modal logic M, each equality $M = M^* \oplus \boldsymbol{\tau}$, where $M^* \in [\mathbf{S4}, \mathbf{Grz}]$ and $\boldsymbol{\tau} \in \mathscr{L}$, is called a τ-decomposition (of M).

Let us draw first observations from Definition 10.2.1.

- In a τ-decomposition $M = M^* \oplus \boldsymbol{\tau}$, the logic $\boldsymbol{\tau} = \tau\rho(M)$; that is, it is unique. Indeed, assume that $\boldsymbol{\tau} = \tau(L)$, for some $L \in$ Ext**Int**. Then $\rho(M) = \rho\tau(L) = L$. Hence $\boldsymbol{\tau} = \tau(L) = \tau\rho(M)$.

The next property will be used in the sequel, e.g., in the proof of Proposition 21.

Proposition 2 (Muravitsky (2006), Proposition 5.1)[5] *If $M = M^* \oplus \boldsymbol{\tau}$ is a τ-decomposition, then the modal component $M^* \subseteq M \cap$ **Grz**. That is, $M \cap$ **Grz** is the greatest modal component of M. Given a modal logic M, the set of its modal components is a convex sublattice of NExt**S4** with the greatest element $M \cap$ **Grz**.*

Problem 1 *Does every* **S4**-*logic possess a least modal component?*

In the sequel, it will be convenient to remember that for any $\boldsymbol{\tau} \in \mathscr{L}$,

$$\mathbf{Grz} \cap \boldsymbol{\tau} = M_0 \cap \boldsymbol{\tau}.$$

Also, the next observation will be useful.

Proposition 3 *Let $\boldsymbol{\tau} \in \mathscr{L}$. Then*

(A) *the following conditions are equivalent:*

$$(a)\ \boldsymbol{\tau} \subseteq \mathbf{Grz}$$
$$(b)\ \boldsymbol{\tau} \subseteq M_0$$
$$(c)\ \boldsymbol{\tau} = \mathbf{S4};$$

(B) *if $M_0 \subseteq \boldsymbol{\tau}$ then $\boldsymbol{\tau} =$* **S5**.

Proof Since $\boldsymbol{\tau} \subseteq$ **S5**, (a) and (b) are equivalent. Also, it is obvious that (c) implies (a). Now, to prove (a) $\Rightarrow$ (c), assume that $\boldsymbol{\tau} \subseteq$ **Grz** and denote $\boldsymbol{\tau} = \tau(L)$. In virtue of property (d) on p. 4, $L = \rho\tau(L) \subseteq \rho(\mathbf{Grz}) = \mathbf{Int}$. Hence $\boldsymbol{\tau} = \mathbf{S4}$.

Now we assume that $M_0 \subseteq \boldsymbol{\tau}$ and, for contradiction, that $\boldsymbol{\tau} \neq$ **S5**. It is well known that lattice Ext**Int** has a pre-top element which the logic of a 3-element Heyting algebra. Let us denote the last logic by by $\mathbf{G}_3$. Since Ext**Int** and $\mathscr{L}$ are isomorphic, the logic $\tau(\mathbf{G}_3)$ is the pre-top element in $\mathscr{L}$. Hence $M_0 \subseteq \boldsymbol{\tau} \subseteq \tau(\mathbf{G}_3)$.

Next, in view of (10.1) and (10.8), we notice that

$$M_0 = \mathbf{S4} + \Box(\Box(\Box(p \to \Box p) \to p) \to p) \vee \Box(\Box\neg\Box\neg\Box q \to \Box q).$$

This, along with the last inclusion, implies that the proper axiom of M_0 above is valid on each **S4**-algebra, the Heyting carcass of which is a 3-element Heyting algebra. Now let us consider the 8-element **S4**-algebra below, (Fig. 10.1) where the squared elements of the algebra are the only open elements.

Let us assign p and q any of the two coatoms over the open atom. Then the additional axiom takes a. Hence $M_0 \not\subseteq \tau(\mathbf{G}_3)$. A contradiction.

---

[5]In Muravitsky (2006), I mistakenly used the term *dense* instead of *convex*.

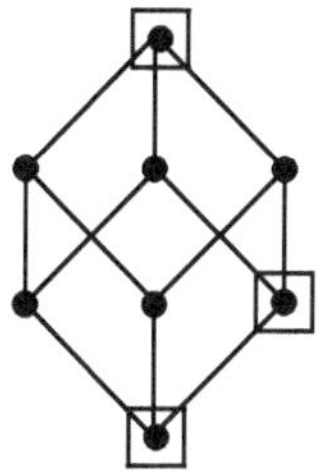

Fig. 10.1 An 8-element **S4**-algebra

Corollary 1 *Let $M = M^* \oplus \boldsymbol{\tau}$ be a τ-decomposition. Then either $M = M^*$ and $\boldsymbol{\tau} = \mathbf{S4}$ or $M = \boldsymbol{\tau}$ or M^* and $\boldsymbol{\tau}$ are incomparable.*

Proof Indeed, if $\boldsymbol{\tau} \subseteq M^*$, then, by virtue of Proposition 3, $\boldsymbol{\tau} = \mathbf{S4}$ and hence $M = M^*$. If $M^* \subseteq \boldsymbol{\tau}$, then, obviously, $M = \boldsymbol{\tau}$. Otherwise, M^* and $\boldsymbol{\tau}$ are incomparable.

We conclude this section with the following examples.

- If $M \in [\mathbf{S4}, \mathbf{Grz}]$, its only τ-decomposition is $M = M \oplus \mathbf{S4}$. Indeed, If $M = M^* \oplus \boldsymbol{\tau}$ is a τ-decomposition, then, according to Proposition 3, $\boldsymbol{\tau} = \mathbf{S4}$ and hence $M = M^*$.
- If $M = \boldsymbol{\tau}$, then $M = M^* \oplus \boldsymbol{\tau}$ is a τ-decomposition of M if and only if $M^* \in [\mathbf{S4}, \mathbf{Grz} \cap \boldsymbol{\tau}]$. (This example will be generalized in Corollary 9.)

10.3 S-Series Slicing of NExtS4

According to Scroggs (1951), the extensions of **S5** constitute a chain of order type $1 + \omega^*$. We denote these logics as follows.

$$\mathbf{S5} = S_0 \subset \ldots \subset S_2 \subset S_1 = \mathbf{S4} + p \to \Box p. \qquad (S\text{-}series)$$

Also, it is well known from Scroggs (1951) that

$$\cap_{n \geqslant 1} S_n = S_0.$$

We remind that logic S_n with $n \geqslant 1$ is the logic of an 2^n-element Boolean algebra with only two open elements, **0** and **1**. Such an algebra, known as a Henle algebra in Dunn and Hardegree (2001), is denoted below by B_n. This logic can also be regarded as the logic of an n-element cluster in the sense of **S4**-frame. About the notion of cluster see Chagrov and Zakharyaschev (1997) or Segerberg (1968, 1971).

Definition 10.3.1 (*S-series slicing*) A logic M belongs to the nth S-slice, $n \geqslant 1$, if $M \subseteq S_n$ and $M \nsubseteq S_{n+1}$. If $M \subseteq S_n$, for all $n \geqslant 1$, that is to say, $M \in [\mathbf{S4}, S_0]$, then M lies in the 0th S-slice. Thus the 0th S-slice is $[\mathbf{S4}, \mathbf{S5}]$. For $n \geqslant 0$, we denote the nth S-slice by $\mathscr{S}_n$.

Proposition 4 $\{\mathscr{S}_n\}_{n<\omega}$ *is a partition of NExt***S4**. *Moreover,*

$$M \in \mathscr{S}_n \Longleftrightarrow M \oplus S_0 = S_n.$$

Proof The first assertion is obvious. Next, if M belongs in the 0th S-slice, then $M \subseteq S_0$, which is equivalent to $M \oplus S_0 = S_0$. The converse is obvious.

Now we assume that $M \subseteq S_n$ and $M \nsubseteq S_{n+1}$, where $n \neq 0$. It is clear that $M \oplus S_0 = S_m$, for some $m \geqslant n$. Obviously, $m \ngeqslant n+1$. Therefore, $M \oplus S_0 = S_n$. The converse is obvious.

Our next goal is to show that each $\mathscr{S}_n$ is an interval (Proposition 5).

Definition 10.3.2 (*logics* K_n) Let χ_n be the characteristic formula of algebra B_n, $n \geqslant 1$. We define $K_n = \mathbf{S4} + \Box\chi_{n+1}$, for $n > 0$, and $K_0 = \mathbf{S4}$.

Proposition 5 *Each S-slice is an interval. For n* $\geqslant 0$, *logic* S_n *is the top of the* nth *slice and logic* K_n *is its bottom.*

Proof The claim is true when $n = 0$. So assume that $n > 0$. Since $\mathrm{B}_{n+1} \not\models \chi_{n+1}$, (Sect. 10.1) $K_n \nsubseteq S_{n+1}$. Next, by virtue of the property (b) of χ_n, since $\mathrm{B}_{n+1} \notin \mathbf{HS}(\mathrm{B}_n)$, $\mathrm{B}_n \models \chi_{n+1}$. Therefore, $K_n \subseteq S_n$. Thus K_n lies in the nth S-slice and hence each logic $M \in [K_n, S_n]$ lies in this slice as well.

Now let a logic M belong to the nth S-slice, $n \geqslant 1$, that is, $M \subseteq S_n$ and $M \nsubseteq S_{n+1}$. The latter implies that there is a formula $\alpha \in M \setminus S_{n+1}$. This in turn implies that $\mathrm{B}_{n+1} \not\models \alpha$. Hence, in view of the property (c) of χ_{n+1}, $\mathbf{S4} + \alpha \vdash \chi_{n+1}$. The latter implies that $K_n \subseteq M$ and hence $K_n \subseteq M \subseteq S_n$.

Definition 10.3.3 (*splitting pair;* cf. McKenzie (1972) Let $\mathscr{K}$ be a sublattice of NExt**S4**. Given two **S4**-logics M and N in $\mathscr{K}$, we call $\langle M, N\rangle$ a splitting pair in $\mathscr{K}$, if for any M' in $\mathscr{K}$, either $M \subseteq M'$ or $M' \subseteq N$.

As an easy consequence of the last proposition we get the following.

Corollary 2 *For each n* $\geqslant 1$, $\langle K_n, S_{n+1}\rangle$ *is a splitting pair in NExt***S4**.

Now for a moment we will focus on $\mathscr{S}_1$.

Proposition 6 $K_1 = \mathbf{S4.1}$; *that is* $\mathscr{S}_1 = [\mathbf{S4.1}, S_1]$.

Proof We prove that **S4.1** is the least logic in the 1st S-slice by showing that

$$\mathbf{S4.1} \subseteq M \Longleftrightarrow M \in \mathscr{S}_1.$$

However, first we observe that **S4.1** $\not\subseteq S_2$. It is so because, as it is well known that formula $\Box\Diamond p \rightarrow \Diamond\Box p$ "prohibits" maximal clusters see Segerberg (1968). But logic S_2 is determined by a 2-element cluster; cf. Scroggs (1951). Thus $\mathbf{S4.1} \in \mathscr{S}_1$.

Now, it suffices to show that if $\mathbf{S4.1} \not\subseteq M$ then $M \subseteq S_2$ and hence $M \notin \mathscr{S}_1$.

So, we assume that an **S4**-algebra B validates M and for some $a \in$ B, $\Box\Diamond a \nleq \Diamond\Box a$. Now we define $\Box$-filter F generated by element $\Box\Diamond a$ and consider quotient algebra B$/F$. Let **0** and **1** be the bottom and top elements of B, respectively. We notice $\Box\Diamond(a/F) = \mathbf{1}/F$. Next we define on B$/F$ the $\Box$-filter G generated by $\Box\neg\Box(a/F)$ and quotient algebra B$/F/G$. Filter G is proper. Indeed, if it were otherwise, we would have $\Box\neg\Box(a/F) \leqslant \mathbf{0}/F$, that is, $\neg\Box\neg\Box a \in F$, which means $\Box\Diamond a \leqslant \Diamond\Box a$. A contradiction.

Element $\Box\neg\Box(a/F/G)$ is the top element of the algebra B$/F/G$ and we denote this element by $\mathbf{1}^*$. Then element $\Diamond\Box(a/F/G)$ is the bottom of this algebra and we denote it by $\mathbf{0}^*$. Also, we denote $a^* = a/F/G$.

It is clear that $\mathbf{1}^* = \mathbf{1}/F/G = \Box\Diamond(a/F/G)$ and $\mathbf{0}^* < \mathbf{1}^*$, since filter G is proper.

Next we notice that $\mathbf{1}^* = \Box\neg\Box a^* = \Box\neg\Box\neg a^*$, which implies that $\Box a^* = \Box\neg a^* = \mathbf{0}^*$. This, in conjunction with the inequality $\mathbf{0}^* < \mathbf{1}^*$, implies the compound inequality $\mathbf{0}^* < a^*, \neg a^* < \mathbf{1}^*$ and also the non-equality $a^* \neq \neg a^*$. Thus we conclude that the elements $\{\mathbf{0}^*, a^*, \neg a^*, \mathbf{1}^*\}$ form a 4-element **S4**-algebra with only two open elements, which is a subalgebra of B$/F/G$. It is well known that such an algebra is adequate for S_2. Since all formulas of M are valid on this subalgebra, $M \subseteq S_2$.

Below we define the logics of M-series, which have been introduced in Muravitsky (2006), though in a different way. So for any $n \geqslant 0$, we define:

$$M_n = S_n \cap \mathbf{Grz}.$$

These logics belong to the interval $[M_0, \mathbf{Grz}]$ and are ordered as follows:

$$\mathbf{Grz} \cap \mathbf{S5} = M_0 \subset \ldots \subset M_2 \subset M_1 = \mathbf{Grz} \qquad (M\text{-}series)$$

This can be obtained from the following.

Proposition 7 (cf. Muravitsky (2006), Theorem 6.1) *Given $\boldsymbol{\tau} \in \mathscr{L}$, the maps*

$$f : [M_0 \oplus \boldsymbol{\tau}, M_1 \oplus \boldsymbol{\tau}] \longrightarrow [S_0, S_1] : M \mapsto M \oplus S_0$$

and

$$g : [S_0, S_1] \longrightarrow [M_0 \oplus \boldsymbol{\tau}, M_1 \oplus \boldsymbol{\tau}] : N \mapsto N \cap (M_1 \oplus \boldsymbol{\tau})$$

establish inverse isomorphisms between $[M_0 \oplus \boldsymbol{\tau}, M_1 \oplus \boldsymbol{\tau}]$ and $[S_0, S_1]$.

Also, we note that

$$\textstyle\bigcap_{n\geqslant 1} M_n = M_0, \tag{10.10}$$

(Muravitsky (2006), Corollary 7.2) and also that for any $n \geqslant 0$,

$$M_n = S_n \cap M_1 \tag{10.11}$$

and

$$S_n = M_n \oplus S_0; \tag{10.12}$$

cf. Muravitsky (2006), Proposition 7.1. Furthermore, the logics M_n fill the entire interval $[M_0, M_1]$; cf. Muravitsky (2006), Theorem 6.1. Also, we observe the following: For any $n \geqslant 0$,

$$M_n \cap S_l = M_l, \tag{10.13}$$

whenever $M_l \subseteq M_n$ or $S_l \subseteq S_n$. We prove (10.13) with help of (10.11) as follows.

Assuming that $M_l \subseteq M_n$, we have:

$$M_n \cap S_l = (M_1 \cap M_n) \cap S_l = M_n \cap (M_1 \cap S_l) = M_n \cap M_l = M_l.$$

Now, if $S_l \subseteq S_n$, we receive:

$$M_n \cap S_l = (M_1 \cap S_n) \cap S_l = M_1 \cap (S_l \cap S_n) = M_1 \cap S_l = M_l.$$

Now we turn to logics K_n. In the next proposition we use the property that NExt**S4** is a Heyting algebra with a relative pseudo-complement $M \Rrightarrow N$ which exists for any two modal logics M and N.

Proposition 8 *For any* $n \geqslant 0$, $K_n \subseteq M_n$.

Proof Given $n \geqslant 0$, we show that $M_n \in \mathscr{S}_n$ and then apply Proposition 5. Indeed, the proposition is obviously true for $n = 0$. Next, we assume that $n > 0$. According to (10.11), $M_n \subseteq S_n$. Therefore, it suffices to prove that $M_n \nsubseteq S_{n+1}$. For contradiction, assume that $M_n \subseteq S_{n+1}$, that is $S_n \cap \mathbf{Grz} \subseteq S_{n+1}$. Regarding NExt**S4** as a Heyting algebra, we derive that $\mathbf{Grz} \subseteq S_n \Rrightarrow S_{n+1}$. It is obvious that $S_n \Rrightarrow S_{n+1} = S_{n+1}$. A contradiction.

Proposition 9 *For any* $n \geqslant 1$, $K_{n+1} \subset K_n$.

Proof We note that $K_{n+1} \neq K_n$, because otherwise we would have $K_n \subseteq S_{n+1}$.

Now we prove that $K_{n+1} \subseteq K_n$. Assuming, for contradiction, that $K_{n+1} \nsubseteq K_n$, we will first prove that $K_n \cap S_{n+1} \subset S_{n+2}$.

For contradiction, assume that $K_n \cap S_{n+1} \nsubseteq S_{n+2}$. Therefore, $K_n \cap S_{n+1}$ lies in $\mathscr{S}_{n+1}$. This implies that $K_{n+1} \subseteq K_n \cap S_{n+1}$. It yields $K_{n+1} \subseteq K_n$, which contradicts the main premise. Thus $K_n \cap S_{n+1} \subseteq S_{n+2}$. However, the equality $K_n \cap S_{n+1} = S_{n+2}$ is not possible, because, with help of Proposition 8, it would lead us to the inclusions $S_{n+2} \subseteq K_n \subseteq M_n$. However, by virtue of Proposition 16, the interval $[M_n, S_n]$, has the cardinality of continuum. Therefore, there would exist infinitely many logics between S_{n+2} and S_n. Next we notice that $K_n \cap S_{n+1} = K_n \cap S_{n+2}$. Indeed,

$K_n \cap S_{n+1} = K_n \cap S_{n+1} \cap S_{n+2} = K_n \cap S_{n+2}$. Also, it is obvious that $K_n \oplus S_{n+1} = K_n \oplus S_{n+2} = S_n$. Thus we have obtained a pentagon $\{K_n, S_n, S_{n+1}, S_{n+2}, K_n \cap S_{n+1}\}$. A contradiction with a well-known property of distributive lattices.

Corollary 3 *For any* $n \geqslant 1$, $\mathbf{S4} \subset K_n$.

The next property is natural to expect.

Proposition 10 $\cap_{n\geqslant 1} K_n = \mathbf{S4}$.

Proof For contradiction, assume that $\mathbf{S4} \subset \cap_{n\geqslant 1} K_n$. Then there is a formula $\alpha \in \cap_{n\geqslant 1} K_n$ and an **S4**-algebra B such that $\mathrm{B} \not\models \alpha$. According to McKinsey and Tarski (1944), without loss of generality, we can count B finite. Since $\alpha \in \cap_{n\geqslant 1} K_n$, then for every $n \geqslant 1$, $\chi_{n+1} \vdash \alpha$. The latter implies that for every $n \geqslant 1$, $\mathrm{B} \not\models \chi_{n+1}$. Therefore, by virtue of the property (b) of χ_{n+1}, for every $n \geqslant 1$, $\mathrm{B}_{n+1} \in \mathbf{HS}(\mathrm{B})$. Therefore, B must be infinite. A contradiction.

For any $n \geqslant 1$, we denote

$$T_n = K_n \cap S_0$$

and derive from Proposition 10 the following.

Corollary 4 $\cap_{n\geqslant 1} T_n = \mathbf{S4}$.

Logics T_n will be used in Sect. 10.7. Here we prove the following property which leads to curious facts about **S4**, **S5**, **Grz**, and M_0.

Proposition 11 *Let* $\boldsymbol{\tau} \in \mathscr{L}$ *and* $\boldsymbol{\tau} \neq \mathbf{S4}$. *Then for any* $n \geqslant 1$, $\mathbf{S4} \subset K_n \cap \boldsymbol{\tau}$ *and* $\mathbf{S4} \subset T_n \cap \boldsymbol{\tau}$. *In particular,* $\mathbf{S4} \subset T_n$.

Proof We prove the first inclusion first. Suppose $\boldsymbol{\tau} = \mathbf{S4} + \Sigma^{\mathbf{t}}$ and $\mathbf{Int} + \Sigma$ is a proper consistent extension of **Int**. Then there is an assertoric formula $A \in \Sigma$, which is refutable at a root of a finite rooted **I**-frame (W, R, a), where R is a partial order and a is the root. We can count that W contains more than one element. If we regard (W, R, a) as an **S4**-frame, we can claim that it refutes formula $A^{\mathbf{t}}$ at a. Now let b be a maximal element of W with respect to R. Then, we obtain the **S4**-frame (W', R', a) by replacing b with an $(n + 1)$-element cluster. Formula $A^{\mathbf{t}}$ is refutable on the new frame at a. The formula $\Box\chi_{n+1}$ is refutable on it too, because χ_{n+1} is refutable in the cluster and hence $\Box\chi_{n+1}$ is refuted at a. Thus the formula $\Box\chi_{n+1} \vee' A^{\mathbf{t}}$ is refuted at a. However, $\Box\chi_{n+1} \vee' A^{\mathbf{t}} \in K_n \cap \boldsymbol{\tau}$. Thus $\mathbf{S4} \subset K_n \cap \boldsymbol{\tau}$. Next, if we assume (for contradiction) that $\mathbf{S4} = T_n \cap \boldsymbol{\tau}$, then we get: $\mathbf{S4} = T_n \cap \boldsymbol{\tau} = (K_n \cap S_0) \cap \boldsymbol{\tau} = K_n \cap \boldsymbol{\tau}$.

Corollary 5 *Let* $\boldsymbol{\tau} \neq \mathbf{S4}$. *Then the following inclusions and equality hold:*

(a) $\neg(\mathbf{Grz} \cap \boldsymbol{\tau}) \subseteq \mathbf{S5}$;
(b) $\neg\boldsymbol{\tau} \subseteq \mathbf{S5}$;
(c) $\neg\mathbf{S5} = \mathbf{S4}$;
(d) $\neg M_0 \subset \mathbf{S5}$.

Proof First we prove (a); the inclusion (b) can be proved analogously.

For contradiction, suppose $\neg(\mathbf{Grz} \cap \boldsymbol{\tau})$ lies in the nth S-slice, for some $n \geqslant 1$. Then $K_n \subseteq \neg(\mathbf{Grz} \cap \boldsymbol{\tau})$, which implies that $K_n \cap \mathbf{Grz} \cap \boldsymbol{\tau} \subseteq \mathbf{S4}$. The latter in turn implies that $K_n \cap \boldsymbol{\tau} = \mathbf{S4}$. A contradiction. Thus $\neg(\mathbf{Grz} \cap \boldsymbol{\tau})$ belongs in the 0th S-slice.

Now we turn to (c). According to (b), for any $\boldsymbol{\tau} \neq \mathbf{S4}$, $\boldsymbol{\tau} \oplus \neg\boldsymbol{\tau} \subseteq \mathbf{S5}$. Then **S5** is a dense element in NExt**S4** regarded as a Heyting algebra. (See Rasiowa and Sikorski (1970), Theorem IV.5.3.) Therefore, $\neg\mathbf{S5} = \mathbf{S4}$.

Finally, we prove (d). For contradiction, assume that $\neg M_0$ belongs to an nth S-slice, for some $n \geqslant 1$; that is $K_n \subseteq \neg M_0$. The latter implies that $K_n \cap M_0 \subseteq \mathbf{S4}$. It in turn implies that $K_n \cap \mathbf{Grz} \cap \mathbf{S5} = K_n \cap \mathbf{S5} \subseteq \mathbf{S4}$. Hence, by virtue of (c) above, $K_n \subseteq \mathbf{S4}$. A contradiction. (Proposition 11) Thus $\neg M_0 \subseteq \mathbf{S5}$. However, the equality $\neg M_0 = \mathbf{S5}$ is impossible, since $\mathbf{S4} \subset M_0 \subset \mathbf{S5}$. (See, e.g., Muravitsky (2006), Proposition 4.2.)

Further, we note that

$$K_n \oplus S_l = S_n, \tag{10.14}$$

whenever $S_l \subseteq S_n$. Indeed, we notice that, if $n = 0$, then $l = 0$ and hence (10.14) is obvious. Next we assume that $n \geqslant 1$. Then we have $S_l \subseteq K_n \oplus S_l \subseteq S_n$. Thus logic $K_n \oplus S_l$ is one of the logics in the S-series. If it were the case that $K_n \oplus S_l \subseteq S_{n+1}$, then we would have $K_n \subseteq S_{n+1}$, which is not true.

10.4 M-Series Slicing of [S4, Grz]

The partition of [**S4**, **Grz**] we are going to discuss in this section has been defined in Muravitsky (2006). Here we will examine it from a different viewpoint.

Definition 10.4.1 (*M-series slicing*) A logic M from [**S4**, **Grz**] belongs to the nth E-slice if and only if M is in the nth S-slice. In other words, the nth E-slice equals $[K_n, S_n] \cap [\mathbf{S4}, \mathbf{Grz}]$. We denote the nth E-slice by $\mathscr{E}_n$, for any $n \geqslant 0$.

Proposition 12[6] *For any* $M \in [\mathbf{S4}, \mathbf{Grz}]$ *and* $n \geqslant 0$, *the following conditions are equivalent:*

(a) $M \in \mathscr{E}_n$;
(b) $M \in [K_n, M_n]$;
(c) $M \oplus M_0 = M_n$.

For any $n \geqslant 1$, *each of (a) – (c) is equivalent to:*

(d) $M \subseteq M_n$ *and* $M \nsubseteq M_{n+1}$.

[6]This proposition answers in the affirmative the question of Problem 9.2 in Muravitsky (2006).

Proof It is clear that if $n = 0$, (a) – (c) are equivalent. So we assume that $n \geqslant 1$.

The equivalence (a) and (b) can be easily obtained with help of (10.11).

To prove (b) $\Rightarrow$ (c) we assume that $M \in [K_n, M_n]$. Then $K_n \oplus M_0 \subseteq M \oplus M_0 \subseteq M_n \oplus M_0 = M_n$. It suffices to show that $K_n \oplus M_0 = M_n$. Indeed, if it were the case that $K_n \oplus M_0 \subseteq M_{n+1}$, we would have that $K_n \subseteq S_{n+1}$.

Proving (c) $\Rightarrow$ (d), we assume, for contradiction, that $M \subseteq M_{n+1}$. Then, we derive, by using (c):

$$M_{n+1} = M \oplus M_{n+1} = M \oplus M_0 \oplus M_{n+1} = M_n \oplus M_{n+1} = M_n.$$

A contradiction.

Finally, proving (d) $\Rightarrow$ (a), we will show that $M \in \mathscr{S}_n$. It is clear that $M \subseteq S_n$. Assume, for contradiction, that $M \subseteq S_{n+1}$. Applying (10.13), we get:

$$M = M \cap M_n \subseteq S_{n+1} \cap M_n = M_{n+1},$$

which contradicts the premise.

Next, we receive the following.

Corollary 6 $\{\mathscr{E}_n\}_{n \geqslant 0}$ *is a partition of* [**S4**, **Grz**].

Proof follows by (d) of Proposition 12 and (10.10).

Corollary 7 *For any* $n \geqslant 1$, $\langle K_n, M_{n+1}\rangle$ *is a splitting pair in* [**S4**, **Grz**].

The next proposition will allow us to give examples of modal logics having infinitely many modal components.

Proposition 13 *Let* $n \geqslant 0$. *If a modal logic* M *lies in* $\mathscr{S}_n$, *then any of its modal components* M^* *belongs to* $\mathscr{E}_n$. *Conversely, for any modal logic* $M^* \in \mathscr{E}_n$ *and any* $\boldsymbol{\tau} \in \mathscr{L}$, *the logic* $M^* \oplus \boldsymbol{\tau}$ *lies in* $\mathscr{S}_n$.

Proof Suppose $M \in \mathscr{S}_n$, that is $K_n \subseteq M \subseteq S_n$. By virtue of (10.13), $K_n \subseteq M \cap$ **grz** $\subseteq S_n \cap M_1 = M_n$. Now let us take any modal component M^* of M, that is, in particular, $M = M^* \oplus \boldsymbol{\tau}$. If it were that $K_l \subseteq M^* \subseteq M_l$ for $l > n$ or $l = 0$, then we would have $K_l \subseteq M \subseteq S_n$. A contradiction. Thus $K_n \subseteq M^* \subseteq M_n$, that is, according to Proposition 12, $M^* \in \mathscr{E}_n$.

Next assume that $K_n \subseteq M^* \subseteq M_n$. Then $K_n \subseteq M^* \oplus \boldsymbol{\tau} \subseteq M_n \oplus \boldsymbol{\tau} \subseteq S_n \oplus \boldsymbol{\tau} = S_n$, for $\boldsymbol{\tau} \subseteq S_0 \subseteq S_n$.

Corollary 8 *For any* $n \geqslant 0$, *all modal logics of* $\mathscr{E}_n$ *are the modal components of the logic* S_n.

Proof Let $M^* \in \mathscr{E}_n$, that is $K_n \subseteq M^* \subseteq M_n$. First, we note that $\tau\rho(S_n) = S_0$. Thus $K_n \oplus S_0 \subseteq M^* \oplus S_0 \subseteq S_n \oplus S_0$. By virtue of (10.14), $M^* \oplus S_0 = S_n$.

Corollary 9 *For any* $n \geqslant 0$ *and any* $\boldsymbol{\tau} \in \mathscr{L}$, *the interval* $[K_n, K_n \oplus (\mathbf{Grz} \cap \boldsymbol{\tau})]$ *contains all modal components of logic* $K_n \oplus \boldsymbol{\tau}$.

Proof First we notice that $\tau\rho(K_n \oplus \boldsymbol{\tau}) = \boldsymbol{\tau}$, $(K_n \oplus \boldsymbol{\tau}) \cap \mathbf{Grz} = K_n \oplus (\mathbf{Grz} \cap \boldsymbol{\tau})$, and $K_n \oplus (\mathbf{Grz} \cap \boldsymbol{\tau}) \oplus \boldsymbol{\tau} = K_n \oplus \boldsymbol{\tau}$. Then we apply Proposition 13.

We will use the following equality below.

$$K_n \oplus M_l = M_n, \tag{10.15}$$

whenever $M_l \subseteq M_n$. This equality can be proven similarly to (10.14), only with a reference to the M-series instead of S-series.

For the sake of completeness, we conclude this section with some results about the extensions of M_0.

Definition 10.4.2 (*slices* $\mathscr{M}_n$) We define $\mathscr{M}_n = \mathscr{S}_n \cap \mathrm{NExt}M_0$.

We note that $\mathscr{M}_1 = \mathrm{NExt}\mathbf{Grz}$. The next proposition is similar to Proposition 12.

Proposition 14 *For any $M \in NExtM_0$ and $n \geqslant 0$, the following conditions are equivalent:*

(a) $M \in \mathscr{M}_n$;
(b) $M \in [M_n, S_n]$;
(c) $M \oplus S_0 = S_n$.

For any $n \geqslant 1$, each of (a) – (c) is equivalent to:

(d) $M \subseteq S_n$ and $M \nsubseteq S_{n+1}$.

Proof It is clear that if $n = 0$, (a) – (c) are equivalent. So we assume that $n \geqslant 1$.

The equivalence (a) and (b) can be easily obtained with help of (10.15).

To prove (b) $\Rightarrow$ (c) we assume that $M \in [M_n, S_n]$. Then $M_n \oplus S_0 \subseteq M \oplus S_0 \subseteq S_n \oplus S_0 = S_n$. In view of (10.12), we get (c).

Proving (c) $\Rightarrow$ (d), we assume, for contradiction, that $M \subseteq S_{n+1}$. Then, we derive, by using (10.12) and (c):

$$S_{n+1} = M \oplus S_{n+1} = M \oplus S_0 \oplus M_{n+1} = S_n \oplus M_{n+1} = S_n.$$

A contradiction.

Finally, proving (d) $\Rightarrow$ (a), we will show that $M \in \mathscr{S}_n$. It is given that $M \subseteq S_n$. Assume, for contradiction, that $M \subseteq S_{n+1}$. Applying (10.13), we receive:

$$M = M \cap M_n \subseteq S_{n+1} \cap M_n = M_{n+1},$$

which contradicts the premise.

Proposition 15 (Muravitsky (2006), Theorem 7.4) *$\{\mathscr{M}_n\}_{n \geqslant 0}$ is a partition of $NExtM_0$.*

Corollary 10 *For any $n \geqslant 1$, $\langle M_n, S_{n+1}\rangle$ is a splitting pair in NExtM_0.*

The following proposition shows that NExt**S4** contains an infinite series of sublattices, each of which is isomorphic to Ext**Int** and one of them is NExt**Grz**.

Proposition 16 (Muravitsky (2006), Theorem 7.3) *For any $n \geqslant 1$, $\mathscr{M}_n = [M_n, S_n]$ is isomorphic to Ext**Int**.*

The next proposition is a generalization of what we discussed in Sect. 10.2 regarding axiomatization of all extensions of **Grz**.

Proposition 17 (Muravitsky (2006), Corollary 6.3) *Given $M \in \mathscr{M}_n$, $M = M_n \oplus \tau\rho(M)$.*

Let us consider $M \in \mathscr{M}_n$ for $n \geqslant 1$. According to Proposition 17,

$$M = M_n + \Lambda^{\mathbf{t}} = M_n \oplus (\mathbf{S4} + \Lambda^{\mathbf{t}}),$$

for some set Λ of assertoric formulas. In view of (10.2), $\rho(M) = \mathbf{Int} + \Lambda$. Thus we have arrived at that which is a generalization of the observation (10.7).

Corollary 11 *For any $n \geqslant 1$,*

$$\mathbf{Int} + \Lambda \vdash A \Longleftrightarrow M_n + \Lambda^{\mathbf{t}} \vdash A^{\mathbf{t}}.$$

10.5 Congruences of [S4, Grz]

In this section we will take a look at Problem 1 from an abstract viewpoint.

Definition 10.5.1 (*relation $\theta_{\boldsymbol{\tau}}$*) Given $\boldsymbol{\tau} \in \mathscr{L}$, we define a relation on [**S4**, **Grz**] as follows*:*

$$\langle M_1^*, M_2^*\rangle \in \theta_{\boldsymbol{\tau}} \Longleftrightarrow M_1^* \oplus \boldsymbol{\tau} = M_2^* \oplus \boldsymbol{\tau}.$$

Let us denote

$$\mathscr{E} = \langle [\mathbf{S4}, \mathbf{Grz}], \cap, \oplus\rangle.$$

That is, $\mathscr{E}$ is a sublattice of NExt**S4**. As usual, by

$$\mathbf{Con}\mathscr{E}$$

we denote the lattice of the congruences of $\mathscr{E}$.

Proposition 18 *For any $\boldsymbol{\tau} \in \mathscr{L}$, $\theta_{\boldsymbol{\tau}} \in$ **Con**$\mathscr{E}$. Moreover, $\mathfrak{L} = \langle\{\theta_{\boldsymbol{\tau}} \mid \boldsymbol{\tau} \in \mathscr{L}\}, \subseteq\rangle$ is a sublattice of **Con**$\mathscr{E}$.*

Proof Let us fix $\boldsymbol{\tau} \in \mathscr{L}$. It is clear that $\theta_{\boldsymbol{\tau}}$ is an equivalence.

Now assume that for modal logics $M^*, M_1^*, N^*, N_1^* \in [\mathbf{S4}, \mathbf{Grz}]$, the equalities $M^* \oplus \boldsymbol{\tau} = M_1^* \oplus \boldsymbol{\tau}$ and $N^* \oplus \boldsymbol{\tau} = N_1^* \oplus \boldsymbol{\tau}$ hold. Then, we obtain:

$$(M^* \oplus N^*) \oplus \boldsymbol{\tau} = (M^* \oplus \boldsymbol{\tau}) \oplus (N^* \oplus \boldsymbol{\tau}) = (M_1^* \oplus \boldsymbol{\tau}) \oplus (N_1^* \oplus \boldsymbol{\tau}) = (M_1^* \oplus N_1^*) \oplus \boldsymbol{\tau}$$

and

$$(M^* \cap N^*) \oplus \boldsymbol{\tau} = (M^* \oplus \boldsymbol{\tau}) \cap (N^* \oplus \boldsymbol{\tau}) = (M_1^* \oplus \boldsymbol{\tau}) \cap (N_1^* \oplus \boldsymbol{\tau}) = (M_1^* \cap N_1^*) \oplus \boldsymbol{\tau}.$$

Now we show that $\mathfrak{L}$ is a sublattice of $\mathbf{Con}\mathscr{E}$. We remind that $\mathscr{L}$ is a sublattice of NExt**S4** (Sect. 10.1) and observe that for any $\boldsymbol{\tau}_1, \boldsymbol{\tau}_2 \in \mathscr{L}$,

$$\boldsymbol{\tau}_1 \subseteq \boldsymbol{\tau}_2 \Longrightarrow \theta_{\boldsymbol{\tau}_1} \subseteq \theta_{\boldsymbol{\tau}_2}. \tag{10.16}$$

Proving the equality $\theta_{\boldsymbol{\tau}_1 \cap \boldsymbol{\tau}_2} = \theta_{\boldsymbol{\tau}_1} \cap \theta_{\boldsymbol{\tau}_2}$, we see that (10.16) implies the inclusion $\theta_{\boldsymbol{\tau}_1 \cap \boldsymbol{\tau}_2} \subseteq \theta_{\boldsymbol{\tau}_1} \cap \theta_{\boldsymbol{\tau}_2}$. Now let $\langle M^*, N^* \rangle \in \theta_{\boldsymbol{\tau}_1} \cap \theta_{\boldsymbol{\tau}_2}$, that is $M^* \oplus \boldsymbol{\tau}_1 = N^* \oplus \boldsymbol{\tau}_1$ and $M^* \oplus \boldsymbol{\tau}_2 = N^* \oplus \boldsymbol{\tau}_2$. Then, we obtain:

$$M^* \oplus (\boldsymbol{\tau}_1 \cap \boldsymbol{\tau}_2) = (M^* \oplus \boldsymbol{\tau}_1) \cap (M^* \oplus \boldsymbol{\tau}_2) = (N^* \oplus \boldsymbol{\tau}_1) \cap (N^* \oplus \boldsymbol{\tau}_2) = N^* \oplus (\boldsymbol{\tau}_1 \cap \boldsymbol{\tau}_2),$$

that is $\langle M^*, N^* \rangle \in \theta_{\boldsymbol{\tau}_1 \cap \boldsymbol{\tau}_2}$.

Next we prove that the join of $\theta_{\boldsymbol{\tau}_1}$ and $\theta_{\boldsymbol{\tau}_2}$ equals $\theta_{\boldsymbol{\tau}_1 \oplus \boldsymbol{\tau}_2}$. Assume that $\langle M^*, N^* \rangle \in \theta_{\boldsymbol{\tau}_1 \oplus \boldsymbol{\tau}_2}$, that is $M^* \oplus \boldsymbol{\tau}_1 \oplus \boldsymbol{\tau}_2 = N^* \oplus \boldsymbol{\tau}_1 \oplus \boldsymbol{\tau}_2$, which implies

$$M^* \oplus (\mathbf{Grz} \cap \boldsymbol{\tau}_1) \oplus (\mathbf{Grz} \cap \boldsymbol{\tau}_2) = N^* \oplus (\mathbf{Grz} \cap \boldsymbol{\tau}_1) \oplus (\mathbf{Grz} \cap \boldsymbol{\tau}_2). \tag{10.17}$$

Using (10.17), we obtain:

$$\begin{aligned}
&M^* \oplus \boldsymbol{\tau}_1 = \mathbf{Grz} \cap (M^* \oplus \boldsymbol{\tau}_1) \oplus \boldsymbol{\tau}_1 = M^* \oplus (\mathbf{Grz} \cap \boldsymbol{\tau}_1) \oplus \boldsymbol{\tau}_1,\\
&M^* \oplus (\mathbf{Grz} \cap \boldsymbol{\tau}_1) \oplus \boldsymbol{\tau}_2 = \mathbf{Grz} \cap (M^* \oplus (\mathbf{Grz} \cap \boldsymbol{\tau}_1) \oplus \boldsymbol{\tau}_2) \oplus \boldsymbol{\tau}_2,\\
&\mathbf{Grz} \cap (M^* \oplus (\mathbf{Grz} \cap \boldsymbol{\tau}_1) \oplus \boldsymbol{\tau}_2) \oplus \boldsymbol{\tau}_2 = M^* \oplus (\mathbf{Grz} \cap \boldsymbol{\tau}_1) \oplus (\mathbf{Grz} \cap \boldsymbol{\tau}_2) \oplus \boldsymbol{\tau}_2,\\
&N^* \oplus (\mathbf{Grz} \cap \boldsymbol{\tau}_1) \oplus (\mathbf{Grz} \cap \boldsymbol{\tau}_2) \oplus \boldsymbol{\tau}_2 = \mathbf{Grz} \cap (N^* \oplus (\mathbf{Grz} \cap \boldsymbol{\tau}_1) \oplus \boldsymbol{\tau}_2) \oplus \boldsymbol{\tau}_2,\\
&\mathbf{Grz} \cap (N^* \oplus (\mathbf{Grz} \cap \boldsymbol{\tau}_1) \oplus \boldsymbol{\tau}_2) \oplus \boldsymbol{\tau}_2 = N^* \oplus (\mathbf{Grz} \cap \boldsymbol{\tau}_1) \oplus \boldsymbol{\tau}_2,\\
&N^* \oplus (\mathbf{Grz} \cap \boldsymbol{\tau}_1) \oplus \boldsymbol{\tau}_1 = N^* \oplus \boldsymbol{\tau}_1.
\end{aligned}$$

Thus we have:

$$\begin{aligned}
&\langle M^*, M^* \oplus (\mathbf{Grz} \cap \boldsymbol{\tau}_1) \rangle \in \theta_{\boldsymbol{\tau}_1},\\
&\langle M^* \oplus (\mathbf{Grz} \cap \boldsymbol{\tau}_1), M^* \oplus (\mathbf{Grz} \cap \boldsymbol{\tau}_1) \oplus (\mathbf{Grz} \cap \boldsymbol{\tau}_2) \rangle \in \theta_{\boldsymbol{\tau}_2},\\
&\langle M^* \oplus (\mathbf{Grz} \cap \boldsymbol{\tau}_1) \oplus (\mathbf{Grz} \cap \boldsymbol{\tau}_2), N^* \oplus (\mathbf{Grz} \cap \boldsymbol{\tau}_1) \oplus (\mathbf{Grz} \cap \boldsymbol{\tau}_2) \rangle \in \theta_{\boldsymbol{\tau}_2},\\
&\langle N^* \oplus (\mathbf{Grz} \cap \boldsymbol{\tau}_1) \oplus (\mathbf{Grz} \cap \boldsymbol{\tau}_2), N^* \oplus (\mathbf{Grz} \cap \boldsymbol{\tau}_1) \rangle \in \theta_{\boldsymbol{\tau}_2},\\
&\langle N^* \oplus (\mathbf{Grz} \cap \boldsymbol{\tau}_1), N^* \rangle \in \theta_{\boldsymbol{\tau}_1}.
\end{aligned}$$

By virtue of Burris and Sankappanavar (1981), Chap. 1, Theorem 4.6, and Chap. 2, Theorem 5.3, $\langle M^*, N^* \rangle$ belongs to the join of $\theta_{\boldsymbol{\tau}_1}$ and $\theta_{\boldsymbol{\tau}_2}$, which in turn, according to (10.16), is included in $\theta_{\boldsymbol{\tau}_1 \oplus \boldsymbol{\tau}_2}$.

In Proposition 19 we will prove that $\mathfrak{L}$ and Ext**Int** are isomorphic, which demonstrates another presence of the latter, an implicit one. We start with a lemma, in the proof of which we will need the formula

$$\mathbf{grz} = \Box(\Box(\Box(p \to \Box p) \to p) \to p).$$

It is not difficult to see that $\mathbf{Grz} = \mathbf{S4} + \mathbf{grz}$.

Lemma 1 *For any* $\boldsymbol{\tau}_1, \boldsymbol{\tau}_2 \in \mathscr{L}$,

$$\boldsymbol{\tau}_1 \neq \boldsymbol{\tau}_2 \Rightarrow \mathbf{Grz} \cap \boldsymbol{\tau}_1 \neq \mathbf{Grz} \cap \boldsymbol{\tau}_2.$$

Proof Let us take two distinct logics from $\mathscr{L}$, $\boldsymbol{\tau}_1 \neq \boldsymbol{\tau}_2$. For contradiction, we assume that $\mathbf{Grz} \cap \boldsymbol{\tau}_1 = \mathbf{Grz} \cap \boldsymbol{\tau}_2$.

Further, we remind that we can represent the first two logics as follows: $\boldsymbol{\tau}_1 = [L_1^{\mathbf{t}}]$ and $\boldsymbol{\tau}_2 = [L_2^{\mathbf{t}}]$, where L_1 and L_2 are two distinct logics from Ext**Int**. Accordingly, we have:

$$\mathbf{Grz} \cap \boldsymbol{\tau}_1 = \left[\{B^{\mathbf{t}} \vee' \mathbf{grz} \mid B \in L_1\}\right],$$

$$\mathbf{Grz} \cap \boldsymbol{\tau}_2 = \left[\{B^{\mathbf{t}} \vee' \mathbf{grz} \mid B \in L_2\}\right].$$

Let $A \in L_1 \setminus L_2$. Then, by assumption, $A^{\mathbf{t}} \vee' \mathbf{grz} \in \mathbf{Grz} \cap \boldsymbol{\tau}_1$. Since $A \notin L_2$, there is a subdirectly irreducible Heyting algebra $\mathfrak{A}$ that validates the logic L_2 and refutes A. Let us denote $L = L(\mathrm{A})$, that is L is the intermediate logic of the algebra A. Now let B be any **S4**-algebra, the Heyting carcass of which is isomorphic to A. It is clear that B validates the logic $\mathbf{Grz} \cap \boldsymbol{\tau}_2$ and hence, by assumption, it validates $\mathbf{Grz} \cap \boldsymbol{\tau}_1$. Therefore, B validates $A^{\mathbf{t}} \vee' \mathbf{grz}$. But B refutes $A^{\mathbf{t}}$, for its Heyting carcass A refutes A. Since the algebra B is subdirectly irreducible and validates $A^{\mathbf{t}} \vee' \mathbf{grz}$, it must validate **grz**. Because of the choice of the algebra B, we conclude that $\mathbf{Grz} \subseteq \tau(L)$. By virtue of Proposition 3, $\tau(L) = \mathbf{S5}$ and hence $L = \rho\tau(L) = \rho(\mathbf{S5}) = \mathbf{Cl}$. The latter implies that A is a Boolean algebra and hence A is not a classical tautology. However, A being in L_1 must be a classical tautology.

Proposition 19 *The map* $\boldsymbol{\tau} \mapsto \theta_{\boldsymbol{\tau}}$ *establishes a partial order isomorphism between* $\mathscr{L}$ *and* $\mathfrak{L}$.

Proof First we show that for any $\boldsymbol{\tau}_1, \boldsymbol{\tau}_2 \in \mathscr{L}$,

$$\boldsymbol{\tau}_1 \neq \boldsymbol{\tau}_2 \Longrightarrow \theta_{\boldsymbol{\tau}_1} \neq \theta_{\boldsymbol{\tau}_2}.$$

Indeed, assume that $\boldsymbol{\tau}_1 \neq \boldsymbol{\tau}_2$. Then, according to Lemma 1, $\mathbf{Grz} \cap \boldsymbol{\tau}_1 \neq \mathbf{Grz} \cap \boldsymbol{\tau}_2$. Assume that $\mathbf{Grz} \cap \boldsymbol{\tau}_1 \nsubseteq \mathbf{Grz} \cap \boldsymbol{\tau}_2$, that is $\mathbf{grz} \cap \boldsymbol{\tau}_1 \notin [\mathbf{S4}, \mathbf{Grz} \cap \boldsymbol{\tau}_2]$. As has been observed at the end of Sect. 10.2, the modal components of $\boldsymbol{\tau}_1$ and $\boldsymbol{\tau}_2$ form the intervals [**S4**, **Grz** $\cap\, \boldsymbol{\tau}_1$] and [**S4**, **Grz** $\cap\, \boldsymbol{\tau}_2$], respectively. Since these sets are the congruence classes of $\theta_{\boldsymbol{\tau}_1}$ and $\theta_{\boldsymbol{\tau}_2}$, respectively, $\langle \mathbf{S4}, \mathbf{Grz} \cap \boldsymbol{\tau}_1\rangle \in \theta_{\boldsymbol{\tau}_1}$ and $\langle \mathbf{S4}, \mathbf{Grz} \cap \boldsymbol{\tau}_1\rangle \notin \theta_{\boldsymbol{\tau}_2}$. Now it remains to apply (10.16).

A given a modal logic $M \in [\mathbf{S4}, \mathbf{Grz}]$ and $\boldsymbol{\tau} \in \mathcal{L}$, we denote by

$$M/\theta_{\boldsymbol{\tau}}$$

the congruence class with respect to $\theta_{\boldsymbol{\tau}}$, which contains M. Further, a given modal logic M, we define

$$\varphi(M) = (M \cap \mathbf{Grz})/\theta_{\tau\rho(M)}.$$

Proposition 20 *Given a modal logic M, $\varphi(M)$ is the set of all modal components of M.*

Proof Indeed, we observe: for any $M^* \in [\mathbf{S4}, \mathbf{Grz}]$,

$$M^* \in \varphi(M) \Leftrightarrow M^* \oplus \tau\rho(M) = (M \cap \mathbf{Grz}) \oplus \tau\rho(M) \Leftrightarrow M^* \oplus \tau\rho(M) = M.$$

Now we define: for any $\boldsymbol{\tau} \in \mathcal{L}$,

$$\psi(\boldsymbol{\tau}) = \{\varphi(M) \mid \boldsymbol{\tau} \subseteq M \subseteq \mathbf{Grz} \oplus \boldsymbol{\tau}\}.$$

Corollary 12 $\psi(\mathbf{S4}) = \{\{M^*\} \mid M^* \in [\mathbf{S4}, \mathbf{Grz}]\}$ *and* $\psi(\mathbf{S5}) = \{\mathscr{E}_n\}_{n \geqslant 0}$.

Proof The first equality follows from Proposition 20 and the first example on p. 7. The second equality is true due to Proposition 20 and Corollary 8.

Our next goal is to prove Proposition 21. We shall do it by using the following two lemmas.

Lemma 2 *For any $\boldsymbol{\tau} \in \mathcal{L}$, $[\mathbf{S4}, \mathbf{Grz}] = \cup\{\varphi(M) \mid \boldsymbol{\tau} \subseteq M \subseteq \mathbf{Grz} \oplus \boldsymbol{\tau}\}$*

Proof Let us fix $\boldsymbol{\tau} \in \mathcal{L}$. Since each $\varphi(M) \subseteq [\mathbf{S4}, \mathbf{Grz}]$, (Proposition 20) the $\supseteq$-inclusion is true. Now assume that $M^* \in [\mathbf{S4}, \mathbf{Grz}]$. We define: $M = M^* \oplus \boldsymbol{\tau}$. Then $M \in [\boldsymbol{\tau}, \mathbf{Grz} \oplus \boldsymbol{\tau}]$ and, by virtue of Proposition 20, $M^* \in \varphi(M)$.

Lemma 3 *For any $\boldsymbol{\tau} \in \mathcal{L}$, $\langle M_1^*, M_2^*\rangle \in \theta_{\boldsymbol{\tau}}$ if and only if there is $M \in [\boldsymbol{\tau}, \mathbf{Grz} \oplus \boldsymbol{\tau}]$ such that $M_1^*, M_2^* \in \varphi(M)$.*

Proof Let us fix $\boldsymbol{\tau} \in \mathcal{L}$. Also, let us fix $M_1^*, M_2^* \in [\mathbf{S4}, \mathbf{Grz}]$. First assume that $M_1^* \oplus \boldsymbol{\tau} = M_2^* \oplus \boldsymbol{\tau}$. We denote: $M = M_1^* \oplus \boldsymbol{\tau}$. It is obvious that $M \in [\boldsymbol{\tau}, \mathbf{Grz} \oplus \boldsymbol{\tau}]$. Also, according to Proposition 20, $M_1^*, M_2^* \in \varphi(M)$.

Next, let $M_1^*, M_2^* \in \varphi(M)$, for some $M \in [\boldsymbol{\tau}, \mathbf{Grz} \oplus \boldsymbol{\tau}]$. It is clear that $\tau(M) = \boldsymbol{\tau}$. By virtue of Proposition 20, $M_1^* \oplus \boldsymbol{\tau} = M_2^* \oplus \boldsymbol{\tau}$, that is $\langle M_1^*, M_2^*\rangle \in \theta_{\boldsymbol{\tau}}$.

Proposition 21 *For any $\boldsymbol{\tau} \in \mathcal{L}$, $\psi(\boldsymbol{\tau})$ is the set of all congruence classes of $\theta_{\boldsymbol{\tau}}$.*

Proof According to its definition and Lemmas 2 and 3, $\psi(\boldsymbol{\tau})$ contains all congruence classes of θ_{τ} and nothing else.

Now, collecting together pieces of our knowledge about the modal components of a modal logic, one can say that for any $\boldsymbol{\tau} \in \mathscr{L}$, each congruence class of θ_{τ} is a convex sublattice with a top element. Thus Problem 1 can be reformulated as follows.

Problem 2 *Is it true that for any $\boldsymbol{\tau} \in \mathscr{L}$, each congruence class of θ_{τ} is an interval?*

In connection with the question of Problem 2, we note that not all congruence of **Con**$\mathscr{E}$ lies in $\mathfrak{L}$. Indeed, by virtue of Propositions 19, 21 and Corollary 12, the congruence classes of any congruence of $\mathfrak{L}$ partition each of $\mathscr{E}_n$. Therefore, by virtue of Proposition 9, the congruence under the filter generated by the set $\{K_n\}_{n\geqslant 1}$ does not belong to $\mathfrak{L}$. Also, not all congruence classes of the last congruence of **Con**$\mathscr{E}$ are intervals; for instance, the filter in question is not. In Sect. 10.8 we will discuss conditions on $\boldsymbol{\tau}$ for the θ_{τ} congruence class containing **Grz** to be an interval.

We close this section with a proposition which could be proven earlier, but now we have a convenient notation for it. Namely, we will use the map φ according to its meaning in Proposition 20.

Proposition 22 *Let $M = M^* \oplus \boldsymbol{\tau}$ be a τ-decomposition. If M is finitely axiomatizable, then there is no chain $\{N_i\}_{i<\omega}$ of logics $N_i \in$ [**S4**, **Grz**] such that $N_i \subseteq N_{i+1}$ with each $N_i \notin \varphi(M)$ and $\cup\{N_i\}_{i<\omega} = M^*$. Conversely, if such a chain of logics does not exist and, in addition, the assertoric component $\boldsymbol{\tau}$ is finitely axiomatizable, then M is finitely axiomatizable.*

Proof First suppose that M is finitely axiomatizable. For contradiction, assume that such a chain of logics N_i exists. It is clear that $N_i \oplus \boldsymbol{\tau} \subseteq M^* \oplus \boldsymbol{\tau}$. If it were that $N_i \oplus \boldsymbol{\tau} = M^* \oplus \boldsymbol{\tau}$, then, by virtue of Proposition 20, N_i would belong to $\varphi(M)$. Thus $N_i \oplus \boldsymbol{\tau} \subset M^* \oplus \boldsymbol{\tau}$. However, we observe that $\cup\{N_i \oplus \boldsymbol{\tau}\}_{i<\omega} = \cup\{N_i\}_{i<\omega} \oplus \boldsymbol{\tau} = M^* \oplus \boldsymbol{\tau} = M$. Therefore, according to Lemmon (1965), Theorem 1, M is not finitely axiomatizable.

Now assume that there is no chain of logics N_i with the indicated properties and $\boldsymbol{\tau}$ is finitely axiomatizable but (for contradiction) M is not. Then, by virtue of Lemmon (1965), Theorem 1, there is a chain $\{P_i\}_{i<\omega}$ of modal logics P_i such that $P_i \subseteq P_{i+1}$ and $\cup\{P_i\}_{i<\omega} = M$. We define: $N_i = P_i \cap M^*$. It is obvious that $N_i \subseteq N_{i+1}$. Also, we observe that

$$N_i \oplus \boldsymbol{\tau} = (P_i \cap M^*) \oplus \boldsymbol{\tau} = (P_i \oplus \boldsymbol{\tau}) \cap (M \oplus \boldsymbol{\tau}) = (P_i \oplus \boldsymbol{\tau}) \cap M = P_i \oplus \boldsymbol{\tau}.$$

Denoting $\boldsymbol{\tau}_i = \tau\rho(P_i)$, we notice that $\boldsymbol{\tau}_i \subseteq \boldsymbol{\tau}_{i+1}$. Using (f) of Sect. 10.1, we derive:

$$\boldsymbol{\tau}_i \oplus \boldsymbol{\tau} \subseteq P_i \oplus \boldsymbol{\tau} = N_i \oplus \boldsymbol{\tau} = M.$$

The latter implies that $\boldsymbol{\tau}_i \subseteq \boldsymbol{\tau}$. If it were that $\boldsymbol{\tau}_i = \boldsymbol{\tau}$, then we would have: $P_i = P_i \oplus \boldsymbol{\tau}_i = P_i \oplus \boldsymbol{\tau} = M$, which is false. Thus $\boldsymbol{\tau}_i \subset \boldsymbol{\tau}$, for each $i < \omega$. On the other hand, using (g) and (h) of Sect. 10.1, we obtain:

$$\bigcup\{\boldsymbol{\tau}_i\}_{i<\omega} = \bigvee\{\boldsymbol{\tau}_i\}_{i<\omega} = \bigvee\{\tau\rho(P_i)\}_{i<\omega} = \tau\rho(\bigvee\{P_i\}_{i<\omega}) = \tau\rho(\bigcup\{P_i\}_{i<\omega}) = \tau\rho(M) = \boldsymbol{\tau}.$$

According to Lemmon (1965), Theorem 1, this is a contradiction with the premise that $\boldsymbol{\tau}$ is finitely axiomatizable.

The last proposition suggests the following question.

Problem 3 *Is it true that the finite axiomatizability of a modal logic M implies the finite axiomatizability of its assertoric component $\tau\rho(M)$?*

10.6 Operation $d(X, Y)$ in NExtS4

The first objective of this section is to show that the question of Problem 1 can be reduced (in Proposition 29) to the definability of a function, $d(X, Y)$, defined below, for logics in the interval [**S4**, **Grz**]. However, our nearest goal is Proposition 27. We begin with the following observation.

Proposition 23 *Let M be any logic in NExt***S4***. Then for any logic $Z \subseteq$* **Grz***, the following conditions are equivalent:*

$$\text{(a) } M = Z \oplus \tau\rho(M);$$
$$\text{(b) } M \cap \mathbf{Grz} = Z \oplus (\tau\rho(M) \cap \mathbf{Grz}).$$

Proof The implication (a) $\Rightarrow$ (b) is true because of distributivity.

Proving (b) $\Rightarrow$ (a), we assume that $M \cap \mathbf{Grz} = Z \oplus (\tau\rho(M) \cap \mathbf{Grz})$. Then we have:

$$\tau\rho(M) \oplus (M \cap \mathbf{Grz}) = Z \oplus (\tau\rho(M) \cap \mathbf{Grz}) \oplus \tau\rho(M).$$

Focusing on the left-hand side of the last equality, we have:

$$\tau\rho(M) \oplus (M \cap \mathbf{Grz}) = (\tau\rho(M) \oplus M) \cap (\tau\rho(M) \oplus \mathbf{Grz}) = M \cap (\tau\rho(M) \oplus \mathbf{Grz}) = M.$$

The right-hand side, obviously, equals $Z \oplus \tau\rho(M)$.

Definition 10.6.1 (*operation $d(X, Y)$*) Given two logics X and Y, a logic $C \subseteq X$ is called the difference of the subtraction of Y from X, if for any logic Z, the following equivalence holds:

$$C \subseteq Z \subseteq X \Longleftrightarrow X = Z \oplus Y. \tag{10.18}$$

If such C exists for given X and Y, it is unique. (See below.) We denote it by $d(X, Y)$.

Uniqueness: Suppose there are two logics C_1 and C_2 which, given X and Y, satisfy the following:

$$C_1 \subseteq Z \subseteq X \Longleftrightarrow X = Z \oplus Y, \text{ for any } Z,$$

and

$$C_2 \subseteq Z \subseteq X \Longleftrightarrow X = Z \oplus Y, \text{ for any } Z.$$

First, we obtain that

$$C_1 \subseteq C_1 \subseteq X \Longleftrightarrow X = C_1 \oplus Y$$

and

$$C_2 \subseteq C_2 \subseteq X \Longleftrightarrow X = C_2 \oplus Y,$$

from which we derive

$$X = C_1 \oplus Y \text{ and } X = C_2 \oplus Y.$$

On the other hand, we have:

$$C_1 \subseteq C_2 \subseteq X \Longleftrightarrow X = C_1 \oplus Y$$

and

$$C_2 \subseteq C_1 \subseteq X \Longleftrightarrow X = C_2 \oplus Y.$$

Thus we receive that

$$C_1 \subseteq C_2 \text{ and } C_2 \subseteq C_1.$$

It is clear that operation $d(X, Y)$ is partial, not total. For instance, $d(X, Y)$ is undefined when $X \subset Y$. However, in the sequel we will need to consider only cases when $Y \subseteq X$.[7] For these cases we have the following two observations.

Proposition 24 *Given $Y \subseteq X$, if $d(X, Y)$ is defined, then for any Z, $Z \oplus Y = X$ implies $d(X, Y) \subseteq Z$.*

Proof follows directly from (10.18).

Proposition 25 *Let $Y \subseteq X$. Then $d(X, Y)$ is defined if and only if the set $\{Z \mid Z \oplus Y = X\}$ has a least element.*

Proof Let us denote $\mathcal{M} = \{Z \mid Z \oplus Y = X\}$. We note that $\mathcal{M} \neq \emptyset$, for $X \in \mathcal{M}$.

Suppose $d(X, Y)$ is defined. Let $\{X_i\}_{i<\gamma}$ be a $\subset$-descending chain with $X_i \subseteq X$ and $X_i \oplus Y = X$ for all $i < \gamma$. Then, by virtue of Proposition 24, $d(X, Y) \subseteq \bigcap_{i<\gamma} X_i \subseteq X$ and hence $d(X, Y) \oplus Y \subseteq \bigcap_{i<\gamma} X_i \oplus Y \subseteq X \oplus Y$. Thus $X \subseteq \bigcap_{i<\gamma} X_i \oplus Y \subseteq X$. This implies that $\mathcal{M}$ satisfies the condition of the Kuratowski-Zorn lemma and, therefore, $\mathcal{M}$ has a minimal element $C_0 \subseteq X$. Let $C_1 \subseteq X$ be another minimal element in $\mathcal{M}$. We notice that

$$(C_0 \cap C_1) \oplus Y = (C_0 \oplus Y) \cap (C_1 \oplus Y) = X.$$

[7]There are examples, when $Y \subset X$ but $d(X, Y)$ is undefined. For instance, $\mathbf{Dum} \subset M_0$, where $\mathbf{Dum} = \mathbf{S4} + \Diamond\Box p \to (\Box(\Box(p \to \Box p) \to p) \to p)$, and $d(M_0, \mathbf{Dum})$ is undefined.

It means $C_0 \cap C_1 \in \mathscr{M}$; that is $\mathscr{M}$ has a least element.

Next assume $\mathscr{M}$ has a least element, say C. Then $C \subseteq X$. Also, if $C \subseteq Z \subseteq X$ then $X = C \oplus Y \subseteq Z \oplus Y \subseteq X \oplus Y = X$. On the other hand, if $Z \oplus Y = X$ then $Z \subseteq X$ and $Z \in \mathscr{M}$. The latter implies that $C \subseteq Z$. We conclude that, by definition, $C = d(X, Y)$.

Next we define

$$d^*(M) = d(M, \tau\rho(M)).$$

We note that if $d^*(M)$ is defined, then $d^*(M) \subseteq M$; also, for any logic Z,

$$d^*(M) \subseteq Z \subseteq M \Longleftrightarrow M = Z \oplus \tau\rho(M). \tag{10.19}$$

The next proposition is in essence another definition of the function $d^*(X)$.

Proposition 26 *Given a logic M, let a logic $C \subseteq M$ satisfy the equivalence*

$$C \subseteq Z \subseteq M \Longleftrightarrow M = Z \oplus \tau\rho(M),$$

for any logic Z. Then $d^*(M)$ is defined and equals C.

Proof By Definition 10.6.1, $C = d(M, \tau\rho(M))$. Thus $d^*(M)$ exists and equals C.

Next we observe the following.

Proposition 27 *Given a logic M, if $d^*(M)$ exists, then $d^*(M) \subseteq$ **Grz** and $d^*(M)$ is the least modal component of M. Conversely, if a logic M has the least modal component, then $d^*(M)$ is defined and equals this least modal component of M.*

Proof Suppose $d^*(M)$ is defined. Replacing Z with $M \cap \mathbf{Grz}$ in (10.19), we get:

$$d^*(M) \subseteq M \cap \mathbf{Grz} \subseteq M \Longleftrightarrow M = (M \cap \mathbf{Grz}) \oplus \tau\rho(M),$$

the right-hand side of which is true. Hence $d^*(M) \subseteq \mathbf{Grz}$. Using (10.19) one more time, we see that $M = d^*(M) \oplus \tau\rho(M)$ and hence $d^*(M)$ is the least modal component of M.

Now let M^* be the least modal component of M. Assume that $M^* \subseteq Z \subseteq M$. Then, obviously, $M^* \subseteq Z \cap \mathbf{Grz} \subseteq M \cap \mathbf{grz}$ and $Z \oplus \tau\rho(M) \subseteq \mathbf{Grz} \oplus \tau\rho(M)$. Next, we obtain:

$$\begin{aligned} M &= (Z \cap \mathbf{Grz}) \oplus \tau\rho(M) \quad \text{(Proposition 2.)} \\ &= (Z \oplus \tau\rho(M)) \cap (\mathbf{Grz} \oplus \tau\rho(M)) = Z \oplus \tau\rho(M). \end{aligned}$$

Now let $M = Z \oplus \tau\rho(M)$. It is obvious that $Z \subseteq M$ and hence $Z \oplus \tau\rho(M) \subseteq \mathbf{Grz} \oplus \tau\rho(M)$. Also, $M = (Z \cap \mathbf{Grz}) \oplus \tau\rho(M)$. Indeed, we have:

$$(Z \cap \mathbf{Grz}) \oplus \tau\rho(M) = (Z \oplus \tau\rho(M)) \cap (\mathbf{Grz} \oplus \tau\rho(M)) = M.$$

Thus we obtain that $M^* \subseteq Z \cap \mathbf{Grz} \subseteq Z \subseteq M$. By virtue of Proposition 26, we conclude that $M^* = d^*(M)$.

Following Kleene (1952), Sect. 63, for two partial functions $F(x)$ and $G(x)$, we write $F(x) \simeq G(x)$ when for any element a, both $F(a)$ and $G(a)$ are defined at a and have equal values, or both are undefined at a. We note that relation $\simeq$ is transitive.

Proposition 28 *For any* $M \in NExt\mathbf{S4}$, $d^*(M) \simeq d(M \cap \mathbf{Grz}, \tau\rho(M) \cap \mathbf{Grz})$.

Proof Let $d^*(M)$ be defined. We prove that for any $Z \in \mathrm{NExt}\mathbf{S4}$,

$$d^*(M) \subseteq Z \subseteq M \cap \mathbf{Grz} \Longleftrightarrow M \cap \mathbf{Grz} = Z \oplus (\tau\rho(M) \cap \mathbf{Grz}).$$

If $Z \nsubseteq M \cap \mathbf{Grz}$ then both parts of the equivalence above are false. Therefore, we will be assuming that $Z \subseteq M \cap \mathbf{Grz}$.

Suppose $d^*(M) \subseteq Z \subseteq M \cap \mathbf{Grz}$. Then, certainly, $d^*(M) \subseteq Z \subseteq M$ and hence, by (10.19), $M = Z \oplus \tau\rho(M)$ which, by distributivity, implies $M \cap \mathbf{Grz} = Z \oplus (\tau\rho(M) \cap \mathbf{Grz})$.

Now assume that the last equality holds. Then, by virtue of Proposition 23, we conclude that $M = Z \oplus \tau\rho(M)$. Thus Z is a modal component of M. According to Proposition 27, $d^*(M) \subseteq Z$.

Now suppose $C = d(M \cap \mathbf{Grz}, \tau\rho(M) \cap \mathbf{Grz})$ is defined. Then for any Z,

$$C \subseteq Z \subseteq M \cap \mathbf{Grz} \Longleftrightarrow M \cap \mathbf{Grz} = Z \oplus (\tau\rho(M) \cap \mathbf{Grz}).$$

If we substitute $M \cap \mathbf{Grz}$ for Z, we see that the right-hand side of this substitution instance is true, which implies that $C \subseteq M \cap \mathbf{Grz}$. Next, we substitute C for Z in the above equivalence. According to Proposition 24, the left-hand side of this substitution instance is true, which implies that $M \cap \mathbf{Grz} = C \oplus (\tau\rho(M) \cap \mathbf{Grz})$. According to Proposition 23, $M = C \oplus \tau\rho(M)$; that is, C is a modal component of M. Now let a logic Z be another modal component of M, that is $M = Z \oplus \tau\rho(M)$. Then, by virtue of Proposition 23, $M \cap \mathbf{Grz} = Z \oplus (\tau\rho(M) \cap \mathbf{Grz})$. The latter, in particular, implies that $C \subseteq Z$. Thus we have proved that C is the least modal component. Applying Proposition 27, we conclude $d^*(M)$ exists and equals C.

The next proposition is the last step in the reduction we talked about in the beginning of this section.

Proposition 29 *Let* $M = M^* \oplus \boldsymbol{\tau}$ *be a* τ*-decomposition. Then* $d^*(M) \simeq d(M^*, \boldsymbol{\tau} \cap M^*)$.

Proof Suppose $d^*(M)$ is defined and equals C. By virtue of Proposition 28,

$$C = d(M \cap \mathbf{Grz}, \boldsymbol{\tau} \cap \mathbf{Grz}).$$

Since $M \cap \mathbf{Grz} = M^* \oplus (\boldsymbol{\tau} \cap \mathbf{Grz})$, then, according to Proposition 24, $C \subseteq M^*$. Then, we obtain:

$$C \oplus (M^* \cap \boldsymbol{\tau}) = (C \oplus M^*) \cap (C \oplus \boldsymbol{\tau}) = M^* \cap (C \oplus \boldsymbol{\tau}) = M^* \cap \mathbf{Grz} \cap (C \oplus \boldsymbol{\tau})$$
$$= M^* \cap (C \oplus (\mathbf{Grz} \cap \boldsymbol{\tau})) = M^* \cap (M \cap \mathbf{Grz}) = M^*.$$

Next, assume $C' \oplus (M^* \cap \boldsymbol{\tau}) = M^*$, for some $C' \subseteq \mathbf{Grz}$. Then we have:

$$C' \oplus (\mathbf{Grz} \cap \boldsymbol{\tau}) = C' \oplus (M^* \cap \boldsymbol{\tau}) \oplus (\mathbf{Grz} \cap \boldsymbol{\tau}) = M^* \oplus (\mathbf{Grz} \cap \boldsymbol{\tau})$$
$$= (M^* \oplus \mathbf{Grz}) \cap (M^* \oplus \boldsymbol{\tau}) = \mathbf{Grz} \cap M.$$

We conclude that $C \subseteq C'$ and hence $C = d(M^*, M^* \cap \boldsymbol{\tau})$.

Now we assume that $d(M^*, M^* \cap \boldsymbol{\tau})$ is defined and equals D. Thus $M^* = D \oplus (M^* \cap \boldsymbol{\tau})$. The latter implies:

$$D \oplus (\mathbf{Grz} \cap \boldsymbol{\tau}) = D \oplus (M^* \cap \boldsymbol{\tau}) \oplus (\mathbf{Grz} \cap \boldsymbol{\tau}) = M^* \oplus (\mathbf{Grz} \cap \boldsymbol{\tau})$$
$$= (M^* \oplus \mathbf{Grz}) \cap (M^* \oplus \boldsymbol{\tau}) = \mathbf{Grz} \cap M.$$

Finally, assume that $D' \oplus (\mathbf{Grz} \cap \boldsymbol{\tau}) = M \cap \mathbf{Grz}$, for some $D' \subseteq \mathbf{Grz}$. First we notice that

$$(D \cap D') \oplus (\mathbf{Grz} \cap \boldsymbol{\tau}) = (D \oplus (\mathbf{Grz} \cap \boldsymbol{\tau})) \cap (D' \oplus (\mathbf{Grz} \cap \boldsymbol{\tau})) = M \cap \mathbf{Grz}.$$

Furthermore, we observe:

$$(D \cap D') \oplus (M^* \cap \boldsymbol{\tau}) = (D \cap D') \oplus (\mathbf{Grz} \cap \boldsymbol{\tau} \cap M^*)$$
$$= ((C \cap D') \oplus (\mathbf{Grz} \cap \boldsymbol{\tau})) \cap ((D \cap D') \oplus M^*)$$
$$= M \cap \mathbf{Grz} \cap M^* = M^*.$$

From the last we conclude that $D \subseteq D'$ and hence $D = d(M \cap \mathbf{Grz}, \mathbf{Grz} \cap \boldsymbol{\tau})$. According to Proposition 28, $D = d^*(M)$.

We conclude this section with a necessary condition for the definability of $d^*(X)$.

Corollary 13 *Let $M = M^* \oplus \boldsymbol{\tau}$ be a $\boldsymbol{\tau}$-decomposition. If for any modal logic Z, the equality $Z \oplus (M^* \cap \boldsymbol{\tau}) = M^*$ implies $[M^* \setminus \boldsymbol{\tau}] \subseteq Z$, then $d^*(M) = [M^* \setminus \boldsymbol{\tau}]$.*

Proof Let us denote $M = M^* \oplus \boldsymbol{\tau}$. Then, we obtain:

$$[M^* \setminus \boldsymbol{\tau}] \oplus (M^* \cap \boldsymbol{\tau}) = ([M^* \setminus \boldsymbol{\tau}] \oplus M^*) \cap ([M^* \setminus \boldsymbol{\tau}] \oplus \boldsymbol{\tau})$$
$$= M^* \cap M \quad \text{(Muravitsky 2006, Proposition 5.3)}$$
$$= M^*.$$

This implies that $d(M^*, M^* \cap \boldsymbol{\tau}) = [M^* \setminus \boldsymbol{\tau}]$. By virtue of Proposition 29, we receive that $d^*(M^* \oplus \boldsymbol{\tau}) = [M^* \setminus \boldsymbol{\tau}]$.

10.7 Embeddings

Definition 10.7.1 (*maps h^n, h_n, and h_{nl}*) We define $h^n : M \mapsto M \cap S_n$, where $n \geqslant 0$. Now for any $l \geqslant n \geqslant 1$, h_{nl} is defined as h^l restricted to $[K_n, S_n]$. Also, for each $n \geqslant 0$, we define the map h_n as h^0 restricted to $[K_n, S_n]$.

Thus in Definition 10.7.1, we have defined the maps of these types:

$$h_{nl} : \mathscr{S}_n \longrightarrow \mathscr{S}_l, \text{ where } l \geqslant n, \text{ and } h_n : \mathscr{S}_n \longrightarrow \mathscr{S}_0.$$

First we want to make an observation which will be used later on. Namely, for any logic M of the nth S-slice,

$$\rho(h_n(M)) = \rho(M). \tag{10.20}$$

Indeed, we have:

$$\rho(h_n(M)) = \rho(M \cap S_0) = \rho(M) \cap \rho(S_0) = \rho(M) \cap \mathbf{Cl} = \rho(M).$$

Definition 10.7.2 (*maps s^n, s_n, and s_{ln}*) We define $s^n : M \mapsto M \oplus K_n$, where $n \geqslant 0$. Now for any $l \geqslant n \geqslant 1$, s_{ln} is defined as s^n restricted to $[K_l, S_l]$. Also, for any $n \geqslant 0$, we define s_n as s^n restricted $[K_0, S_0]$.

Definition 10.7.2 defines maps of these types:

$$s_{ln} : \mathscr{S}_l \longrightarrow \mathscr{S}_n, \text{ where } l \geqslant n, \text{ and } s_n : \mathscr{S}_0 \longrightarrow \mathscr{S}_n.$$

We note that the maps h_{nl}, h_n, s_{ln} and s_n are monotone.

Proposition 30 *Given $l \geqslant n \geqslant 1$, the map h_{nl} is a lattice embedding of the* nth *S-slice into the* lth *S-slice. Given $n \geqslant 0$, the map h_n is a lattice embedding of the* nth *S-slice onto $[T_n, S_0]$. Moreover, the maps s_{ln}, restricted to $[K_n \cap S_l, S_l]$, and s_n, restricted to $[T_n, S_0]$, are the inverses of h_{nl} and h_n, respectively.*

Proof Let $l \geqslant n \geqslant 1$ and $M \in [K_n, S_n]$. Then, with help of (10.14), we observe:

$$s_{ln}(h_{nl}(M)) = (M \cap S_l) \oplus K_n = (M \oplus K_n) \cap (S_l \oplus K_n) = M \cap S_n = M.$$

Also, for any $M \in [K_n \cap S_l, S_l]$, with help of (10.13), we have:

$$h_{nl}(s_{ln}(M)) = (M \oplus K_n) \cap S_l = (M \cap S_l) \oplus (K_n \cap S_l) = M.$$

Remembering that h_{nl} and s_{ln} are monotone bijections, by virtue of Burris and Sankappanavar (1981), Chap. 1, Theorem 2.3, we conclude that they are lattice embeddings that are inverses of one another. Then, using h_n and s_n instead of h_{nl} and s_{ln}, respectively, we prove the second part of the proposition.

Definition 10.7.3 (*maps h_n^* and s_n^**) The map h_n restricted to the nth E-slice, $\mathscr{E}_n$, is denoted by h_n^*. The map s_n restricted to the 0th E-slice is denoted by s_n^*.

Proposition 31 *For any $n \geqslant 1$, $[T_n, M_0]$ is an isomorphic image of the* nth *E-slice with respect to h_n^*. Moreover, map s_n^* restricted to $[T_n, M_0]$ is the inverse of h_n^*.*

Proof follows straightforward from Definition 10.4.1 and Proposition 30.

The next proposition shows that the definability of $d^*(M)$ at $M \in$ NExt**S4** is reduced to the definability of this operation at a corresponding logic belonging to $\mathscr{S}_0$.

Proposition 32 *Let M be a logic of the nth S-slice, $n \geqslant 0$. If $d^*(M)$ is defined then $d^*(h_n(M))$ is also defined, in which case $d^*(h_n(M)) = h_n(d^*(M))$. Conversely, if $d^*(h_n(M))$ is defined then $d^*(M)$ is defined as well, in which case $d^*(M) = s_n(d^*(h_n(M)))$. In other words, $d^*(h_n(M)) \simeq h_n(d^*(M))$ and $d^*(M) \simeq s_n(d^*(h_n(M)))$.*

Proof First, we observe, with help of (10.20), that

$$\tau\rho(M) = \tau\rho(h_n(M)).$$

Second, according to Proposition 30, the mappings h_n and s_n, restricted to $[T_n, S_0]$, are inverses of one another.

Now we assume that $d^*(M)$ is defined, that is, for any logic Z, the equivalence (10.19) holds. By virtue of Propositions 27 and 13, $d^*(M)$ belongs to the nth S-slice.[8] We want to prove that for any logic Z,

$$h_n(d^*(M)) \subseteq Z \subseteq h_n(M) \Leftrightarrow h_n(M) = Z \oplus \tau\rho(h_n(M)).$$

Assume that $h_n(d^*(M)) \subseteq Z \subseteq h_n(M)$. All the three terms of the last inequality belong to $[T_n, S_0]$. Further, since s_n is monotone, applying it to each term of the last inequality, by virtue of Propositions 30, we have $d^*(M)) \subseteq s_n(Z) \subseteq M$. Then, by (10.19), $M = s_n(Z) \oplus \tau\rho(M)$. Therefore, $h_n(M) = (s_n(Z) \oplus \tau\rho(M)) \cap S_0 = (s_n(Z) \cap S_0) \oplus \tau\rho(M) = h_n(s_n(Z)) \oplus \tau\rho(M) = Z \oplus \tau\rho(h_n(M))$.

Now we assume that $h_n(M) = Z \oplus \tau\rho(h_n(M))$. Then, according to Proposition 30, $M = (Z \oplus K_n) \oplus \tau\rho(h_n(M)) = s_n(Z) \oplus \tau\rho(M)$. By (10.19), we have $d^*(M) \subseteq s_n(Z) \subseteq M$. Applying h_n to each term of the last inequality, we get $h_n(d^*(M)) \subseteq Z \subseteq h_n(M)$. Thus, by virtue of Proposition 26, $d^*(h_n(M))$ is defined and equals $h_n(d^*(M))$.

Next assume that $d^*(h_n(M))$ is defined. That is, for any logic Z,

$$d^*(h_n(M)) \subseteq Z \subseteq h_n(M) \Leftrightarrow h_n(M) = Z \oplus \tau\rho(h_n(M)).$$

We will prove that for any Z,

[8] Actually, $d^*(M)$ lies in the nth M-slice which is a proper subset of the nth S-slice.

$$s_n(d^*(h_n(M))) \subseteq Z \subseteq M \Leftrightarrow M = Z \oplus \tau\rho(M).$$

So, we suppose that $s_n(d^*(h_n(M))) \subseteq Z \subseteq M$. Since $d^*(h_n(M))$ belongs to the 0th M-slice, $s_n(d^*(h_n(M)))$ lies in the nth M-slice and hence Z belongs also to the nth S-slice. Then, we apply h_n to all three terms of the last inequality to obtain $(d^*(h_n(M)) \oplus K_n) \cap S_0 = d^*(h_n(M)) \oplus (K_n \cap S_0) \subseteq h_n(Z) \subseteq h_n(M)$. Therefore, $d(h_n(M)) \subseteq h_n(Z) \subseteq h_n(M)$, which in turn implies that $h_n(M) = h_n(Z) \oplus \tau\rho$ $(h_n(M))$. Applying s_n to both sides of the latter, we get $M = (h_n(Z) \oplus K_n) \oplus \tau\rho(h_n(M)) = s_n(h_n(Z)) \oplus \tau\rho(M)$. Therefore, $M = Z \oplus \tau\rho(M)$.

Now assume that $M = Z \oplus \tau\rho(M)$. Since M belongs to the nth S-slice, $Z \subseteq S_n$. If it were the case that $Z \subseteq S_{n+1}$, we would have that $M = Z \oplus \tau\rho(M) \subseteq S_{n+1} \oplus S_0 = S_{n+1}$. Therefore, Z also belongs to the nth S-slice. Now we apply h_n to both sides of the first equality to obtain: $h_n(M) = (Z \oplus \tau\rho(M)) \cap S_0 = (Z \cap S_0) \oplus \tau\rho(h_n(M)) = h_n(Z) \oplus \tau\rho(M)$. By premise, it follows that $d^*(h_n(M) \subseteq h_n(Z) \subseteq h_n(M)$. Then, we apply s_n to the three terms of the latter to derive $s_n(d^*(h_n(M))) \subseteq Z \subseteq M$. Thus $d^*(M)$ is defined and equals $s_n(d^*(h_n(M)))$.

In view of Propositions 27 and 32, Problem 1 can be reformulated equivalently as follows.

Problem 4 *Is it true that $d^*(M)$ is defined for any $M \in \mathscr{S}_0 =$ [**S4**, **S5**]?*

Now we also observe the following.

Proposition 33 *Let $M_1^*, M_2^* \in \mathscr{E}_n$ and $\boldsymbol{\tau} \in \mathscr{L}$. Then*

$$\langle M_1^*, M_2^* \rangle \in \theta_{\boldsymbol{\tau}} \Longleftrightarrow \langle h_n^*(M_1^*), h_n^*(M_2^*) \rangle \in \theta_{\boldsymbol{\tau}}.$$

Proof Indeed, suppose that $M_1^* \oplus \boldsymbol{\tau} = M_2^* \oplus \boldsymbol{\tau}$. Then $(M_1^* \cap M_0) \oplus (\boldsymbol{\tau} \cap M_0) = (M_2^* \cap M_0) \oplus (\boldsymbol{\tau} \cap M_0)$ and hence $(M_1^* \cap M_0) \oplus \boldsymbol{\tau} = (M_2^* \cap M_0) \oplus \boldsymbol{\tau}$, that is $\langle h_n^*(M_1^*), h_n^*(M_2^*) \rangle \in \theta_{\boldsymbol{\tau}}$.

Now assume that $(M_1^* \cap S_0) \oplus \boldsymbol{\tau} = (M_2^* \cap S_0) \oplus \boldsymbol{\tau}$. This implies that $K_n \oplus (M_1^* \cap S_0) \oplus \boldsymbol{\tau} = K_n \oplus (M_2^* \cap S_0) \oplus \boldsymbol{\tau}$. By distributivity and (10.14), we obtain that $M_1^* \oplus \boldsymbol{\tau} = M_2^* \oplus \boldsymbol{\tau}$, that is $\langle M_1^*, M_2^* \rangle \in \theta_{\boldsymbol{\tau}}$.

Thus, according to Proposition 33, the question of Problem 2 can be reduced to the following.

Problem 5 *Is it true that for any $\boldsymbol{\tau} \in \mathscr{L}$, each congruence class of $\theta_{\boldsymbol{\tau}}$ restricted to $\mathscr{E}_0 \times \mathscr{E}_0$ is an interval?*

10.8 Some Definability Conditions

In this section we consider a few conditions of the definability of $d^*(M_0 \oplus \boldsymbol{\tau})$. The following consideration leads to the question, whether $d^*(M_0 \oplus \boldsymbol{\tau})$ is defined.

Let $M \in \mathrm{NExt}M_0$. Assume that $M \in \mathscr{M}_n$. Then, according to Proposition 17, $M = M_n \oplus \boldsymbol{\tau}$, for some $\boldsymbol{\tau} \in \mathscr{L}$. Further, by virtue of Proposition 32, $d^*(M_n \oplus \boldsymbol{\tau}) \simeq s_n(d^*(h_n(M_n \oplus \boldsymbol{\tau})))$. However, $h_n(M_n \oplus \boldsymbol{\tau}) = M_0 \oplus \boldsymbol{\tau}$. Therefore, $d^*(M_n \oplus \boldsymbol{\tau})$ is defined if and only if $d^*(M_0 \oplus \boldsymbol{\tau})$ is. In particular, given $\boldsymbol{\tau} \in \mathscr{L}$, the congruence class of θ_τ containing **Grz** is an interval if and only if $d^*(M_0 \oplus \boldsymbol{\tau})$ is defined.

We start with a more general case, when a logic $M \in \mathscr{S}_0$.

Proposition 34 *Let* $M = M^* \oplus \boldsymbol{\tau}$ *be a* τ*-decomposition, where* $M^* \in \mathscr{E}_0$. *If* $\mathbf{Grz} \cap \boldsymbol{\tau} \subseteq M^*$, *then* $d^*(M) \simeq d(M^*, \mathbf{Grz} \cap \boldsymbol{\tau})$.

Proof Since $M^* \in \mathscr{E}_0$, then $\mathbf{Grz} \cap \boldsymbol{\tau} = M^* \cap (\mathbf{Grz} \cap \boldsymbol{\tau}) = M^* \cap (M_0 \cap \boldsymbol{\tau}) = M^* \cap \boldsymbol{\tau}$. Then, we apply Proposition 29.

Proposition 35 *Let* $M = M^* \oplus \boldsymbol{\tau}$ *be a* τ*-decomposition, where* $M^* \in \mathscr{E}_0$. *If* C *is a relative complement of* $M^* \cap \boldsymbol{\tau}$ *in* $[\mathbf{S4}, M^*]$, *then* $d^*(M)$ *is defined and equals* C.

Proof Suppose $C \oplus (M^* \cap \boldsymbol{\tau}) = M^*$ and $C \cap (M^* \cap \boldsymbol{\tau}) = \mathbf{S4}$. If for some modal logic $Z \subseteq M^*$, $Z \oplus (M^* \cap \boldsymbol{\tau}) = M^*$, then, by distributivity, $(C \cap Z) \oplus (M^* \cap \boldsymbol{\tau}) = M^*$. Therefore, $C \cap Z = C$, that is $C \subseteq Z$. This implies that $C = d(M^*, M^* \cap \boldsymbol{\tau})$. It remains to apply Proposition 29.

Next we will consider some conditions for the definability of $d^*(M_0 \oplus \boldsymbol{\tau})$. We begin with the following observations. First, for any $\boldsymbol{\tau} \in \mathscr{L}$,

$$\mathbf{Grz} \cap \neg(\mathbf{Grz} \cap \boldsymbol{\tau}) = \mathbf{Grz} \cap \neg\boldsymbol{\tau}. \tag{10.21}$$

Indeed, we have: $\mathbf{Grz} \cap \neg(\mathbf{Grz} \cap \boldsymbol{\tau}) = \mathbf{Grz} \cap (\mathbf{Grz} \cap \boldsymbol{\tau} \Rrightarrow \mathbf{S4}) = \mathbf{Grz} \cap \neg\boldsymbol{\tau}$.

Second, if $\boldsymbol{\tau} \neq \mathbf{S4}$, then $\neg\boldsymbol{\tau} \subseteq \mathbf{S5}$ (Corollary 5) and hence

$$M_0 \cap \boldsymbol{\tau} = \mathbf{Grz} \cap \boldsymbol{\tau} \text{ and } M_0 \cap \neg\boldsymbol{\tau} = \mathbf{Grz} \cap \neg\boldsymbol{\tau}; \tag{10.22}$$

also, similarly to (10.21), we have:

$$M_0 \cap \neg(M_0 \cap \boldsymbol{\tau}) = M_0 \cap \neg\boldsymbol{\tau}. \tag{10.23}$$

According to (10.22), logics $M_0 \cap \boldsymbol{\tau}$ and $\mathbf{Grz} \cap \boldsymbol{\tau}$, on the one hand, and $M_0 \cap \neg\boldsymbol{\tau}$ and $\mathbf{Grz} \cap \neg\boldsymbol{\tau}$, on the other, will be used interchangeably. The next lemma is rather technical.

For two **S4**-logics M and N, we denote

$$M \bowtie N \Longleftrightarrow M \nsubseteq N \text{ and } N \nsubseteq M,$$

that is M and N are incomparable with respect to $\subseteq$. Remembering that the $\bowtie$ relation is not transitive, we will use

$$M \bowtie N \bowtie K \text{ to mean } M \bowtie N \text{ and } N \bowtie K.$$

Certainly, the relation $\bowtie$ is symmetric.

Lemma 4 *Let* $\boldsymbol{\tau} \neq \mathbf{S4}$ *and let* Z *satisfy the equation* $Z \oplus (\mathbf{Grz} \cap \boldsymbol{\tau}) = M_0$. *Then* $\mathbf{Grz} \cap \neg\boldsymbol{\tau} \subseteq Z$, *or equivalently* $M_0 \cap \neg\boldsymbol{\tau} \subseteq Z$.

Proof We prove the last inequality, which is easier to do. For this, we have to consider several cases. In all cases we assume that $\mathbf{S4} \subset M_0 \cap \neg\boldsymbol{\tau}$, since otherwise the conclusion is trivial.

Cases: $Z = M_0$ or $\boldsymbol{\tau} = \mathbf{S5}$. The statement is obvious for the first premise. For the second, it is true, since $\neg\mathbf{S5} = \mathbf{S4}$. (See Corollary 5.)

Case: $\boldsymbol{\tau} \neq \mathbf{S4}$, $\boldsymbol{\tau} \neq \mathbf{S5}$, and $Z \neq M_0$. By premise, $Z \subseteq M_0$. Therefore, $Z \subset M_0$. This condition implies (with help of (10.22) and Proposition 3) that $Z \bowtie M_0 \cap \boldsymbol{\tau}$. Next we notice that $M_0 \cap \boldsymbol{\tau} \subset M_0$. If it were otherwise, we would get that $M_0 \subseteq \boldsymbol{\tau}$, which implies that $\boldsymbol{\tau} = \mathbf{S5}$. Also, we observe that $\mathbf{S4} \subset \mathbf{Grz} \cap \boldsymbol{\tau}$. If it were the case that $\mathbf{Grz} \cap \boldsymbol{\tau} = \mathbf{S4}$, we would get that $\mathbf{Grz} \subseteq \neg\boldsymbol{\tau}$. A contradiction with Corollary 5. Thus also $\mathbf{S4} \subset M_0 \cap \boldsymbol{\tau}$. Let us consider two cases. First, $M_0 \cap (\boldsymbol{\tau} \oplus \neg\boldsymbol{\tau}) = (M_0 \cap \boldsymbol{\tau}) \oplus (M_0 \cap \neg\boldsymbol{\tau}) = M_0$. This indicates that $M_0 \cap \neg\boldsymbol{\tau}$ is a relative complement of $M_0 \cap \boldsymbol{\tau}$ in $[\mathbf{S4}, M_0]$. By virtue of Proposition 35, $d^*(M_0 \oplus \boldsymbol{\tau}) = M_0 \cap \neg\boldsymbol{\tau}$ and hence $M_0 \cap \neg\boldsymbol{\tau} \subseteq Z$. Next, we assume that $M_0 \cap (\boldsymbol{\tau} \oplus \neg\boldsymbol{\tau}) \subset M_0$. We observe that $M_0 \cap \boldsymbol{\tau} \bowtie M_0 \cap \neg\boldsymbol{\tau}$; also, $Z \not\subset M_0 \cap \neg\boldsymbol{\tau}$. For contradiction, assume that $Z \bowtie M_0 \cap \neg\boldsymbol{\tau}$. We call ii the *Main Assumption*. We collect all current assumptions in one block.

$$\left.\begin{array}{l}\boldsymbol{\tau} \neq \mathbf{S4},\ \boldsymbol{\tau} \neq \mathbf{S5},\ Z \subset M_0,\ M_0 \cap (\boldsymbol{\tau} \oplus \neg\boldsymbol{\tau}) \subset M_0 \\ M_0 \cap \boldsymbol{\tau} \bowtie Z \bowtie M_0 \cap \neg\boldsymbol{\tau},\ M_0 \cap \boldsymbol{\tau} \bowtie M_0 \cap \neg\boldsymbol{\tau}.\end{array}\right\} \qquad (10.24)$$

Now we observe that

$$\mathbf{S4} \subset Z \cap \boldsymbol{\tau}.$$

Indeed, if it were that $Z \cap \boldsymbol{\tau} = \mathbf{S4}$, we would derive

$$\begin{array}{l} Z \cap \boldsymbol{\tau} = \mathbf{S4} \Rightarrow Z \cap M_0 \cap \boldsymbol{\tau} = \mathbf{S4} \Rightarrow Z \subseteq \neg(M_0 \cap \boldsymbol{\tau}) \Rightarrow Z \subseteq M_0 \cap \neg(M_0 \cap \boldsymbol{\tau}) \\ \qquad \Rightarrow Z \subseteq M_0 \cap \neg\boldsymbol{\tau} \quad \text{(using (10.3)).}\end{array}$$

Also, we obtain that

$$\mathbf{S4} \subset Z \cap \neg\boldsymbol{\tau}.$$

Indeed, for contradiction, assume that $Z \cap \neg\boldsymbol{\tau} = \mathbf{S4}$. Then, on the one hand, we would have $Z \cap M_0 \cap (\boldsymbol{\tau} \oplus \neg\boldsymbol{\tau}) = Z \cap M_0 \cap \boldsymbol{\tau} = Z \cap \boldsymbol{\tau}$ and, on the other, by virtue of (10.22), $M_0 = Z \oplus (M_0 \cap \boldsymbol{\tau}) \subseteq Z \oplus (M_0 \cap (\boldsymbol{\tau} \oplus \neg\boldsymbol{\tau})) \subseteq Z \oplus M_0 = M_0$. Hence $Z \oplus (M_0 \cap \boldsymbol{\tau}) = Z \oplus M_0 \cap (\boldsymbol{\tau} \oplus \neg\boldsymbol{\tau}) = M_0$. This implies that the logics $Z \cap \boldsymbol{\tau}$, Z, M_0, $M_0 \cap (\boldsymbol{\tau} \oplus \neg\boldsymbol{\tau})$, and $M_0 \cap \boldsymbol{\tau}$ form a pentagon.

Next we prove that

$$Z \cap (\boldsymbol{\tau} \oplus \neg\boldsymbol{\tau}) \subset Z. \qquad (10.25)$$

For contradiction, assume that $Z \cap (\boldsymbol{\tau} \oplus \neg\boldsymbol{\tau}) = Z$. Since, by premise, $Z \subseteq \mathbf{Grz}$, we have: $Z \subseteq \mathbf{Grz} \cap (\boldsymbol{\tau} \oplus \neg\boldsymbol{\tau})$. Using distributivity and (10.22), we get: $Z \oplus (\mathbf{Grz} \cap$

$\boldsymbol{\tau}) \subseteq Z \oplus (\mathbf{Grz} \cap (\boldsymbol{\tau} \oplus \neg\boldsymbol{\tau})) = \mathbf{Grz} \cap (\boldsymbol{\tau} \oplus \neg\boldsymbol{\tau}) = M_0 \cap (\boldsymbol{\tau} \oplus \neg\boldsymbol{\tau}) \subset M_0$. A contradiction.

Our next observation is that

$$Z \cap (\boldsymbol{\tau} \oplus \neg\boldsymbol{\tau}) \bowtie M_0 \cap \boldsymbol{\tau}.$$

To refute the inclusion $M_0 \cap \boldsymbol{\tau} \subseteq Z \cap (\boldsymbol{\tau} \oplus \neg\boldsymbol{\tau})$, we notice that it implies that $M_0 \cap \boldsymbol{\tau} \subseteq Z$, from which, with help of (10.22), we deduce: $M_0 = Z \oplus (\mathbf{Grz} \cap \boldsymbol{\tau}) = Z \oplus (M_0 \cap \boldsymbol{\tau}) = Z \subset M_0$. Otherwise, for contradiction, we assume that $Z \cap (\boldsymbol{\tau} \oplus \neg\boldsymbol{\tau}) \subset M_0 \cap \boldsymbol{\tau}$. Then, we observe:

$$Z \cap (\boldsymbol{\tau} \oplus \neg\boldsymbol{\tau}) = Z \cap (\boldsymbol{\tau} \oplus \neg\boldsymbol{\tau}) \cap M_0 \cap \boldsymbol{\tau} = Z \cap M_0 \cap \boldsymbol{\tau}.$$

Since $M_0 \cap \boldsymbol{\tau} \subset M_0 \cap (\boldsymbol{\tau} \oplus \neg\boldsymbol{\tau})$, using (10.25), we conclude that the logics $Z \cap (\boldsymbol{\tau} \oplus \neg\boldsymbol{\tau})$, Z, M_0, $M_0 \cap (\boldsymbol{\tau} \oplus \neg\boldsymbol{\tau})$, and $M_0 \cap \boldsymbol{\tau}$ form a pentagon.

Next, we note that, according to (10.25) and (10.24), $Z \cap (\boldsymbol{\tau} \oplus \neg\boldsymbol{\tau}) \subseteq M_0 \cap (\boldsymbol{\tau} \oplus \neg\boldsymbol{\tau})$ and hence $Z \cap (\boldsymbol{\tau} \oplus \neg\boldsymbol{\tau}) \oplus (M_0 \cap \boldsymbol{\tau}) \subseteq M_0 \cap (\boldsymbol{\tau} \oplus \neg\boldsymbol{\tau})$. Thus we have two alternatives: either

$$Z \cap (\boldsymbol{\tau} \oplus \neg\boldsymbol{\tau}) \oplus (M_0 \cap \boldsymbol{\tau}) \subset M_0 \cap (\boldsymbol{\tau} \oplus \neg\boldsymbol{\tau}) \tag{10.26}$$

or

$$Z \cap (\boldsymbol{\tau} \oplus \neg\boldsymbol{\tau}) \oplus (M_0 \cap \boldsymbol{\tau}) = M_0 \cap (\boldsymbol{\tau} \oplus \neg\boldsymbol{\tau}). \tag{10.27}$$

If (10.26) is the case, then the logics $Z \cap (\boldsymbol{\tau} \oplus \neg\boldsymbol{\tau})$, Z, M_0, $M_0 \cap (\boldsymbol{\tau} \oplus \neg\boldsymbol{\tau})$, and $(Z \cap \neg\boldsymbol{\tau}) \oplus (M_0 \cap \boldsymbol{\tau})$ form a pentagon. If (10.27) is the case, then the logics **S4**, $Z \cap \neg\boldsymbol{\tau}$, $M_0 \cap \neg\boldsymbol{\tau}$, $M_0 \cap (\boldsymbol{\tau} \oplus \neg\boldsymbol{\tau})$, and $M_0 \cap \boldsymbol{\tau}$ form a pentagon. Thus the Main Assumption is false, which implies that $M_0 \cap \neg\boldsymbol{\tau} \subseteq Z$.

The following proposition follows immediately from Lemma 4.

Proposition 36 *Let* $\boldsymbol{\tau} \neq \mathbf{S4}$. *If* $d^*(M_0 \oplus \boldsymbol{\tau})$ *is defined, then* $\mathbf{Grz} \cap \neg\boldsymbol{\tau} \subseteq d^*(M_0 \oplus \boldsymbol{\tau})$, *or equivalently* $M_0 \cap \neg\boldsymbol{\tau} \subseteq d^*(M_0 \oplus \boldsymbol{\tau})$.

Proof According to Proposition 29, $d^*(M_0 \oplus \boldsymbol{\tau}) \simeq d(M_0, M_0 \cap \boldsymbol{\tau})$. Then, we apply (10.22) and Lemma 4.

The two propositions below do not require Lemma 4. However, the first proposition gives conditions for $d^*(M_0 \oplus \boldsymbol{\tau}) = \mathbf{Grz} \cap \neg\boldsymbol{\tau}$.

Proposition 37 *Let* $\boldsymbol{\tau} \neq \mathbf{S4}$. *Then the following conditions are equivalent.*

(a) $M_0 \subseteq \boldsymbol{\tau} \oplus \neg\boldsymbol{\tau}$;
(b) $\mathbf{Grz} \cap \neg\boldsymbol{\tau}$ *is a relative complement of* $\mathbf{Grz} \cap \boldsymbol{\tau}$ *in* $\mathscr{E}_0$;
(c) $d^*(M_0 \oplus \boldsymbol{\tau}) = \mathbf{Grz} \cap \neg\boldsymbol{\tau}$;
(d) $M_0 = \mathbf{Grz} \cap (\boldsymbol{\tau} \oplus \neg\boldsymbol{\tau})$.

Proof First we prove that (a) $\Rightarrow$ (b) $\Rightarrow$ (c) $\Rightarrow$ (a). Indeed, suppose $M_0 \subseteq \boldsymbol{\tau} \oplus \neg\boldsymbol{\tau}$. With help of (10.22), we derive that $(\mathbf{Grz} \cap \boldsymbol{\tau}) \oplus (\mathbf{Grz} \cap \neg\boldsymbol{\tau})$, which in turn implies (b). Then, according to Proposition 35, $d^*(M_0 \oplus \boldsymbol{\tau}) = \mathbf{Grz} \cap \neg\boldsymbol{\tau}$. Finally, the latter, by virtue of (10.22) and Proposition 29, implies that $(M_0 \cap \boldsymbol{\tau}) \oplus (M_0 \cap \neg\boldsymbol{\tau}) = M_0$, that is (a).

Now we show the equivalence of (a) and (d). Let $M_0 \subseteq \boldsymbol{\tau} \oplus \neg\boldsymbol{\tau}$, that is $M_0 = M_0 \cap (\boldsymbol{\tau} \oplus \neg\boldsymbol{\tau})$. By distributivity with help of (10.22), we obtain (d). On the other hand, it is obvious that (d) implies (a).

Proposition 38 *Let* $\boldsymbol{\tau} \in \mathscr{L}$ *and* $\boldsymbol{\tau} \neq \mathbf{S5}$. *If* $d^*(M_0 \oplus \boldsymbol{\tau})$ *is defined, then there is* $n \geqslant 1$ *such that* $M_0 \nsubseteq T_n \oplus \boldsymbol{\tau}$.

Proof We prove the contrapositive implication. For this, assume that for all $n \geqslant 1$, $M_0 \subseteq T_n \oplus \boldsymbol{\tau}$. Then for all $n \geqslant 1$, $T_n \oplus (M_0 \cap \boldsymbol{\tau}) = M_0$. By virtue of Corollary 4 and the premise that $\boldsymbol{\tau} \neq \mathbf{S5}$, the set $\{Z \mid Z \oplus (M_0 \cap \boldsymbol{\tau}) = M_0\}$ does not have a least element. According to Proposition 25, it implies that $d(M_0, M_0 \cap \boldsymbol{\tau})$ is undefined. It remains to apply Proposition 29.

10.9 Conclusion

As we have seen, the idea of τ-decomposition of **S4**-logics sheds light on the structure of NExt**S4** as a distributive lattice and, to some extent, as a Heyting algebra. The initial point of this research is the Embedding Theorem owing to Gödel, McKinsey, and Tarski. Further development of the idea of embedding includes such contributors as Dummett, Lemmon, Grzegorczyk, Maksimova, Rybakov, Blok, and Esakia. Although we leave the question of Problem 1 open, we hope that it will soon find its solution, for, at least in relation to mathematics, *dies diem docet.*

References

Balbes, R., & Dwinger, P. (1974). *Distributive lattices*. Columbia: University of Missouri Press.

Blok, W. (1976). Varieties of interior algebras. Ph.D. thesis, University of Amsterdam.

Burris, S., & Sankappanavar, H. (1981). *A course in universal algebra* (Vol. 78), Graduate texts in mathematics. New York: Springer.

Chagrov, A., & Zakharyaschev, M. (1997). *Modal logic* (Vol. 35), Oxford logic guides. New York: The Clarendon Press, Oxford University Press.

Citkin, A. (2013). Jankov formula and ternary deductive term. *Proceedings of the Sixth Conference on Topology, Algebra, and Category in Logic (TACL 2013)* (pp. 48–51). Nashville: Vanderbilt University.

Citkin, A. (2014). Characteristic formulas 50 years later (An algebraic account). arXiv:1407.583v1.

Dummett, M., & Lemmon, E. (1959). Modal logics between S4 and S5. *Zeitschrift für Mathematische Logik und Grundlagen der Mathematik, 5*, 250–264.

Dunn, J., & Hardegree, G. (2001). *Algebraic methods in philosophical logic* (Vol. 41), Oxford logic guides. New York: The Clarendon Press, Oxford University Press.

Esakia, L. (1976). O modalnykh "naparnikakh" superintuicionistskikh logik, *Abstracts of the 7th All-Union Symposium on Logic and Methodology of Science, Kiev* (pp. 135–136). [On modal "counterparts" of superintuitionistic logics].
Esakia, L. (1979). In Mikhailov, A. I. (Ed.), *O mnogoobrazii algebr Grzegorczyka* (pp. 257–287), Studies in non-classical logics and set theory. Moscow: Nauka. [Translation: On the variety of Grzegorczyk algebras. *Selecta Mathematica Sovietica 3*, pp. 343–366.].
Fine, K. (1974). An ascending chain of S4 logics. *Theoria*, *40*(2), 110–116.
Gabbay, D., & Maksimova, L. (2005). *Interpolation and definability* (Vol. 46), Oxford logic guides. Oxford: The Clarendon Press, Oxford University Press.
Gödel, K. (1933). Eine Interpretation des intuitionistischen Aussagenkalküls. *Ergebnisse eines Mathematischen Kolloquiums* (Vol. 4, pp. 39–40). [Translation is available in Gödel (1986)].
Gödel, K. (1986). In S. Feferman (Ed.), *Collected works, Vol. 1, publications 1929–1936*, With a preface by Solomon Feferman. New York: The Clarendon Press, Oxford University Press.
Grätzer, G. (1978). *General lattice theory* (1st ed.). Berlin: Akademie Verlag.
Grätzer, G. (1979). *Universal algebra* (2nd ed.). New York: Springer.
Grzegorczyk, A. (1967). Some relational systems and the associated topological spaces. *Fundamenta Mathematicae*, *60*, 223–231.
Hosoi, T. (1969). On intermediate logics. II. Journal of the Faculty of Science. *University of Tokyo Sect. I*, *16*, 1–12.
Hughes, G., & Cresswell, M. (1996). *A new introduction to modal logic*. London: Routledge.
Jankov, V. (1969). Konyunktino nerazlozhimye formuly v propozicional'nom ischislenii. *Izvestia Akademii Nauk SSSR*, Ser. Mat. *33*, 18–38. [Transation: Conjunctively indecomposable formulas in propositional calculi, *Mathematics of the USSR–Izvestiya 3*, pp. 17–37.].
Lemmon, E. (1965). Some results on finite axiomatizability in modal logic. *Notre Dame Journal of Formal Logic*, *6*(4), 301–308.
Kleene, S. (1952). *Introduction to metamathematics*. New York: D. Van Nostrand Co., Inc.
Maksimova, L. (1975). Modalnye logiki konechnykh sloev. *Algebra i Logika14*(3), 304–319. [Transation: Modal logics of finite layers, *Algebra and Logic 14*, pp. 188–197.].
Maksimova, L., & Rybakov, V. (1974). Reshetka normalnykh modalnykh logik. *Algebra i Logika*, *13*(2), 188–216. [Transation: The lattice of normal modal logics, *Algebra and Logic 13*, pp. 105–122.].
McKenzie, R. (1972). Equational bases and nonmodular lattice varieties. *Transactions of the American Mathematical Society*, *174*, 1–43.
McKinsey, J., & Tarski, A. (1944). The algebra of topology. *Annals of Mathematics*, *45*, 141–191.
McKinsey, J., & Tarski, A. (1948). Some theorems about the sentential calculi of Lewis and Heyting. *Journal of Symbolic Logic*, *13*(1), 1–15.
Muravitsky, A. (2006). The embedding theorem: its further development and consequences. Part 1. *Notre Dame Journal of Formal Logic*, *47*(4), 525–540.
Rasiowa, H., & Sikorski, R. (1970). *The Mathematics of Metamathematics* (3rd ed., Vol. 41), Monografie Matematyczne. Warsaw: PWN.
Rautenberg, W. (1979). *Klassische und nichtclassische Aussagenlogik* (Vol. 22), Logik und Grundlagen der Mathematik. Braunschweig: Friedr. Vieweg & Sohn.
Rautenberg, W. (1980). Splitting lattices of logics. *Archiv für. Mathematische Logik und Grundlagenforschung*, *20*(3–4), 155–159.
Scroggs, S. (1951). Extensions of the Lewis system S5. *Journal of Symbolic Logic*, *16*, 112–120.
Segerberg, K. (1968). Decidability of S4.1. *Theoria*, *34*, 7–20.
Segerberg, K. (1971). *An Essay in Classical Modal Logic* (Vol. 1-3), Filosofiska Föreningen och Filosofiska Institutionen vid Uppsala Universitet. Uppsala: Filosofiska Studier.

Chapter 11
Linear Temporal Logic with Non-transitive Time, Algorithms for Decidability and Verification of Admissibility

Vladimir V. Rybakov

I dedicate this paper to Larisa Maksimova, a bright mathematician with outstanding contributions to mathematical logic and my former PhD adviser, with whom for many years I have shared a close and friendly relationship.

Abstract We investigate linear temporal logic $\mathcal{LTL}_{NT}$ with non-transitive time and operators NEXT and UNTIL, as well as some possible interpretations of logical knowledge operators in this context. We assume time to be non-transitive, linear and discrete, the former being a major innovative part of this paper. We provide motivation for our approach along with ideas of why we might want to consider time to be non-transitive, and comment on possible interpretations of logical knowledge operators. The main result of Sect. 11.5 is a solution for decidability problem for $\mathcal{LTL}_{NT}$, for which we describe in details the decision algorithm. In Sect. 11.6 we introduce non-transitive linear temporal logic $\mathcal{LTL}_{NT}(m)$ with uniform bound (m) for non-transitivity. There we also solve the admissibility problem for $\mathcal{LTL}_{NT}(m)$, that is we provide an algorithm for verifying admissibility of inference rules in $\mathcal{LTL}_{NT}(m)$. Last section contains description of remaining interesting open problems.

11.1 Introduction

The conception of time is a very fundamental one, though even now it still does not have a definitive and precise definition. We cannot even observe or feel it except as by means of some changes in our environment (including the work of time-measuring devices). Anyway, prosaically, it works fine in CS, e.g. for interpretation of spread the set of check points in computational runs, etc.; temporal logic has found an important

V. V. Rybakov (✉)
Institute of Mathematics and Informatics, Siberian Federal University Krasnoyarsk, and (part time) Institute of Informatics Systems (Siberian Division RAS), av. Svobodnui 79, Krasnoyarsk 660041, Russia
e-mail: Vladimir_Rybakov@mail.ru

S. Odintsov (ed.), *Larisa Maksimova on Implication, Interpolation, and Definability*, Outstanding Contributions to Logic 15, https://doi.org/10.1007/978-3-319-69917-2_11

application in formal verification, where it is used to state requirements of hardware or software systems.

Historically, investigations of temporal logic in mathematical/philosophical logic based on modal systems were originated by Arthur Prior in late 1950s. Since then temporal logic has been a very active area of mathematical logic, information sciences, AI and CS (cf. e.g. Gabbay et al. 1994; Gabbay and Hodkinson 1990; 1995).

In mathematical logic and CS, a popular model of time is the so called linear temporal logic or linear-time temporal logic — a temporal modal logic with modalities referring to time. For example, linear temporal logic $\mathcal{LTL}$ (with Until and Next) proved to be very useful (cf. Manna and Pnueli 1992, 1995; Vardi 1995, 1998): it was used for analyzing protocols of computations, checking consistency, etc. The decidability and satisfiability problems for $\mathcal{LTL}$, the main problems were in focus of investigation and were successfully resolved (see references above).

The conception of knowledge, and especially the one formulated by means of multi-agent approaches, is also a popular investigation topic in logic and computer science. Its various aspects, including interactions and autonomy, effects of cooperation, etc. were successfully investigated (cf. e.g.. Wooldridge and Lomuscio 2000; Woolridge 2003; Woolridge et al. 2006; Lomuscio and Michaliszyn 2013; Balbiani and Vakarelov 2012). In particular, in Babenyshev and Rybakov (2010) a multi-agent logic with distances was suggested and studied, and its satisfiability problem was solved. In Rybakov (2011, 2012) a conception of *chance discovery* within multi-agent approach was considered; a logic modeling uncertainty of agent's views was investigated in McLean and Rybakov (2013). A representation of interactions between agents was suggested in Rybakov (2009a, b) as a dual of an elegant notion of the *common knowledge* suggested and profoundly developed in Fagin et al. (1995). Historically, common knowledge was formalized and technically developed in 1990s in a series of papers lately summarized in Fagin et al. (1995). At the core of this approach, as well as of majority of those introduced later for working with logical knowledge operators, is agent's knowledge represented by $S5$-like modalities.

In general, attempts to model knowledge by means of symbolic logic can likely be dated to the end of 1950s. Hintikka's 1962 book *Knowledge and Belief* (Hintikka (1962)) was very likely the first book-length work, in which the use of modalities to capture the semantics of knowledge was suggested. Since then knowledge representation and reasoning about knowledge in logical framework became very popular areas of research. Thus, for example, modal and multi-modal logics were used for formalizing agent's reasoning in Balbiani and Vakarelov (2012); Vakarelov (2005); Fagin et al. (1995); Rybakov (2003, 2009a). Some up-to-date studies of knowledge and beliefs in terms of modal logic can be found in Halpern et al. (2009). Modern approaches to knowledge frequently use the notion of justification in terms of epistemic logic (cf. e.g. Artemov 2006; Artemov and Nogina 2005).

In this paper, we will investigate linear temporal logic $\mathcal{LTL}_{NT}$ with non-transitive time and operators next **N** and until **U**, and discuss some possible interpretations for logical knowledge operators in this setting. The use of non-transitive linear time is a major innovative part of this paper. We will consider **N** and **U** to be directed to the past and time to be discrete, linear, non-transitive and also directed to the past.

In Sect. 11.2 we will recall some preliminary facts and notation concerning the standard (transitive) linear temporal logic $\mathcal{LTL}$. In Sect. 11.3 we will explain the motivation behind our approach, give reasons as to why it is reasonable to consider time to be non-transitive and introduce some useful and appealing interpretations. In Sect. 11.4 we will formally introduce linear non-transitive temporal logic $\mathcal{LTL}_{NT}$ and develop a technique necessary for solving the main problem, which is the decidability problem for $\mathcal{LTL}_{NT}$. In Sect. 11.5 we solve this problem: we show that $\mathcal{LTL}_{NT}$ is decidable and, hence, that the satisfiability problem for it is also decidable. We will obtain and give a detailed description of the decision algorithm. Let us remark that non-transitivity of time makes our algorithm significantly more complicated than that for $\mathcal{LTL}$. In Sect. 11.6 we will introduce non-transitive linear temporal logic $\mathcal{LTL}_{NT}(m)$ with uniform bounds for non-transitivity. Section 11.7 is dedicated to describing some remaining open problems.

11.2 Preliminary Definitions and Notation

In this paper we investigate a temporal logic with non-transitive time, which is its major innovation point, and operators *until* and *next*. Let us remark that temporal logics with non-transitive time were already considered before, yet it seems that there were no prior studies dedicated to linear temporal logics with operators until and next based on non-transitive time. We begin with a short recollection of notation and definitions concerning standard linear temporal logic $\mathcal{LTL}$. The reader familiar with $\mathcal{LTL}$ may skip this section and go straight to the next one.

The language of the Linear Temporal Logic ($\mathcal{LTL}$ in the sequel) extends the language of Boolean logic with operators **N** (*next*) and **U** (*until*). Formulas of $\mathcal{LTL}$ are built up from a set $Prop$ of atomic propositions (or propositional letters) using Boolean operations, an unary operation **N** and a binary operation **U**. Later on for some extensions of $\mathcal{LTL}$ we will also introduce a binary operation **S** (*since*) and a unary operation **Pr** (*previous*). Intuitively, formula $\mathbf{N}\varphi$ means that the statement φ holds at the next time point (state); formula $\mathbf{Pr}\varphi$ means that the statement φ was true at the time point immediately before the current one; formula $\varphi\mathbf{U}\psi$ means that φ will hold until ψ will hold first time; formula $\varphi\mathbf{S}\psi$ means that φ holds since the point in which ψ became true. Semantics for $\mathcal{LTL}$ consists of *infinite transition systems (runs, computations)*, which are formally represented by linear Kripke structures based on the set of natural numbers.

An *infinite linear Kripke structure* (or simply a *Kripke structure*) is a tuple $\mathcal{M} := \langle N, \leqslant, \text{Next}, V\rangle$, where N is the set of all natural numbers (for some extensions of $\mathcal{LTL}$ we can consider the set of all integers Z instead); $\leqslant$ is the standard ordering relation on N, Next is a binary relation, such that a Next b holds iff $b = a + 1$; V is a valuation of some subset S of $Prop$, which assigns truth values to elements of S. Formally, for any $p \in S$ we have $V(p) \subseteq N$ and $V(p)$ represents the set of all $n \in N$

where p is true (w.r.t. V). Elements of N are called possible *states* (or worlds), $\leqslant$ is called the *transition* relation (which is linear in our case), and V can be interpreted as a *labeling* operation of states with atomic propositions. A tuple $\langle N, \leqslant, \text{Next}\rangle$ is called a *Kripke frame* and for brevity will be denoted by $\mathscr{N}$.

The truth values of any Kripke structure $\mathscr{M}$ can be extended from propositions in S to arbitrary formulas constructed using these propositions.

Definition 1 Computational rules for logical operations ($a \in N$):

- $(\mathscr{M}, a) \Vdash_V p \;\Leftrightarrow\; a \in V(p)$ for $p \in S$;
- $(\mathscr{M}, a) \Vdash_V (\varphi \wedge \psi) \;\Leftrightarrow\; (\mathscr{M}, a) \Vdash_V \varphi \wedge (\mathscr{M}, a) \Vdash_V \psi$
- $(\mathscr{M}, a) \Vdash_V \neg\varphi \;\Leftrightarrow\; not[(\mathscr{M}, a) \Vdash_V \varphi]$;
- $(\mathscr{M}, a) \Vdash_V \mathbf{N}\varphi \;\Leftrightarrow\; [[(a \text{ Next } b) \Rightarrow (\mathscr{M}, b) \Vdash_V \varphi]]$;
- $(\mathscr{M}, a) \Vdash_V \mathbf{Pr}\,\varphi \;\Leftrightarrow\; [[(b \text{ Next } a) \Rightarrow (\mathscr{M}, b) \Vdash_V \varphi]]$;
- $(\mathscr{M}, a) \Vdash_V (\varphi \mathbf{U} \psi) \;\Leftrightarrow$

$$\exists b\,[(a \leqslant b) \wedge ((\mathscr{M}, b) \Vdash_V \psi) \wedge \forall c\,[(a \leqslant c < b) \;\Rightarrow\; (\mathscr{M}, c) \Vdash_V \varphi]];$$

- $(\mathscr{M}, a) \Vdash_V (\varphi \mathbf{S} \psi) \;\Leftrightarrow$

$$\exists b\,[(b \leqslant a) \wedge ((\mathscr{M}, b) \Vdash_V \psi) \wedge \forall c\,[(a \leqslant c < b) \Rightarrow (\mathscr{M}, c) \Vdash_V \varphi]].$$

For a Kripke structure $\mathscr{M} := \langle N, \leqslant, \text{Next}, V\rangle$ and a formula φ with letters from the domain of V we say that φ is *valid* in $\mathscr{M}$ and denote it by $\mathscr{M} \Vdash \varphi$ if for any state b in $\mathscr{M}$ (that is $b \in N$) formula φ is true at b (that is $(\mathscr{M}, b) \Vdash_V \varphi$).

Definition 2 The linear temporal logic $\mathscr{LTL}$ is the set of all formulas which are valid in all infinite temporal linear Kripke structures $\mathscr{M}$ based on $\mathscr{N}$ with standard $\leqslant$ and Next.

Modal operators $\Box$ (necessary) and $\Diamond$ (possible) can be defined using temporal ones as follows: $\Diamond p := \top \mathbf{U} p$, $\Box p := \neg\Diamond\neg p$.

11.3 Motivation and the Conception of Knowledge from the Temporal Perspective

Our motivation is based on ideas suggested in preprints Rybakov (2014a, b). Here we extend this approach and, in particular, give complete proofs for some statements given in these preprints.

It is easy to accept that knowledge is not absolute and depends on opinions of individuals (agents), who may accept a statement as either definitely true or not, on the one hand, and on what we actually consider as true knowledge, on the other hand.

Firstly, we would like to look at this from the temporal perspective. Some evident and trivial observations are that

(i) human beings remember (at least some of) theirs past, but
(ii) they do not know future at all (rather they could surmise what is going to happen in some immediate proximity of time);
(iii) individual memory tells us that time was linear in the past (though there is a chance that it might only be this way in our perception).

Therefore it seems meaningful to look for interpretations of knowledge in linear temporal logic with accessibility relations and operator next directed to the past.

Now, we would like to discuss several approaches to defining knowledge operations. Here we will use unary logical operations K_i for logical knowledge operations. Let us remind once more, that we will consider models for $\mathscr{LTL}$ with both *next* and $\leqslant$ directed to the past.

(i) Simple approach: a knowledge is something that was once discovered to be true and remained true since then.

$$(\mathscr{M}, a) \Vdash_V K_1\varphi \Leftrightarrow \exists b\, [a \leqslant b \wedge ((\mathscr{M}, b+1) \nVdash \varphi) \wedge \forall c\, (a \leqslant c \leqslant b \Rightarrow (\mathscr{M}, c) \Vdash_V \varphi]]$$

At first glance, it seems to be a rather plausible interpretation. The bigger b is (that is the further it is in the past), the more reasonable it is to consider φ as a knowledge. But if $a = b$ this definition actually tells us nothing, it then admits one-day knowledge, which is definitely not good.

(ii) Rigid approach: something is knowledge if it always held true.

$$(\mathscr{M}, a) \Vdash_V K_2\varphi \Leftrightarrow (\mathscr{M}, a) \Vdash_V \neg(\top \mathbf{U} \neg\varphi)$$

This interpretation is perfectly fine, though it is too rigid, since it presupposes that we know all of the past (and besides it does not take into account that any knowledge has been obtained only at a particular time point).

(iii) Parametric approach: a knowledge is something that held true ever since some specified event happened.

$$(\mathscr{M}, a) \Vdash_V K_\psi\varphi \Leftrightarrow (\mathscr{M}, a) \Vdash_V \varphi \mathbf{U} \psi$$

In this interpretation φ has been true ever since some event happened in the past (which is modeled by formula ψ being true at a state). Thus, as soon as ψ became true, φ also became true and held true since. Here we use the standard *until*. Formula ψ may be chosen with any desirable meaning, so that a *knowledge* is something that held since ψ.

(iv) Agent-based approach: knowledge as voted truth.

This is a very well developed area, cf. Fagin et al. (1995) as well as more recent publications e.g. Rybakov (2003, 2009a). Here, though, we would like to look at it from a different standpoint. Before, knowledge operations (agents knowledge) of this sort were considered simply as unary logical operations K_i interpreted as $S5$-modalities, and knowledge operations were introduced via vote of agents, etc. We would like to suggest here a somewhat very simple yet rather fundamental interpretation, which does not seem to have been considered before. We assume that every agent has his own valuation on the frame $\mathcal{N}$. So, we have n agents, n valuations V_i and thus, as earlier, we have truth values w.r.t. V_i of any propositional letter p_j at any world $a \in \mathcal{N}$. From applications standpoint, V_i corresponds to agents information about truth of p_j (they may be different). So, V_i is just individual *information*.

How can information be transformed into *local* knowledge? One way is as a result of a vote (in a sense close to what has been considered before): we consider a new valuation V, w.r.t. which p_i is true at a if majority (with specified confidence) of agents believes that p_i is true at a. Then we obtain a model with single (standard) valuation V. After that we can apply any of the already known approaches. But we could also extend valuations V_i for complex formulas first and only then consider a knowledge valuation V, which says that an arbitrary formula φ is true at a w.r.t. V if it is true at a w.r.t. majority of V_i (with specified confidence level). Besides that we could also use a more temporal feature as in the following approach.

(v) Agent-based approach: knowledge as a resolution at the evaluation state.

Here we suggest an interpretation which is somewhat similar to (iv). Again we have valuations V_i for every agent i, but then we put

$$(\mathcal{M}, a) \Vdash_V K\varphi \ \Leftrightarrow \forall i[(\mathcal{M}, a) \Vdash_{V_i} \Diamond\varphi \wedge \Box[\neg\varphi \to \mathbf{N}\neg\varphi].$$

In this case, if we will then allow using nested knowledge operations K in formulas together with several valuations V_i for agents information as well as the obtained valuation V (for all cases, in which we evaluate $K\varphi$, no matter for which agent, i.e. V_i) a decision procedure for the logic based on this approach is not known. We think that it is an interesting open question.

Summarizing these observations, we think that the linear temporal logic is a very promising tool for making subtle distinctions between different definitions of logical knowledge operations. In the sequel, our approach for various kinds of logical knowledge operations (interpreted as above) will be based on the assumption that *knowledge is a true fact, which was observed once, was widely acknowledged to be true in the past, and remained true since then.* However we would like to abolish the requirement for time to be transitive (and then we will base our approach on suggested modification).

Why Time Might be Non-transitive ?

View (i). *Individual being view.* If we interpret time in the past as a line of all events which individuals remember, and time flow as the chain of events which individual remembers as they pass to each other, the things are clear. We do not know and do not remember all of which our ancestors knew.

View (ii). *Computational view.* Inspections of protocols for computations are limited by time resources and have non-uniform length (besides, at any point of inspection, verification may refer to some old stored protocols). Therefore, if we interpret our models as reflecting verification of computations, then the number of check points is finite, yet not all of them might be available at a given time point.

View (iii). *Agent as admin.* We may consider states (worlds of our model) as checkpoints of network states, which a number of admins (agents) are allowed to inspect. Any admin is allowed to inspect some of the previous states, but only within his/hers areas of responsibility (due to security reasons or some other reasons). So then, the accessibility is not transitive again. Admin (a1) can inspect a state and then admin (a2), responsible for this state (either the same one or some other), also has some states available for inspection to him. In general (a1) cannot inspect all states, which (a2) can.

View (iv). *Agent as user.* We can consider states of a model as the content of web pages available to users, with accessibility relation connecting any page with those that can be reached from it using links. Some page may only be available to those individuals, who possess the password — these users can then continue surfing by passing from this page to other available from it and etc. Clearly, in such an interpretation surfing looks like it will generate a non-transitive relation. Further, if we interpret web-surfing as time-steps the accessibility will be intransitive.

View (v). *From the standpoint of collecting knowledge.*

In human perception, an individual could only observe events and record knowledge in some finite interval of time in the past. The time in the past feels linear and we can only remember finite amount of information. If we consider the furthest time point in the past, of which we remember something, there was an individual, whose collected information, again, was limited to some interval of time, and so forth. Thus, from this standpoint the time in the past, generally speaking does not look to be transitive.

View (vi). *Collected knowledge as voted knowledge.* Here the picture is similar to the case (v) above, but we may consider knowledge as a collection of facts which only a majority (not necessarily all) of experts consider to be true (assert, are positive about). In the past the voted opinion of experts on one matter could differ at different time points. Besides time intervals which experts remember might be very diverse. Therefore, time again looks non-transitive from this standpoint.

In the next section we will suggest some non-transitive linear models, which are based on these observation about knowledge and non-transitivity of time.

11.4 Linear Temporal Logic Based on Non-transitive Time: Introduction and Preliminary Techniques

We start with the formal definition of the logic itself and then we will turn to interpretations of the above-mentioned logical knowledge operators. Let us denote by N the set of all natural numbers. For $a, b \in N$, such that $a < b$, put $[a, b] := \{c \in \mathcal{N} \mid a \leqslant c \leqslant b\}$, $[a, b) := [a, b] \setminus \{b\}$, $(a, b) := [a, b) \setminus \{a\}$ and for $a, b \in N$ put $(a \text{ Next } b) \Leftrightarrow (b = a + 1)$.

Definition 3 A *linear non-transitive possible-worlds frame* (henceforth, simply a *frame*) is a tuple $\mathcal{F} := \langle N, \leqslant, \text{Next}, \bigcup_{i \in N}\{R_i\}\rangle$ such that for some fixed $X \subset N$ we have $N = \bigcup_{i \in X}[i, m_i)$, where m_i denotes the smallest element of X with $i < m_i$, and intervals $[i, m_i)$, $[j, m_j)$ have empty intersections for different $i, j \in X$. For any $j \in N$ by R_j we will denote the standard linear order on $[j, m_i]$ for $i \in X$ and $j \in [i, m_i)$. For any $i \in X$ put $t(i) := m_i$. We will call X the *set of nodes of* $\mathcal{F}$ and denote it by $Node(\mathcal{F})$.

Notice that we consider $(\bigcup_{i \in N}\{R_i\})$ not as a binary relation, but as a countable set of finite binary relations R_i. As before, we can define a *model* $\mathcal{M}$ over $\mathcal{F}$ by adding a valuation V on $\mathcal{F}$ and extending it on all formulas as above, except for **U** and **N**, for which the computational clauses are defined as follows:

Definition 4 For any $a \in N$ put

$$(\mathcal{M}, a) \Vdash_V (\varphi\, \mathbf{U}\, \psi) \quad \Leftrightarrow$$

$$\exists b\, [(a R_a b) \wedge ((\mathcal{M}, b) \Vdash_V \psi) \wedge \forall c\, [(a \leqslant c < b) \Rightarrow (\mathcal{M}, c) \Vdash_V \varphi]];$$

$$(\mathcal{M}, a) \Vdash_V \mathbf{N}\varphi \quad \Leftrightarrow \quad \forall b\, [(a \text{ Next } b) \;\Rightarrow\; (\mathcal{M}, b) \Vdash_V \varphi].$$

Definition 5 Logic $\mathcal{LTL}_{NT}$ is the set of all formulas which are valid in any model $\mathcal{M}$.

The relation $\langle \bigcup_{i \in N}\{R_i\}\rangle$ is evidently non-transitive, despite the fact that any R_i is linear and transitive. States related by R_i (or, more precisely, the whole interval $[i, m_i]$ itself) may be interpreted as points of time which agent i remembers.

At any time point $i + k$ it might be a new agent, who observes this point, or the previous one, still we will refer to him as $i + k$ anyway. Relation R_j for $j \in [i, m_i]$ is contained in R_i, the interpretation being that agent j remembers the past to the same extent that i does, despite the fact that agent j is situated at an earlier time point, i then is the more knowledgeable of the two.

Therefore, in this approach we may interpret $i \in N$ as a time point and $[i, m_i]$ as a time interval available for the agent responsible for verification/reasoning at the time point i. The name of the agent in this case is anonymous, it is simply the one who is responsible for time point i.

Relying on this interpretation we can consider *interpretations of various aspects of knowledge* discussed above as well as some of theirs extensions.

Example 1

$$(\mathcal{M}, a) \Vdash_V K\varphi \;\Leftrightarrow\; (\mathcal{M}, a) \Vdash_V [\mathbf{N}^m \varphi] \wedge [\mathbf{N}^{m+1} \neg\varphi] \wedge [\varphi\, \mathbf{U}\, \neg\varphi].$$

Here $K\varphi$ means that knowledge coded by φ has been obtained exactly m 'years' ago: φ did not hold $m + 1$ years ago but it did held m years ago and it held ever since.

$$(\mathcal{M}, a) \Vdash_V K_1\varphi \;\Leftrightarrow\; (\mathcal{M}, a) \Vdash_V \Box\neg\varphi \;\wedge \Diamond(\neg\varphi \wedge \mathbf{N}(K\varphi)).$$

$K_1\varphi$ means that φ was wrong in all observable (from a) past, but before that there has been a time interval of length m, when φ was true (it was, so to say, a local temporal knowledge).

$$(\mathcal{M}, a) \Vdash_V K_2\varphi \Leftrightarrow (\mathcal{M}, a) \Vdash_V \bigwedge_{i \leqslant k} \Box^i \neg\varphi \wedge \Diamond^k(\neg\varphi \wedge \mathbf{N}(K\varphi)).$$

Here $K_2\varphi$ says that φ was wrong in last k 'remembered' intervals of time, but before that it has been a local knowledge for some time interval of length m.

Even based on these simple examples, it is easy to imagine wide possibilities for expressing properties of knowledge from temporal perspectives, which can be obtained by assuming that time is non-transitive.

Now we turn to our main topic, that is solving decidability and satisfiability problems for our logic $\mathcal{LTL}_{NT}$. It can be immediately seen that, since our accessibility relation is not transitive, standard techniques of solving these problems do not work here. Thus such techniques, as rarefication (as e.g. in Rybakov 2009a, b), filtration, or direct usage of automatons technique, would not readily work here. Besides that non-transitivity does not even allow us to transform formulas to more suitable and transparent forms. This is a consequence of the fact that it is impossible to say by means of formulas that two formulas are equivalent at all time points (we cannot use $\Box$ for this, since time is intransitive). Therefore, we will need to do some preliminary work to avoid nested operations.

We will rework a technique we have already used many times before (e.g. in Rybakov 2009b), which consists of reducing formulas to rules and then converting them into reduced forms. This approach simplifies all proofs a lot, since it allows us to consider very simple and uniform formulas without nested temporal operations.

Recall that a (sequential) *inference rule* is an expression

$$\mathbf{r} := \frac{\varphi_1(x_1, \ldots, x_n), \ldots, \varphi_l(x_1, \ldots, x_n)}{\psi(x_1, \ldots, x_n)},$$

where $\varphi_1(x_1, \ldots, x_n), \ldots, \varphi_l(x_1, \ldots, x_n)$ and $\psi(x_1, \ldots, x_n)$ are formulas constructed out of letters (variables) $x_1, \ldots, x_n$. Meaning of $\mathbf{r}$ is that formula $\psi(x_1, \ldots, x_n)$ (which is called the conclusion of $\mathbf{r}$) follows (logically follows) from assumptions $\varphi_1(x_1, \ldots, x_n), \ldots, \varphi_l(x_1, \ldots, x_n)$.

Definition 6 A rule $\mathbf{r}$ is said to be *valid* in a model $\mathcal{M}$ if the following holds:

$$\forall a\, [(\mathcal{M}, a) \Vdash_V \bigwedge_{1 \leqslant i \leqslant l} \varphi_i] \Rightarrow \forall a\, [(\mathcal{M}, a) \Vdash_V \psi].$$

Otherwise we say that $\mathbf{r}$ is *refuted* in $\mathcal{M}$ or that it is *refuted in* $\mathcal{M}$ *by* V and write $\mathcal{M} \nVdash_V \mathbf{r}$. A rule $\mathbf{r}$ is *valid* in a frame $\mathcal{F}$ (notation $\mathcal{F} \Vdash \mathbf{r}$) if it is valid in any model based on $\mathcal{F}$.

We can transform any formula φ into a rule $(x \to x)/\varphi$ and employ a technique of reduced normal forms for inference rules. We start with a self-evident

Lemma 1 *An arbitrary formula φ is a theorem of $\mathcal{LTL}_{NT}$ (that is $\varphi \in \mathcal{LTL}_{NT}$) iff a rule $(x \to x)/\varphi$ is valid in any frame $\mathcal{F}$.*

Definition 7 A rule $\mathbf{r}$ is said to be in *reduced normal form* if $\mathbf{r} = \varepsilon/x_1$, where

$$\varepsilon := \bigvee_{1 \leqslant j \leqslant l} \left[\bigwedge_{1 \leqslant i \leqslant n} x_i^{t(j,i,0)} \wedge \bigwedge_{1 \leqslant i \leqslant n} (\mathbf{N}x_i)^{t(j,i,1)} \wedge \bigwedge_{1 \leqslant i,k \leqslant n, i \neq k} (x_i \mathbf{U} x_k)^{t(j,i,k,1)} \right],$$

$t(j, i, z), t(j, i, k, z) \in \{0, 1\}$ and $\alpha^0 := \alpha$, $\alpha^1 := \neg\alpha$ for any formula α.

Definition 8 A rule $\mathbf{r_{nf}}$ in reduced normal form is said to be a *normal reduced form for a rule* $\mathbf{r}$ if for any frame $\mathcal{F}$ for $\mathcal{LTL}_{NT}$ we have

$$\mathcal{F} \Vdash \mathbf{r} \;\Leftrightarrow\; \mathcal{F} \Vdash \mathbf{r_{nf}}.$$

Theorem 1 *There exists a (single) exponential time algorithm, which given a rule* $\mathbf{r}$ *constructs its normal reduced form* $\mathbf{r_{nf}}$.

Similar statements for various logics have been proved by us quite a while ago, for instance, it is literally the same as the proof for $\mathcal{LTL}$ itself (cf. Lemma 5 in Babenyshev and Rybakov 2011). We will provide a proof for a similar statement for another logic in Sect. 11.6, in fact the proof for both cases is the same.

Summarizing, a formula φ is a theorem of $\mathcal{LTL}_{NT}$ iff a rule $r := (p \to p)/\varphi$ is valid in all frames (cf. Lemma 1) iff its reduced form $\mathbf{r_{nf}}$ is valid in all frames (cf. Theorem 1). Thus, from now on we can consider only rules in reduced normal forms.

11.5 Deciding Algorithm

First, we need to modify slightly models for $\mathcal{LTL}_{NT}$ introduced earlier.

Consider a frame $\mathcal{F} := \langle N, \leqslant, \text{Next}, (\bigcup_{i \in N}\{R_i\})\rangle$ and recall the definitions of a function t and a set $X = Node(\mathcal{F})$ as introduced in Definition 3. For any number k and any $i \in Node(\mathcal{F})$ let us denote by $t^k(i)$ the result of applying t to i k times. For convenience sake, we will sometimes write $\text{Next}(a) = b$ instead of a Next b. For any natural numbers r and g with $r > g \geqslant t^2(0)$ we define a frame $\mathcal{F}(N(r, g))$ based on the initial interval of $\mathcal{F}$:

$$\mathcal{F}(N(r, g)) := \langle N(r, g), \leqslant, \text{Next}, (\bigcup_{i \in N}\{R_i\})\rangle,$$

where the base set $N(r, g)$ of this frame is $[0, t^{r+1}(0))$, that is

$$N(r,g) := [0, t(0)] \cup [t(0), t^2(0)] \cup \cdots \cup [t^g(0), t^{g+1}(0)] \cup \cdots \cup [t^r(0), t^{r+1}(0)),$$

with $\leqslant$, Next and R_i (for $i \in [0, t^{r+1}(0))$) defined exactly they were defined for $\mathscr{F}$, except that (i) $\text{Next}(t^{r+1}(0) - 1) := t^g(0)$ and (ii) $j R_j t^g(0)$ for any $j \in [t^r(0), t^{r+1}(0))$.

A valuation V on such $\mathscr{F}(N(r,g))$ is defined as before and it is extended to formulas with **U** and **N** as earlier.

Lemma 2 *If a rule* $\mathbf{r_{nf}}$ *in reduced normal form is refuted in some frame* $\mathscr{F}$ *then* $\mathbf{r_{nf}}$ *can be refuted in a model over* $\mathscr{F}(N(r,g))$ *for some* $r, g \in N$, *such that the cardinality of* $\mathscr{F}(N(r,g))$ *is effectively computable from the size of the rule* $\mathbf{r_{nf}}$ *and is at most* $l! \cdot l_1$, *where* $l = l_1 + 1$ *and* l_1 *is the number of disjuncts in* $\mathbf{r_{nf}}$.

Proof Let $\mathbf{r_{nf}} = \varepsilon/x_1$ be a rule in reduced normal form, such that $\varepsilon = \bigvee_{1 \leqslant j \leqslant l} \theta_j$ and

$$\theta_j = \left[\bigwedge_{1 \leqslant i \leqslant n} x_i^{t(j,i,0)} \wedge \bigwedge_{1 \leqslant i \leqslant n} (\mathbf{N}\, x_i)^{t(j,i,1)} \wedge \bigwedge_{1 \leqslant i,k \leqslant n, i \neq k} (x_i \mathbf{U} x_k)^{t(j,i,k,1)} \right].$$

Also, consider a frame

$$\mathscr{F} := \langle N, \leqslant, \text{Next}, (\bigcup_{i \in N} \{R_i\}) \rangle$$

as above, where $\mathbf{r_{nf}}$ is refuted by a valuation V.

Our first step is to show that $\mathbf{r_{nf}}$ can be refuted in a frame $\mathscr{F}_1$, such that all distances $m_i - i$ for $i \in Node(\mathscr{F}_1)$ are effectively bounded.

For a frame refuting $\mathbf{r_{nf}}$ by a valuation V and $i \in N$ denote by $\theta(i)$ a unique disjunct θ_j from $\mathbf{r_{nf}}$, which is valid at i w.r.t. V. Recall that $Node(\mathscr{F}) := X$ is the set of nodes of $\mathscr{F}$ and that for any $k \in X$ we have $t(k) := m_k$, while $t^r(k)$ denotes the result of applying t to k r times.

Lemma 3 *There exists a frame* $\mathscr{F}_1$ *refuting* $\mathbf{r_{nf}}$ *by a valuation* V, *such that* $m_i - i \leqslant l$ *for all* $i \in Node(\mathscr{F}_1)$, *where* l *is the number of disjuncts in the premise of* $\mathbf{r_{nf}}$.

Proof Take the original model based on $\mathscr{F}$ which refutes $\mathbf{r_{nf}}$ by V and consider interval $[0, t(0))$ in it. Put $a = \text{Next}(0)$ and take the greatest $g(a)$ in $(a, t(0))$ (if any) such that $\theta(a) = \theta(g(a))$. If there is no such $g(a)$ consider $[\text{Next}(0), t(0))$ instead. Delete all numbers from $[a, g(a))$, thus putting $g(a)$ next to 0. Let valuation V on the remaining worlds of the resulting frame be the same as before. Then, the suggested modification will not affect truth values of sub-formulas of $\mathbf{r_{nf}}$ w.r.t. V.

Verification of this fact is a short standard routine procedure, we only need to check preservation of truth values of formulas with **U** and that is trivial: if formula $x_i \mathbf{U} x_j$ is disproved at 0 by virtue of a world in $[a, g(a))$, then it must be disproved by virtue of a world in $[g(a), t(0)]$, since $\theta(a) = \theta(g(a))$). Hence, the resulting model will also refute $\mathbf{r_{nf}}$.

Next, consider an interval $[g(a), t(0))$ and repeat the same transformation for it choosing $g(g(a))$ as above. We repeat the same rarefication procedure until it terminates. It will terminate in at most l steps and we will obtain a frame, such that the first interval $[0, t(0)]$ in it has at most l elements and which preserves truth values w.r.t. V of all sub-formulas from the rule $\mathbf{r_{nf}}$. Consider now any interval

$$[t^k(0), t^{k+1}(0)] \subset \mathscr{F}$$

of our frame which does not satisfy conditions of this lemma. We apply to the interval $[t^k(0), t^{k+1}(0)]$ exactly the same transformation as for $[0, t(0)]$ above. This transformation, as we observed earlier, will not affect truth values of sub-formulas of $\mathbf{r_{nf}}$ at any world of the new frame. By this rarefication procedure for $\mathscr{F}$ we conclude that Lemma 3 holds. □

Now, having a model $\mathscr{F}_1$ based on a frame as in Lemma 3, which refutes $\mathbf{r_{nf}}$, we do the following. Moving from $t(0)$ upward, starting with the second interval $[t(0), t(t(0))]$, we find first two intervals of the form $[t^s(0), t^{s+1}(0)]$ with empty intersection, such that strings of the form $\theta(t^s(0)), \ldots, \theta(t^{s+1}(0))$ for them coincide. Let

$$[t^s(0), t^{s+1}(0)] \text{ and } [t^{s+r}(0), t^{s+r+1}(0)]$$

be these lowermost intervals ($s + r \geqslant s + 1$).

Then we do the following: we delete all points of interval $[t^{s+r}(0), \infty)$ from $\mathscr{F}_1$ and appoint $[t^s(0), t^{s+1}(0)]$ to be the next interval after $[t^{s+r-1}(0), t^{s+r}(0))$: that is put $\text{Next}(t^{s+r}(0) - 1) := t^s(0)$ and for any $j \in [t^{s+r-1}(0), t^{s+r}(0))$ put $j R_j t^s(0))$. We preserve the valuation V on remaining worlds as before. Denote the resulting frame by $\mathscr{F}_2$ and the resulting model by $\mathscr{M}_2$, respectively.

Lemma 4 *Truth values of all sub-formulas from* $\mathbf{r_{nf}}$ *are the same at any world in* $\mathscr{M}_2$ *w.r.t.* V *as they were in* $\mathscr{F}_1$ *w.r.t.* V.

Proof Since we did not change intervals of transitivity $[k, t(k)]$ in $\mathscr{M}_1$, this is almost immediate and is proved by a routine verification.

Application of the last lemma completes the proof of Lemma 2. □

Lemma 5 *Any rule* $\mathbf{r_{nf}}$ *in reduced normal form, which is refuted in a model described in Lemma 2, is not valid in* $\mathscr{LTL}_{NT}$, *i.e. there is a standard frame* $\mathscr{F}$ *refuting* $\mathbf{r_{nf}}$.

Proof This is a simple modification of the standard unraveling technique: it is sufficient to roll the cyclic part of the frame towards the future. Even though our accessibility relation is non-transitive, the method still works. Just in case, we give below some sufficient details. Consider a frame $\mathscr{F}$ as in Lemma 2 refuting $\mathbf{r_{nf}}$ by a valuation V, where

$$\mathscr{F}(N(r, g)) := \langle N(r, g), \leqslant, \text{Next}, (\bigcup_{i \in N}\{R_i\})\rangle,$$

for $r > g \geqslant t^2(0)$ and $N(r, g) = [0, t^{r+1}(0))$, that is

$$N(r, g) := [0, t(0)] \cup [t(0), t^2(0)] \cup \cdots \cup [t^g(0), t^{g+1}(0)] \cup \cdots \cup [t^r(0), t^{r+1}(0)).$$

Recall that relations R_i and Next are defined the standard way, except (i) Next$(t^{r+1}(0) - 1) := t^g(0)$ and (ii) for any $j \in [t^r(0), t^{r+1}(0))$ we have $jR_jt^g(0)$. Now, it will suffice to stretch $\mathcal{F}(N(r, g))$ to the future. Consider a frame

$$\mathcal{F}(N(r, g, \infty)) := \langle W, \leqslant, \text{Next}, (\bigcup_{i \in N}\{R_i\})\rangle,$$

where W consists of all elements of $N(r, g)$:

$$N(r, g) := [0, t(0)] \cup [t(0), t^2(0)] \cup \cdots \cup [t^g(0), t^{g+1}(0)] \cup \cdots \cup [t^r(0), t^{r+1}(0))$$

as well as of an infinite ascending sequence of intervals

$$[t^g(0), t^{g+1}(0)] \cup \cdots \cup [t^r(0), t^{r+1}(0)),$$

adjoined to $[0, t^{r+1}(0))$. Here we do not transfer the relation Next from $\mathcal{F}(N(r, g))$, but assume instead that it behaves the standard way (which, nonetheless, is in accordance with how it behaved in $\mathcal{F}(N(r)))$. The valuation V is transferred to $\mathcal{F}(N(r, \infty))$ and is defined to be the same for copies of elements as it has been defined for original elements. It is a trivial routine computation to verify that all worlds and their copies will have the same truth values for sub-formulas of $\mathbf{r_{nf}}$ in $\mathcal{F}(N(r, \infty))$ as they had in $\mathcal{F}(N(r, g))$. Lemma is proved. □

Using Lemmas 1, 2, 5 and Theorem 1 we immediately derive

Theorem 2 *Logic $\mathcal{LTL}_{NT}$ is decidable and the satisfiability problem for it is also decidable. More specifically, for any formula we can compute, whether it is satisfiable and, if the answer is positive, compute a valuation, which satisfies this formula in a finite model of the form $\mathcal{F}(N(r, g))$.*

We currently cannot answer the question on recognizing admissibility in $\mathcal{LTL}_{NT}$, but we were able to solve this problem for some restricted version of our logic, and its extension. Next section is devoted to this problem.

11.6 Logics with Uniform Bound of Intransitivity and the Admissibility Problem

In this section we will consider a variation of $\mathcal{LTL}_{NT}$ — its extension, generated by models with uniformly bound measure of non-transitivity.

Definition 9 A non-transitive possible-worlds linear frame $\mathcal{F}$ with measure of intransitivity m ($m \in N$) is a frame:

$$\mathcal{F} := \langle N, \leqslant, \text{Next}, (\bigcup_{i \in N}\{R_i\})\rangle,$$

defined as in Definition 3 of Sect. 11.4, such that additionally for all $i \in X$ we have $m_i - i = m$. Recall, that $X = Node(\mathcal{F})$ is the set of nodes of $\mathcal{F}$.

Notice, that then $Node(\mathcal{F})$ consists of all numbers divisible by m, that is of all numbers of the form $m \cdot j$ for $j \in N$. For any set of letters P we can define a valuation V on $\mathcal{F}$ and to extend V to all formulas built up from letters in P the same way as we did for $\mathcal{LTL}_{NT}$.

Definition 10 Logic $\mathcal{LTL}_{NT}(m)$ is the set of all formulas which are valid at any model $\mathcal{M}$ with the measure of intransitivity m.

It seems reasonable to consider and to discuss such logic, since we may put a limitation on the size of time intervals that agents (experts) may introspect in the future (or can remember in the past).

Recall that the *temporal degree* of a formula φ is the largest number of nested occurrences of temporal operations **U** and **N**. More formally, temporal degree of a propositional letter is zero, i.e. $td(p) := 0$; temporal degree of a formula whose main logical operation is a boolean operation is the maximum of temporal degrees of its main components, i.e. if $\varphi := \varphi_1 \star \varphi_2$, were $\star$ is a binary boolean operation, then $td(\varphi) := max\{td(\varphi_1), td(\varphi_2)\}$, and $tg(\neg\varphi) := td(\varphi)$; for $\varphi := \varphi_1 \mathbf{U} \varphi_2$ put $td(\varphi) := max\{td(\varphi_1), td(\varphi_2) + 1\}$; also put $td(\mathbf{N}\varphi) := td(\varphi) + 1$.

Our first immediate and simple observation about $\mathcal{LTL}_{NT}(m)$ is

Proposition 1 *Logic $\mathcal{LTL}_{NT}(m)$ is decidable.*

The proof is trivial since to verify, whether a formula of temporal degree k is a theorem of $\mathcal{LTL}_{NT}(m)$, we only need to check its validity on the initial part of a frame, consisting only of $k + 1$ subsequent intervals of length m each.

Despite that, the question of recognizing admissible rules in $\mathcal{LTL}_{NT}(m)$ is not trivial. For a rule

$$\mathbf{r} := \frac{\varphi_1(x_1, \ldots, x_n), \ldots, \varphi_l(x_1, \ldots, x_n)}{\psi(x_1, \ldots, x_n)},$$

the definition of validity in a model or a frame is the same as before (bearing in mind different structures of frames and models). Again, we can transform any formula φ into a rule $x \to x/\varphi$ and, using such rules, determine, whether a formula is a theorem as follows: formula φ is a theorem of $\mathcal{LTL}_{NT}(m)$ (that is $\varphi \in \mathcal{LTL}_{NT}(m)$) iff the rule $(x \to x/\varphi)$ is valid in any frame $\mathcal{F}$. Recall that

Definition 11 A rule $r := \varphi(x_1, \ldots, x_n), \ldots, \varphi_m(x_1, \ldots, x_n)/\psi(x_1, \ldots, x_n)$ is said to be *admissible* in a logic L if for all formulas $\alpha_1, \ldots, \alpha_n$

$$\left[\left(\bigwedge_{1 \leqslant i \leqslant m} \varphi_i(\alpha_1, \ldots, \alpha_n)\right) \in L\right] \Longrightarrow [\psi(\alpha_1, \ldots, \alpha_n) \in L].$$

Reduced normal forms for rules play an even bigger role in solving the admissibility problem than in solving the decidability problem. The definition for reduced normal forms for rules is the same as in Definition 7 of Sect. 11.4.

Definition 12 Given a rule $\mathbf{r_{nf}}$ in reduced normal form $\mathbf{r_{nf}}$ is said to be a *normal reduced form for* $\mathbf{r}$ in logic $\mathcal{LTL}_{NT}(m)$ if for any frame $\mathcal{F}$ w.r.t. $\mathcal{LTL}_{NT}(m)$:

$$\mathcal{F} \Vdash \mathbf{r} \;\Leftrightarrow\; \mathcal{F} \Vdash \mathbf{r_{nf}}.$$

It is clear that we can restrict ourselves to consider only rules with a single premise, since any rule $\alpha_1, \ldots, \alpha_n/\beta$ is equivalent w.r.t. both derivability and admissibility to a rule $\alpha_1 \wedge \cdots \wedge \alpha_n/\beta$.

Theorem 3 *There exists a (single) exponential time algorithm, which, given a rule* $\mathbf{r}$, *constructs its normal reduced form* $\mathbf{r_{nf}}$ *in* $\mathcal{LTL}_{NT}(m)$.

Proof We draft the proof here. It will suffice to use a general algorithm described in e.g. Lemma 5 in Babenyshev and Rybakov (2011) for our specific language. Let $\mathbf{r} = \alpha/\beta$ be an inference rule. Denote by $Sub(\mathbf{r})$ the set of all subformulas of formulas in $\mathbf{r}$. We need a set of variables $Z = \{z_\gamma \mid \gamma \in Sub(r)\}$, not appearing in $\mathbf{r}$. Let us consider a rule in the *intermediate form*:

$$\mathbf{r_{if}} = z_\alpha \wedge \bigwedge_{\gamma \in Sub(\mathbf{r}) \setminus Var(\mathbf{r})} (z_\gamma \leftrightarrow \gamma^\sharp)/z_\beta,$$

where

$$\gamma^\sharp = \begin{cases} z_\delta * z_\epsilon, & \text{if } \gamma = \delta * \epsilon \textbf{ for } * \in \{\wedge, \vee, \rightarrow, \mathbf{U}\}. \\ *z_\delta, & \text{if } \gamma = *\delta \text{ for } * \in \{\neg, \mathbf{N}\}. \end{cases}$$

Rules $\mathbf{r}$ and $\mathbf{r_{if}}$ are definably equivalent.

Indeed, suppose $\mathcal{M}$ is a model of $\mathcal{LTL}_{NT}(m)$ based on frame $\mathcal{F}$ over natural numbers with a valuation V, such that $\mathcal{M} \nVdash_V \mathbf{r}$. Then $\mathcal{M} \Vdash_V \alpha$ and there exists an element $w \in \mathcal{F}$, such that $(\mathcal{M}, w) \nVdash_V \beta$. Let $V_1 : Z \rightarrow 2^N$ be a valuation defined by $V_1(z_\gamma) := V(\gamma)$. It is straightforward to show that $\mathcal{M} \Vdash_{V_1} z_\alpha \wedge \bigwedge\{z_\gamma \leftrightarrow \gamma^\sharp \mid \gamma \in Sub(\mathbf{r}) \setminus Var(\mathbf{r})\}$. In addition, $(\mathcal{M}, w) \nVdash_{V_1} z_\beta$.

For the other direction, suppose that for a similar model $\mathcal{M}$ with a valuation $V_1 : Z \rightarrow 2^N$ and $w \in \mathcal{F}$ we have $\mathcal{M} \Vdash_{V_1} z_\alpha \wedge \bigwedge\{z_\gamma \leftrightarrow \gamma^\sharp \mid \gamma \in Sub(\mathbf{r}) \setminus Var(\mathbf{r})\}$ and $(\mathcal{M}, w) \nVdash_{V_1} z_\beta$. Define $V : Var(\mathbf{r}) \rightarrow 2^N$ by $V(x_i) = V_1(z_{x_i})$. Then clearly $V(\gamma) = V_1(z_\gamma)$ for all $\gamma \in Sub(r)$. Thus $\mathcal{M} \Vdash_V \alpha$, $(\mathcal{M}, w) \nVdash_V \beta$ and hence $\mathcal{M} \nVdash_V \mathbf{r}$.

Finally, we transform the premise of $\mathbf{r_{if}}$ into a perfect disjunctive normal form over primitives of the form x_i, $\mathbf{N}x_i$ and $x_i\mathbf{U}x_j$. This requires not more than exponential time depending on the number of primitives, i.e. on the number of subformulas of the original rule (the same as for reduction of a boolean formula to the perfect disjunctive normal form). □

The reduced normal form for a rule $\mathbf{r}$, obtained in the last theorem, is uniquely defined and will be denoted in the sequel by $\mathbf{r_{nf}}$. In the formulation of the theorem below $\mathcal{F}_1$ and $\mathcal{F}_1(n)$ are models over some frames of the form specified in Definition 9 above, that is, they have all the relations and satisfy all the properties listed in this definition. Since these are frames for $\mathcal{LTL}_{NT}(m)$, we have $Node(\mathcal{F}_1) = Node(\mathcal{F}_1(n)) = m \cdot N$, where $m \cdot N := \{m \cdot j \mid j \in N\}$.

Theorem 4 *If a rule* $\mathbf{r_{nf}} = \bigvee_{1 \leqslant j \leqslant l} \varphi_j / x_1$ *in reduced normal form is not admissible in* $\mathcal{LTL}_{NT}(m)$ *then there are frames* $\mathcal{F}_1$ *and* $\mathcal{F}_1(n)$ *with valuations called* V_1 *in both cases such that*

(i) $\mathcal{F}_1 := \langle N, \leqslant, \text{Next}, (\bigcup_{i \in N} R_i) \rangle$ *and for some* $a_1 \in Node(\mathcal{F}_1)$ *we have* $N = [0, a_1] \cup [a_1 + 1, \infty)]$ *and*

(a) *all letters from* $\mathbf{r_{nf}}$ *have the same truth values w.r.t.* V_1 *at all worlds in* $[a_1 + 1, \infty)$*;*
(b) *there is* j *such that* $(\mathcal{F}_1, n) \Vdash_{V_1} \varphi_j$ *for all* $n \in [a_1 + 1, \infty)$*;*
(c) *for each* $n \in N$ *there is* j *such that* $(\mathcal{F}_1, n) \Vdash_{V_1} \varphi_j$ *and we denote a unique such* φ_j *by* $\theta(n)$*;*
(d) *the size of* $[0, a_1]$ *is at most* $m \cdot l^m$ *and* $\mathcal{F}_1, b \nVdash_{V_1} x_1$ *for some* $b \in [0, m)$.

(ii) *For* a_1 *specified above and for all* $n \in [0, a_1] \cap (\mathcal{F}_1)$ *the frame* $\mathcal{F}_1(n)$ *with valuation* V_1 *is such that, for some* $n_1 \in N$*, the upper segment* $[n_1, \infty)$ *of* $\mathcal{F}_1(n)$ *coincides with the segment* $[n, \infty)$ *of* $\mathcal{F}_1$ *(coincides as model). Beside that, we have*

(a) *for all* $m_1 \in N$ *there is* j *such that* $(\mathcal{F}_1(n), m_1) \Vdash_{V_1} \varphi_j$*. Again, we denote a unique such* φ_j *by* $\theta(m_1)$*. Generally speaking* $\theta(m_1)$ *may be different from the one in (i) above, but for worlds in* $[n_1, \infty)$ *in* $\mathcal{F}_1(n)$ *we can assume that they are the same as for corresponding worlds in* $[n, \infty)$ *in* $\mathcal{F}_1$*;*
(b) *the initial segment* $[0, n_1]$ *of* $\mathcal{F}_1(n)$ *contains interval* $[0, 2m]$ *for* $m \in Nodes(\mathcal{F}_1(n)$*, situated before* $[n_1, \infty)$ *which has properties:*
(c) *there is* $j \in N$ *such that for all* $n_2 \in [0, 2m]$ *we have* $(\mathcal{F}_1(n), n_2) \Vdash_{V_1} \varphi_j$ *and* $\theta(n_2) = \varphi_j = \theta(a_1 + 1)$*, where* $\theta(a_1 + 1)$ *defined as in (i) above;*
(d) *the size of* $[0, n_1]$ *in* $\mathcal{F}_1(n)$ *is at most* $m \cdot l^m + 2m$.

Proof Put $\mathbf{r_{nf}} = \phi / x_1$, where

$$\phi := \bigvee_{1 \leqslant j \leqslant l} \left[\bigwedge_{1 \leqslant i \leqslant n} x_i^{t(j,i,0)} \wedge \bigwedge_{1 \leqslant i \leqslant n} (\mathbf{N} x_i)^{t(j,i,1)} \wedge \bigwedge_{1 \leqslant i,k \leqslant n, i \neq k} (x_i \mathbf{U} x_k)^{t(j,i,k,1)} \right]$$

and let $x_i \to \varepsilon(x_i)$ be a substitution of formulas $\varepsilon(x_i)$ for x_i in $\mathbf{r_{nf}}$, such that after extending this substitution to formulas we get

$$\varepsilon(\phi) \in \mathcal{LTL}_{NT}(m) \text{ and } \varepsilon(x_1) \notin \mathcal{LTL}_{NT}(m).$$

Put

$$\varphi_j := \bigwedge_{1\leqslant i\leqslant n} x_i^{t(j,i,0)} \wedge \bigwedge_{1\leqslant i\leqslant n} (\mathbf{N}x_i)^{t(j,i,1)} \wedge \bigwedge_{1\leqslant i,k\leqslant n, i\neq k} (x_i\mathbf{U}x_k)^{t(j,i,k,1)}.$$

Then there is a frame

$$\mathscr{F} := \langle N, \leqslant, \text{Next}, (\bigcup_{i\in N}\{R_i\})\rangle,$$

for $\mathscr{LTL}_{NT}(m)$ with $N = \bigcup_{i\in Node(\mathscr{F})}[i, i+m]$, (cf. Definition 9) with a valuation V such that $(\mathscr{F}, b) \nVdash_V \varepsilon(x_1)$ for some $b \in [0, m)$.

Let k be the maximum of temporal degrees of formulas $\varepsilon(x_i)$ for all i and k_1 be the smallest number in $Node(\mathscr{F})$ strictly bigger than k. Now we modify valuation V in such a way that all formulas $\varepsilon(x_i)$ become true at all worlds from the interval $[(k_1+1)\cdot m, \infty)$.

This modification of V does not change truth values of formulas $\varepsilon(x_i)$ at b (which can be immediately inferred using an argument on temporal degrees of formulas), and therefore the resulting model also disproves (invalidates) formula $\varepsilon(x_1)$ at b. At the same time, since by our assumption $\varepsilon(\phi) \in \mathscr{LTL}_{NT}(m)$, we have

$$\forall n_1 \in N\, \exists j : (\mathscr{F}, n_1) \Vdash_V \varepsilon(\varphi_j).$$

We will denote a unique such φ_j by $\theta(n_1)$.

Next, for any $n \in \mathscr{F}$ with $n \in [0, (k_1+1)\cdot m]$ we introduce a new model $\mathscr{F}(n)$ for $\mathscr{LTL}_{NT}(m)$ such that (i) the initial segment $[0, (k_1+3)\cdot m + x)$ of it is the initial segment of some frame for $\mathscr{LTL}_{NT}(m)$ and (ii) it is followed by an infinite segment $[n, \infty)$ from $\mathscr{F}$ (considered again as a frame, names of states are reset so that the base set is again the set of all natural numbers). Then $\mathscr{F}(n)$ is a frame for our logic, i.e. the size of all intransitivity intervals is exactly m (that is for $n \in Node(\mathscr{F})$ we have $t(n) - n = m$). We denote by $f(g)$ the world in $[n, \infty)$ from $\mathscr{F}(n)$, which corresponds to g from $[n, \infty)$ in $\mathscr{F}$. Thus, $f([n, \infty))$ is the upper interval of $\mathscr{F}_1(n)$ adjoined to $[0, (k_1+3)\cdot m + x)$.

We define the valuation V of model $\mathscr{F}(n)$ for elements in $[f(n), \infty)$ to be the same as for the corresponding elements of $\mathscr{F}$ in $[n, \infty)$, and we assume all letters from all formulas $\varepsilon(x_i)$ to be true w.r.t. V at all worlds from the initial interval $[0, (k_1+3)\cdot m + x)$ of $\mathscr{F}(n)$.

Then again from $\varepsilon(\phi) \in \mathscr{LTL}_{NT}(m)$ we infer

$$\forall n_1 \in N\, \exists j : (\mathscr{F}(n), n_1) \Vdash_V \varepsilon(\varphi_j).$$

We will denote a unique such φ_j by $\theta(n_1, n)$.

Lemma 6 *(i) There exists j such that for all $n \in [(k_1+1)\cdot m), \infty)$ we have $(\mathscr{F}, n) \Vdash_V \varepsilon(\varphi_j)$.*

(ii) $\forall \mathscr{F}(n)$, there is j such that $(\mathscr{F}(n), n_1) \Vdash_V \varepsilon(\varphi_j)$ for all $n_1 \in [0, (k_1+2)\cdot m)$ where φ_j is the same as in (i) (from the formulation of our theorem).

Proof Item (i) follows immediately from the fact that all letters occurring in $\varepsilon(x_i)$ are true at all worlds from the interval $[(k_1+1)\cdot m, \infty)$ in $\mathscr{F}$, while (ii) is obtained by a standard argument using temporal degrees of formulas and the fact that all letters occurring in $\varepsilon(x_i)$ are true at all elements in $[0, (k_1+3)\cdot m + x)$ of $\mathscr{F}(n)$. □

Now, we modify models based on $\mathscr{F}$ and $\mathscr{F}(n)$ by replacing valuation V with V_1 such that $V_1(x_i) := V(\varepsilon(x_i))$ for all i. Then we can reformulate the lemma above the following way

Lemma 7 *(i) There is j such that for all $n_1 \in [(k_1+1)\cdot m, \infty)$ we have: $(\mathscr{F}, n_1) \Vdash_{V_1} \varphi_j$;*
(ii) For all $\mathscr{F}(n)$ there is j such that $(\mathscr{F}(n), n_1) \Vdash_{V_1} \varphi_j$ for all $n_1 \in [0, (k_1+2)\cdot m)$, where φ_j is the same as in (i).

We are now ready to apply the rarefication technique. First we clean up the frame $\mathscr{F}$. To do so starting with 0 we consider all intervals of the form $[i, i+m]$ for $i \in Node(\mathscr{F})$ and find the largest $j \in Node(\mathscr{F})$ (if any) such that $j < (k_1+1)\cdot m$ and strings $\theta(j), \ldots, \theta(j+m)$ and $\theta(i), \ldots, \theta(i+m)$ coincide. Then we delete all intermediate intervals between $[i, i+m]$ and $[j, j+m]$ and replace $[j, j+m]$ with $[i, i+m_i]$ rewriting all relations and the valuation accordingly.

It is clear that this modification does not affect truth values of formulas and for the resulting model $\mathscr{F}_1$ an analog of statement (i) from Lemma 7 will hold. Besides, if we denote by N_1 the base set of $\mathscr{F}_1$ then

$$\forall n_1 \in N_1\, \exists j : (\mathscr{F}_1, n_1) \Vdash_{V_1} \varphi_j,$$

since (see above) we had $\forall n_1 \in N\, \exists j : (\mathscr{F}, n_1) \Vdash_V \varepsilon(\varphi_j)$. We apply this procedure for $i \in Node(\mathscr{F})$ as long as possible. After it terminates (in at most l^m steps), the initial segment of the resulting model $\mathscr{F}_1$ situated before $[(k_1+1)\cdot m, \infty)$ will have the size at most $m \cdot l^m$. This completes the proof of (i) from the statement of our theorem, if we put $a_1 + 1 := (k_1+1)\cdot m$ (considered as worlds and not as numbers).

For the second item we apply a similar procedure to a model $\mathscr{F}(n)$ starting with the interval $[2m, 3m] \subset [0, (k_1+3)\cdot m)$ and proceeding with all intervals situated before $f(n)$. By the same reasoning it will terminate in a finite amount of steps (at most l^m) and the resulting model $\mathscr{F}_1(n)$ will have a segment of at most $m \cdot l^m + 2m$ elements situated before $f(n)$. And again this procedure does not affect truth values of formulas, and therefore in resulting models $\mathscr{F}_1(n)$ the analogs of (i) and (ii) from Lemma 7 hold, thus for all $n_1 \in \mathscr{F}_1(n)$ there is j such that $(\mathscr{F}_1(n), n_1) \Vdash_{V_1} \varphi_j$. □

Theorem 5 *Consider a rule $\mathbf{r_{nf}} = \bigvee_{1 \leqslant j \leqslant l} \varphi_j / x_1$ in normal reduced form. If there are models $\mathscr{F}_1$ and $\mathscr{F}_1(n)$ for all $n \in N$ with valuations for letters in $\mathbf{r_{nf}}$, such that $\mathscr{F}_1$ and $\mathscr{F}(n)$ satisfy all conditions of Theorem 4, then $\mathbf{r_{nf}}$ is not admissible in $\mathscr{L}\mathscr{T}\mathscr{L}_{NT}(m)$.*

Proof For any $a \in \mathscr{F}_1 \setminus [a_1+2m+2, \infty)$ we fix a unique letter p_a. Let b_a be the smallest world from $Node(\mathscr{F}_1) \cap \mathscr{F}_1 \setminus [a_1+2m+2, \infty)$ strictly bigger than a (if any). For each world a from the set $\mathscr{F}_1 \setminus [a_1+2m+2, \infty)$ put

$$\varphi_0(a) := p_a \wedge \left[\bigwedge_{c\in[0,a_1+2m+1]} \ \bigwedge_{0\leqslant x\leqslant a_1+2m+1-a} \mathbf{N}^x (p_c \to \bigwedge_{d\in[0,a_1+2m+1],\, c\neq d} \neg p_d) \right] \wedge$$

$$\left[\bigwedge_{b>a, b\leqslant a_1+2m+1} \mathbf{N}^{b-a} p_b \right] \wedge \left[\bigwedge_{a\leqslant y\leqslant b_a} \Diamond p_y \right] \wedge \neg\Diamond p_{\mathrm{Next}(b_a)},$$

in case both b_a and $p_{\mathrm{Next}(b_a)}$ exist; if b_a exist, but $p_{\mathrm{Next}(b_a)}$ does not, we replace $a \leqslant y \leqslant b_a$ in the second to last conjunct with $a \leqslant y \leqslant a_1 + 2m + 1$ and remove the last conjunct; if b_a does not exists we remove last two conjuncts from the definition of $\varphi_0(a)$ above.

$$\varphi(a) := \varphi_0(a) \wedge \left[\bigwedge_{a<c\leqslant a_1+2m+1} \mathbf{N}^{c-a} \varphi_0(c) \right] \wedge$$

$$\bigwedge_{0\leqslant t\leqslant a_1+2m+1-a} \Box^t \left[p_{a_1+2m+1} \to \bigwedge_{k\in[0,m\cdot l^m+2m+m], c\in[0,a_1+2m+1]} \mathbf{N}^k \neg p_c) \right]$$

(recall the estimate $m \cdot l^m + 2m$ from item (d) of Theorem 4). Formula $\varphi(a)$ in a sense describes the path (run) of length $a_1 + 2m + 1 - a$ in $\mathscr{F}_1$, by pointing out the exact locations of nodes on this path using formulas

$$\left[\bigwedge_{a\leqslant y\leqslant b_a} \Diamond p_y \right] \wedge \neg\Diamond p_{\mathrm{Next}(b_a)}.$$

Let $\varphi(\mathscr{F}_1)$ be the set of all formulas $\varphi(a)$ for $a \in \mathscr{F}_1$. Now we choose a substitution for letters of our rule $\bigvee_{1\leqslant j\leqslant l} \varphi_j / x_1$ as follows. For any disjunct φ_j from the premise of our rule denote by φ_j^+ the set of all letters (variables) that occur positively in φ_j while not being nested under temporal operators. Recall that $\theta(a)$ denotes some disjunct φ_j of our rule, chosen in subitem (c) of item (i) of Theorem 4 for any $a \in \mathscr{F}_1 \setminus [a_1 + 2m + 2, \infty)$. Let Θ be the set of all such $\theta(a)$. Recall also that $\theta(b)$ is chosen in subitem (a) of item (ii) of Theorem 4 for any $b \in \mathscr{F}_1(n) \setminus \mathscr{F}_1$. Let

$$\varepsilon(x_i) := \bigvee_{x_i\in\varphi_j^+, \varphi_j = \theta(a), \theta(a)\in\Theta} \varphi(a) \vee \alpha(x_i) \vee \beta(x_i), \text{ where}$$

$$\alpha(x_i) := \left[\bigwedge_{\varphi(a)\in\varphi(\mathcal{F}_1)} \neg\varphi(a)\right] \wedge \left[\bigvee_{n\in[0,a_1]\subseteq\mathcal{F}_1, n\in\mathcal{F}_1, b\in\mathcal{F}_1(n)\setminus\mathcal{F}_1, x_i\in\theta^+(b)}\right.$$

$$\left.(\mathbf{N}^{n-b}\varphi(n) \wedge \neg \bigvee_{t\,<\,n-b,\ a\in[0,a_1+2m+1]\subseteq\mathcal{F}_1} \mathbf{N}^t\varphi(a))\right];$$

$$\beta(x_i) := \gamma(x_i) \wedge \left[\bigwedge_{\varphi(a)\in\varphi(\mathcal{F}_1)} \neg\varphi(a)\right] \wedge \neg \left(\bigvee_{n\in[0,a_1]\subseteq\mathcal{F}_1, n\in\mathcal{F}_1, b\in\mathcal{F}_1(n)\setminus\mathcal{F}}\right.$$

$$\left.(\mathbf{N}^{n-b}\varphi(n) \wedge \neg \bigvee_{t\,<\,n-b,\ a\in[0,a_1+2m+1]\subseteq\mathcal{F}_1} \mathbf{N}^t\varphi(a))\right), \text{ where}$$

$$\gamma(x_i) = \begin{cases} \top, & \text{if } x_i \in \theta^+(a_1 + 2m + 1); \\ \bot, & \text{otherwise.} \end{cases}$$

We will show now using this substitution that $\mathbf{r_{nf}}$ is not admissible in $\mathcal{LTL}_{NT}$ (m). To do so we first consider a frame $\mathcal{F}_1$ from Theorem 4 with its original valuation V_1. Consider a valuation V_2 for letters p_a in $\mathcal{F}_1$ defined as follows: $V_2(p_a) := \{a\}$ for all $a \in [0, a_1 + 2m + 1] \subseteq \mathcal{F}_1$.

Using definition of formulas $\varepsilon(x_i)$ above by a straightforward computation it can be immediately seen that

$$\forall a \in \mathcal{F}_1 \text{ if } a \leqslant a_1 + 2m + 1, \text{ then } (\mathcal{F}_1, a) \Vdash_{V_1} x_i \;\Leftrightarrow\; (\mathcal{F}_1, a) \Vdash_{V_2} \varepsilon(x_i).$$

Thus $(\mathcal{F}_1, b) \nVdash_{V_2} \varepsilon(x_1)$ (where $b \in [0, m)$ is the state from $\mathcal{F}_1$ chosen in item (i) of Theorem 4) and consequently $\varepsilon(x_1) \notin \mathcal{LTL}_{NT}(m)$. To complete the proof of our theorem we need

Lemma 8 *The substitution ε unifies the premise of $\mathbf{r_{nf}}$ in $\mathcal{LTL}_{NT}(m)$, that is $\varepsilon(\bigvee_{1\leqslant j\leqslant l}\varphi_j) \in \mathcal{LTL}_{NT}(m)$.*

Indeed, consider an arbitrary model $\mathcal{M}$ of our logic with a valuation V_2 for all letters p_a.

Step I. Assume first that for some $b \in \mathcal{M}$ and for some $\varphi(a)$, where $a \in \mathcal{F}_1 \setminus [a_1 + 2m + 2, \infty)$, we have $(\mathcal{M}, b) \Vdash_{V_2} \varphi(a)$. If, then, we consider the segment $[b, b + (a_1 + 2m + 1 - a)]$ in $\mathcal{M}$, then from definition of $\varphi(a)$ it follows that at any state in it an appropriate formula $\varphi(c)$ is true w.r.t. V_2. Recall, that conjuncts

$$\left[\bigwedge_{a\leqslant y\leqslant b_a} \Diamond p_y\right] \wedge \neg\Diamond p_{\text{Next}(b_a)},$$

in $\varphi(a)$ are responsible for locating nodes in $\mathscr{M}$ following b at exactly the same distances as they were after a in $\mathscr{F}_1$ and having exactly the same distances to the upper nodes in $[b, b + (a_1 + 2m + 1 - a)]$ as they had in $\mathscr{F}_1$. As a result we infer using the definition of $\varepsilon(x_i)$ that for all such b and all such $a \in \mathscr{F}_1$

$$(\mathscr{M}, b) \Vdash_{V_2} \varepsilon(x_i) \;\Leftrightarrow\; (\mathscr{F}_1, a) \Vdash_{V_1} x_i.$$

And hence, since $(\mathscr{F}_1, a) \Vdash_{V_1} \theta(a)$, we have

$$(\mathscr{M}, b) \Vdash_{V_2} \varepsilon(\theta(a))$$

which follows from the definition of $\varphi(a)$ (which describes in details the run after a in $\mathscr{F}_1$), the fact that $\theta(a)$ does not have nested temporal operations and subitems (a) from (i) and (c) from (ii) of Theorem 4.

Step II. Suppose now that for some b_1 and $a \in \mathscr{F}_1 \setminus [a_1 + 2m + 2, \infty)$ we have $(\mathscr{M}, b_1) \nVdash_{V_2} \varphi(a)$. Beside that assume (see the definition of $\alpha(x_i)$ above) that

$$(\mathscr{M}, b_1) \Vdash_{V_2} \left[\bigvee_{n\in[0,a_1]\subseteq\mathscr{F}_1, n\in\mathscr{F}_1, b\in\mathscr{F}_1(n)\setminus\mathscr{F}_1, x_i\in\theta^+(b)} \right.$$

$$\left. (\mathbf{N}^{n-b}\varphi(n) \wedge \neg \bigvee_{t\,<\,n-b,\; a\in[0,a_1+2m+1]\subseteq\mathscr{F}_1} \mathbf{N}^t\varphi(a))\right];$$

In this case, there is a path leading from b_1 to $b_1(n)$, where $\varphi(n)$ is true in $\mathscr{M}$ w.r.t. V_2, such that the length of this path is $n - b$ for some b, where

$$b \in \mathscr{F}_1(n) \setminus \mathscr{F}_1$$

(see formula above), and there is no shorter path in $\mathscr{M}$ where some formula $\varphi(a)$ is true.

Then truth values of all formulas $\varepsilon(x_i)$ at b_1 w.r.t. V_2 coincide with truth values of letters x_i at the corresponding b in the model $\mathscr{F}_1(n)$ w.r.t. V_1. The same will hold for all elements after b_1, which satisfy the same conditions as b_1. Besides, each of such elements (including b_1 itself) has a path leading to $b_1(n)$ with interval of transitivity of length m, and the interval from $b_1(n)$ to the next node is (as the model w.r.t. V_2) the same as the interval from n to the next node after it in $\mathscr{F}_1(n)$ w.r.t. V_1.

Therefore, since for all $b \in \mathscr{F}_1(n) \setminus \mathscr{F}_1$ we have $(\mathscr{F}_1(n), b) \Vdash_{V_1} \theta(b)$, using Step I above with assumption $(\mathscr{M}, b) \Vdash_{V_2} \varphi(a)$, items (ii) and (a) in (ii) of Theorem 4 we obtain

$$(\mathcal{M}, b_1) \Vdash_{V_2} \varepsilon(\theta(b)).$$

Step III. Consider now the final case (see definition of $\beta(x_i)$ above), when for some b_2 and for $a \in \mathcal{F}_1 \setminus [a_1 + 2m + 2, \infty)$ we have $(\mathcal{M}, b_2) \nVdash_{V_2} \varphi(a)$ and

$$(\mathcal{M}, b_2) \Vdash_{V_2} \neg \left[\bigvee_{n\in[0,a_1]\subseteq\mathcal{F}_1, n\in\mathcal{F}_1, b\in\mathcal{F}_1(n)\setminus\mathcal{F}_1, x_i\in\theta^+(b)} \right.$$

$$\left. (\mathbf{N}^{n-b}\varphi(n) \wedge \neg \bigvee_{t\,<\,n-b,\ a\in[0,a_1+2m+1]\subseteq\mathcal{F}_1} \mathbf{N}^t\varphi(a)) \right],$$

which means that for $n \in \mathcal{F}_1$ such that $n \leqslant a_1$ we cannot reach from b_2 by $\mathbf{N}$ any world, where some $\varphi(n)$ is true w.r.t. V_2, in the same number of steps as it would require to reach n from any world in $\mathcal{F}_1(n) \setminus \mathcal{F}_1$ in $\mathcal{F}_1(n)$.

Let us consider a closest w.r.t. $\mathbf{N}$ successor b_3 of b_2 such that from b_3 we can reach by $\mathbf{N}$ some world c, where some $\varphi(n)$ is true w.r.t. V_2, in the same number of steps it would take to reach some world in $\mathcal{F}_1(n) \setminus \mathcal{F}_1$. Then this number of steps is at least n_1 by item (ii) of Theorem 4 and the interval $[0, n_1]$ in $\mathcal{F}_1(n)$ satisfies all properties described in item (ii) of Theorem 4 (in particular, property (c)).

Therefore, all worlds from the interval $[b_2, b_2 + 2m]$ have the same truth values for formulas $\varepsilon(x_i)$ w.r.t. V_2 as 0 had w.r.t. V_1 in $\mathcal{F}_1(n)$ (see (c) from Theorem 4, definition of formulas $\beta(x_i)$ and Steps I and II above).

Thus, since $(\mathcal{F}_1(n), 0) \Vdash_{V_1} \theta(a_1 + 1)$ and $(\mathcal{F}_1, a_1 + 1) \Vdash_{V_1} \theta(a_1 + 1)$ (see items (i) and (ii) of Theorem 4), we have $(\mathcal{M}, b) \Vdash_{V_2} \varepsilon(\theta(a_1 + 1))$, which concludes the proof of our theorem. □

Using Theorems 4 and 5 we immediately obtain:

Theorem 6 *There is an algorithm verifying admissibility of inference rules in* $\mathcal{LTL}_{NT}(m)$.

We were not able to solve the admissibility problem for logic $\mathcal{LTL}_{NT}$ itself yet. Theorem 6 provides a solution of this problem only for its extension $\mathcal{LTL}_{NT}(m)$. To illustrate it, we conclude this section with a brief comparison of the introduced non-transitive linear temporal logics.

Proposition 2 $\mathcal{LTL}_{NT} \subset \mathcal{LTL}_{NT}(m)$ *and* $\mathcal{LTL}_{NT}(m+1) \neq \mathcal{LTL}_{NT}(m)$.

The proof is trivial and immediately follows from definitions and the following observation

$$\left(\bigwedge_{i\leqslant m} \mathbf{N}^i p \to \Box p \right) \in \mathcal{LTL}_{NT}(m), \quad \text{but}$$

$$\left(\bigwedge_{i\leqslant m}\mathbf{N}^i p \rightarrow \Box p\right) \notin \mathcal{LTL}_{NT}(m+1) \text{ and } \left(\bigwedge_{i\leqslant m}\mathbf{N}^i p \rightarrow \Box p\right) \notin \mathcal{LTL}_{NT}.$$

Proposition 3 $\mathcal{LTL} \nsubseteq \mathcal{LTL}_{NT}$ *and* $\mathcal{LTL} \nsubseteq \mathcal{LTL}_{NT}(m)$ *for all* m.

This is obvious, since $\Box p \rightarrow \Box\Box P \in \mathcal{LTL}$.

Proposition 4 $\mathcal{LTL}_{NT}(m) \nsubseteq \mathcal{LTL}$ *for all* m.

This one is also obvious, since $(\bigwedge_{i\leqslant m}\mathbf{N}^i p \rightarrow \Box p) \in \mathcal{LTL}_{NT}(m)$.

11.7 Open Problems

There are many interesting open problems concerning the suggested framework, for example, problems for multi-agent temporal logics described in Sect. 11.3 (see multi-agent approaches (vi) and (v)). Problems of axiomatizing $\mathcal{LTL}_{NT}$ and $\mathcal{LTL}_{NT}(m)$ are open. It also seems reasonable to extend our approach to logics with linear non-transitive time which extends both backwards and forwards. Another problem is developing a technique of recognizing admissible rules of $\mathcal{LTL}_{NT}$ and for a version of $\mathcal{LTL}_{NT}(m)$ in which non-transitivity is bound by m but not uniformly (not always equal to m). To extend the suggested framework to include multi-agent case, in which any $n \in N$ would be represented by a cluster (circle) with agent knowledge relations K_i, would also be quite interesting.

References

Artemov, S. (2006). Justified common knowledge. *Theoretical Computer Science*, *357*, 4–22.

Artemov, S., & Nogina, E. (2005). Introducing justification into epistemic logic. *Journal of Logic and Computation*, *15*, 1059–1073.

Babenyshev, S., & Rybakov, V. (2011). Linear temporal logic LTL: Basis for admissible rules. *Journal of Logic and Computation*, *21*, 157–177.

Balbiani, P., & Vakarelov, D. (2002). A modal logic for indiscernibility and complementarity in information systems. *Fundamenta Informaticae*, *50*, 243–263.

Belardinelli, F., & Lomuscio, A. (2012). Interactions between knowledge and time in a first-order logic for multi-agent systems: Completeness results. *Journal of Artificial Intelligence Research*, *45*, 1–45.

Fagin, R., Halpern, J., Moses, Y., & Vardi, M. (1995). *Reasoning about knowledge*. Cambridge: MIT Press.

Friedman, H. (1975). One hundred and two problems in mathematical logic. *Journal of Symbolic Logic*, *40*, 113–130.

Gabbay, D. M., & Hodkinson, I. M. (1990). An axiomatization of the temporal logic with until and since over the real numbers. *Journal of Logic and Computation*, *1*, 229–260.

Gabbay, D. M., & Hodkinson, I. M. (1995). Temporal logic in context of databases. In J. Copeland (Ed.), *Logic and reality, essays on the legacy of Arthur prior* (pp. 89–120). Oxford: Oxford University Press.

Gabbay, D. M., Hodkinson, I. M., & Reynolds, M. A. (1994). *Temporal logic: mathematical foundations and computational aspects* (Vol. 1). Oxford: Clarendon Press.

Halpern, J., Samet, D., & Segev, E. (2009). Defining knowledge in terms of belief: The modal logic perspective. *The Review of Symbolic Logic*, *2*, 469–487.

Hintikka, J. (1962). *Knowledge and belief: An Introduction to the logic of the two notions*. Ithaca: Cornell University Press.

Lomuscio, A., & Michaliszyn, J. (2013). An epistemic Halpern–Shoham logic. In *Proceedings of the 23rd International Joint Conference on Artificial Intelligence (IJCAI13)* (pp.1010–1016). Beijing: AAAI Press.

Manna, Z., & Pnueli, A. (1992). *The temporal logic of reactive and concurrent systems: Specification*. New York: Springer.

Manna, Z., & Pnueli, A. (1995). *Temporal verification of reactive systems: Safety*. Berlin: Springer.

McLean, D., & Rybakov, V. (2013). Multi-agent temporary logic $TS4^U_{K_n}$ based at non-linear time and imitating uncertainty via agents interaction, *in Artificial Intelligence and Soft Computing 2013, Conference Proceedings* (pp. 375–384). Heidelberg: Springer.

Rybakov, V. V. (2003). Refined common knowledge logics or logics of common information. *Archive for mathematical Logic*, *42*, 179–200.

Rybakov, V. V. (2005). Logical consecutions in discrete linear temporal logic. *Journal of Symbolic Logic*, *70*, 1137–1149.

Rybakov, V. V. (2008). Linear temporal logic with until and next, logical consecutions. *Annals of Pure and Applied Logic*, *155*, 32–45.

Rybakov, V. (2009a). Logic of knowledge and discovery via interacting agents. Decision algorithm for true and satisfiable statements. *Information Sciences*, *179*, 1608–1614.

Rybakov, V. (2009b). Linear temporal logic LTL_{K_n} extended by multi-agent logic K_n with interacting agents. *Journal of logic and Computation*, *19*, 989–1017.

Rybakov V. V. (2011). Chance discovery and unification in linear modal logic. In *Knowledge-Based and Intelligent Information and Engineering Systems 2011, Proceedings* (pp. 478–485). Heidelberg: Springer.

Rybakov, V. V. (2012). Logical analysis for chance discovery in multi-agents' environment. In *Knowledge-Based and Intelligent Information and Engineering Systems 2011, Proceedings* (pp. 1593–1601). Heidelberg: Springer.

Rybakov, V. V. (2014a). A note on parametrized knowledge operations in temporal logic, from arXiv:1405.0559v1.

Rybakov, V. V. (2014b). Linear non-transitive temporal logic, knowledge operations, algorithms for admissibility, from arXiv:1406.2783.

Rybakov, V. V. (2014c). Non-transitive linear temporal logic and logical knowledge operations. Submitted, December 2014.

Rybakov, V., & Babenyshev. S. (2010). Multi-agent logic with distances based on linear temporal frames. In *Artificial Intelligence and Soft Computing 2010, Conference Proceedings* (pp. 337–344). Heidelberg: Springer.

Vakarelov, D. (2005). A modal characterization of indiscernibility and similarity relations in Pawlak's information systems. In D. Slezak et al (Eds.), *Rough Sets, Fuzzy Sets, Data Mining, and Granular Computing: 10th International Conference, RSFDGrC 2005, Regina, Canada, August 31–September 3, 2005, Proceedings, Part I* (pp. 12–22). Heidelberg: Springer.

Vardi, M. (1995). An automata-theoretic approach to linear temporal logic. In *Banff Higher Order Workshop* (pp. 38–266), from http://citeseer.ist.psu.edu/vardi96automatatheoretic.html.

Vardi, M. Y. (1998). Reasoning about the past with two-way automata, In K. G. Larsen, S. Skyum, & G. Winskel (Eds.), *Automata, Languages and Programming, 25th International Colloquium, ICALP'98, Aalborg, Denmark July 13–17, 1998, Proceedings* (pp. 628–641). Heidelberg: Springer.

Wooldridge, M. (2003). An Automata-theoretic approach to multi-agent planning, *in Proceedings of the First European Workshop on Multi-agent Systems (EUMAS 2003)*, Oxford: Oxford University.

Wooldridge, M., & Lomuscio, A. (2000). Multi-agent VSK logic. In M. Ojeda-Aciego, I. P. de Guzman, G. Brewka, & L. M. Pereira (Eds.), *Logics in Artificial Intelligence, European Workshop, JELIA 2000 Malaga, Spain, September 29–October 2, 2000 Proceedings* (pp. 300–312). Heidelberg: Springer.

Wooldridge, M., Huget, M.-P., Fisher, M., & Parsons, S. (2006). Model checking multi-agent systems: The MABLE language and its applications. *International Journal on Artificial Intelligence Tools*, *15*, 195–225.

Chapter 12
Segerberg Squares of Modal Logics and Theories of Relation Algebras

Valentin Shehtman

Abstract The paper studies two-dimensional modal logics with additional connectives (so-called Segerberg squares) and can be regarded as a continuation of Shehtman (Russian Mathematical Surveys, 67(4):721–778, 2012). It gives a new simpler proof of the finite model property of minimal Segerberg squares using bisimulation games. It proves the square finite model property for Segerberg squares of polymodal **T** and **D**. It also constructs a faithful embedding of Segerberg squares in the equational theory of relation algebras.

12.1 Introduction

In several works of Larisa Maksimova from the 1970–80s local tabularity of modal and intermediate logics was studied Maksimova (1975, 1981). This subject caused interest of other modal logicians from former Soviet Union, and the research was continued by several scientists: Sergey Mardaev, Alex Citkin, Revaz Grigolia, Guram and Nick Bezhanishvili. The author also got involved in these studies after he realized the importance of local tabularity for two-dimensional modal logics (essentially in Gabbay and Shehtman 1998). The recent progress in the field is due to bisimulation games. For a more detailed exposition of this topic we address the reader to Shehtman (2018); here we concentrate on a special kind of two-dimensional modal logics — Segerberg squares.

These logics are squares (in the sense of Gabbay and Shehtman 1998) with additional connectives corresponding to the diagonal symmetry and projections onto the diagonal, horizontal and vertical. This set of modal connectives first appeared in Krister Segerberg's paper Segerberg (1973). A formal definition of Segerberg squares was

V. Shehtman (✉)
Kharkevich Institute for Information Transmission Problems, RAS, Moscow, Russia
e-mail: shehtman@nestcape.net

V. Shehtman
National Research University Higher School of Economics, Moscow, Russia

V. Shehtman
Moscow State University, Moscow, Russia

S. Odintsov (ed.), *Larisa Maksimova on Implication, Interpolation, and Definability*, Outstanding Contributions to Logic 15, https://doi.org/10.1007/978-3-319-69917-2_12

given in our paper Shehtman (2012) (and its preliminary version Shehtman 2011). The Segerberg square of **S5** was (implicitly) studied in Segerberg (1973), and basic results on Segerberg squares of other logics were obtained in Shehtman (2012).

The completeness theorem in that paper gave axiomatizations for Segerberg squares of complete Horn axiomatizable logics. However, the finite model property and its stronger version, the square fmp (i.e., completeness w.r.t. squares of finite frames), was proved only for Segerberg squares of minimal polymodal logics $\mathbf{K}_N$. There was also a construction of embedding of Segerberg squares in classical first-order logic.

The present paper can be regarded as a continuation to Shehtman (2012); the main new results are the following.

- A simpler proof of the fmp for the Segerberg square of $\mathbf{K}_N$ using bisimulation games.
- A proof of the conjecture from Shehtman (2012) that the fmp and also the square fmp hold for Segerberg squares of the logic of reflexive frames **T** and the logic of serial frames **D**.
- A faithful embedding of the Segerberg square of $\mathbf{K}_N$ in the equational theory of relation algebras.

As the proofs of completeness and the square fmp in Shehtman (2012) also used games (so-called 'rectification games'), it turns out that game-theoretic arguments are deeply incorporated in proofs of all main model-theoretic results on Segerberg squares.

Let us now give a more detailed description of the contents.

To make this paper more self-contained we repeat some of the material from Shehtman (2012), with slight changes in the notation. In Sect. 12.2 we recall basic definitions and facts from modal logic. Sections 12.3 and 12.4 discuss the definition and axiomatization of Segerberg squares from Shehtman (2012). Section 12.5 is about bisimulation games; we explain how to find the modal depth of a modal logic using games and as an example we calculate the modal depth of the logics $\mathbf{K}_N + \Box^m \bot$. In Sect. 12.6 by the same method we prove finiteness of modal depth of their Segerberg squares, which implies their local tabularity, and thus the fmp of the Segerberg square of $\mathbf{K}_N$.

In Sect. 12.7 we recall the result from Shehtman (2012) on the square fmp for the Segerberg square of $\mathbf{K}_N$.

In Sect. 12.8 we prove consequences of the results of Shehtman (2012) for relation algebras. The idea of the proofs is the same as in papers Gabbay and Shehtman (2000, 2002) on products, but now we need more of technical details (some of them are moved to the Appendix).

Section 12.9 studies Segerberg squares of polymodal **T**; we show that they are embeddable in Segerberg squares of polymodal **K** and also enjoy the square fmp. In Sect. 12.10 we use a similar method for Segerberg squares of polymodal **D**.

In Sect. 12.11 we recall the embedding of Segerberg squares in classical first-order theories and prove the corresponding consequences for polymodal **T** and **D**.

The conclusion briefly discusses open problems and further research directions.

12.2 Preliminaries

12.2.1 Modal Formulas and Logics

We consider *N-modal formulas* built from a countable set of proposition letters $PL := \{p_1, p_2, \ldots\}$, the classical connectives $\to$, $\bot$, and the unary modal connectives $\Box_1, \ldots, \Box_N$; other connectives ($\wedge$, $\vee$, $\neg$, $\top$, $\leftrightarrow$, $\Diamond_i$) are derived. $\mathscr{L}_N$ denotes the set of all these formulas.

We also use the 'unmarked' box and diamond:

$$\Box A := \Box_1 A \wedge \ldots \wedge \Box_N A, \ \Diamond A := \Diamond_1 A \vee \ldots \vee \Diamond_N A.$$

As usual, $\Box^m$ denotes the m-th iteration of $\Box$ (which is empty for $m = 0$); similarly, for $\Diamond^m$, $\Box_i^m$, $\Diamond_i^m$.

For a sequence of indices $\alpha = i_1 \ldots i_r \in \{1, \ldots, n\}^r$ put

$$\Diamond_\alpha := \Diamond_{i_1} \ldots \Diamond_{i_r}, \ \Box_\alpha := \Box_{i_1} \ldots \Box_{i_r}.$$

If α is a void sequence, then $\Diamond_\alpha$, $\Box_\alpha$ are also void.

$\mathscr{L}_N \lceil k$ denotes the set of all N-modal formulas built from the finite set of proposition letters $PL\lceil k := \{p_1, p_2, \ldots, p_k\}$; they are called *k-formulas*. A modal formula without proposition letters is called *closed*.

The *length* $|A|$ of a modal formula A is the number of occurrences of logical connectives in A.

A *(normal) N-modal logic* is a subset of $\mathscr{L}_N$ containing the classical tautologies, the axioms $\Box_i(p_1 \to p_2) \to (\Box_i p_1 \to \Box_i p_2)$, which is closed under Substitution, Modus Ponens, and $\Box$-introduction ($A/\Box_i A$). The *k-restriction* of a modal logic L is $L\lceil k := L \cap \mathscr{L}_N\lceil k$. These sets $L\lceil k$ are called *k-weak modal logics*.

$\mathbf{K}_N$ denotes the minimal N-modal logic; $\mathbf{K} := \mathbf{K}_1$.

For an N-modal logic L and a set of N-modal formulas Φ, $L + \Phi$ denotes the smallest N-modal logic containing $L \cup \Phi$; we say that Φ *axiomatizes* $L + \Phi$ over L.

For a formula A, $A \in L$ is also denoted by $L \vdash A$. We will deal with particular modal logics:

$$\mathbf{D} := \mathbf{K} + \Diamond\top, \ \mathbf{T} := \mathbf{K} + \Box p \to p, \ \mathbf{SL} := \mathbf{K} + \Diamond p \leftrightarrow \Box p,$$
$$\mathbf{S5} := \mathbf{T} + \Box p \to \Box\Box p + \Diamond\Box p \to p, \ \mathbf{K.t} := \mathbf{K}_2 + \Diamond_1\Box_2 p \to p + \Diamond_2\Box_1 p \to p.$$

In the case of $\mathbf{K.t}$ we prefer the notations $\Box_{-1}$, $\Diamond_{-1}$ to $\Box_2$, $\Diamond_2$.

Definition 12.2.1 The *modal depth* $md(A)$ of a modal formula A is defined by induction:

$$md(\bot) = md(p_i) = 0, \ md(A \to B) = \max(md(A), md(B)),$$
$$md(\Box_j A) = md(A) + 1.$$

Definition 12.2.2 For a modal logic L and a formula A in its language we define the *modal depth of A in L*:

$$md_L(A) := min\{md(B) \mid L \vdash A \leftrightarrow B\}.$$

The *modal depth of L* is

$$md(L) := \begin{cases} max\{md_L(A) \mid A \text{ is in the language of } L\} & \text{if it exists,} \\ \infty & \text{otherwise.} \end{cases}$$

Definition 12.2.3 The *fusion* of an N_1-modal logic L_1 and an N_2-modal logic L_2 is the $(N_1 + N_2)$-modal logic

$$L_1 * L_2 := \mathbf{K}_{N_1+N_2} + L_1 + L_2^{+N_1},$$

where $L_2^{+N_1}$ is obtained from L_2 by renaming all the modalities $\Box_j$ into $\Box_{j+N_1}$.

For a modal logic L, the N-tuple fusion $\underbrace{L * \ldots * L}_{N}$ is denoted by L_N. In particular, for the logic $\mathbf{K.t}_N$ we use the modalities $\Diamond_i, \Box_i$, where $i = \pm 1, \ldots, \pm N$; it is axiomatized by the formulas $\Diamond_i \Box_i p \to p$ for all these i.

12.2.2 Kripke Frames and Models

An *(N-modal) (Kripke) frame* is a tuple $F = (W, R_1, \ldots, R_N)$, where $W \neq \varnothing$, $R_i \subseteq W \times W$.

For a frame $(W, R_1, \ldots, R_N)$ we use the standard notation

$$R_i(x) := \{y \in W \mid x R_i y\};$$

R denotes the union $R_1 \cup \ldots \cup R_N$. If uRv we say that v is a *successor* of u, u is a *predecessor* of v.

A *Kripke model over F* is a pair $M = (F, \theta)$, where $\theta : PL \longrightarrow 2^W$ is a *valuation*. Similarly, a *k-weak Kripke model* over F is $M = (F, \theta)$, where $\theta : PL\lceil k \longrightarrow 2^W$ is a *k-valuation*.

We say that a formula A is *in the language* of a Kripke model M if A is N-modal, and also A is a k-formula if M is k-weak.

$M, x \vDash A$ (or $x \vDash A$, for short) denotes that a formula A in the language of a Kripke model M is true at a point x. The definition is standard, in particular,

$$M, x \vDash \Box_i A \text{ iff } \forall y \in R_i(x)\ M, y \vDash A.$$

Then

$$M, x \vDash \Diamond_i A \text{ iff } \exists y \in R_i(x)\ M, y \vDash A,$$

$$M, x \vDash \Box A \text{ iff } \forall y \in R(x)\ M, y \vDash A,$$

$$M, x \vDash \Diamond A \text{ iff } \exists y \in R(x)\ M, y \vDash A.$$

A *submodel* of a Kripke model is its restriction to some subset. A submodel M' of M is called *reliable* if $M, x \vDash A \Leftrightarrow M', x \vDash A$ for any modal formula A and x in M'.

A formula A is *true in a Kripke model* M (in symbols: $M \vDash A$) if it is true at every world of M. A is *valid in a frame* F (in symbols: $F \vDash A$) if it is true in every Kripke model over F. A set of formulas Γ is *valid* in F (in symbols: $F \vDash \Gamma$) if every $A \in \Gamma$ is valid in F. In the latter case we also say that F is a Γ*-frame.* Similarly, Γ is *true in a model* M ($M \vDash \Gamma$) if every $A \in \Gamma$ is true.

The *logic of* an N-modal frame F (or the logic *determined by* F) is $\mathbf{L}(F) := \{A \in \mathscr{L}_N \mid F \vDash A\}$. The logic *determined by* a class of N-modal frames $\mathscr{C}$ is $\mathbf{L}(\mathscr{C}) := \bigcap\{\mathbf{L}(F) \mid F \in \mathscr{C}\}$. Logics determined by classes of frames are called *(Kripke) complete.* A logic determined by some class of finite (respectively, at most countable) frames is said to have the *finite model property (fmp)* (respectively, the *countable frame property*).

It is well known that every finitely axiomatizable modal logic with the fmp is decidable (*Harrop theorem*).

For a modal logic Λ, the class of all Λ-frames is called its *frame variety* and denoted by $\mathbf{V}(\Lambda)$. If L is complete, $\Lambda = \mathbf{L}(\mathbf{V}(\Lambda))$. For a set of N-modal formulas Γ we also use the notation

$$\mathbf{V}(\Gamma) := \{F \mid F \text{ is an } N\text{-frame}, F \vDash \Gamma\};$$

then $\mathbf{V}(\Gamma) = \mathbf{V}(\mathbf{K}_N + \Gamma)$ (soundness theorem). The notation $\mathbf{V}(\{A\})$ is abbreviated to $\mathbf{V}(A)$.

A modal formula A is called *elementary* if the class is $\mathbf{V}(A)$ elementary, i.e., it is the class $Mod(\alpha)$ of all models of some classical first-order formula α in the signature with only binary predicates and equality. In this case we say that α *corresponds* to A. Analogously, a modal logic is called *elementary* if its frame variety is elementary. A modal logic Λ is called Δ*-elementary* if its variety $\mathbf{V}(\Lambda)$ is Δ-elementary i.e., is a class of all models of some first-order theory in the corresponding signature (or equivalently, if $\mathbf{V}(\Lambda)$ is an intersection of elementary classes).

Lemma 12.2.4 (Cf. Gabbay and Shehtman 1998, Proposition 5.4).
Every complete Δ-elementary modal logic has the countable frame property.

Definition 12.2.5 A *p-morphism* from a frame $F = (W, R_1, \dots, R_N)$ onto a frame $F' = (W', R'_1, \dots, R'_N)$ is a surjective map $f : W \longrightarrow W'$ satisfying the conditions (for any $i \leqslant N$)

- $xR_iy \Longrightarrow f(x)R'_if(y)$ (*monotonicity*),
- $f(x)R'_iz \Longrightarrow \exists y\,(xR_iy\ \&\ f(y) = z)$ (the *lift property*).

A p-morphism from a Kripke model $M = (F, \theta)$ onto $M' = (F', \theta')$ is a p-morphism from F onto F' satisfying the condition

$$\theta(p) = f^{-1}(\theta'(p))$$

for any $p \in PL$ (or $PL\lceil k$ if the models are k-weak).

$f : F \twoheadrightarrow F'$ denotes that f is a p-morphism from F onto F'; the same notation is used for Kripke models.

Lemma 12.2.6 *Let F, F' be N-modal Kripke frames, M, M' Kripke models over them (maybe both k-weak), A an N-modal formula, and $f : F \twoheadrightarrow F'$. Then*

(i) $\mathbf{L}(F) \subseteq \mathbf{L}(F')$.
(ii) *If A is closed, then $F \vDash A$ iff $F' \vDash A$.*
(iii) *If $f : M \twoheadrightarrow M'$, then for any x in F, for any A in the language of M $M, x \vDash A$ iff $M', f(x) \vDash A$.*

Definition 12.2.7 The *disjoint sum* of a family of frames $F_j = (W_j, R_{1j}, \dots, R_{Nj})$, where $j \in J$, is the frame $\bigsqcup_{j\in J} F_j := (W, R_1, \dots, R_N)$, where

$$W = \bigsqcup_{j\in J} W_j := \bigcup_{j\in J} (W_j \times \{j\}),$$

$$(x, j)R_i(y, j') \text{ iff } j = j'\ \&\ xR_{ij}y.$$

The *disjoint sum* of a family of Kripke models $M_j = (W_j, R_{1j}, \dots, R_{Nj}, \theta_j)$, for $j \in J$, is the model $\bigsqcup_{j\in J} M_j := (\bigsqcup_{j\in J} F_j, \theta)$, where

$$\theta(p) = \bigsqcup_{j\in J} \theta_j(p)$$

for every $p \in PL$.

Lemma 12.2.8 $\mathbf{L}\left(\bigsqcup_{j\in J} F_j\right) = \bigcap_{j\in J} \mathbf{L}(F_j)$.

12.2.3 Paths and Frame Depth

Definition 12.2.9 Let $F = (W, R_1, \ldots, R_N)$ be a frame, $u, v \in W$, $m \geqslant 1$. A *path of length* m from u to v is a sequence $(u_0, j_0, u_1, \ldots, j_{m-1}, u_m)$ such that $u = u_0$, $v = u_m$ and for all $i < m$, $u_i R_{j_i} u_{i+1}$.

A singleton sequence (u) is the *path of length 0* (from u to u).

Definition 12.2.10 The *depth* of a point x in a frame F (denoted by $d(x)$) is the maximum of lengths of paths in F beginning from x (if this maximum exists), or ∞ otherwise.

The depth of x w.r.t. to the relation R_i (denoted by $d_i(x)$) is the depth of x in the frame (W, R_i).

The depth of a frame F (denoted by $d(F)$) is the maximal depth of its points (if it exists) and ∞ otherwise.

Lemma 12.2.11 *Let $F = (W, R_1, \ldots, R_N)$ be a frame, $n \geqslant 1$. Then (in any model over F), for any $x \in W$*

$$d(x) < n \text{ iff } x \vDash \Box^n \bot,$$
$$d(x) \geqslant n \textit{ iff } x \vDash \Diamond^n \top,$$
$$d(x) = n \textit{ iff } x \vDash \Diamond^n \top \wedge \Box^{n+1} \bot.$$

The proof of the first claim is by induction, cf. Gabbay and Shehtman (1998), Lemma 9.2; the others follow trivially.

Definition 12.2.12 A point x in a frame F is *path accessible* from a point u if there exists a path from u to x. A *cone with root u in a frame F* (in symbols, $F{\uparrow}u$) is the restriction of F to the set of all points path accessible from u. Similarly we define a cone $M{\uparrow}u$ in a Kripke model M.

A frame F is *rooted with root u* if $F = F{\uparrow}u$; similarly for a Kripke model.

If $F = F{\uparrow}u$ is rooted, then it is clear that $d(F) = d(u)$, since any path from $x \neq u$ can be extended by adding a path from u to x at the beginning.

Lemma 12.2.13 *(1)* $\mathbf{L}(F) = \bigcap_{u \in F} \mathbf{L}(F{\uparrow}u)$.

(2) $M{\uparrow}u$ is a reliable submodel of M.

Definition 12.2.14 Let F, G be N-modal Kripke frames. A *reduction* from F to G is a p-morphism from a cone of F onto G. If G is rooted and there exists a reduction from F to G, we say that F is *reducible* to G. If G is arbitrary we also say that F is reducible to G if it is reducible to every cone in G.

Lemmas 12.2.6 and 12.2.13 imply

Lemma 12.2.15 *If F is reducible to G, then* $\mathbf{L}(F) \subseteq \mathbf{L}(G)$.

Definition 12.2.16 An N-modal frame F with root u is called an *N-tree* if for any point v in F there exists a unique path from u to v. The length of this path is called the *height* of v and denoted by $h(v)$.

Note that the depth of a tree is the maximal height of its points (if it exists), or ∞ otherwise. The depth of a tree F is also called its *height* and denoted by $h(F)$.

Definition 12.2.17 The *full standard (intransitive irreflexive) tree* is the frame $T_\omega := (\omega^*, \sqsubset)$, where ω^* is the set of all finite sequences of non-negative integers, and

$$\alpha \sqsubset \beta \text{ iff } \exists m\ \beta = \alpha m.$$

The *full standard N-tree* is $T_{\omega,N} := (\omega^*, \sqsubset_1, \ldots, \sqsubset_N)$, where

$$\alpha \sqsubset_i \beta \text{ iff } \exists m (\beta = \alpha m \ \&\ m \equiv i\ (mod\ N));$$

Definition 12.2.18 A subset $V \subseteq T_{\omega,N}$ is called *stable* if

$$\forall \alpha, \beta, i\ (\alpha \sqsubset_i \beta \ \&\ \beta \in V \Rightarrow \alpha \in V).$$

A *standard N-tree* is a stable subframe of $T_{\omega,N}$, i.e., a restriction of $T_{\omega,N}$ to a stable subset.

Note that the height $h(\alpha)$ of any point $\alpha \in \omega^*$ of a standard tree is the length of α; it is also denoted by $|\alpha|$.

Lemma 12.2.19 *Every at most countable N-tree is isomorphic to a standard N-tree.*

In this paper we usually deal with at most countable trees, with few exceptions.

Definition 12.2.20 Let F be a tree. It gives rise to a *truncated tree of height m*: $F^{(m)} := F \upharpoonright \{x \mid h(x) \leqslant m\}$.

The next lemma is well-known, but still we recall its proof.

Lemma 12.2.21 (Truncation lemma) *Let $F \subseteq T_{\omega,N}$ be a standard tree, $F^- = F^{(m)}$, M a Kripke model over F, M^- its restriction to F^-. Then for any N-modal formula A, for any $x \in F^-$, $md(A) + |x| \leqslant m$ implies*

$$M, x \vDash A \textit{ iff } M^-, x \vDash A.$$

Proof By induction on $|A|$ we prove the claim for all x.

If $A \in PL$, the claim holds by the definition of M^-.

The cases $A = \bot$, $A = B \to C$ are obvious.

Let $A = \Box_i B$, $md(A) + |x| \leqslant m$. Then

$$M, x \vDash A \Leftrightarrow \forall y \in M (x \sqsubset_i y \Rightarrow M, y \vDash B),$$
$$M^-, x \vDash A \Leftrightarrow \forall y \in M (x \sqsubset_i y \Rightarrow M^-, y \vDash B),$$

since the same points are accessible from x in M and M^- (provided $|x| < m$, as in our case).

Note that $x \sqsubset_i y$ implies $|y| \leqslant |x| + 1$. Therefore $md(B) + |y| \leqslant md(B) + 1 + |x| = md(A) + |x| \leqslant m$, and so we can apply the IH to B. ⊠

Remark Lemma 12.2.21 as well as its two-dimensional analogue (Lemma 12.6.6) and another version (Lemma 12.10.4) can also be proved using bisimulation games for two Kripke models (and Theorem 12.5.5 for this case). The use of games is not so crucial here, and we leave this argument as an exercise for the reader.

Recall the standard unravelling construction (cf. Gabbay and Shehtman 1998).

Definition 12.2.22 Let $F = (W, R_1, \ldots, R_N)$ be a cone with root u. The *unravelling* of F is the frame $F^\sharp = (W^\sharp, R_1^\sharp, \ldots, R_N^\sharp)$, in which $W^\sharp$ is the set of all paths from u to points in F, and $\alpha R_i^\sharp \beta$ iff $\beta = (\alpha, i, x)$ for some $x \in W$.

Remark In a rooted frame there may be several roots, so strictly speaking, the frame $F^\sharp$ depends on the choice of u. However, we do not include u in the notation.

Lemma 12.2.23 *$F^\sharp$ is a tree. The map π sending every path to its endpoint is a p-morphism $F^\sharp \twoheadrightarrow F$.*

Unravelling is extended in an obvious way to Kripke models: if $M = (F, \theta)$ is a Kripke model over a rooted frame, we define its unravelling $M^\sharp := (F^\sharp, \theta^\sharp)$, where for any $q \in PL$, $\theta^\sharp(q) := \pi^{-1}(\theta(q))$. Then clearly $\pi : M^\sharp \twoheadrightarrow M$.

12.2.4 Canonical Models

Recall the main properties of canonical frames and models.

Definition 12.2.24 The *canonical frame* for an N-modal logic (maybe weak) L is $F_L = (W_L, R_{1,L}, \ldots, R_{N,L})$, where W_L is the set of all maximal L-consistent sets of formulas in the language of L;

$$xR_{i,L}y \text{ iff for any } A, \quad \Box_i A \in x \text{ implies } A \in y.$$

The *canonical model* for L is $M_L = (F_L, \theta_L)$, where

$$\theta_L(p_i) = \{x \mid p_i \in x\}.$$

Theorem 12.2.25 (Canonical model theorem) *For any formula A in the language of L,*

(1) $M_L, x \vDash A$ iff $A \in x$;
(2) $M_L \vDash A$ iff $A \in L$.

Corollary 12.2.26 *Every canonical model is* distinguishable*: if for any A,* $M_L, x \vDash A \Leftrightarrow M_L, y \vDash A$*, then* $x = y$.

Definition 12.2.27 A modal logic L is called *canonical* if $F_L \vDash L$ (or equivalently, $L = \mathbf{L}(F_L)$) and *weakly canonical* if $F_{L\lceil k} \vDash L$ for any finite k.

12.2.5 Tabularity, Local Tabularity, Modal Depth

Definition 12.2.28 An N-modal logic L is called *locally tabular* if for any finite k there exist finitely many N-modal k-formulas up to equivalence in L.

The local tabularity of L is obviously equivalent to the *local finiteness* of the variety of L-algebras, i.e., finiteness of all finitely generated L-algebras, cf. Malcev (1973), Chap. 6, Sect. 14.

Definition 12.2.29 An N-modal logic L is called *tabular* if $L = \mathbf{L}(F)$ for some finite N-modal frame F. L has the *finite model property (fmp)* if it is an intersection of tabular logics.

The following simple facts are well-known:

Proposition 12.2.30 *(1) A modal logic L is locally tabular iff every weak canonical model* $M_{L\lceil k}$ *is finite.*
(2) Every extension of a locally tabular modal logic in the same language is locally tabular.
(3) Every tabular logic is locally tabular.
(4) Every locally tabular logic has the fmp.

Lemma 12.2.31 *For a non-weak modal logic L*

$$md(L) = max\{md(L\lceil k) \mid k \geqslant 0\}.$$

Proof Note that for any k-formula A we have $md_L(A) = md_{L\lceil k}(A)$.

In fact, if $md_L(A) = n$, then $L \vdash A \leftrightarrow B$ for some B of modal depth n. If B is not a k-formula, we can substitute $\bot$ for all extra proposition letters. So we obtain a k-formula B' such that $md(B') = md(B)$ and $L \vdash A \leftrightarrow B'$; thus $md_{L\lceil k}(A) = n$.

Eventually (if L is N-modal)

$$md(L) = max\{md_L(A) \mid A \in \mathscr{L}_N\} =$$
$$max\{max\{md_L(A) \mid A \in \mathscr{L}_N\lceil k\} \mid k \geqslant 0\} = max\{md(L\lceil k) \mid k \geqslant 0\}.$$

⊠

Proposition 12.2.32 *Let L be a (non-weak) modal logic.*

(1) If $L\lceil k$ *is of finite modal depth, then* $L\lceil k$ *is tabular.*

(2) If L is of finite modal depth, then L is locally tabular.
(3) If $L \subseteq L'$ and L, L' are in the same language, then $md(L) \geqslant md(L')$.

Proof (1) There are finitely many k-formulas of modal depth $\leqslant n$ up to equivalence in $\mathbf{K}_N$. So if $L\lceil k$ is of finite modal depth, there are finitely many k-formulas up to equivalence in L, i.e., the Lindenbaum algebra of $L\lceil k$ is finite.

(2) follows from (1) and Lemma 12.2.31. (3) is obvious. ⊠

12.2.6 Products

Definition 12.2.33 (Shehtman 1978) The *product of Kripke frames* $F = (W, R_1, \dots, R_n)$ and $G = (V, S_1, \dots, S_m)$ is defined as follows.

$$F \times G := (W \times V, R_{11}, \dots, R_{n1}, S_{12}, \dots, S_{m2}),$$

where

$$(x, y)R_{i1}(x', y') \Leftrightarrow xR_i x' \ \&\ y = y';$$

$$(x, y)S_{j2}(x', y') \Leftrightarrow x = x' \ \&\ yS_j y'.$$

Definition 12.2.34 Let $\mathscr{C}_1, \mathscr{C}_2$ be classes of respectively n- and m-frames. Their *product* is defined as the class of $(n+m)$-frames

$$\mathscr{C}_1 \times \mathscr{C}_2 := \{F_1 \times F_2 \mid F_1 \in \mathscr{C}_1,\ F_2 \in \mathscr{C}_2\}.$$

Definition 12.2.35 Let Λ_1, Λ_2 be an n-modal and m-modal logic respectively, such that $\mathbf{V}(\Lambda_1), \mathbf{V}(\Lambda_2) \neq \varnothing$. Their *product* is the $(n+m)$-modal logic

$$\Lambda_1 \times \Lambda_2 := \mathbf{L}(\mathbf{V}(\Lambda_1) \times \mathbf{V}(\Lambda_2)).$$

Definition 12.2.36 The *commutative join* of an n-modal logic Λ_1 and an m-modal logic Λ_2 is obtained from their fusion $\Lambda_1 * \Lambda_2$ by adding the axioms

$$(Com_{ij})\quad \Box_i \Box_{n+j} p \leftrightarrow \Box_{n+j} \Box_i p \ \text{(commutation axioms)}$$

and

$$(CR_{ij})\quad \Diamond_i \Box_{n+j} p \to \Box_{n+j} \Diamond_i p \ \text{(Church–Rosser axioms)}$$

for $1 \leqslant i \leqslant n,\ 1 \leqslant j \leqslant m$.

Logics Λ_1, Λ_2 are called *product-matching*, if $\Lambda_1 \times \Lambda_2 = [\Lambda_1, \Lambda_2]$.

One can easily see that these axioms are elementary; their varieties are defined by the following conditions (Fig. 12.1).

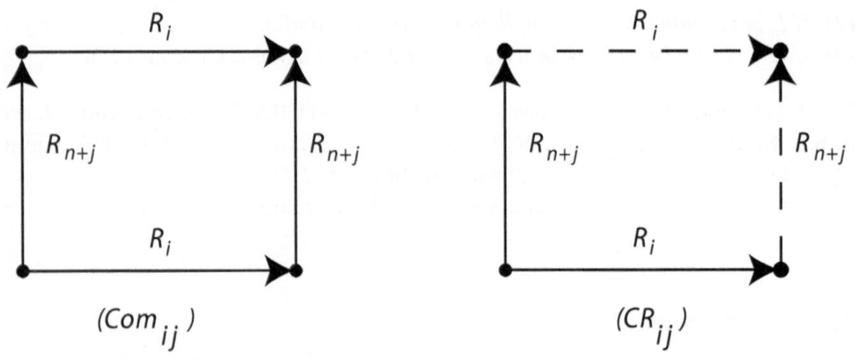

Fig. 12.1 Semantic conditions for commutation and Church–Rosser axioms

$(W, R_1, \dots, R_n) \vDash Com_{ij}$ iff $R_i \circ R_{n+j} = R_{n+j} \circ R_i$.
$(W, R_1, \dots, R_n) \vDash CR_{ij}$ iff $R_i^{-1} \circ R_{n+j} \subseteq R_{n+j} \circ R_i^{-1}$.

Recall that the frames of $\mathbf{SL} = \mathbf{K} + \Box p \leftrightarrow \Diamond p$ are of the form (W, R), where R is a graph of a function $W \longrightarrow W$ (we will usually identify such a function with its graph). There is also the following completeness result Segerberg (1967):

Proposition 12.2.37 $\mathbf{SL} = \mathbf{L}(\omega, S)$, *where* S *is the successor function* $(S(x) = x + 1)$.

Now we recall the definition of a Horn axiomatizable modal logic from Gabbay et al. (2003). We consider the classical first-order language with the binary predicates $R_1, \dots, R_N$ and equality.

Definition 12.2.38 A *Horn sentence* is a classical first-order sentence of the form $\overline{\forall}(\varphi \rightarrow R_i(x, y))$, where $\overline{\forall}$ is the universal closure, φ is a conjunction of atomic formulas, x, y are different variables. A set of Horn sentences is called a *Horn theory*.

Definition 12.2.39 Let $F_1 = (W_1, S_1, \dots, S_N)$, $F_2 = (W_2, S_1', \dots, S_N')$ be N-modal frames. F_2 is called a *weak extension of* F_1 (in symbols: $F_1 \subseteqq F_2$) if $W_1 = W_2$ and for all i, $S_i \subseteq S_i'$.

The next lemma allows us to construct the "Horn closure", the minimal weak extension of a frame satisfying a certain Horn theory; for a proof see e.g. Gabbay and Shehtman (1998).

Lemma 12.2.40 *Let* $F = (W, \rho_1, \dots, \rho_N)$ *be a Kripke frame,* $\Gamma = \{\psi_k \mid k \in I\}$ *a Horn theory,*

$$\psi_k = \overline{\forall}(\varphi_k \rightarrow R_{i_k}(x, y)).$$

Then there exists a frame F_Γ^+ *such that*

(1) $F \subseteqq F_\Gamma^+$;
(2) $F_\Gamma^+ \vDash \Gamma$;
(3) If $G \vDash \Gamma$ *and* $f : F \longrightarrow G$ *is a monotonic map, then the same* f *is also a monotonic map* $F_\Gamma^+ \longrightarrow G$.

Definition 12.2.41 A modal formula is called *Horn* if it corresponds to a universal Horn sentence. A modal logic is *Horn axiomatizable* if it is obtained from the minimal logic by adding closed or Horn axioms.

Typical examples of Horn formulas are pseudo-transitive formulas from Gabbay and Shehtman (1998).

Definition 12.2.42 A modal formula is called *pseudo-transitive* if it is of the form $\Diamond_\alpha \Box_k p \to \Box_\beta p$, where $p \in PL$, α, β are sequences of indices. A *PTC-logic* (from 'pseudo-transitive or closed') is a modal logic axiomatized by a set of formulas that are pseudo-transitive or closed.

Proposition 12.2.43 *(1) A pseudo-transitive formula* $A = \Diamond_\alpha \Box_k p \to \Box_\beta p$ *corresponds to the Horn sentence*

$$\varphi = \forall x \forall v (\exists u (u R_\alpha x \wedge u R_\beta v) \to R_k(x, v)),$$

where

$$R_{i_1 \dots i_r} := R_{i_1} \circ \dots \circ R_{i_r}.$$

(2) Every PTC-logic is canonical.

Here $u R_\alpha x$, $u R_\beta v$ are abbreviations of some existential formulas; thus, in a prenex form φ becomes a Horn proposition. If α is void, R_α means equality.

However, not every Horn modal formula is pseudo-transitive. The simplest counterexample is $\Box(\Box p \to p)$ corresponding to $\overline{\forall}(R(x, y) \to R(y, y))$.

Theorem 12.2.44 (Cf. Gabbay et al. 2003) *Every two complete Horn axiomatizable modal logics are product-matching.*

12.3 Segerberg Squares

Definition 12.3.1 The *Segerberg square* of a Kripke frame $F = (W, R_1, \dots, R_N)$ is the $(2N + 3)$-frame

$$F_{Sg}^2 := (F^2, \sigma_0, \sigma_1, \sigma_2),$$

obtained from $F^2 = F \times F$ by adding the functions

$$\sigma_0 : (x, y) \mapsto (y, x) \text{ (the diagonal symmetry)},$$

$$\sigma_1 : (x, y) \mapsto (y, y) \text{ (the horizontal projection onto the diagonal)},$$

$$\sigma_2 : (x, y) \mapsto (x, x) \text{ (the vertical projection onto the diagonal)}.$$

Definition 12.3.2 The *Segerberg square* of an N-modal logic Λ is the $(2N+3)$-modal logic

$$\Lambda^2_{Sg} := \mathbf{L}(\{F^2_{Sg} \mid F \vDash \Lambda\}).$$

Its connectives are denoted by $\Box_1, \ldots, \Box_N, \blacksquare_1, \ldots, \blacksquare_N, \bigcirc, \bigcirc_1, \bigcirc_2$. If $F = (W, R_1, \ldots, R_N)$ is a Kripke frame, then the corresponding modal formulas are interpreted in a Kripke model over the Segerberg square F^2_{Sg} as follows:

$$(x, y) \vDash \Box_i A \text{ iff } \forall z(xR_i z \Rightarrow (z, y) \vDash A),$$

$$(x, y) \vDash \blacksquare_i A \text{ iff } \forall z(yR_i z \Rightarrow (x, z) \vDash A),$$

$$(x, y) \vDash \bigcirc A \text{ iff } \sigma_0(x, y) \vDash A,$$

$$(x, y) \vDash \bigcirc_1 A \text{ iff } \sigma_1(x, y) \vDash A,$$

$$(x, y) \vDash \bigcirc_2 A \text{ iff } \sigma_2(x, y) \vDash A.$$

Let us recall specific axioms of Segerberg squares proposed in Shehtman (2012). First of all, we should have axioms of **SL** for the circle-connectives $\bigcirc_1, \bigcirc_2, \bigcirc$, since $\sigma_1, \sigma_2, \sigma_0$ are functions. In addition to them, we have the following list.

$(Sg1)$ $\bigcirc^2 p \leftrightarrow p$,
$(Sg2)$ $\bigcirc_1^2 p \leftrightarrow \bigcirc_1 p$,
$(Sg3)$ $\bigcirc_2 p \leftrightarrow \bigcirc\bigcirc_1 p$,
$(Sg4)$ $\bigcirc_1 \bigcirc p \leftrightarrow \bigcirc_1 p$,
$(Sg5)$ $\blacksquare_i p \leftrightarrow \bigcirc \Box_i \bigcirc p$,
$(Sg6)$ $\bigcirc_1 \Box_i (\blacksquare_i p \rightarrow \bigcirc_2 p)$,
$(Sg7)$ $\bigcirc_1 p \rightarrow \Box_i \bigcirc_1 p$,
$(Sg8)$ $\blacksquare_i \bigcirc_1 p \leftrightarrow \bigcirc_1 \blacksquare_i \bigcirc_1 p$.

Here are their classical first-order correspondents.

Lemma 12.3.3 *(1) Let* $G = (V, X_1, \ldots, X_n, Y_1, \ldots, Y_n, f_0, f_1, f_2) \vDash \mathbf{K}_{2n} * \mathbf{SL}_3$. *Then*

$G \vDash (Sg1)$ *iff* $f_0 f_0 = 1_V$,
$G \vDash (Sg2)$ *iff* $f_1 f_1 = f_1$,
$G \vDash (Sg3)$ *iff* $f_1 f_0 = f_2$,
$G \vDash (Sg4)$ *iff* $f_0 f_1 = f_1$,

$G \models (Sg5)$ *iff* $\forall a, b \in V(aX_i b \Leftrightarrow f_0(a) Y_i f_0(b))$ *(provided* $G \models (Sg1)$*)*
$G \models (Sg6)$ *iff* $\forall a, b \in V(f_1(a) X_i b \Rightarrow bY_i f_2(b))$,
$G \models (Sg7)$ *iff* $\forall a, b \in V(aX_i b \Rightarrow f_1(a) = f_1(b))$,
$G \models (Sg8)$ *iff* $\forall a \in V\ f_1[Y_i(a)] = f_1[Y_i(f_1(a))]$.

(2) Axioms $(Sg1) - (Sg8)$ *are valid in Segerberg squares.*

Remark Unlike the composition of relations, we denote the composition of functions $f : U \longrightarrow V$ and $g : V \longrightarrow W$ by gf. 1_V denotes the identity function $V \longrightarrow V$. $f[Z]$ denotes the image of the set Z under the map f.

A routine proof of this lemma is an easy exercise; alternatively, one can apply Sahlqvist theorem.

Let us give some comments on these axioms.

Axiom $(Sg1)$ means that f_0 is an involution on V; in Segerberg squares this is a diagonal symmetry. The fixed points of f_0 are called *weakly diagonal*; in Segerberg squares they constitute the diagonal.

Axiom $(Sg2)$ means that f_1 is an idempotent map from V into itself. So the image of f_1 consists of the fixed points of f_1. They are called *diagonal*.

Axiom $(Sg3)$ allows us to express $\bigcirc_2$ in terms of $\bigcirc_1$ and $\bigcirc$, and similarly, f_2 in terms of f_1, f_0. Later on we will regard $\bigcirc_2$ as derived. $(Sg3)$ also implies that the images of f_2 and of f_1 coincide.

Axiom $(Sg4)$ means that all diagonal points are weakly diagonal. In Segerberg squares the converse also holds, which may not be the case in general.

In a set of the form W^2 the diagonal symmetry, two projections onto the diagonal and the identity map constitute a 4-element monoid. It is presented by two generators σ_0, σ_1 and the equations

$$\sigma_0^2 = 1,\ \sigma_1^2 = \sigma_1,\ \sigma_0\sigma_1 = \sigma_1,$$

exactly corresponding to axioms $(Sg1)$, $(Sg2)$, $(Sg3)$.

Axioms $(Sg5)$ and $(Sg1)$ imply that f_0 is an isomorphism between the relations X_i and Y_i. This allows us to express the vertical modalities in terms of the horizontal ones and the connective $\bigcirc$.

$(Sg5)$ also means that the relations X_i and Y_i are conjugate by f_0 in the following sense:

$$f_0 \circ X_i \circ f_0 = Y_i$$

(where $\circ$ is the composition of relations), see Fig. 12.2.

$(Sg6)$ means that if b is horizontally (X_i-)accessible from a diagonal point, then some diagonal point is vertically (Y_i-)accessible from b. In Segerberg squares this is clear: $(y, y)X_i(x, y)$ implies $(x, y)Y_i(x, x)$ (Fig. 12.3).

Axiom $(Sg7)$ means that horizontally related points are on the same horizontal, i.e., their f_1-projections are the same.

Axiom $(Sg8)$ means that for any point a, the horizontal of any vertically (Y_i-) accessible point y always contains a point y' that is vertically accessible from the

Fig. 12.2 Semantic condition for (Sg5)

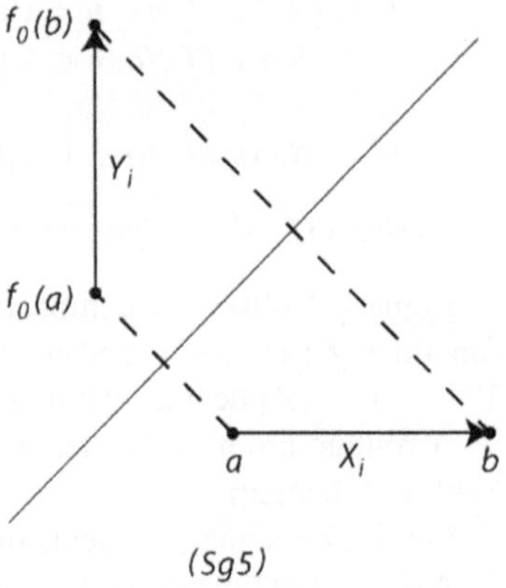

Fig. 12.3 Semantic condition for (Sg6)

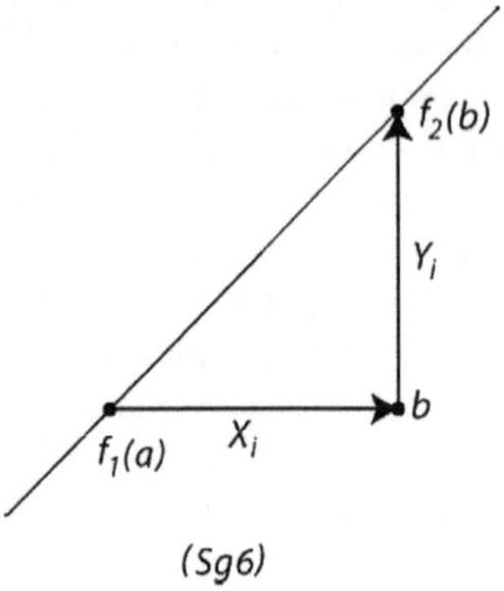

Fig. 12.4 Semantic condition for (Sg8)

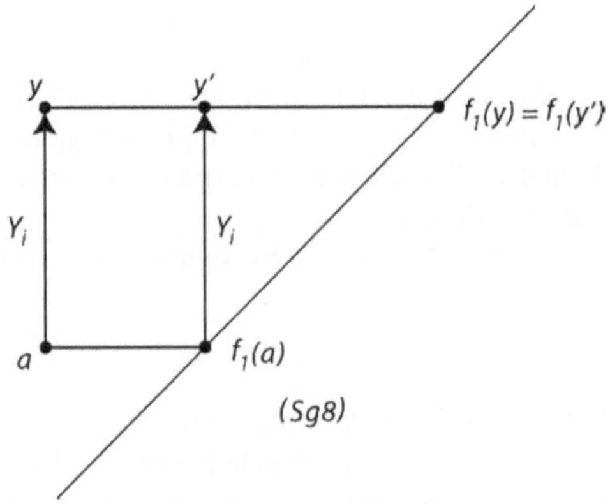

diagonal point $f_1(a)$, and the other way round, the horizontal of any point y' vertically accessible from $f_1(a)$, contains a point y vertically accessible from a (Fig. 12.4).

In the language of the logic $\mathbf{K}_{2N} * \mathbf{SL}_3$ with horizontal and vertical modalities every formula A has a conjugate A° obtained by interchange of vertical and horizontal modalities. A formal definition is by induction:

Definition 12.3.4 $A^\circ := A$ if A is atomic,
$(A \to B)^\circ := A^\circ \to B^\circ$,
$(\Box_i A)^\circ := \blacksquare_i A^\circ$,
$(\blacksquare_i A)^\circ := \Box_i A^\circ$,
$(\bigcirc_1 A)^\circ := \bigcirc_2 A^\circ$,
$(\bigcirc_2 A)^\circ := \bigcirc_1 A^\circ$,
$(\bigcirc A)^\circ := \bigcirc A^\circ$.

Lemma 12.3.5 (Conjugacy principle) *Let L be a $(2N+3)$-modal logic with the connectives $\Box_i, \blacksquare_i, \bigcirc_1, \bigcirc_2, \bigcirc$ containing*

$$\mathbf{K}_{2N} * \mathbf{SL}_3 + \{(Sg1), (Sg3), (Sg4), (Sg5)\}.$$

Then

(1) for any formula $A(p_1, \dots, p_n)$,

$$L \vdash A^\circ \leftrightarrow \bigcirc A(\bigcirc p_1, \dots, \bigcirc p_n);$$

(2) the rule A / A° is admissible in L.

We skip the routine proof, cf. Shehtman (2012). In particular, if a logic L contains axioms $(Sg2) - (Sg8)$, then their conjugates $(Sg2)^\circ - (Sg8)^\circ$ are provable.

In this paper we minimize the set of basic connectives for Segerberg squares and regard the modalities $\Box_i, \bigcirc_1, \bigcirc$ as basic and $\blacksquare_i, \bigcirc_2$ as derived:

$$\blacksquare_i := \bigcirc \Box_i \bigcirc, \ \bigcirc_2 := \bigcirc \bigcirc_1 .$$

Respectively for an N-modal frame $F = (W, R_1, \dots, R_N)$ the frame F^2_{Sg} is regarded as $(N+2)$-modal: $(W^2, R_{11}, \dots, R_{N1}, \sigma_0, \sigma_1)$, and for an N-modal logic Λ we regard Λ^2_{Sg} as $(N+2)$-modal.

12.4 Axiomatization of Segerberg Squares

Definition 12.4.1 For an N-modal logic Λ put

$$[\Lambda, \Lambda]^{\circledast} := [\Lambda, \Lambda] * \mathbf{SL}_3 + \{(Sg1), (Sg2), (Sg4), (Sg6), (Sg7), (Sg8)\}.$$

in the same $(N+2)$-modal language, with the same derived connectives as in the Segerberg square Λ^2_{Sg}.

Note that in this version the axioms ($Sg3$) and ($Sg5$) turn into definitions. Conjugacy principle implies that this axiomatization can be slightly simplified. Viz., Com_{ij} (Definition 12.2.36) can be replaced with the one-way commutation axioms

$$\Box_i \blacksquare_j p \to \blacksquare_j \Box_i p,$$

since the converse implications are their conjugates. We can also eliminate the **SL**-axioms for $\bigcirc_2$.

In many cases the above list of axioms is sufficient; the following result was proved in Shehtman (2012):

Theorem 12.4.2 *If Λ is a complete Horn axiomatizable modal logic, then $\Lambda^2_{Sg} = [\Lambda, \Lambda]^{\circledast}$.*

The proof is based on two lemmas.

Lemma 12.4.3 *If Λ is a complete Horn axiomatizable modal logic, then the logic $[\Lambda, \Lambda]^{\circledast}$ has the countable frame property.*

Lemma 12.4.4 (Main lemma) *Assume that a logic Λ is Horn axiomatizable, $G \vDash [\Lambda, \Lambda]^{\circledast}$, where G is (at most) countable and rooted. Then for some Λ-frame F, there exists a p-morphism $F^2_{Sg} \twoheadrightarrow G$.*

In some cases the completeness Theorem 12.4.2 can be refined: Segerberg squares are determined by trees.

We identify $T_{\omega,N} \sqcup T_{\omega,N}$ (cf. Definition 12.2.7) with $T_{\omega,N} \upharpoonright U$, where U is the set of all sequences from ω^* beginning with 8 or 9.[1]

The completeness proof in Shehtman (2012) yields the following observation (cf. Shehtman 2012, Lemma 4.10).

Lemma 12.4.5 *Let Λ be a logic from Theorem 12.4.2, Γ the corresponding Horn theory, A a formula in the language of Λ. If $\Lambda^2_{Sg} \nvdash A$, then there exists a frame $F \subseteq T_{\omega,N}$ such that $(F^+_\Gamma)^2_{Sg}, (8, 9) \nvDash A$, and in this case $F^+_\Gamma \vDash \Lambda$.*

In this general case F is a disjoint sum of two trees, but sometimes it can be chosen as a single tree:

Theorem 12.4.6 (Cf. Shehtman 2012, Theorem 5.2) *Let $\Lambda = \mathbf{K}_N + \Sigma_1 + \Sigma_2$, where the formulas from Σ_1 are pseudo-transitive of the form $\Box_k p \to \Box_\alpha p$, the formulas from Σ_2 are closed of the form $\Diamond_i \top$. Let*

$$\mathscr{C}_\Lambda := \{F^+_\Gamma \mid F \text{ is a standard } N\text{-tree}, \ F \vDash \Sigma_2\},$$

where Γ is a Horn theory corresponding to Σ_1. Then

$$\Lambda^2_{Sg} = \mathbf{L}(\{G^2_{Sg} \mid G \in \mathscr{C}_\Lambda\}).$$

[1] The choice of 8, 9 is arbitrary; they are used only as markers for two halves of the disjoint union.

There is a further refinement of Theorem 12.4.6.

Proposition 12.4.7 (Cf. Shehtman 2012, Proposition 4.7) *Let Λ be a logic from Theorem 12.4.6, A a formula in its language, $\Lambda^2_{Sg} \nvdash A$. Then $G^2_{Sg}, (8, 9) \not\models A$ for some $G \in \mathscr{C}_\Lambda$.*

In this paper we are especially interested in the logics $\Lambda = \mathbf{K}_N + \Box^r \bot$. We can notice that their Segerberg squares are not determined by squares of trees:

Lemma 12.4.8 *(1) Let F be an N-tree of depth less than r. Then*

$$F^2_{Sg} \models p \wedge \blacklozenge^{r-1}\top \wedge \Diamond^{r-1}\top \to \bigcirc p.$$

(2) $(\mathbf{K}_N + \Box^r \bot)^2_{Sg} \nvdash p \wedge \blacklozenge^{r-1}\top \wedge \Diamond^{r-1}\top \to \bigcirc p$.

Proof (1) If in F^2_{Sg} we have $(x, y) \models \blacklozenge^{r-1}\top \wedge \Diamond^{r-1}\top$, then $d(x) = d(y) = r - 1$ in F. So $x = y$ is the root of F. Then $(x, y) \models p$ implies $(x, y) \models \bigcirc p$.

(2) Let F_0 be the intransitive chain $(\{0, \dots, r-1\}, R)$, where xRy iff $y = x + 1$, and let F the disjoint sum $F_0 \sqcup F_0 = \bigsqcup_{j \in \{1,2\}} F_0$. So according to Definition 12.2.7, F is the union of two cones isomorphic to F_0, with the roots $a := (0, 1)$ and $b := (0, 2)$.

Then $F \models \Box^r \bot$, but $F^2_{Sg} \not\models p \wedge \blacklozenge^{r-1}\top \wedge \Diamond^{r-1}\top \to \bigcirc p$.

In fact, consider a Kripke model over F^2_{Sg}, in which

$$z \models p \text{ iff } z = (a, b).$$

Then we have

$$(a, b) \models p \wedge \blacklozenge^{r-1}\top \wedge \Diamond^{r-1}\top,$$

but $(a, b) \not\models \bigcirc p$, since p holds only at (a, b) and this point is not diagonal. ⊠

Another example is the logic **S5**. In this case the closure of a tree is a frame with a universal relation. The logic of Segerberg squares of these frames was axiomatized in Segerberg (1973). It contains the formula $\Box p \to \bigcirc_1 p$ non-provable in $\mathbf{S5}^2_{Sg}$.

12.5 Bisimulation Games

In this section we recall some facts about bisimulation games. In more detail they are discussed in Shehtman (2018).

Definition 12.5.1 Let $M = (W, R_1, \dots, R_N, \theta)$ be a k-weak Kripke model. By induction on n we define the *n-bisimilarity* (or *n-equivalence*) relation $M, x \equiv_n M, x'$ on W:

- $M, x \equiv_0 M, x'$ iff the same proposition letters from $PL\lceil k$ are true at M, x and M, x'.

- $M, x \equiv_{n+1} M, x'$ iff $M, x \equiv_0 M, x'$ and $(R_i(x) / \equiv_n) = (R_i(x') / \equiv_n)$ for any $i \leqslant N$.

We will use a simpler notation $x \equiv_n x'$ if M is clear from the context.

Definition 12.5.2 Let $M = (W, R_1, \dots, R_N, \theta)$ be a Kripke model, $x, x' \in W$. The *r-round bisimulation game on M with the initial position* (x, x') (denoted by $BG_r(M, x, x')$) is described as follows.

- Every play of the game consists of moves made in turn by two players, the Spoiler ('Abelard') and the Duplicator ('Éloïse'); the Spoiler makes the first move.
- Every position in a play is a pair $(y, y') \in W \times W$; the n-th position is reached after n moves of each player.
- (x, x') is the 0th position.
- If the nth position is (y, y'), then the Spoiler's move can be

(1) a triple $(1, i, z)$, where $1 \leqslant i \leqslant N$ and yR_iz;
(2) a triple $(2, i, z')$, where $1 \leqslant i \leqslant N$ and $y'R'_iz'$.

We say that the moves $(1, i, z)$, $(2, i, z')$ are *of type* i. The response of the Duplicator should be respectively

(1) z' such that $y'R'_iz'$ and $M, z \equiv_0 M, z'$;
(2) z such that yR_iz and $M, z \equiv_0 M, z'$.

The $(n + 1)$th position will be (z, z').

- If the Duplicator cannot respond to the nth move of the Spoiler for some $n \leqslant r$, then the play terminates, and the Spoiler wins.
- If the Spoiler cannot make the next move, then the play terminates, and the Duplicator wins.
- If the play reaches the rth position (i.e., the Duplicator responds to every move of the Spoiler), then the play terminates, and the Duplicator wins.

A *short play* of $BG_r(M, x, x')$ is a sequence of moves made according to the rules, i.e., an unfinished play of this game.

The next lemma is an easy consequence of Definition 12.5.2.

Lemma 12.5.3 *Consider a short play of* $BG_r(M, x, x')$ *of length* $2n$. *For* $1 \leqslant j \leqslant n$ *let* (x_j, x'_j) *be the* j*th position of this play, and put* $x_0 := x$, $x'_0 := x'$. *Suppose the* j*th move of this play is of type* i_j. *Then* $(x_0, i_1, x_1, \dots, i_n, x_n)$ *and* $(x'_0, i_1, x'_1, \dots, i_n, x'_n)$ *are paths in* M.

The paths from the above lemma are called the *play paths* (for the given play).

The next definition is necessary only for formal reasons. Usually the results on games are proved by informal natural arguments.

Definition 12.5.4 A *winning strategy* for the Duplicator in $BG_r(M, x, x')$ is a partial function σ defined on short plays of odd length with the following property.

If $\Gamma = (s_1, d_1, \ldots, s_n, d_n, s_{n+1})$ is a short play such that $d_k = \sigma(s_1, d_1, \ldots, s_k)$ for any $k \leqslant n$ (in particular, if $n = 0$ and s_1 is arbitrary),[2] then $\sigma(\Gamma)$ is an allowed move of the Duplicator, i.e., $(\Gamma, \sigma(\Gamma))$ is also a short play.

Recall the main theorem on bisimulation games, cf. Goranko and Otto (2007), Theorem 32.

Theorem 12.5.5 *For a weak Kripke model M the following conditions are equivalent:*

(1) $M, x \equiv_n M, x'$;
(2) $M, x \vDash A$ iff $M, x' \vDash A$
for any formula A in the language of M of modal depth $\leqslant n$;
(3) the Duplicator has a winning strategy in $BG_n(M, x, x')$.

Proof Let us denote the second relation by $\sim_n$ and the third by $\approx_n$.

(I) The inclusion $\equiv_n \subseteq \approx_n$ is proved by induction. The base is trivial.

For the step, suppose $\equiv_n \subseteq \approx_n, x \equiv_{n+1} y$. To show that $x \approx_{n+1} y$, consider the first Spoiler's move $(1, i, x')$ in a play of $BG_{n+1}(x, y)$ (for moves of type 2 the argument is similar). Since $x \equiv_{n+1} y$, there exists $y' \in R_i(y)$ such that $x' \equiv_n y'$, so $x' \approx_n y'$ by IH. This y' can be the Duplicator's first move, because $x' \approx_n y'$ guarantees the winning strategy for the remaining n moves. Thus $x \approx_{n+1} y$.

(II) The inclusion $\approx_n \subseteq \sim_n$ is also proved by induction. The base is again trivial.

For the step, suppose $\approx_n \subseteq \sim_n, x \approx_{n+1} y$. Now by induction on the length of a formula A of modal depth $\leqslant n+1$ we have to show that $x \vDash A$ iff $y \vDash A$.

The only nontrivial case is $A = \Box_i B$. It is sufficient to show that $x \nvDash A$ implies $y \nvDash A$, because the converse follows by symmetry. So assume $x \nvDash A$; then $x' \nvDash B$ for some $x' \in R_i(x)$. Consider a play of $BG_{n+1}(x, y)$ with the first move $(1, i, x')$. Since $x \approx_{n+1} y$, there is a Duplicator's response $y' \in R_i(x')$ such that $x' \approx_n y'$. By IH $x' \sim_n y'$, so $y' \nvDash B$, since $md(B) = md(A) - 1 \leqslant n$. Therefore $y \nvDash \Box_i B$.

(III) For the inclusion $\sim_n \subseteq \equiv_n$ we also argue by induction, and the base is trivial.

For the step suppose $\sim_n \subseteq \equiv_n, x \sim_{n+1} y$. Consider an arbitrary $x' \in R_i(x)$.

Note that there are finitely many N-modal k-formulas of modal depth $\leqslant n$ up to equivalence in $\mathbf{K}_N$ (this is easily proved by induction on n). So we can define

$$B := \bigwedge \{A \mid x' \vDash A,\ md(A) \leqslant n\}),$$

where the conjuncts are supposed non-equivalent in $\mathbf{K}_N$. Then $md(B) \leqslant n, x \vDash \Diamond_i B$, so $x \sim_{n+1} y$ implies $y \vDash \Diamond_i B$. Hence we obtain $y' \in R_i(y)$ such that $y' \vDash B$. Thus $x' \sim_n y'$; in fact, every formula of modal depth $\leqslant n$ satisfied at x' is equivalent to a conjunct in B, so it is satisfied at y'; the other way round, if $x' \nvDash A$ and $md(A) \leqslant n$, then $md(\neg A) \leqslant n, x' \vDash \neg A$, so $y' \vDash \neg A$. Now by IH we obtain $x' \equiv_n y'$. It follows that $(R_i(x)/\equiv_n) \subseteq (R_i(y)/\equiv_n)$, and the converse follows by symmetry. ⊠

Put $\equiv_\infty := \bigcap_n (\equiv_n)$.

[2]In this case we say that Γ is *controlled by* σ.

Corollary 12.5.6 *(1) The sequence of n-bisimilarities decreases:* $\equiv_n \supseteq \equiv_{n+1}$.
(2) $M, x \equiv_\infty M, y$ *iff* $(M, x \vDash A \Leftrightarrow M, y \vDash A)$ *for any formula A in the language of M (i.e., iff x and y are indistinguishable).*

Proposition 12.5.7 *In every weak Kripke k-model the number of n-bisimilarity classes is finite; it is bounded by a function depending only on n and k.*

Proof Note that there is a finite number of N-modal k-formulas of depth $\leqslant n$ up to equivalence in $\mathbf{K}_N$ — this is easily checked by induction. Alternatively, one can estimate the number $f(n,k)$ of $\equiv_n$-classes for a k-weak Kripke model as follows.

Clearly, $f(0,k) \leqslant 2^k$ as there can be at most 2^k combinations of truth values for k proposition letters.

$f(n+1,k) \leqslant 2^k \cdot 2^{Nf(n,k)}$, since every $\equiv_{n+1}$-class of a point x is determined by its $\equiv_0$-class and the family of the quotient sets $R_i(x)/\equiv_n$ (for all $i \leqslant N$); each $R_i(x)/\equiv_n$ is characterised by the set of $\equiv_n$-classes intersecting $R_i(x)$.

Thus $f(n,k) \leqslant g(n,k)$, where $g(n,k)$ is defined by recursion:

$$g(0,k) = 2^k, \;\; g(n+1,k) = 2^{k+Ng(n,k)}.$$

⊠

Proposition 12.5.8 *(1) If* $\equiv_n = \equiv_{n+1}$, *then* $\equiv_n = \equiv_\infty$.
(2) If in a canonical model $M_{L\lceil k}$ *for some n* $\equiv_n = \equiv_{n+1}$, *then* $M_{L\lceil k}$ *is finite.*

Proof (1) Assuming $\equiv_n = \equiv_{n+1}$, by induction we obtain that $\equiv_n = \equiv_m$ for any $m > n$. In fact, $\equiv_n = \equiv_m$ implies $\equiv_{n+1} = \equiv_{m+1}$, since

$$x \equiv_{m+1} y \text{ iff } x \equiv_0 y \;\&\; \forall i \leqslant N\; (R_i(x)/\equiv_m) = (R_i(y)/\equiv_m),$$

$$x \equiv_{n+1} y \text{ iff } x \equiv_0 y \;\&\; \forall i \leqslant N\; (R_i(x)/\equiv_n) = (R_i(y)/\equiv_n).$$

(2) By (1), $\equiv_n = \equiv_{n+1}$ implies $\equiv_n = \equiv_\infty$; so by Proposition 12.5.7, the number of $\equiv_\infty$-classes is finite. By Corollaries 12.5.6, 12.2.26, $\equiv_\infty$ is the equality relation. Hence $M_{L\lceil k}$ is finite. ⊠

Proposition 12.5.9 *Let L be a (non-weak) modal logic.*

(1) $md(L\lceil k) \leqslant n$ *iff* $\equiv_n = \equiv_{n+1}$ *in* $M_{L\lceil k}$.
(2) $md(L) \leqslant n$ *iff* $\equiv_n = \equiv_{n+1}$ *in every weak canonical model of L.*

Proof (1) (If.) Let us choose a set $\Phi_{k,n}$ of representatives for classes of k-formulas of modal depth $\leqslant n$ modulo equivalence in $\mathbf{K}_N$.

Consider a k-formula A and $M := M_{L\lceil k}$. For every $x \in M$ put $B_x := \bigwedge\{B \in \Phi_{k,n} \mid x \vDash B\}$. Since by Proposition 12.5.8 $\equiv_n = \equiv_{n+1}$ implies $\equiv_n = \equiv_\infty$ and in a canonical model $\equiv_\infty$ is the diagonal relation, it follows that two points in M are equal iff they satisfy the same k-formulas of modal depth $\leqslant n$. In other words, B_x defines x in M.

Therefore $M \vDash A \leftrightarrow \bigvee\{B_x \mid x \vDash A\}$, i.e., $L \vdash A \leftrightarrow \bigvee\{B_x \mid x \vDash A\}$ by the Canonical model theorem. The latter disjunction is of modal depth $\leqslant n$.

'Only if' is a trivial consequence of Theorem 12.5.5 (since $M \models L\lceil k$).

(2) follows from (3) and Lemma 12.2.31. ⊠

Lemma 12.5.10 *Let M be a Kripke model, $x, y \in M$. Suppose $x \equiv_n y$ and the Duplicator has a winning strategy σ in $BG_n(x, y)$ such that every play controlled by σ has at least two repeating positions. Then $x \equiv_{n+1} y$.*

Proof To win the game of length $n + 1$, the Duplicator should follow σ for the first n moves. If she does not win earlier than at the nth round, there must be repeating positions. If the positions (x_i, y_i) and (x_j, y_j) with $i < j$ are the same, we have $x_j = x_i \equiv_{n-i} y_i = y_j$, so anyway $x_j \equiv_{n+1-j} y_j$. Thus she has a got a winning strategy for the rest $n + 1 - j$ moves after the j-th position. ⊠

As an illustration of the game-theoretic method, let us give an alternative proof of local tabularity for logics of finite depth from Gabbay and Shehtman (1998). Actually we will show that these logics are of finite modal depth.

Theorem 12.5.11 $md(\mathbf{K}_N + \Box^n\bot) = n - 1$.

Proof Let M be a weak canonical model of this logic L and let us show that in this model $\equiv_{n-1} = \equiv_n$. So suppose $x \equiv_{n-1} x'$, i.e., the Duplicator has a winning strategy for $BG_{n-1}(M, x, x')$. We claim that this is also a winning strategy for $BG_n(M, x, x')$.

In fact, for the first $n - 1$ Spoiler's moves, the strategy works fine. If the Spoiler does not lose until the $(n - 1)$th move and (y, y') is the $(n - 1)$th position, then by Lemma 12.5.3 there exist paths of length $n - 1$ from x to y and from x' to y'. If the nth move of the Spoiler is $(1, i, z)$ and yRz, then there is a path of length n from x to z, which contradicts Lemma 12.2.11. By the same reason this move cannot be $(2, i, z')$ with $y'Rz'$. So the Spoiler cannot make the nth move, and the Duplicator wins.

Thus by Proposition 12.5.9 $md(L) \leqslant n - 1$. To prove the equality, consider the formula $\Box^{n-1}\bot$. Let us show that it is not L-equivalent to any formula of modal depth $< n - 1$. In fact, consider a Kripke model over an L-frame, which is a disjoint sum of two intransitive chains shown on Fig. 12.5, with all the proposition letters true. Here we have $x \equiv_{n-2} y$, since the Duplicator wins $BG_{n-2}(x, y)$. So x, y satisfy the same formulas of depth $< n - 1$. However, $\Box^{n-1}\bot$ holds only at y; so it is not equivalent to any formula of smaller depth. ⊠

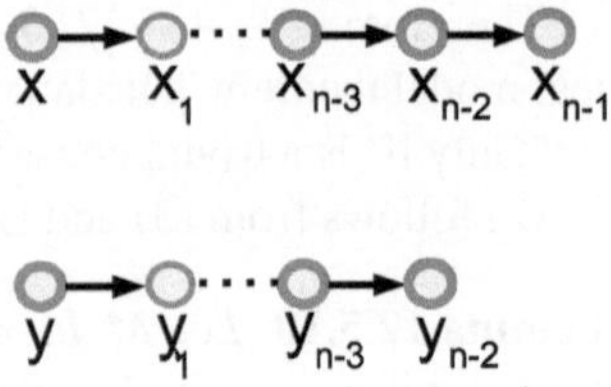

Fig. 12.5 The Duplicator wins $BG_{n-2}(x, y)$

12.6 Finite Modal Depth

In this section we show that the logics $(\mathbf{K}_N + \Box^m \bot)^2_{Sg}$ are of finite modal depth, which implies their local tabularity. The latter fact was already proved in Shehtman (2012), but the proof was technically more difficult than the one proposed here. In its turn, the local tabularity of $(\mathbf{K}_N + \Box^m \bot)^2_{Sg}$ implies the fmp of $(\mathbf{K}_N)^2_{Sg}$.

We use the notation $\Lambda_0 := (\mathbf{K}_N)^2_{Sg}$, $\Lambda_m := (\mathbf{K}_N + \Box^m \bot)^2_{Sg}$.

Definition 12.6.1 Let $F = (W, X_1, \dots, X_N, f_0, f_1)$ be a Λ_0-frame. The X_i are called the *horizontal relations*, and we also define the *vertical relations*

$$xY_iy \text{ iff } f_0(x)X_i f_0(y).$$

Also put $f_2 := f_1 f_0$.

The *horizontal depth* $d_1(x)$ of a point x in F is the depth of x in the frame $F_1 := (W, X_1, \dots, X_N)$. Similarly, the *vertical depth* $d_2(x)$ of x in F is the depth of x in $F_2 := (W, Y_1, \dots, Y_N)$. The *horizontal* (respectively, the *vertical*) *depth* of F is the depth of F_1 (respectively, F_2).

From Lemma 12.2.11 we readily obtain

Lemma 12.6.2 $x \vDash \Box^m\bot \Leftrightarrow d_1(x) \leqslant m - 1$,
$x \vDash \blacksquare^m\bot \Leftrightarrow d_2(x) \leqslant m - 1$.

Lemma 12.6.3 *Under the conditions of Definition 12.6.1:*

(1) $d_2(f_0(x)) = d_1(x)$,
(2) $d_1(f_0(x)) = d_2(x)$,
(3) $d_1(f_1(x)) = d_2(f_1(x)) = d_2(x)$,
(4) $d_1(f_2(x)) = d_2(f_2(x)) = d_1(x)$,
(5) $xX_iy \Rightarrow d_2(x) = d_2(y) \;\&\; d_1(y) \leqslant d_1(x) - 1$,
(6) $xY_iy \Rightarrow d_1(x) = d_1(y) \;\&\; d_2(y) \leqslant d_2(x) - 1$.

Here we assume that $\infty - 1 = \infty$ (although this case is not really used).

We omit the proof, see Lemma 6.5 from Shehtman (2012).

Definition 12.6.4 We define the length of a pair of sequences of natural numbers:

$$|(\alpha, \beta)| := \max(|\alpha|, |\beta|).$$

Definition 12.6.5 The $\Box$-*depth* $d_\Box(A)$ of a modal formulas A in the language of Λ_0 is defined by induction:

$$d_\Box(p_k) = d_\Box(\bot) = 0, \ d_\Box(A \to B) = \max\,(d_\Box(A), d_\Box(B));$$
$$d_\Box(\bigcirc A) = d_\Box(\bigcirc_i A) = d_\Box(A); d_\Box(\blacksquare_j A) = d_\Box(\Box_j A) = d_\Box(A) + 1.$$

Lemma 12.6.6 (Two-dimensional truncation lemma) *Let F be a standard N-tree, $F^- = F^{(m)}$ (Definition 12.2.20), M a Kripke model over F^2_{Sg}, M^- its restriction to $(F^-)^2_{Sg}$. Then for any $\mathscr{L}_N S$-formula A, for any $z \in M^-$, $d_\Box(A) + |z| \leqslant m$ implies*

$$M, z \vDash A \Leftrightarrow M^-, z \vDash A.$$

Proof The proof is by induction on the length of A similar to Lemma 12.2.21. We consider only the nontrivial cases.

Let $A = \bigcirc B, d_\Box(A) + |z| \leqslant m$. Then

$$M, z \vDash A \Leftrightarrow M, \sigma(z) \vDash B; \ \ M^-, z \vDash A \Leftrightarrow M^-, \sigma(z) \vDash B.$$

But $|z| = |\sigma(z)|, \ d_\Box(A) = d_\Box(B)$, so $d_\Box(B) + |\sigma(z)| \leqslant m$, and we can apply the induction hypothesis for B:

$$M, \sigma(z) \vDash B \Leftrightarrow M^-, \sigma(z) \vDash B.$$

Hence the claim for A follows.

Analogously, let $A = \bigcirc_1 B, d_\Box(A) + |z| \leqslant m$. Then

$$M, z \vDash A \Leftrightarrow M, \sigma_1(z) \vDash B; \ \ M^-, z \vDash A \Leftrightarrow M^-, \sigma_1(z) \vDash B.$$

Note that $|(y, y)| \leqslant |(x, y)|$, i.e., $|\sigma_1(z)| \leqslant |z|$. Also, $d_\Box(A) = d_\Box(B)$, Therefore $d_\Box(B) + |\sigma_1(z)| \leqslant d_\Box(A) + |z| \leqslant m$, and again we can apply the IH for B.

Let $A = \Box_i B, d_\Box(A) + |z| \leqslant m$. Then

$$M, z \vDash A \Leftrightarrow \forall t \in M(z \sqsubset_{i1} t \Rightarrow M, t \vDash B),$$
$$M^-, z \vDash A \Leftrightarrow \forall t \in M(z \sqsubset_{i1} t \Rightarrow M^-, t \vDash B),$$

since the same points are accessible from z in M and M^- (provided $|z| < m$, as in our case).

Note that $z \sqsubset_{i1} t$ implies $|t| \leqslant |z| + 1$ (more exactly, $|t| = |z|$ or $|t| = |z| + 1$). Therefore

$$d_\Box(B) + |t| \leqslant d_\Box(B) + 1 + |z| = d_\Box(A) + |z| \leqslant m,$$

and again we can apply the IH for B. ⊠

Lemma 12.6.7 *If $\Lambda_0 \nvdash A$, then $\Lambda_0 + \Box^{d_\Box(A)+1}\bot \nvdash A$.*

Proof So let $\Lambda_0 \nvdash A$. Then by Proposition 12.4.7, $G^2_{Sg}, (8, 9) \nvDash A$ for some standard tree G.

Let $m = d_\Box(A) + 1$, $G^- = G^{(m)}$. Since $d_\Box(A) + |(8,9)| = m$, then by Lemma 12.6.6, $(G^-)^2_{Sg}, (8,9) \nvDash A$. Now note that $(G^-)^2_{Sg}\uparrow(8,9) \vDash \Box^{m+1}\bot$.

In fact,

$$(\alpha, \beta) \sqsubset_{i1} (\gamma, \beta) \Rightarrow |\gamma| = |\alpha| + 1.$$

Therefore, if in the frame $(G^-)^2$ with the relations $\sqsubset_{11}, \dots, \sqsubset_{N1}$ there is a path $(\alpha_1, \beta), \dots, (\alpha_k, \beta)$, then its length is $k - 1 = |\alpha_k| - |\alpha_1| \leqslant m + 1$.

However, if this path is in the generated subframe $(G^-)^2\uparrow(8,9)$ with $\sqsubset_{11}, \dots, \sqsubset_{N1}$, then $|\alpha_1| \neq 0$, so the length of the path does not exceed m. Hence our claim follows by Lemma 12.2.11.

Eventually, by the generation lemma we have

$$(G^-)^2_{Sg}\uparrow(8,9) \vDash \Lambda_0 + \Box^{m+1}\bot$$

and $(G^-)^2_{Sg}\uparrow(8,9) \nvDash A$, and so $\Lambda_0 + \Box^{m+1}\bot \nvdash A$. ☒

Hence we obtain

Lemma 12.6.8 $\Lambda_0 = \bigcap\limits_m \Lambda_m$.

Our next goal is to prove that Λ_m is of finite modal depth. So we examine bisimulation games in weak canonical models of Λ_m. Since Λ_m is axiomatized by Sahlqvist formulas, it is d-persistent, and thus weakly canonical. Now we fix

$$M := M_{\Lambda_m\lceil k}, \quad F := F_{\Lambda_m\lceil k}.$$

We use the notation X_j, Y_j, f_0, f_1, f_2 of relations and functions in F from Definition 12.6.1.

Since $F \vDash \Lambda_0$, we can use Lemma 12.6.3.

The set (for a finite r)

$$S_r := \{x \in M \mid \max(d_1(x), d_2(x)) = r\}$$

is called the *r-th layer* in M. It is partitioned into *sectors*, *horizontal*

$$S_{kr} := \{x \in M \mid d_1(x) = k, \ d_2(x) = r\}$$

and *vertical*

$$S_{rk} := \{x \in M \mid d_1(x) = r, \ d_2(x) = k, \}$$

where $0 \leqslant k \leqslant r$. The sectors S_{rr} are called *diagonal*; they are both horizontal and vertical.

Lemma 12.6.9 *Assume that* $x \equiv_n y$; $x, y \in \bigcup\limits_{r'<r} S_{r'}$ *and* $n > r + 1$. *Then* x, y *are in the same sector.*

Proof First let us show that $d_1(x) = d_1(y)$ if $n > r$. Suppose the contrary, e.g. $d_1(x) > d_1(y)$. Then in the game $BG_n(x, y)$ the Spoiler can make $d_1(y) + 1$ successive X-moves from x. The Duplicator has no response for the last of them, so she loses. Hence $x \not\equiv_n y$ contradicting the assumption of the lemma.

Let us also show that $d_2(x) = d_2(y)$, i.e., $d_1(f_0(x)) = d_1(f_0(y))$ by Lemma 12.6.3(2). Note that $(f_0(x), f_0(y))$ is the first position in the play of $BG_n(x, y)$ if the Spoiler makes the first move from x to $f_0(x)$. So $f_0(x) \equiv_{n-1} f_0(y)$, while $n - 1 > r$ by the assumption of the lemma. Hence $d_1(f_0(x)) = d_1(f_0(y))$ as proved above. ⊠

Lemma 12.6.10 *Assume that $n > 2r + 2$ and x, y are in the same sector of layer r. Then every n-round play of $BG_n(x, y)$ within the layer r has repeating positions.*

Proof Suppose the contrary and consider such a play without repetitions.

`Case 1.` The points x, y are in a vertical sector S_{rk}, with $k < r$. As the first position in the play is in S_r, the first move can be only of f_0-type, so the first position is $(f_0(x), f_0(y)) \in S_{kr}$. Starting from this position we can argue as in Case 2 below.

`Case 2.` x, y are in a horizontal sector S_{kr}. Then (by Lemma 12.6.3(3),(5)) all positions in our play remain in horizontal sectors until the first Spoiler's f_0-move. Suppose this move is made from the i-th position (x_i, y_i). Thus the position (x_{i+1}, y_{i+1}) is in a vertical sector. It follows that $i + 1 = n$ — otherwise to keep the play within S_r we should have $x_{i+2} = f_0(x_{i+1}) = x_i$ as in Case 1, and similarly $y_{i+2} = y_i$, so there would be repeating positions.

Therefore the first $(n - 1)$ Spoiler's moves of the play are of X- or f_1-type.

Now note that $f_1(x_i) = f_1(x)$ for any $i < n$. This easily follows by induction from Lemma 12.3.3 (which is applicable since $F \models \Lambda_0$). In fact, $x_i X_j x_{i+1}$ implies $f_1(x_i) = f_1(x_{i+1})$ by $(Sg7)$, and $x_{i+1} = f_1(x_i)$ implies $f_1(x_i) = f_1(x_{i+1})$ by $(Sg2)$. Similarly we have $f_1(y_i) = f_1(y)$ for any $i < n$.

As the positions do not repeat, it follows that at most one of the first $(n - 1)$ Spoiler's moves is of f_1-type. At the same time every X-move decreases the horizontal depth, so there may be at most r successive Spoiler's X-moves. In total we obtain that $2r + 1 \geqslant n - 1$ contradicting the assumption $n > 2r + 2$. ⊠

Lemma 12.6.11 *Suppose $x, y \in \bigcup_{r' \leqslant r} S_{r'}$ and $x \equiv_n y$ for $n > (r + 1)(r + 2)$. Then every n-round play of the game $BG_n(x, y)$ has repeating positions.*

Proof First note that x, y are in the same sector by Lemma 12.6.9, so we may assume that $x, y \in S_r$. Then we argue by induction on r. The base follows from 12.6.10. So assume $r > 0$.

Suppose there is an n-round play without repetitions. By Lemma 12.6.10 at most $(2r + 2)$ first rounds can keep the position at layer r. Any move away from S_r leads to a smaller layer. In fact, this can be only an X- or f_1-move from a vertical sector S_{rk} (see the proof of the previous lemma); by Lemma 12.6.3(5) an X-move leads from S_{rk} to $S_{r'k}$ with $r' < r$, and obviously an f_1-move leads from S_{rk} to S_{kk}.

So after the play leaves S_r there are $n' \geqslant n - (2r + 2)$ rounds remaining, from a position (x', y') with $x' \equiv_{n'} y'$ at layers $\leqslant r - 1$. Then

$$n' > (r+1)(r+2) - 2(r+1) = (r+1)r.$$

So by IH there must be repetitions in the remaining part of the play. ⊠

Theorem 12.6.12 $md(\Lambda_m) \leqslant m(m+1)+1$.

Proof In a canonical model of Λ_m we have $x \vDash \Box^m\bot \wedge \blacksquare^m\bot$, so $d_1(x), d_2(x) \leqslant m-1$. Put $n := m(m+1)+1$ and suppose $x \equiv_n y$. By Lemma 12.6.11 applied to $r = m-1$, every n-round play of $BG_n(x,y)$ has repeating positions, so by Lemma 12.5.10, $x \equiv_{n+1} y$.

Therefore $md(\Lambda_m) \leqslant n$ by Proposition 12.5.9. ⊠

Hence by Proposition 12.2.32 we obtain

Corollary 12.6.13 *Every logic Λ_m is locally tabular.*

Theorem 12.6.14 *Λ_0 has the fmp.*

Proof By Lemma 12.6.8 and Corollary 12.6.13, Λ_0 is an intersection of locally tabular logics. Now Proposition 12.2.30(4) can be applied. ⊠

From the finite axiomatizability of Λ_0 (Theorems 12.4.2) and 12.6.14 we obtain

Corollary 12.6.15 *Λ_0 is decidable.*

12.7 Square Finite Model Property

In this section we recall the *square fmp* of the logic $(\mathbf{K}_N)^2_{Sg}$ proved in Shehtman (2012). An analogue to this property for usual products is the *product fmp* proved in Gabbay and Shehtman (2000) for some cases.

Definition 12.7.1 Let Λ be an N-modal logic. We say that its Segerberg square Λ^2_{Sg} has the *square finite model property* if

$$\Lambda^2_{Sg} = \mathbf{L}(\{F^2_{Sg} \mid F \vDash \Lambda,\ F \text{ finite }\}).$$

Theorem 12.7.2 *$(\mathbf{K}_N)^2_{Sg}$ has the square fmp.*

The proof of this result in Shehtman (2012) is by a game-theoretic argument. Let us recall the plan of that proof and make some refinement. We again denote $(\mathbf{K}_N)^2_{Sg}$ by Λ_0.

Definition 12.7.3 A Λ_0-frame G is called *principal rooted* if $G = G{\uparrow}u$ for some diagonal point u.

Proposition 12.7.4

$$\Lambda_0 = \mathbf{L}(\{G \mid G \vDash \Lambda_0,\ G \textit{ is finite principal rooted of finite horizontal depth}\}).$$

Moreover, if $\Lambda_0 \nvDash A$, then there exists a finite principal rooted frame G of finite horizontal depth such that for some i, j the formula $\Diamond_i \blacklozenge_j \neg A$ is satisfiable at the root of G.

Proof The first claim is Proposition 7.4 from Shehtman (2012), and the second one can be extracted from its proof. So let us repeat the details.

Let $\Lambda_0 \nvDash A$ for a k-formula A, $m = d_{\Box}(A) + 1$. Then as in the proof of Lemma 12.6.7 we obtain a standard tree G^- of depth $\leqslant m$ such that $(G^-)^2_{Sg}, (8, 9) \nvDash A$. Note that $G^- \vDash \Box^{m+1}\bot$ and $\lambda \sqsubset_i 8,\ \lambda \sqsubset_j 9$ for some i, j.

Now consider the set of formulas[3]

$$\Theta := \{\Diamond_i \blacklozenge_j \neg A\} \cup \{\bigcirc_1 B \leftrightarrow B \mid B \in \mathscr{L}_N S\lceil k\}.$$

By the construction, for some model M over $(G^-)^2_{Sg}$ we have $M, (\lambda, \lambda) \vDash \Theta$. Thus Θ is consistent in the logic $\Lambda_{m+1} = (\mathbf{K}_N + \Box^{m+1}\bot)^2_{Sg}$. Also $\Theta \subseteq \mathscr{L}_N S\lceil k$. Therefore, by the canonical model theorem, for some u we have $M_{\Lambda_{m+1}\lceil k}, u \vDash \Theta$. Then by the generation lemma, $M_{\Lambda_{m+1}\lceil k}\!\uparrow\! u, u \vDash \Diamond_i \blacklozenge_j \neg A$. As we know from Sect. 12.6, Λ_{m+1} is locally tabular, so $F_{\Lambda_{m+1}\lceil k}$ (and $F_{\Lambda_{m+1}\lceil k}\!\uparrow\! u$) is a finite frame of finite horizontal depth.

From the definition of canonical models and the axioms of **SL** for $\bigcirc_1$, in the model $M_{\Lambda_{m+1}\lceil k}$ we have:

$$f_1(u) = \{B \mid \bigcirc_1 B \in u\}.$$

Hence by the construction of Θ we get $f_1(u) = u$, i.e., u is a diagonal point, and the rooted frame $F_{\Lambda_{m+1}\lceil k}\!\uparrow\! u$ is principal. ⊠

Lemma 12.7.5 (Main lemma) *Let G be a finite principal rooted Λ_0-frame of finite horizontal depth r. Then for some finite (standard) N-tree F of depth r there exists a p-morphism $F^2_{Sg} \twoheadrightarrow G$.*

Theorem 12.7.6 $(\mathbf{K}_N)^2_{Sg} = \mathbf{L}(\{F^2_{Sg} \mid F \textit{ is a finite } N\textit{-tree}\})$.
Moreover, if $(\mathbf{K}_N)^2_{Sg} \nvDash A$, then for some finite standard N-tree F, for some $x, y \in \omega$, A is refutable at the point (x, y) of F^2_{Sg}.

Proof By Proposition 12.7.4(1), for some i, j the formula $\Diamond_i \blacklozenge_j \neg A$ is satisfiable at the root u of some finite principal rooted frame G of finite horizontal depth. The p-morphism $F^2_{Sg} \twoheadrightarrow G$ constructed in Lemma 12.7.5 maps the root (λ, λ) to the root u. So by the p-morphism lemma $\Diamond_i \blacklozenge_j \neg A$ is satisfiable at (λ, λ) in F^2_{Sg}, i.e., A is refutable at some (x, y), with $\lambda \sqsubset_i x, \lambda \sqsubset_j y$. ⊠

[3]In Shehtman (2012) there is a misprint in the definition of Θ.

12.8 From Segerberg Squares to Relation Algebras

In this section we extend the results from Gabbay and Shehtman (2000, 2002) connecting products of modal logics with relation algebras.

Consider formulas in the language of $(\mathbf{K}.\mathbf{t}_N)^2_{Sg}$, for which we now choose diamonds as primitives instead of boxes. More exactly, the basic modal connectives for this logic will be $\Diamond_i$, $\blacklozenge_i$ for $i = \pm 1, \ldots, \pm N$, and also $\bigcirc$, $\bigcirc_1$. The language of $(\mathbf{K}.\mathbf{t}_N)^2$ is denoted by $\mathscr{L}_N t$, the language of $(\mathbf{K}_N)^2_{Sg}$ by $\mathscr{L}_N S$, and the language of $(\mathbf{K}.\mathbf{t}_N)^2_{Sg}$ by $\mathscr{L}_N tS$.

For basic facts about relation algebras we address the reader to the contemporary monograph (Hirsch and Hodkinson 2002). Here we recall only few things.

In the signature $(\cup, \cap, -, \mathbf{0}, \mathbf{1}, \circ, ^{-1} \delta)$ relation algebras form a variety denoted by RA and axiomatized by the identities

$$
\begin{aligned}
&(RA1)\ (x \cup y) \circ z = (x \circ z) \cup (y \circ z),\\
&(RA2)\ (x \cup y)^{-1} = x^{-1} \cup y^{-1},\\
&(RA3)\ (x \circ y) \circ z = x \circ (y \circ z),\\
&(RA4)\ x \circ \delta = x,\\
&(RA5)\ (x^{-1})^{-1} = x,\\
&(RA6)\ (x \circ y)^{-1} = y^{-1} \circ x^{-1},\\
&(RA7)\ x^{-1} \circ (-(x \circ y)) \leqslant -y,
\end{aligned}
$$

together with the boolean identities.

The *full relation algebra* $Rel(W)$ over a set W consists of all binary relations on W with the standard boolean operations and also the composition, the converse, and the diagonal relation I_W. Full relation algebras generate the variety RRA of *representable relation algebras*. Both equational theories $Eq(\mathsf{RA})$, $Eq(\mathsf{RRA})$ are undecidable (Tarski 1941), and $Eq(\mathsf{RRA})$ is not finitely axiomatizable Monk (1964).

One can also consider the corresponding varieties generated by finite algebras, RA_f generated by finite relation algebras and RRA_f generated by finite full relation algebras. The corresponding equational theories $Eq(\mathsf{RA}_f)$, $Eq(\mathsf{RRA}_f)$ are also undecidable (Andréka et al. 1997) and not finitely axiomatizable (cf. Hirsch and Hodkinson 2002). Moreover, one can prove that all four mentioned theories are different (cf. Hirsch and Hodkinson 2002 for further details).

However, in Gabbay and Shehtman (2000, 2002) we identified a certain decidable common fragment of these theories obtained by translation from products of modal logics. Let us now extend the translation defined in Gabbay and Shehtman (2002) to $\mathscr{L}_N tS$-formulas.

Definition 12.8.1 Every $\mathscr{L}_N tS$-formula A is translated into a relation algebra term A^∇ in variables $p_1, p_2, \ldots, r_1, \ldots, r_N$ defined by induction:

$$
\begin{aligned}
&p_k^\nabla = p_k;\\
&\bot^\nabla = \mathbf{0};\\
&(A \to B)^\nabla = (-A^\nabla) \cup B^\nabla;\\
&(\Diamond_i A)^\nabla = r_i \circ A^\nabla;\\
&(\Diamond_{-i} A)^\nabla = r_i^{-1} \circ A^\nabla;\\
&(\blacklozenge_i A)^\nabla = A^\nabla \circ r_i^{-1};\\
&(\blacklozenge_{-i} A)^\nabla = A^\nabla \circ r_i;\\
&(\bigcirc_1 A)^\nabla = \mathbf{1} \circ (A^\nabla \cap \delta);\\
&(\bigcirc A)^\nabla = (A^\nabla)^{-1},
\end{aligned}
$$

where $1 \leqslant i \leqslant N$.

$\mathbf{K.t}_N$-frames are also called *N-temporal*; they are characterized by the conditions $R_{-i} = R_i^{-1}$ (for $i = 1, \ldots, N$).

Lemma 12.8.2 *Let $F = (W, R_1, \ldots, R_N, R_1^{-1}, \ldots, R_N^{-1})$ be an N-temporal frame, and consider a Kripke model $M = (F^2_{Sg}, \theta)$ over its Segerberg square. Consider the algebra $Rel(W)$ with an assignment μ sending the relation variables r_i to R_i, p_j to $\theta(p_j)$. Then for any $u, v \in W$ and for any $\mathscr{L}_N S$-formula A*

$$M, (x, y) \vDash A \text{ iff } (x, y) \in |A^\nabla|_\mu,$$

where $|A^\nabla|_\mu$ denotes the value of the term A^∇ in $Rel(W)$ under μ.

Proof Routine, by induction on the length of A, cf. Gabbay and Shehtman (2000), Lemma 9.2. Consider only some cases.

1. $A = \Diamond_{-i} B$. Then

$$M, (x, y) \vDash A \text{ iff } \exists z (x R_i^{-1} z \;\&\; (z, y) \vDash B) \text{ iff } \exists z (x R_i^{-1} z \;\&\; (z, y) \in |B^\nabla|_\mu) \text{ by IH} \\ \text{iff } (x, y) \in R_i^{-1} \circ |B^\nabla|_\mu = |A^\nabla|_\mu.$$

2. $A = \blacklozenge_i B$. Then

$$M, (x, y) \vDash A \text{ iff } \exists z (z R_i^{-1} y \;\&\; (x, z) \vDash B) \text{ iff } \exists z (z R_i^{-1} y \;\&\; (x, z) \in |B^\nabla|_\mu) \text{ by IH} \\ \text{iff } (x, y) \in |B^\nabla|_\mu \circ R_i^{-1} = |A^\nabla|_\mu.$$

3. $A = \bigcirc_1 B$. Then

$$M, (x, y) \vDash A \text{ iff } (y, y) \vDash B) \text{ iff } \exists z ((x, z) \in \mathbf{1} \;\&\; (z, y) \in \delta \;\&\; M, (z, y) \vDash B) \\ \text{iff } \exists z ((x, z) \in \mathbf{1} \;\&\; (z, y) \in \delta \;\&\; (z, y) \in |B^\nabla|_\mu) \text{ by IH } \text{ iff } (x, y) \in \mathbf{1} \circ (\delta \cap |B^\nabla|_\mu) = |A^\nabla|_\mu.$$

4. $A = \bigcirc B$. Then

$$M, (x, y) \vDash A \text{ iff } M, (y, x) \vDash B \text{ iff } (y, x) \in |B^\nabla|_\mu \text{ by IH } \text{ iff } (x, y) \in (|B^\nabla|_\mu)^{-1} = |A^\nabla|_\mu.$$

⊠

Theorem 12.8.3 *For any $\mathscr{L}_N t$-formula A the following conditions are equivalent:*

(i) $\mathbf{K.t}_N^2 \vdash A$;
(ii) $\mathsf{RA} \models A^\triangledown = \mathbf{1}$;
(iii) $\mathsf{RRA} \models A^\triangledown = \mathbf{1}$.

Proof Everything has been already proved in Gabbay and Shehtman (2000, 2002), and it remains to make few remarks.

(i) ⇒ (ii). Cf. the proof of Theorem 9.4 from Gabbay and Shehtman (2000).

(ii) ⇒ (iii) is trivial.

(iii) ⇒ (i). Suppose $\mathbf{K.t}_N^2 \nvdash A$. Then for some N-temporal frame F and for some Kripke model M over F^2 $M \not\models A$. Hence by Lemma 12.8.2, $|A^\triangledown|_\mu \neq \mathbf{1}$ in the full relation algebra over the set of worlds of F. ⊠

Theorem 12.8.4 *For any $\mathscr{L}_N tS$-formula A the following conditions are equivalent:*

(i) $(\mathbf{K.t}_N)^2_{Sg} \vdash A$;
(ii) $\mathsf{RA} \models A^\triangledown = \mathbf{1}$;
(iii) $\mathsf{RRA} \models A^\triangledown = \mathbf{1}$.

Proof Along the same lines as the previous theorem. Again (ii) ⇒ (iii) is trivial, and for (iii) ⇒ (i) one can use Lemma 12.8.2. So it remains to check (i) ⇒ (ii).

For this purpose we first recall the corresponding part of the proof of Theorem 9.4 from Gabbay and Shehtman (2000).

We use the following axiomatization of $\mathbf{K.t}_N^2$.

Axioms

Classical tautologies,
$\Diamond_i(p \vee q) \leftrightarrow \Diamond_i p \vee \Diamond_i q$,
$\blacklozenge_i(p \vee q) \leftrightarrow \blacklozenge_i p \vee \blacklozenge_i q$,
$\neg\Diamond_i\bot$,
$\neg\blacklozenge_i\bot$,
$\Diamond_{-i}\Box_i p \to p$,
$\blacklozenge_{-i}\blacksquare_i p \to p$,
$\Diamond_i\blacklozenge_j p \leftrightarrow \blacklozenge_j\Diamond_i p$.

Inference rules

Substitution,
Modus Ponens,
Replacement : $\vdash A \leftrightarrow B \Rightarrow \vdash \Diamond_i A \leftrightarrow \Diamond_i B, \quad \blacklozenge_i A \leftrightarrow \blacklozenge_i B$.

where $i, j = \pm 1, \dots, \pm N$.

So we have (cf. Gabbay and Shehtman 2000)

(1) If A is an axiom, then $\mathsf{RA} \models A^\triangledown = \mathbf{1}$,

(2) If A/B is an inference rule and $\mathsf{RA} \models A^\nabla = \mathbf{1}$, then $\mathsf{RA} \models B^\nabla = \mathbf{1}$.

To axiomatize the logic $(\mathbf{K.t}_N)^2_{Sg} = [\mathbf{K.t}_N, \mathbf{K.t}_N]^{\circledast}$ we can extend the above axiomatization of $(\mathbf{K.t}_N)^2$ by the Replacement rule for $\bigcirc$ and $\bigcirc_1$, the axioms of **SL** for $\bigcirc$ and $\bigcirc_1$:

$$
\begin{aligned}
&(AS1)\ \bigcirc(p \vee q) \leftrightarrow \bigcirc p \vee \bigcirc q,\\
&(AS2)\ \bigcirc_1(p \vee q) \leftrightarrow \bigcirc_1 p \vee \bigcirc_1 q,\\
&(AS3)\ \bigcirc\neg p \leftrightarrow \neg \bigcirc p,\\
&(AS4)\ \bigcirc_1\neg p \leftrightarrow \neg \bigcirc_1 p.
\end{aligned}
$$

and also $(Sg1)$, $(Sg2)$, $(Sg4)$, $(Sg5), \ldots, (Sg8)$. Then the axiom $(AS3)$ becomes redundant (see Appendix).

The claim (2) for the Replacement rule is checked easily, so we have to prove the claim (1) for the extra axioms.

Let us begin with simple cases.

For $A = (AS1)$ we have $A^\nabla = ((p \cup q)^{-1} \Leftrightarrow (p^{-1} \cup q^{-1}))$, where $\Leftrightarrow$ denotes the boolean equivalence operation. So $A^\nabla = \mathbf{1}$ follows from $(p \cup q)^{-1} = p^{-1} \cup q^{-1}$, which is $(RA2)$.

For $A = (Sg1)$ we have $A^\nabla = ((p^{-1})^{-1} \Leftrightarrow p)$. So $A^\nabla = \mathbf{1}$ follows from $(p^{-1})^{-1} = p$, which is $(RA5)$.

For $(Sg5)$ we use an equivalent dual form: $\blacklozenge_i p \leftrightarrow \bigcirc \Diamond_i \bigcirc p$. For this we need the identity

$$p \circ r_i^{-1} = (r_i \circ p^{-1})^{-1}$$

readily following from $(RA5)$ and $(RA6)$.

The remaining axioms involve the connective $\bigcirc_1$, and the direct proofs of their RA-correspondents are rather tedious. To simplify the job we choose another strategy: we prove the relational algebra analogues for the axioms of a more powerful logic.

Consider the language $\mathscr{L}_N St\exists_1\eth$ obtained from $\mathscr{L}_N St$ by adding the unary modality $\exists_1$ (the 'horizontal existence') and the propositional constant $\eth$ (the 'diagonal'). Their interpretations in square Kripke models are fixed:

$$
\begin{aligned}
&(x, y) \models \exists_1 A \text{ iff } \exists z\, (z, y) \models A,\\
&(x, y) \models \eth \text{ iff } x = y.
\end{aligned}
$$

We also use the derived modality $\forall_1 := \neg\exists_1\neg$. The translation $^\nabla$ is extended to this language as follows:

$$
\begin{aligned}
&(\exists_1 A)^\nabla := \mathbf{1} \circ A^\nabla,\\
&\eth^\nabla := \delta.
\end{aligned}
$$

Consider the logic $(\mathbf{K.t}_N U)^2_{Sg}\delta$ extending $(\mathbf{K.t}_N)^2$ in $\mathscr{L}_N St\exists_1\eth$. We add the above discussed $(AS1)$, $(Sg1)$, $(Sg5)$, and also the following axioms

$(A\exists 1)\ \exists_1(p \vee q) \leftrightarrow \exists_1 p \vee \exists_1 q,$
$(A\exists 2)\ \neg\exists_1\bot,$
$(A\exists 3)\ p \to \exists_1 p,$
$(A\exists 4)\ \exists_1\exists_1 p \to \exists_1 p,$
$(A\exists 5)\ \exists_1\forall_1 p \to p,$
$(A\exists 6)\ \Diamond_i p \to \exists_1 p,$
$(A\exists 7)\ \blacklozenge_i\exists_1 p \leftrightarrow \exists_1\blacklozenge_i p,$
$(A\delta 1)\ \delta \wedge \exists_1(\delta \wedge p) \to p,$
$(A\delta 2)\ \exists_1\delta,$
$(A\delta 3)\ \delta \wedge p \to \bigcirc p,$
$(A\delta 4)\ \Diamond_{-i}\delta \to \blacklozenge_i\delta,$
$(A\delta 5)\ \bigcirc_1 p \leftrightarrow \exists_1(\delta \wedge p).$

for $i = \pm 1, \ldots, \pm N$, and the Replacement rules for $\bigcirc$ and $\exists_1$:

$$\frac{A \leftrightarrow B}{\exists_1 A \leftrightarrow \exists_1 B}.$$

The reader can recognize the axioms of **S5** in $(A\exists 1), \ldots, (A\exists 5)$, while $(A\exists 6)$ states that $\exists_1$ is the existential modality along the first coordinate. $(A\delta 1)$, $(A\delta 2)$ state that along the first coordinate we have a hybrid logic with the nominal δ. The validity of $(A\delta 3)$ and $(A\delta 4)$ in Segerberg squares is almost obvious, and $(A\delta 5)$ is valid by definition. These observations imply

Lemma 12.8.5 *If* $(\mathbf{K.t}_N U)^2_{Sg}\delta \vdash A$ *and* F *is an* N*-temporal Kripke frame, then* $F^2_{Sg} \models A$.

This lemma is stated here only for the reader's convenience; in fact it also follows from the considerations below.

Lemma 12.8.6 *If* $(\mathbf{K.t}_N U)^2_{Sg}\delta \vdash A$*, then* $\mathsf{RA} \models A^\nabla = \mathbf{1}$.

Proof In view of Theorem 12.8.3 it remains to check the claim for the extra axioms $(A\exists 1), \ldots, (A\exists 6), (A\delta_1), \ldots, (A\delta 4)$.

$(A\exists 1)^\nabla = \mathbf{1}$ is equivalent to the standard identity $\mathbf{1} \circ (p \cup q) = \mathbf{1} \circ p \cup \mathbf{1} \circ q$ (see Appendix).

For $(A\exists 2)$ we use the standard identity $\mathbf{1} \circ \mathbf{0} = \mathbf{0}$.

$(A\exists 3)$ is equivalent to $p \leqslant \mathbf{1} \circ p$. This inequality is obtained from $\delta \leqslant \mathbf{1}$ by composing with p from the right and using $\delta \circ p = p$ (see Appendix).

For $(A\exists 4)$ compose the inequality $\mathbf{1} \circ \mathbf{1} \leqslant \mathbf{1}$ with p from the right and use the associativity $(RA3)$.

For $(A\exists 5)$ use the standard identity $\mathbf{1}^{-1} = \mathbf{1}$ (see Appendix) and apply $(RA7)$:

$$\mathbf{1}^{-1} \circ (-(\mathbf{1} \circ (-p))) \leqslant -(-p) = p.$$

For $(A\exists 6)$ compose the inequality $r_i \leqslant \mathbf{1}$ with p from the right.

For $(A\exists 7)$ we need the identity

$$(\mathbf{1} \circ p) \circ r_i^{-1} = \mathbf{1} \circ (p \circ r_i^{-1})$$

following from $(RA3)$.

For $(A\delta 1)$ we have to check the inequality

$$(*) \qquad \delta \cap (\mathbf{1} \circ (\delta \cap p)) \leqslant p.$$

To show this, first note that

$$(**) \qquad x \cap (\mathbf{1} \circ y) \leqslant x \circ y^{-1} \circ y.$$

In fact the latter inequality is a ${}^{\nabla}$-translation of the temporal logic theorem (see Appendix):

$$p \wedge \blacklozenge^{-1}\top \to \blacklozenge^{-1}\blacklozenge p.$$

Now by applying $(**)$ to $x = \delta,\ y = \delta \cap p$ we have

$$\delta \cap (\mathbf{1} \circ (\delta \cap p)) \leqslant \delta \circ (\delta \cap p)^{-1} \circ (\delta \cap p) = (\delta \cap p)^{-1} \circ (\delta \cap p) \leqslant \delta^{-1} \circ (\delta \cap p) = \delta \cap p \leqslant p.$$

This proves $(*)$.

For $(A\delta 2)$ we use $\mathbf{1} \circ \delta = \mathbf{1}$ following from $(RA4)$.

For $(A\delta 3)$ we have to check

$$\delta \cap p \leqslant p^{-1},$$

This follows from the standard property of relation algebras (see Appendix)

$$(* * *) \qquad x \leqslant \delta \Rightarrow x = x^{-1},$$

which we apply to $x = \delta \cap p$ and then note that $(\delta \cap p)^{-1} \leqslant p^{-1}$.

For $(A\delta 4)$ we have to check

$$r_i^{-1} \circ \delta = \delta \circ r_i^{-1}.$$

This follows from $(RA4)$ and its consequence (see Appendix):

$$\delta \circ x = x.$$

Finally for $A = (A\delta 5)$ the identity $A^{\nabla} = \mathbf{1}$ holds in RA by definition. ⊠

Now to complete the proof of the theorem we need to show

Lemma 12.8.7 $(\mathbf{K.t}_N U)^2_{Sg}\delta \vdash A$ *for* $A = (AS2), (AS4), (Sg2), (Sg4), (Sg6), (Sg7), (Sg8)$.

Proof 1. $(AS2)$ is equivalent to

$$\exists_1(\partial \wedge (p \vee q)) \leftrightarrow \exists_1(\partial \wedge p) \vee \exists_1(\partial \wedge q).$$

To prove this formula in $(\mathbf{K}.\mathbf{t}_N U)^2_{Sg}$ one can apply the axiom $(A\exists 1)$, the boolean distributivity, and the Replacement rule.

2. $(AS4)$ is equivalent to

$$\exists_1(\partial \wedge \neg p) \leftrightarrow \neg\exists_1(\partial \wedge p).$$

To prove the left-to-right implication, we first apply $(A\delta 1)$:

$$\exists_1(\partial \wedge \neg p) \rightarrow (\partial \rightarrow \neg p).$$

Hence by **S5** we obtain

$$\exists_1(\partial \wedge \neg p) \rightarrow \forall_1(\partial \rightarrow \neg p).$$

Then note that $\forall_1(\partial \rightarrow \neg p)$ is equivalent to $\neg\exists_1(\partial \wedge p)$.

For the converse implication we apply the axiom $(A\delta 2)$:

$$\exists_1 \partial,$$

hence by classical logic and Replacement for $\exists_1$ we obtain

$$\exists_1((\partial \wedge p) \vee (\partial \wedge \neg p)),$$

and then by $(A\exists_1)$

$$\exists_1(\partial \wedge p) \vee \exists_1(\partial \wedge \neg p),$$

which implies

$$\neg\exists_1(\partial \wedge p) \rightarrow \exists_1(\partial \wedge \neg p).$$

3. $(Sg2)$ is equivalent to

$$\exists_1(\partial \wedge p) \leftrightarrow \exists_1(\partial \wedge \exists_1(\partial \wedge p)).$$

For the left-to right direction note that

$$\partial \wedge p \rightarrow \exists_1(\partial \wedge p)$$

follows from $(A\exists 3)$; hence we have

$$\partial \wedge p \rightarrow \partial \wedge \exists_1(\partial \wedge p)$$

by classical logic, and next

$$\exists_1(\partial \wedge p) \to \exists_1(\partial \wedge \exists_1(\partial \wedge p)).$$

by the monotonicity of $\exists_1$ (admissible in **S5**).

For the converse implication we apply $(A\delta 1)$:

$$\partial \wedge \exists_1(\partial \wedge p) \to p.$$

Hence by classical logic

$$\partial \wedge \exists_1(\partial \wedge p) \to \partial \wedge p,$$

and then by the monotonicity of $\exists_1$

$$\exists_1(\partial \wedge \exists_1(\partial \wedge p)) \to \exists_1(\partial \wedge p).$$

4. $(Sg4)$ is equivalent to

$$\exists_1(\partial \wedge p) \leftrightarrow \exists_1(\partial \wedge \bigcirc p).$$

For the left-to-right implication we apply $(A\delta 3)$

$$\partial \wedge p \to \bigcirc p,$$

hence by classical logic

$$\partial \wedge p \to \partial \wedge \bigcirc p,$$

and so by the monotonicity of $\exists_1$

$$\exists_1(\partial \wedge p) \to \exists_1(\partial \wedge \bigcirc p).$$

For the converse implication we substitute $\bigcirc p$ for p:

$$\exists_1(\partial \wedge \bigcirc p) \to \exists_1(\partial \wedge \bigcirc \bigcirc p),$$

and then apply $(Sg1)$

$$\bigcirc \bigcirc p \leftrightarrow p$$

and Replacement.

5. $(Sg6)$

$$\bigcirc_1 \Box_i(\blacksquare_i p \to \bigcirc_2 p)$$

is equivalent to

$$\exists_1(\partial \wedge \Box_i(\blacksquare_i p \to \bigcirc_2 p)).$$

To obtain this we first prove

(♯) $$\partial \to \Box_i(\blacksquare_i p \to \bigcirc_2 p).$$

By **K.t** we have

$$\partial \to \Box_i \Diamond_{-i} \partial,$$

and by $(A\delta 4)$ and the monotonicity for $\Box_i$

$$\Box_i \Diamond_{-i} \partial \to \Box_i \blacklozenge_i \partial.$$

Thus

(♯1) $$\partial \to \Box_i \blacklozenge_i \partial.$$

Now let us show

(♯2) $$\blacklozenge_i \partial \wedge \blacksquare_i p \to \bigcirc_2 p.$$

Note that by **K** we have

$$\blacklozenge_i \partial \wedge \blacksquare_i p \to \blacklozenge_i(\partial \wedge p).$$

So (♯2) follows from

(♯3) $$\blacklozenge_i(\partial \wedge p) \to \bigcirc_2 p.$$

By the definition of $\bigcirc_2$ and $(Sg5)$ this is equivalent to

(♯4) $$\bigcirc \Diamond_i \bigcirc (\partial \wedge p) \to \bigcirc \exists_1 (\partial \wedge p).$$

To show the latter we first note that

$$\bigcirc \Diamond_i \bigcirc (\partial \wedge p) \to \bigcirc \exists_1 \bigcirc (\partial \wedge p)$$

by $(A\exists 6)$ and the admissible monotonicity rule for $\bigcirc$. So (♯4) reduces to

(♯5) $$\bigcirc \exists_1 \bigcirc (\partial \wedge p) \to \bigcirc \exists_1 (\partial \wedge p).$$

To obtain (♯5) it is sufficient to prove

(♯6) $$\partial \wedge p \to \bigcirc(\partial \wedge p).$$

and use monotonicity and $(Sg1)$. But (♯6) is equivalent to

$$\partial \wedge \partial \wedge p \to \bigcirc(\partial \wedge p),$$

following readily from $(A\delta 3)$. This completes the proof of (♯2).

Note that (♯2) is equivalent to

$$\blacklozenge_i \eth \rightarrow (\blacksquare_i p \rightarrow \bigcirc_2 p);$$

hence by monotonicity

$$(\sharp 7) \qquad \Box_i \blacklozenge_i \eth \rightarrow \Box_i (\blacksquare_i p \rightarrow \bigcirc_2 p).$$

Now (♯7) and (♯1) imply (♯).

Finally note that by classical logic (♯) implies

$$\eth \rightarrow \eth \wedge \Box_i (\blacksquare_i p \rightarrow \bigcirc_2 p).$$

and by monotonicity

$$\exists_1 \eth \rightarrow \exists_1 (\eth \wedge \Box_i (\blacksquare_i p \rightarrow \bigcirc_2 p)).$$

Here the premise is $(A\delta 2)$ and the consequent is equivalent to $(Sg6)$ as we have noticed above.

6. $(Sg7)$ is equivalent to

$$\exists_1 (\eth \wedge p) \rightarrow \Box_i \exists_1 (\eth \wedge p).$$

This reduces to

$$\exists_1 (\eth \wedge p) \rightarrow \forall_1 \exists_1 (\eth \wedge p).$$

and

$$\forall_1 \exists_1 (\eth \wedge p) \rightarrow \Box_i \exists_1 (\eth \wedge p).$$

The first implication follows by **S5** and the second is a substitution instance of

$$\forall_1 p \rightarrow \Box_i p$$

following from $(A\exists 6)$ by contraposition.

7. $(Sg8)$ is equivalent to

$$\blacklozenge_i \bigcirc_1 p \leftrightarrow \bigcirc_1 \blacklozenge_i \bigcirc_1 p.$$

By $(A\delta 5)$ $\bigcirc_1 p$ is equivalent to a formula beginning with $\exists_1$, so the above formula follows from

$$(\ddagger) \qquad \blacklozenge_i \exists_1 p \leftrightarrow \bigcirc_1 \blacklozenge_i \exists_1 p.$$

By $(A\exists 7)$ and Replacement this is equivalent to

(‡1)$\exists_1 \blacklozenge_i p \leftrightarrow \bigcirc_1 \exists_1 \blacklozenge_i p$, (‡1)

so it suffices to prove

(‡2)$\exists_1 p \leftrightarrow \bigcirc_1 \exists_1 p$.(‡2)

The left-to-right-direction follows from two implications

$$\exists_1 p \rightarrow \forall_1 \exists_1 p$$

and

$$\forall_1 \exists_1 p \rightarrow \bigcirc_1 \exists_1 p.$$

The first one is obtained by **S5**. For the second one it suffices to prove

(‡3) $$\forall_1 p \rightarrow \bigcirc_1 p,$$

which is equivalent to

(‡4)$\forall_1 p \rightarrow \exists_1(\eth \wedge p)$.(‡4)

To prove this note that

$$\exists_1 \eth \rightarrow (‡4)$$

follows by **K**, while $\exists_1 \eth$ is $(A\delta 2)$.

The right-to-left implication in (‡2) is equivalent to

$$\exists_1(\exists_1 p \wedge \eth) \rightarrow \exists_1 p.$$

The latter is obtained in two steps:

$$\exists_1(\exists_1 p \wedge \eth) \rightarrow \exists_1 \exists_1 p$$

(by classical logic and monotonicity) and

$$\exists_1 \exists_1 p \rightarrow \exists_1 p$$

(by **S5**). ⊠

⊠

Theorem 12.8.8 *For any $\mathscr{L}_N S$-formula A the following conditions are equivalent:*

(i) $(\mathbf{K}_N)^2_{Sg} \vdash A$;
(ii) $\mathsf{RA} \models A^\triangledown = \mathbf{1}$;
(iii) $\mathsf{RRA} \models A^\triangledown = \mathbf{1}$,
(iv) $\mathsf{RA}_f \models A^\triangledown = \mathbf{1}$,

(v) $\mathsf{RRA}_f \vDash A^\nabla = \mathbf{1}$.

Proof $(\mathbf{K}_N)^2_{Sg}$ is contained in $(\mathbf{K.t}_N)^2_{Sg}$, so the implication $(i) \Rightarrow (ii)$ follows from the previous theorem.

The implications $(ii) \Rightarrow (iii)$, $(ii) \Rightarrow (iv)$, $(iii) \Rightarrow (v)$, $(iv) \Rightarrow (v)$ are trivial.

Finally $(v) \Rightarrow (i)$ follows from the square fmp of $(\mathbf{K}_N)^2_{Sg}$. In fact, by the square fmp, if $(\mathbf{K}_N)^2_{Sg} \nvdash A$, then A is refuted in a Kripke model M over F^2_{Sg} for some finite N-modal frame F; it can also be regarded as an N-temporal frame. Then by Lemma 12.8.2, $Rel(W)$ refutes $A^\nabla = \mathbf{1}$, where W is the set of worlds of F; therefore $\mathsf{RRA}_f \nvDash A^\nabla = \mathbf{1}$. ⊠

Corollary 12.8.9 *An identity of the form $A^\nabla = B^\nabla$, where A, B are $\mathscr{L}_N S$-formulas, holds over* RA *iff it holds over* RRA *iff it holds over* RA_f *iff it holds over* RRA_f.

The fragment of Eq(RA) consisting of these identities is decidable.

Proof In fact, $A^\nabla = B^\nabla$ iff $(A \leftrightarrow B)^\nabla = \mathbf{1}$. So we can apply Theorem 12.8.8. ⊠

12.9 Adding the Reflexivity Axiom

In this section we consider Segerberg squares of polymodal $\mathbf{T}$ and show that they are reducible to Segerberg squares of polymodal $\mathbf{K}$.

Let us first recall the embedding of $\mathbf{T}_N$ in $\mathbf{K}_N$.

Definition 12.9.1 The translation $A \mapsto A^\odot$ from $\mathscr{L}_N$ to $\mathscr{L}_N$ is defined by induction:

$$p_j^\odot := p_j, \ \bot^\odot := \bot,$$
$$(A \to B)^\odot := A^\odot \to B^\odot,$$
$$(\Box_i A)^\odot := \Box_i A^\odot \wedge A^\odot.$$

Definition 12.9.2 For a Kripke frame $F = (W, R_1, \dots, R_N)$ let $F^\odot := (W, R_1^\odot, \dots, R_N^\odot)$, where $R_i^\odot := R_i \cup I_W$ is the reflexive closure of R_i (I_W is the diagonal relation). For a Kripke model $M = (F, \varphi)$ put $M^\odot := (F^\odot, \varphi)$.

Proposition 12.9.3 *For any N-modal formula A, frame F and Kripke model M*

$$M \vDash A^\odot \textit{ iff } M^\odot \vDash A,$$
$$F \vDash A^\odot \textit{ iff } F^\odot \vDash A.$$

Proof By induction on the length of A it follows that for any world x

$$M, x \vDash A^\odot \text{ iff } M^\odot, x \vDash A.$$

E.g. in the case $A = \Box_i B$:

$$M, x \vDash A^\odot \text{ iff } \forall y \in R_i^\odot(x)\ M, y \vDash B^\odot \text{ iff } \forall y \in R_i^\odot(x)\ M^\odot, y \vDash B \text{ (by IH) iff } M^\odot, x \vDash A.$$

Hence

$$M \vDash A^{\odot} \text{ iff } M^{\odot} \vDash A.$$

As valuations in F and $F^{\odot}$ are the same, it follows that

$$F \vDash A^{\odot} \text{ iff } F^{\odot} \vDash A.$$

⊠

Recall that $\mathbf{T}_N$ is determined by reflexive N-trees (i.e., reflexive closures of N-trees).

Proposition 12.9.4 *For any N-modal formula A*

$$\mathbf{T}_N \vdash A \textit{ iff } \mathbf{K}_N \vdash A^{\odot}.$$

Proof By Proposition 12.9.3 we have

$$\mathbf{K}_N \vdash A^{\odot} \text{ iff (for any } N\text{-tree } F \ \ F \vDash A^{\odot}) \text{ iff (for any } N\text{-tree } F \ \ F^{\odot} \vDash A)$$
$$\text{iff (for any reflexive } N\text{-tree } F \ \ F \vDash A) \text{ iff } \mathbf{T}_N \vdash A.$$

⊠

Now let us apply the same method to $(\mathbf{T}_N^2)_{Sg}$.

Definition 12.9.5 The translation $A \mapsto A^{\odot}$ is extended to $\mathscr{L}_N S$ by the following items

$$(\bigcirc A)^{\odot} := \bigcirc A^{\odot},$$
$$(\bigcirc_1 A)^{\odot} := \bigcirc_1 A^{\odot}.$$

Definition 12.9.6 For a $\mathbf{K}_N * \mathbf{SL}_2$-frame $F = (W, R_1, \ldots, R_N, f_0, f_1)$ put $F^{\odot} := (W, R_1^{\odot}, \ldots, R_N^{\odot}, f_0, f_1)$; for a Kripke model $M = (F, \varphi)$ put $M^{\odot} := (F^{\odot}, \varphi)$.

Proposition 12.9.7 *For any $\mathscr{L}_N S$-formula A, $\mathbf{K}_N * \mathbf{SL}_2$-frame F and Kripke model M over F*

$$M \vDash A^{\odot} \textit{ iff } M^{\odot} \vDash A,$$
$$F \vDash A^{\odot} \textit{ iff } F^{\odot} \vDash A.$$

Proof As in Proposition 12.9.3, by induction we show that

$$M, x \vDash A^{\odot} \text{ iff } M^{\odot}, x \vDash A.$$

The proof is the same, with trivial new cases $A = \bigcirc B$, $A = \bigcirc_1 B$. E.g. for $A = \bigcirc B$ we have

$$M, x \vDash A^{\odot} \text{ iff } M, f_0(x) \vDash B^{\odot} \text{ iff } M^{\odot}, f_0(x) \vDash B \text{ (by IH) iff } M^{\odot}, x \vDash A.$$

⊠

Theorem 12.9.8 *For any N-modal formula A*

$$(\mathbf{T}^2_N)_{Sg} \vdash A \textit{ iff } (\mathbf{K}^2_N)_{Sg} \vdash A^{\odot}.$$

Proof We have on the one hand, by Proposition 12.9.7 and Theorem 12.4.6

$$(\mathbf{K}^2_N)_{Sg} \vdash A^{\odot} \text{ iff (for any } N\text{-tree } F \;\; F^2_{Sg} \vDash A^{\odot}) \text{ iff (for any } N\text{-tree } F \;\; (F^2_{Sg})^{\odot} \vDash A).$$

On the other hand,

$$(\mathbf{T}^2_N)_{Sg} \vdash A \text{ iff (for any reflexive } N\text{-tree } F \; F^2_{Sg} \vDash A) \text{ iff (for any } N\text{-tree } F \; (F^{\odot})^2_{Sg}) \vDash A).$$

Now note that

$$(F^2_{Sg})^{\odot} = (F^{\odot})^2_{Sg}.$$

In fact, let $F = (W, R_1, \dots, R_N)$. Then

$$F^{\odot} = (W, R_1^{\odot}, \dots, R_N^{\odot}),\;\; (F^{\odot})^2_{Sg} = (W^2, R_{11}^{\odot}, \dots, R_{N1}^{\odot}, \sigma_0, \sigma_1),$$

where

$$(x, y)R_{i1}^{\odot}(x', y') \text{ iff } x R_i^{\odot} x' \;\&\; y = y' \text{ iff } (x R_i x' \vee x = x') \;\&\; y = y'$$

(see Definition 12.2.33);

$$(F^2_{Sg})^{\odot} = (W^2, R_{11}, \dots, R_{N1}, \sigma_0, \sigma_1)^{\odot} = (W^2, (R_{11})^{\odot}, \dots, (R_{N1})^{\odot}, \sigma_0, \sigma_1),$$

where

$$(x, y)(R_{i1})^{\odot}(x', y') \text{ iff } x R_i x' \;\&\; y = y' \vee x = x' \;\&\; y = y',$$

and it is obvious that $R_{i1}^{\odot} = (R_{i1})^{\odot}$. ⊠

Theorem 12.9.9 $(\mathbf{T}^2_N)_{Sg}$ *has the square fmp, and moreover,*

$$(\mathbf{T}^2_N)_{Sg} = \mathbf{L}(\{F^2_{Sg} \mid F \textit{ is a finite reflexive } N\textit{-tree}\}).$$

Proof The inclusion $\subseteq$ is obvious.

For the converse, suppose $(\mathbf{T}^2_N)_{Sg} \nvdash A$; then by the previous theorem $(\mathbf{K}^2_N)_{Sg} \nvdash A^{\odot}$. By Theorem 12.7.6 $F^2_{Sg} \nvDash A^{\odot}$ for some finite N-tree F; hence by Proposition 12.9.7 $(F^2_{Sg})^{\odot} \nvDash A$, i.e., $(F^{\odot})^2_{Sg} \nvDash A$ (see the proof of Theorem 12.7.6). ⊠

12.10 Adding the Seriality Axiom

Now let us consider Segerberg squares of polymodal **D**; we will also reduce them to Segerberg squares of polymodal **K**.

Again we begin with an embedding of $\mathbf{D}_N$ in $\mathbf{K}_N$ (probably not so well-known as in the case of **T**).

Definition 12.10.1 Let $F = (W, R_1, \dots, R_N)$ be an N-tree of height m. We define its *top-reflexive closure* $F^\Diamond := (W, R_1^\Diamond, \dots, R_N^\Diamond)$, where

$$xR_i^\Diamond y := xR_i y \vee x = y \,\&\, h(x) = m.$$

Lemma 12.10.2 *Let F be an N-tree of finite height m, $M = (F, \varphi)$ a Kripke model over F, $M^\Diamond = (F^\Diamond, \varphi)$. Then for any $A \in \mathscr{L}_N$, for any $x \in F$, $md(A) + h(x) \leqslant m$ implies*

$$M, x \vDash A \Leftrightarrow M^\Diamond, x \vDash A.$$

Proof By induction on the length of A, similar to Lemma 12.2.21. Note that adding reflexivity for the top points does not influence the truth of proposition letters. ⊠

For an $\mathscr{L}_N$-formula A put

$$\Box^{<n} A := \bigwedge_{k=0}^{n-1} \Box^k A, \ \Diamond^{<n} A := \bigvee_{k=0}^{n-1} \Diamond^k A.$$

Proposition 12.10.3 *For any $\mathscr{L}_N$-formula A*

$$\mathbf{D}_N \vdash A \textit{ iff } \mathbf{K}_N \vdash \Box^{<md(A)}(\Diamond_1\top \wedge \ldots \wedge \Diamond_N\top) \to A.$$

Proof 'If' is trivial, since $\mathbf{D}_N \vdash \Box^n(\Diamond_1\top \wedge \ldots \wedge \Diamond_N\top)$.

To prove 'only if' suppose $\mathbf{K}_N \nvdash \Box^{<md(A)}(\Diamond_1\top \wedge \ldots \wedge \Diamond_N\top) \to A$, $md(A) = m$. Then $\Box^{<m}(\Diamond_1\top \wedge \ldots \wedge \Diamond_N\top) \to A$ is refuted at the root λ of a standard N-tree F. So there is a model M over F such that $M, \lambda \vDash \Box^{<m}(\Diamond_1\top \wedge \ldots \wedge \Diamond_N\top) \wedge \neg A$. Note that

$$md(\Box^{<m}(\Diamond_1\top \wedge \ldots \wedge \Diamond_N\top) \wedge \neg A) = m.$$

So by the truncation Lemma 12.2.21 we obtain

$$M^-, \lambda \vDash \Box^{<m}(\Diamond_1\top \wedge \ldots \wedge \Diamond_N\top) \wedge \neg A,$$

where M^- is the restriction of M to $F^- = F^{(m)}$, and next by Lemma 12.10.2

$$(M^-)^\Diamond, \lambda \vDash \Box^{<m}(\Diamond_1\top \wedge \ldots \wedge \Diamond_N\top) \wedge \neg A.$$

Finally note that the frame $(F^-)^\Diamond$ is serial (for all its relations):

$$\lambda \vDash \Box^{<m}(\Diamond_1 \top \wedge \ldots \wedge \Diamond_N \top)$$

implies that every sequence of length $< m$ has $\sqsubset_i$-successors, while the top points of length m are reflexive by the definition of the top-reflexive closure. Therefore $\mathbf{D}_N \nvDash A$. ⊠

Now we return to Segerberg squares. If F is an N-tree and $(x, y) \in F^2$, we define

$$h(x, y) := \max(h(x), h(y)).$$

Lemma 12.10.4 *Let F be an N-tree of finite height m, $M = (F^2_{Sg}, \varphi)$ a Kripke model over F^2_{Sg}, $M^\Diamond = ((F^\Diamond)^2_{Sg}, \varphi)$.*

Then for any $A \in \mathcal{L}_N S$, for any $z \in F^2$, $d_\Box(A) + h(z) \leqslant m$ implies

$$M, z \vDash A \Leftrightarrow M^\Diamond, z \vDash A.$$

Proof By induction, the same as in the two-dimensional truncation Lemma 12.6.6. ⊠

Lemma 12.10.5 *Let F be an N-tree with root u such that*

$$F, u \vDash \Box^{<m}(\Diamond_1 \top \wedge \ldots \wedge \Diamond_N \top),$$

(1) If $h(F) = m$, then $F^\Diamond \vDash \mathbf{D}_N$.
(2) If $F, u \vDash \Box^{m+1}\bot$, then $h(F) = m$.

Proof (1) Since $u \vDash \Box^k(\Diamond_1 \top \wedge \ldots \wedge \Diamond_N \top)$ for $k < m$, we have $x \vDash \Diamond_1 \top \wedge \ldots \wedge \Diamond_N \top$ for any x of height $< m$. Now in $F^\Diamond$ we still have $x \vDash \Diamond_i \top$ for $h(x) < m$, and the points of height m become reflexive for all relations, so we obtain $F^\Diamond \vDash \mathbf{D}_N$.

(2) $u \vDash \Box^{m+1}\bot$ means that $h(F) \leqslant m$ (Lemma 12.2.11). As we have noted, $x \vDash \Diamond_1 \top \wedge \ldots \wedge \Diamond_N \top$ for $h(x) < m$, so x has successors for all relations of F. Now by induction it follows that there exist points of any height $\leqslant m$, thus $h(F) = m$. ⊠

Theorem 12.10.6 *For any $\mathcal{L}_N S$-formula A*

$$(\mathbf{D}_N)^2_{Sg} \vdash A \text{ iff } (\mathbf{K}_N)^2_{Sg} \vdash \Box^{<d_\Box(A)}(\Diamond_1 \top \wedge \ldots \wedge \Diamond_N \top) \wedge \blacksquare^{<d_\Box(A)}(\blacklozenge_1 \top \wedge \ldots \wedge \blacklozenge_N \top) \to A.$$

Proof Along the same lines as Proposition 12.10.3.

For 'if' note again that $\mathbf{D}_N \vdash \Box^n(\Diamond_1 \top \wedge \ldots \wedge \Diamond_N \top)$.

To prove 'only if' suppose $d_\Box(A) = m$,

$$(\mathbf{K}_N)^2_{Sg} \nvdash \Box^{<m}(\Diamond_1 \top \wedge \ldots \wedge \Diamond_N \top) \wedge \blacksquare^{<m}(\blacklozenge_1 \top \wedge \ldots \wedge \blacklozenge_N \top) \to A.$$

Then by Proposition 12.4.7 there is a standard N-tree F and a model M over F^2_{Sg} such that

$$M, (8,9) \models \Box^{<m}(\Diamond_1\top \wedge \ldots \wedge \Diamond_N\top) \wedge \blacksquare^{<m}(\blacklozenge_1\top \wedge \ldots \wedge \blacklozenge_N\top) \wedge \neg A.$$

By Lemma 12.6.6

$$M^-, (8,9) \models \Box^{<m}(\Diamond_1\top \wedge \ldots \wedge \Diamond_N\top) \wedge \blacksquare^{<m}(\blacklozenge_1\top \wedge \ldots \wedge \blacklozenge_N\top) \wedge \neg A,$$

where M^- is the restriction of M to $(F^-)^2$ (and now $F^- := F^{(m+1)}$); Lemma 12.6.6 is applicable, since

$$d_\Box(\Box^{<m}(\Diamond_1\top \wedge \ldots \wedge \Diamond_N\top) \wedge \blacksquare^{<m}(\blacklozenge_1\top \wedge \ldots \wedge \blacklozenge_N\top) \wedge \neg A) = m$$

and $|(8,9)| = 1$. Then by Lemma 12.10.4

$$(M^-)^\Diamond, (8,9) \models \Box^{<m}(\Diamond_1\top \wedge \ldots \wedge \Diamond_N\top) \wedge \blacksquare^{<m}(\blacklozenge_1\top \wedge \ldots \wedge \blacklozenge_N\top) \wedge \neg A.$$

Now

$$(8,9) \models \Box^{<m}(\Diamond_1\top \wedge \ldots \wedge \Diamond_N\top)$$

implies

$$8 \models \Box^{<m}(\Diamond_1\top \wedge \ldots \wedge \Diamond_N\top)$$

in F^- (and $F^-{\uparrow}8$). So by Lemma 12.10.5(1), $(F^-{\uparrow}8)^\Diamond = (F^-)^\Diamond{\uparrow}8 \models \mathbf{D}_N$.

Similarly,

$$(8,9) \models \blacksquare^{<m}(\blacklozenge_1\top \wedge \ldots \wedge \blacklozenge_N\top)$$

implies $(F^-)^\Diamond{\uparrow}9 \models \mathbf{D}_N$.

To make the whole frame $(F^-)^\Diamond$ serial, we add reflexivity (under all relations) to the remaining part of $(F^-)^\Diamond$ and obtain a model M' over a frame G^2_{Sg}, with G validating $\mathbf{D}_N$. By the generation lemma this transformation of $(M^-)^\Diamond$ into M' does not affect the truth value of A at $(8,9)$, and thus $G^2_{Sg} \not\models A$. Therefore $(\mathbf{D}_N)^2_{Sg} \nvdash A$. ⊠

Theorem 12.10.7 $(\mathbf{D}_N)^2_{Sg}$ *has the square fmp.*

Proof Suppose $(\mathbf{D}_N)^2_{Sg} \nvdash A$, $d_\Box(A) = m$. Then by the (trivial part of) the Theorem 12.10.6

$$(\mathbf{K}_N)^2_{Sg} \nvdash \Box^{<m}(\Diamond_1\top \wedge \ldots \wedge \Diamond_N\top) \wedge \blacksquare^{<m}(\blacklozenge_1\top \wedge \ldots \wedge \blacklozenge_N\top) \to A.$$

For the latter implication B we have $d_\Box(B) = m$. Hence by Lemma 12.6.7

$$(\mathbf{K}_N)^2_{Sg} + \Box^{m+1}\bot \nvdash B,$$

which implies (since $(\mathbf{K}_N)^2_{Sg} + \Box^{m+1}\bot \vdash \blacksquare^{m+1}\bot$)

$$(\mathbf{K}_N)^2_{Sg} \nvdash \Box^{m+1}\bot \wedge \blacksquare^{m+1}\bot \wedge \Box^{<m}(\Diamond_1\top \wedge \ldots \wedge \Diamond_N\top) \wedge \blacksquare^{<m}(\blacklozenge_1\top \wedge \ldots \wedge \blacklozenge_N\top) \to A.$$

So by Theorem 12.7.6 there is a finite standard N-tree F, successors x, y of its root λ and a model M over F^2_{Sg} such that

$$M, (x, y) \vDash \Box^{m+1}\bot \wedge \blacksquare^{m+1}\bot \wedge \Box^{<m}(\Diamond_1\top \wedge \ldots \wedge \Diamond_N\top) \wedge \blacksquare^{<m}(\blacklozenge_1\top \wedge \ldots \wedge \blacklozenge_N\top) \wedge \neg A.$$

Now

$$(x, y) \vDash \Box^{m+1}\bot \wedge \Box^{<m}(\Diamond_1\top \wedge \ldots \wedge \Diamond_N\top)$$

implies

$$x \vDash \Box^{m+1}\bot \wedge \Box^{<m}(\Diamond_1\top \wedge \ldots \wedge \Diamond_N\top)$$

in F (and $F{\uparrow}x$). So by Lemma 12.10.5 $h(F{\uparrow}x) = m$ and $(F{\uparrow}x)^{\Diamond} \vDash \mathbf{D}_N$ Similarly,

$$(x, y) \vDash \blacksquare^{m+1}\bot \wedge \blacksquare^{<m}(\blacklozenge_1\top \wedge \ldots \wedge \blacklozenge_N\top)$$

implies $h(F{\uparrow}y) = m$ and $(F{\uparrow}y)^{\Diamond} \vDash \mathbf{D}_N$.

By Lemma 12.10.4 we also obtain

$$M^{\Diamond}, (x, y) \vDash \neg A.$$

(the lemma is applicable as we have $|(x, y)| = 1$ and $h(F) = m + 1$, so $d_{\Box}(A) + |(x, y)| = h(F)$).

Now as in the previous theorem we transform the frame $F^{\Diamond}$ into a serial frame G by making all points beyond $F{\uparrow}x$ and $F{\uparrow}y$ reflexive under all relations. Respectively we obtain a model M' over G^2_{Sg} refuting A. ⊠

12.11 Some Consequences for Classical First-Order Theories

In Shehtman (2012) we defined a translation from Segerberg squares to a certain fragment of classical first-order logic. The results from the previous two sections allows us to extend it to first-order theories of reflexive or serial relations with binary predicates.

Let us briefly recall the details. We consider the first-order language (with equality) $\mathscr{L}1_N$ with the binary predicate letters $R_1, \ldots, R_N$; $\mathscr{L}1^2_N$ is the expansion of $\mathscr{L}1_N$ by a countable set of extra binary predicate letters $P_1, P_2, \ldots$

Definition 12.11.1 Every $\mathscr{L}_N S$-formula A translates into a $\mathscr{L}1^2_N$-formula $A^2(x_1, x_2)$ with (maybe dummy) parameters x_1, x_2:

$(p_i)^2(x_1, x_2) := P_i(x_1, x_2)$,
$(\bot)^2(x_1, x_2) := \bot$,
$(A \to B)^2(x_1, x_2) := A^2(x_1, x_2) \to B^2(x_1, x_2)$,
$(\Box_i A)^2(x_1, x_2) := \forall y\ (R_i(x_1, y) \to A^2(y, x_2))$,

$(\bigcirc A)^2(x_1, x_2) := A^2(x_2, x_1),$
$(\bigcirc_1 A)^2(x_1, x_2) := A^2(x_2, x_2),$

where y is a new variable that does not occur in $A^2(x_1, x_2)$.

Proposition 12.11.2 *Let Λ be a Kripke complete Δ-elementary N-modal logic, with $\mathbf{V}(\Lambda) = Mod(\Theta)$ for a first-order theory Θ and suppose that Λ^2_{Sg} has the square fmp. Then for any $\mathscr{L}_N S$-formula A the following conditions are equivalent:*

(1) $\Lambda^2_{Sg} \vdash A$,
(2) $\Theta \vdash \forall x_1 \forall x_2 A^2(x_1, x_2)$ *in classical first-order logic,*
(3) $\forall x_1 \forall x_2 A^2(x_1, x_2)$ *is true in any finite model of* Θ *(in the language* $\mathscr{L}1^2_N$*).*

Proof First note that for a frame $F = (W, \rho_1, \dots, \rho_N)$, there is a bijection between Kripke models over M over F^2_{Sg} and $\mathscr{L}1^2_N$-structures of the form $\mu = (F, \Pi_1, \Pi_2, \dots)$, where Π_i is an interpretation of P_i, such that

$$(*) \qquad M, (a, b) \vDash A \text{ iff } \mu \vDash A^2(a, b)$$

for any proposition letter A. Then the same equivalence holds for any $\mathscr{L}_N S$-formula A; this is easily proved by induction.

(1) $\Rightarrow$ (2). Suppose $\Lambda^2_{Sg} \vdash A$, and let $\mu = (F, \Pi_1, \Pi_2, \dots)$ be an $\mathscr{L}1^2_N$-structure, where F is an N-frame, Π_i are binary predicates interpreting P_i and $\mu \vDash \Theta$. Then $F \in Mod(\Theta)$, so $F \vDash \Lambda$, $F^2_{Sg} \vDash \Lambda^2_{Sg}$, and thus $M \vDash A$ for any Kripke model M over F^2_{Sg}. Hence $\mu \vDash \forall x_1 \forall x_2 A^2(x_1, x_2)$ by (*). Now (2) follows by Gödel completeness theorem.

(2) $\Rightarrow$ (3) follows by soundness of classical logic.

(3) $\Rightarrow$ (1). Suppose $\forall x_1 \forall x_2 A^2(x_1, x_2)$ holds in any finite model of Θ, and let F be a finite frame from $\mathbf{V}(\Lambda) = Mod(\Theta)$. For any Kripke model M over F^2_{Sg} consider the corresponding $\mathscr{L}1^2_N$-structure μ; then $\mu \vDash \Theta$, so $\mu \vDash \forall x_1 \forall x_2 A^2(x_1, x_2)$, and hence $M \vDash A$ by (*). Thus $F^2_{Sg} \vDash A$, and therefore $\Lambda^2_{Sg} \vdash A$ by the square fmp. ⊠

Applying this to the logics from the previous two sections we obtain

Theorem 12.11.3 *(1) Let Ref_N be the classical first-order theory of N reflexive relations and countably many binary predicates. Then for any A $\in \mathscr{L}_N S$*

$$(\mathbf{T}_N)^2_{Sg} \vdash A \textit{ iff } Ref_N \vdash \forall x_1 \forall x_2 A^2(x_1, x_2) \textit{ iff}$$
$$\forall x_1 \forall x_2 A^2(x_1, x_2) \textit{ is true in any finite model of } Ref_N.$$

(2) Let Ser_N be the classical first-order theory of N serial relations and countably many binary predicates. Then for any A $\in \mathscr{L}_N S$

$$(\mathbf{D}_N)^2_{Sg} \vdash A \textit{ iff } Ser_N \vdash \forall x_1 \forall x_2 A^2(x_1, x_2) \textit{ iff}$$
$$\forall x_1 \forall x_2 A^2(x_1, x_2) \textit{ is true in any finite model of } Ser_N.$$

12.12 Conclusion

This paper gives answers to some questions from Shehtman (2012): it proves the square fmp for Segerberg squares of polymodal **T** and **D**, and also establishes connection between relation algebras and Segerberg squares.

Another new thing is application of bisimulation games to the proofs of local tabularity of truncated fragments, which gives an alternative and (as we think) a simpler proof of the fmp for $\mathbf{K}^2_{Sg}$. This method can be applied to other kinds of logics, see the forthcoming paper Shehtman (2018).

Here are new open questions.

1. Theorem 12.6.12 gives an upper bound for the modal depth of truncated squares Λ_m. What is the precise value of this depth? This question does not seem difficult, one should only look closer at winning strategies, cf. Shehtman (2018).

2. Do there exist similar results for truncated fragments of $(\mathbf{T}_N)^2_{Sg}$ and $(\mathbf{D}_N)^2_{Sg}$? We suppose the answer is positive and the proof is again not very difficult.

3. However, our method obviously fails for Segerberg squares of minimal temporal logics, and even for their fragment, $\mathbf{KB}^2_{Sg}$. Is this logic decidable? What are the properties of corresponding truncated fragments? Here is a related question: are the truncated fragments of $\mathbf{KB}^2$ locally tabular?

4. Theorem on the square fmp can probably be extended to Segerberg squares of the logics $\mathbf{K}_N + \Box^m\bot$, and moreover, to Segerberg squares of any consistent logic of the form $\mathbf{K}_N + \Box^m\bot + C$, where C is closed. In general, if Λ is axiomatizable by a single closed formula, even the properties of Λ^2 are unknown. So there are questions: is Λ^2 decidable for a consistent Λ with a single closed axiom? Does it have the fmp?

Acknowledgements I would like to thank the anonymous referee for useful comments on the first version of the manuscript.
The research presented in this paper was done in part within the framework of the Basic Research Program at National Research University Higher School of Economics and was partially supported within the framework of a subsidy by the Russian Academic Excellence Project 5-100. It was also supported by the RFBR project 16-01-00615 and by the Russian president project NSh-9091.2016.1.

Appendix

In this part we prove a simple syntactic fact from modal logic and recall the proof of some standard identities in relation algebras (for the readers who are less familiar with the subject).

Proposition 1 $\mathbf{K} + \Diamond^2 p \leftrightarrow p \vdash \Box p \leftrightarrow \Diamond p$, *and thus*
$\mathbf{K} + \Diamond^2 p \leftrightarrow p \vdash \neg\Diamond p \leftrightarrow \Diamond\neg p$.

Proof In this logic we have $\Diamond^2\top \leftrightarrow \top$, i.e., $\Diamond^2\top$. So it contains **D** and proves $\Box p \to \Diamond p$.

For the converse we first obtain $\Box^2 p \leftrightarrow p$ (by substituting $\neg p$ for p in the axiom). So $p \to \Box^2 p$, and by substitution, $\Diamond p \to \Box^2 \Diamond p$. From $\Box p \to \Diamond p$ by substitution and monotonicity we have $\Box^2 \Diamond p \to \Box \Diamond^2 p$. Thus $\Diamond p \to \Box \Diamond^2 p$, and by applying the axiom, $\Diamond p \to \Box p$. ⊠

Proposition 2 *The following identities and quasiidentities hold in* **RA***:*

(1) $x \leqslant y \Rightarrow x^{-1} \leqslant y^{-1}$,
(2) $\mathbf{1}^{-1} = \mathbf{1}$,
(3) $z \circ (x \cup y) = (z \circ x) \cup (z \circ y)$,
(4) $\delta^{-1} = \delta$,
(5) $\delta \circ x = x$,
(6) $x \leqslant y \Rightarrow z \circ x \leqslant z \circ y$,
(7) $x \circ \mathbf{0} = \mathbf{0}$,
(8) $x \cap (\mathbf{1} \circ y) \leqslant x \circ y^{-1} \circ y$,
(9) $x \leqslant y \Rightarrow x \circ z \leqslant y \circ z$,
(10) $x \leqslant \delta \Rightarrow x = x^{-1}$.

Proof (1) $x \leqslant y$ implies $y = x \cup y$, and so $y^{-1} = (x \cup y)^{-1} = x^{-1} \cup y^{-1}$ by $(RA2)$. Hence $x^{-1} \leqslant y^{-1}$.

(2) $\mathbf{1}^{-1} \leqslant \mathbf{1}$ implies $(\mathbf{1}^{-1})^{-1} \leqslant \mathbf{1}^{-1}$ by (1), and so $\mathbf{1} \leqslant \mathbf{1}^{-1}$ by $(RA5)$.

(3) From $(RA1)$, $(RA2)$, $(RA6)$ we have:

$$z^{-1} \circ (x^{-1} \cup y^{-1}) = (z^{-1} \circ x^{-1}) \cup (z^{-1} \circ y^{-1}).$$

Now replace x, y, z by their converses and use $(RA5)$.

(4) $\delta^{-1} \circ \delta = \delta^{-1}$ by $(RA4)$; hence

$$(\delta^{-1} \circ \delta)^{-1} = (\delta^{-1})^{-1},$$

and so

$$\delta^{-1} \circ \delta = \delta,$$

by $(RA6)$ and $(RA5)$. Thus $\delta^{-1} = \delta$.

(5) $(RA5)$ and $(RA6)$ imply $\delta^{-1} \circ x^{-1} = x^{-1}$. Then replace x by x^{-1} and use (4) and $(RA5)$.

(6) Replace y by $x \cup y$ and apply (3).

(7) Obviously,

$$\mathbf{0} \leqslant -(x \circ \mathbf{1}).$$

Hence by (6) and $(RA7)$

$$x^{-1} \circ \mathbf{0} \leqslant x^{-1} \circ (-(x \circ \mathbf{1})) \leqslant -\mathbf{1} = \mathbf{0}.$$

Now replace x by x^{-1}.

(8) As mentioned in the proof of Lemma 12.8.6, this is a translation of the **K.t**-theorem

$$p \wedge \blacklozenge^{-1}\top \rightarrow \blacklozenge^{-1}\blacklozenge p.$$

To prove the latter, we first obtain

$$p \wedge \blacklozenge^{-1}\top \rightarrow \blacksquare^{-1}\blacklozenge p \wedge \blacklozenge^{-1}\top$$

from a temporal axiom, and then apply the **K**-theorem

$$\blacksquare^{-1}q \wedge \blacklozenge^{-1}\top \rightarrow \blacklozenge^{-1}q.$$

(9) Similar to (6)

(10) Apply (5), (9) and (8) for $y = \delta$:

$$x = x \cap (\delta \circ x) \leqslant x \cap (\mathbf{1} \circ x) \leqslant x \circ x^{-1} \circ x.$$

So for $x \leqslant \delta$ we have by monotonicity and (6), (9)

$$x \leqslant \delta \circ x^{-1} \circ \delta \leqslant x^{-1}.$$

Hence by taking the converses we obtain

$$x^{-1} \leqslant x,$$

and thus $x^{-1} = x$. ⊠

References

Andréka, H., Givant, V., & Németi, I. (1997). *Decision problems for equational theories of relation algebras* (Vol. 126), Memoirs of the American Mathematical Society. Providence: American Mathematical Society.

Gabbay, D., & Shehtman, V. (1998). Products of modal logics, part 1. *Logic Journal of the IGPL*, *6*, 73–146.

Gabbay, D., & Shehtman, V. (2000). Products of modal logics, part 2: Relativised quantifiers in classical logic. *Logic Journal of the IGPL*, *8*, 165–210.

Gabbay, D., & Shehtman, V. (2002). Products of modal logics, part 3: Products of modal and temporal logics. *Studia Logica*, *72*, 157–183.

Gabbay, D. M., Kurucz, A., Wolter, F., & Zakharyaschev, M. (2003). *Many-dimensional modal logics: Theory and applications*. Amsterdam: Elsevier.

Goranko, V., & Otto, M. (2007). Model theory of modal logic. In P. Blackburn, J. Van Benthem, & F. Wolter (Eds.), *Handbook of modal logic* (pp. 249–329). Amsterdam: Elsevier.

Hirsch, R., & Hodkinson, I. (2002). *Relation algebras by games*. Amsterdam: Elsevier.

Maksimova, L. L. (1975). Modal logics of finite slices. *Algebra and logic*, *14*(3), 304–319.

Maksimova, L. L. (1981). O lokalno konechnykh mnogoobraziyakh psevdobulevykh algebr. In *6th All-Union Algebraical Conference (abstracts)*, No. 1, Leningrad, pp. 99–100, [On locally finite varieties of pseudoboolean algebras].
Malcev, A. I. (1973). *Algebraic systems*. Berlin: Springer.
Monk, D. (1964). On representable relation algebras. *Michigan Mathematical Journal*, *11*, 207–210.
Segerberg, K. (1967). On the logic of tomorrow. *Theoria*, *33*, 45–52.
Segerberg, K. (1973). Two-dimensional modal logic. *Journal of Philosophical Logic*, *2*, 77–96.
Shehtman, V. B. (1978). Two-dimensional modal logics. *Mathematical Notes*, *23*, 417–424.
Shehtman, V. (2011). O kvadratakh modalnykh logik s dopolnitelnymi svyazkami. *Works of Steklov Mathematical Institute (Trudy)*, *274*, 343–351. [On squares of modal logics with additional connectives].
Shehtman, V. B. (2012). Squares of modal logics with additional connectives. *Russian Mathematical Surveys*, *67*(4), 721–778.
Shehtman, V. B. (2018). Bisimulation games in modal and intuitionistic logic, *in preparation*.
Tarski, A. (1941). On the calculus of relations. *Journal of Symbolic Logic*, *6*, 73–89.

Chapter 13
On Algebraisation of Superintuitionistic Predicate Logics

Dmitry Tishkovsky

Abstract Algebraisation is a process of translating the syntax and deductive properties of given logic to an algebraic language. While propositional logics fit various algebraisation frameworks reasonably well, algebraisation of first-order logics has many difficulties. Recently, a class of quasicylindric algebras was introduced and investigated. It was proved that each superintuitionistic predicate logic is strongly complete with respect to a variety of quasicylindric algebras. In this paper, we prove that the semantics of pseudo-boolean models of Rasiowa and Sikorski and the Kripke semantics is subsumed by the semantics of quasicylindric algebras. We also expand results obtained by Larisa Maksimova on algebraisation of non-classical propositional logics to the case of first-order superintuitionistic logics. We consider such deductive properties of first-order superintuitionistic logics as the Beth property and the projective Beth property, the Craig interpolation property, the disjunctive and existential properties. We formulate algebraic equivalents which correspond to these properties in the language of varieties of quasicylindric algebras and establish equivalences of the logical properties and their algebraic counterparts.

13.1 Introduction

It is known that there are infinitely many logics. The more so, we have to embrace the fact that there are uncountable families of logics. For example, the set of all intermediate logics has a cardinality of continuum (Jankov 1968). Fortunately, there are criteria which help researchers to choose between logics. For instance, a decidable logic is usually preferred over an undecidable one if both logics can be used to approach given set of problems. A logic with decidability problem in a lower complexity class will be favoured more than a logic with higher complexity.

Apart from computational properties of logics there are properties which characterise provability relations of logics. Because such logical properties reflect the basic characteristics of logics, investigating them is very important. Examples of

D. Tishkovsky (✉)
School of Computer Science, The University of Manchester, Manchester, UK
e-mail: dmitry@cs.man.ac.uk

S. Odintsov (ed.), *Larisa Maksimova on Implication, Interpolation, and Definability*, Outstanding Contributions to Logic 15, https://doi.org/10.1007/978-3-319-69917-2_13

such properties include variants of Craig interpolation and Beth definability properties (Maksimova 1986, 1995, 1998), disjunctive and existential property, as well as admissibility of derivation rules (Rybakov 1997).

Formalisms of non-classical logics can be very different in terms of syntax and properties of their relations of logical consequence. Families of modal and description logics, temporal and dynamic logics, fuzzy logics, paraconsistent logics, logics of conditionals, etc. are orthogonal each to others and allow reasoning about aspects of applications from very different perspectives. Because of such an extreme diversity of families of non-classical logics it is very difficult to develop a theory of each logical family from only the logical syntax and the logical consequence relation. This is one of main reasons why logicians try to supply semantical formalisms for logics they investigate. Semantics is, in fact, a translation of the language of given logic to some other formalism and appropriate analogues of Gödel completeness theorem for the logic ensure that the consequence relation of the logic is an invariant of this translation.

It is essential that semantics have a well-developed theory and provides an easier way of reasoning about some aspects of the logic. As a consequence, many logical semantics are represented in formalisms of Model Theory or Universal Algebra.

Representing logical formalisms in the language of Universal Algebra and studying them by algebraic methods form the area of Algebraic Logic. Algebraic Logic takes its start from the work by Tarski (1935) where Tarski introduced a notion of a *formula algebra*. He showed that the logical equivalence on the set of all propositional formulae induces a congruence on the formula algebra. He proved that the quotient of the formula algebra by this congruence forms a Boolean algebra and all the classical tautologies coincide with formulae which are equivalent to the top element of the algebra. Such algebras are now called *algebras of Lindenbaum-Tarski* and play a central role in Algebraic Logic.

The Tarski method have been further applied to algebraisation of modal, superintuitionistic and other non-classical logics (Rasiowa and Sikorski 1963; Pinter 1973; Rasiowa 1974; Pinter 1975; Andréka et al. 1984; Blok and Pigozzi 1989; Németi 1991; Tishkovsky 1999; Czelakowski 2001; Czelakowski and Pigozzi 2004; Czelakowski and Jansana 2014) and Tarski lead a wide research program on algebraisation of first-order logic (Henkin et al. 1971).

Important contributions to Algebraic Logic have been made by Larisa Maksimova. In her famous works (e.g. Maksimova 1977) on Craig interpolation properties she proved an equivalence of the interpolation property of a superintuitionistic propositional logic and the amalgamation property for a corresponding variety of Heyting algebras. This allowed to establish that the number of superintuitionistic propositional logics with an interpolation property is seven. This result, for almost forty years, remains a perfect example of application of Algebraic Logic in the area of non-classical logics and is an inspiration of other researchers working in the area (see e.g. Hoogland 1997; Madarász 1995; Tishkovsky 2001). The result was further expanded to many other logics such as modal and temporal logics, and other logical properties were considered (see Gabbay and Maksimova 2005).

Our aims in this paper is twofold. First, we will take a look at algebraisation of superintuitionistic first-order logics from the point of view of other semantics for these logics. And, second, we are going to show how to lift the results on equivalence of algebraic and deductive properties from a propositional language to its extension with first-order quantifiers.

The structure of the paper is as follows. In Sect. 13.2 we give definitions and describe notations of standard notions used in the paper. Section 13.3 presents basic fact on quasicylindric algebras (Tishkovsky 1999). We present unpublished results on interrelation of the sematics of quasicylindric algebras with an algebraic semantics of Rasiowa and Sikorski (1963) and the Kripke semantics of possible worlds in Sect. 13.4. In Sect. 13.5, we consider various deductive properties of superintuitionistic predicate logics and formulate their algebraic equivalents in the language of quasicylindric algebras. We conclude with future directions of research on this topic in Sect. 13.6.

13.2 Definitions and Notations

Let, as usual, ω be the smallest infinite ordinal, that is, $\omega \stackrel{\text{def}}{=} \{0, 1, \dots, n, \dots\}$ be the set of all natural numbers (finite ordinals) and $\aleph_0$ be the cardinality of the ordinal ω.

We consider first-order languages that contain neither functional symbols nor equal-sign. Each set of subject variables of any considered language is assumed to be equal to a *fixed* countable set $\mathrm{Var} = \{x_i \mid i < \omega\}$. Since under these conditions any first-order language is defined by its set of predicate symbols, we use same notation for a first-order language and its set of predicate symbols. Let P be a set of predicate symbols. A mapping $\# : P \to \omega$ maps each p from P to its arity, that is each p from P is $\#p$-ary predicate symbol. Denote by $\mathrm{For}(P)$ the set of all formulae of the language P. For any formula A let $\mathrm{FV}(A)$ be the set of all free variables of the formula A and $P(A)$ be the set of all predicate symbols occurred in the formula A. We denote by $s^i_j A$ the result of replacement of all free occurrences of the variable x_i in A for the variable x_j provided that this replacement is admissible, that is there are no free occurrences of x_i under quantifiers on x_j in A. For any formulae A and B the expression $A \leftrightarrow B$ is an abbreviation of the formula $(A \to B) \wedge (B \to A)$.

Let Pr be a *fixed* set of predicate symbols that contain a countable set of n-ary predicate symbols for each natural number n. A *superintuitionistic logic* is a set of formulae in the language Pr which contains intuitionistic predicate logic and is closed under the substitution rule (see e.g. Tishkovsky 1999), *modus ponens* and the generalisation rule: $A \vdash \forall x_i A$ ($A \in \mathrm{For}(\mathrm{Pr})$ and $i < \omega$). In this paper, we consider only superintuitionistic logics and usually omit the word 'superintuitionistic'.

Let L be a logic, P be a set of predicate symbols. Assume that a set T of formulae in the language P is closed under modus ponens and the generalization rule and contain all substitutional instances in the language P of formulae of the logic L. We call any such set T an *L-theory* in the language P.

It is not difficult to see that the set of all substitutional instances of formulae of any logic L in the language P is the smallest L-theory in the language P. We denote this theory by $L(P)$. Clearly, every logic L is itself a smallest L-theory in the language Pr.

Let L be a logic, P be some first-order language and $\Gamma \cup \{A\} \subseteq \mathrm{For}(P)$. We write $\Gamma \vdash_L A$ and say that 'A is derivable from Γ in L' iff the formula A belongs to the smallest L-theory in the language P which contains Γ. If the set Γ is empty, we write $L \vdash A$ and say that 'A is derivable in L'.

Formulae A and B are *congruent* (we write: $A \sim B$) iff A and B can be obtained one from the other by applying the well-known principle of replacement of bounded variables (see Ershov and Palyutin 1986).

For a formulae A and a sequence of different predicate symbols $p_0, \ldots, p_m$, we write $A[p_0, \ldots, p_m]$ iff all the predicate symbols of A can be found among $p_0, \ldots, p_m$. If q_l is a predicate symbol of the same arity as p_l then the formula obtained from A by replacement of the symbol p_l by q_l is denoted by $A[p_1, \ldots, p_{l-1}, q_l, p_{l+1}, \ldots, p_m]$.

Following a set-theoretic notation we denote a set of all functions from a set V to a set U by U^V.

Given an algebra $\mathscr{A}$ we denote by $|\mathscr{A}|$ the underlying set of the algebra $\mathscr{A}$.

13.3 Quasicylindric Algebras

We denote by σ the similarity type $\langle \wedge, \vee, \top, \bot, \neg, \rightarrow, \forall_i, \exists_i, s^i_j \rangle_{i,j<\omega}$ where $\top$, $\bot$ are propositional constants, $\neg, \forall_i, \exists_i, s^i_j$ are unary and $\wedge, \vee, \rightarrow$ are binary operation symbols. An algebra $\mathscr{A} \overset{\text{def}}{=} \langle A, \sigma \rangle$ is *quasicylinric algebra* iff $\langle A, \wedge, \vee, \rightarrow, \neg, \top, \bot \rangle$ is a Heyting algebra and all the following groups of equalities are valid in $\mathscr{A}$ for all $i, j, k, l < \omega$:

(Q1) $s^i_i X = X$
(Q2) $\forall_i \exists_i X = \exists_i X$
(Q3) $s^i_j \forall_i X = \forall_i X$
(Q4) $\exists_i s^i_j X = s^i_j X \ (i \neq j)$
(Q5) $\exists_i \bot = \bot$
(Q6) $s^i_j \exists_k X = \exists_k s^i_j X \ (k \neq i, j)$
(Q7) $s^i_j \forall_k X = \forall_k s^i_j X \ (k \neq i, j)$
(Q8) $s^i_j s^k_i X = s^k_j s^i_j X$
(Q9) $s^i_j s^k_l X = s^k_l s^i_j X \ (k \neq i, j; l \neq i)$
(Q10) $s^i_j (X \rightarrow Y) = s^i_j X \rightarrow s^i_j Y$
(Q11) $s^i_j (X \wedge Y) = s^i_j X \wedge s^i_j Y$
(Q12) $s^i_j (X \vee Y) = s^i_j X \vee s^i_j Y$
(Q13) $s^i_j \neg X = \neg s^i_j X$
(Q14) $\forall_i (X \rightarrow Y) \leqslant (\forall_i X \rightarrow \forall_i Y)$
(Q15) $\forall_i X \leqslant s^i_j X$
(Q16) $\forall_i (X \rightarrow Y) \leqslant (\exists_i X \rightarrow \exists_i Y)$
(Q17) $s^i_j X \leqslant \exists_i X$
(Q18) $\forall_i \top = \top$

The expression $X \leqslant Y$ is an abbreviation for the equation $X \rightarrow Y = \top$.

For each element a of a quasicylindric algebra we define a set Δa by $\Delta a \overset{\text{def}}{=} \{i < \omega \mid \forall_i a \neq a\}$.

Lemma 1 (Tishkovsky 1999, Statement 5.2) *For all elements a, b of arbitrary quasicylindric algebra the following equalities and inclusions hold* ($i, j < \omega$)*:*

(1) $\Delta\top = \Delta\bot = \varnothing$;
(2) $\Delta(a \to b) \subseteq \Delta a \cup \Delta b$;
(3) $\Delta(a \wedge b) \subseteq \Delta a \cup \Delta b$;
(4) $\Delta(a \vee b) \subseteq \Delta a \cup \Delta b$;
(5) $\Delta\forall_i a \subseteq \Delta a \setminus \{i\}$;
(6) $\Delta\exists_i a \subseteq \Delta a \setminus \{i\}$;
(7) $\Delta s^i_j a \subseteq (\Delta a \setminus \{i\}) \cup \{j\}$.

This lemma implies the following statement.

Proposition 1 *Let* $n \leqslant \omega$ *and* $\langle A, \sigma\rangle$ *be a quasicylindric algebra. The set* $A^{(n)} \stackrel{\text{def}}{=} \{a \in A \mid \mathsf{Card}(\Delta a) < n\}$ *is closed under all the operations of the algebra* $\langle A, \sigma\rangle$*. That is,* $\langle A^{(n)}, \sigma\rangle$ *is a subalgebra of the algebra* $\langle A, \sigma\rangle$*.*

An element a is *finitary* iff Δa is finite. A quasicylindric algebra is *locally finitary* algebra iff all its elements are finitary. For arbitrary class K of algebras we denote by $K^{(\omega)}$ a class of all locally finitary algebras from the class K.

By Proposition 1, the set of all finitary elements of arbitrary quasicylindric algebra is a locally finitary subalgebra of the algebra. We call this subalgebra as *the largest locally finitary subalgebra* of the algebra.

An *valuation* of formulae in algebra $\mathscr{A}$ is a mapping h of first-order formulae into the base set of the algebra $\mathscr{A}$ such that for all formulae A and B and for all $i, j < \omega$ the following conditions are satisfied:

(1) $h\top = \top, h\bot = \bot$;
(2) $h\neg A = \neg hA$;
(3) $h(A \vee B) = hA \vee hB$;
(4) $h(A \wedge B) = hA \wedge hB$;
(5) $h(A \to B) = hA \to hB$;
(6) $h\forall x_i A = \forall_i hA$;
(7) $h\exists x_i A = \exists_i hA$;
(8) if the variable replacement $s^i_j A$ is admissible then $hs^i_j A = s^i_j hA$.

We say that a formula A is *true* under the valuation h iff hA is the greatest element $\top$ of the algebra $\mathscr{A}$. Let K be a class of quasicylindric algebras. A formula A is a *semantic consequence* of the set Γ of formulae with respect to the class K (we write: $\Gamma \models_K A$ and $\models_K A$ if Γ is empty), iff for every algebra from the class K and for every valuation h of formulae in this algebra if all the formulae from Γ are true under h then A is also true under h. If K is a singleton $\{\mathscr{A}\}$ and $\models_K A$ then the formula A is said to be *valid* in the algebra $\mathscr{A}$. A logic L in some language P is *complete with respect to the class* K (or *characterised by the class* K) iff $L = \{A \mid A \in \mathrm{For}(P) \text{ and } \models_K A\}$. A logic L is *strongly complete with respect to the class* K, iff for each formula A and for each set Γ of formulae the equivalence $\Gamma \vdash_L A \iff \Gamma \models_K A$ holds. It is clear that if L is strongly complete with respect to K then L is complete with respect to K.

Lemma 2 (Tishkovsky 1999, Statement 5.3) *Let h be a valuation of formulae into some quasicylindric algebra. For each formula A if $FV(A) = \{x_{i_0}, \dots, x_{i_n}\}$ then $\Delta h A \subseteq \{i_0, \dots, i_n\}$.*

Theorem 5.7 in Tishkovsky (1999) remains true if it is formulated as follows.

Lemma 3 *Assume that, for every n-ary p from P, $y_1^p, \dots, y_n^p$ is a fixed sequence of different variables and $At \stackrel{\text{def}}{=} \{p(y_1^p, \dots, y_n^p) \mid p \in P\}$.*
Let $\mathscr{A}$ be a quasicylindric algebra and h_0 be a mapping from the set At to the base set of the algebra $\mathscr{A}$.

If $\Delta h_0 p(x_{i_1}, \dots, x_{i_n}) \subseteq \{i_1, \dots, i_n\}$ for all $p(x_{i_1}, \dots, x_{i_n})$ from At then h_0 can be uniquely extended to a valuation h of formulae into the algebra $\mathscr{A}$.

Let T be an intuitionistic theory in a language P. For an arbitrary formula A from $\mathrm{For}(P)$ we let $\|A\| \stackrel{\text{def}}{=} \{B \in \mathrm{For}(P) \mid T \ni A \equiv B\}$. Let $F(T) \stackrel{\text{def}}{=} \{\|A\| | A \in \mathrm{For}(P)\}$. The operations of the similarity type σ on $F(T)$ are defined as follows ($i, j < \omega$):

(1) $\top = \|\top\|$ and $\bot = \|\bot\|$;
(2) $\neg\|A\| = \|\neg A\|$;
(3) $\|A\| \to \|B\| = \|A \to B\|$;
(4) $\|A\| \wedge \|B\| = \|A \wedge B\|$;
(5) $\|A\| \vee \|B\| = \|A \vee B\|$;
(6) $\forall_i \|A\| = \|\forall x_i A\|$;
(7) $\exists_i \|A\| = \|\exists x_i A\|$;
(8) $s_j^i \|A\| = \|s_j^i B\|$ where B is a formula which is congruent to the formula A and such that the variable replacement $s_j^i B$ is admissible.

It is easy to prove that the above definition does not depend on representatives of equivalence classes. The algebra $\mathbf{F}(T) \stackrel{\text{def}}{=} \langle F(T), \sigma \rangle$ is called *Lindenbaum–Tarski algebra* of the theory T. A Lindenbaum–Tarski algebra of a logic L in a language P is the Lindenbaum–Tarski algebra of the theory $L(P)$. Note that the mapping $h_T : \mathrm{For}(P) \to F(T)$, which is defined for all $A \in \mathrm{For}(P)$ by the equation $h_T A = \|A\|$ is a valuation of formulae in $\mathbf{F}(T)$. Hence, as it follows from Lemma 2, any Lindenbaum–Tarski algebra is locally finitary.

The main theorem which connects quasicylindric algebras and first-order superintuitionistic logics is the following.

Theorem 1 *Let L be a superintutionistic predicate logic and V_L be the class of all quasicylindric algebras such that all the formulae from L are valid in any algebra from V_L.*

Then V_L is a variety and L is strongly complete with respect to the class $V_L^{(\omega)}$ of all locally finitary quasicylindric algebras from V_L.

13.4 Embeddings of Semantics

In this section, we consider the two most known semantics for superintuitionistic predicate logics: the semantics of pseudo-boolean models of Rasiowa and Sikorski (1963) and the Kripke semantics. We prove that the algebraic semantics of quasicylindric algebras subsumes these semantics. This is established by building quasicylindric algebras from valuation of formulae in given semantical structures.

13.4.1 Algebraic Models of Rasiowa and Sikorski

The semantics of pseudo-boolean models of Rasiowa and Sikorski (1963) provides an interesting alternative to algebraic semantics. In these models, the propositional part of the language is treated purely algebraically within a complete Heyting algebra. The quantifiers and the individual variables, however, are dealt as in Model Theory with use of interpretation of the variables in a domain set. Within the semantics, the existential and universal formula becomes an infinite disjunction or, respectively, an infinite conjunction of all values of the formula under all interpretations of the quantified variable.

A *pseudo-boolean model* (Rasiowa and Sikorski 1963) is a pair $\langle M, D\rangle$ where M is a complete Heyting algebra and D is a non-empty set. A mapping $v : \mathrm{Var} \to D$ is called *variable interpretation* in the model $\langle M, D\rangle$. A *formula valuation* in the model $\langle M, D\rangle$ is a function φ which maps every n-ary predicate symbol p to some function from D^n to M.

Let v be a variable interpretation and φ be a formula valuation in the model $\langle M, D\rangle$. The *value* $\varphi(A[v])$ of the formula A under interpretation v in the model $\langle M, D\rangle$ is defined by induction on the length of A:

(1) $\varphi(\bot[v]) \stackrel{\text{def}}{=} \bot$ and $\varphi(\top[v]) \stackrel{\text{def}}{=} \top$;

(2) $\varphi(p(y_1, \dots, y_n)[v]) \stackrel{\text{def}}{=} \varphi(p)(vy_1, \dots, vy_n)$ for any atomic formula $p(y_1, \dots, y_n)$;

(3) $\varphi((A \wedge B)[v]) \stackrel{\text{def}}{=} \varphi(A[v]) \wedge \varphi(B[v])$;

(4) $\varphi((A \vee B)[v]) \stackrel{\text{def}}{=} \varphi(A[v]) \vee \varphi(B[v])$;

(5) $\varphi((A \to B)[v]) \stackrel{\text{def}}{=} \varphi(A[v]) \to \varphi(B[v])$;

(6) $\varphi(\forall x_i A[v]) \stackrel{\text{def}}{=} \bigwedge\{\varphi(A[w]) \mid wx_j = vx_j \text{ whenever } i \neq j\}$;

(7) $\varphi(\exists x_i A[v]) \stackrel{\text{def}}{=} \bigvee\{\varphi(A[w]) \mid wx_j = vx_j \text{ whenever } i \neq j\}$.

A formula A is *valid* in the model $\langle M, D\rangle$ if $\varphi(A[v]) = \top$ for every variable interpretation v and every formula valuation φ.

A (superintuitionistic predicate) logic L is *complete* with respect to a class K of pseudo-boolean models if L coincide with the set of all formulae (in the language Pr) which are valid in all models from K. It is known that the cardinality of the set of all logics which are not complete with respect to any class of pseudo-boolean models is at least $2^{\aleph_0}$ (Ono 1973).

Let $\langle M, D\rangle$ be a fixed pseudo-boolean model. Let V be a set of all variable interpretations of the model. On the set M^V, it is straightforward to define the following algebraic operations from the similarity type σ. For all $f, g \in M^V$ and $v \in V$ we define: $\bot_v \overset{\text{def}}{=} \bot$, $\top_v \overset{\text{def}}{=} \top$, $(\neg f)_v \overset{\text{def}}{=} \neg f_v$, $(f \wedge g)_v \overset{\text{def}}{=} f_v \wedge g_v$, $(f \vee g)_v \overset{\text{def}}{=} f_v \vee g_v$, $(f \to g)_v \overset{\text{def}}{=} f_v \to g_v$, $(\forall_i f)_v \overset{\text{def}}{=} \bigwedge\{f_w \mid w \in V, wx_j = vx_j \text{ unless } i = j\}$, $(\exists_i f)_v \overset{\text{def}}{=} \bigvee\{f_w \mid w \in V, wx_j = vx_j \text{ unless } i = j\}$. In order to define the operations s^i_j we introduce the following definitions.

For every sequence of variables $\bar{y} = \langle y_i \mid i < \gamma\rangle$ (with $\gamma \leqslant \omega$) and any variable interpretation v we denote by $v\bar{y}$ the sequence $\langle vy_i \mid i < \gamma\rangle$ of all values of variables from the sequence $\bar{y}$ under interpretation v.

A *(functional) form* is any pair $\langle F, \bar{y}\rangle$ where $\bar{y}$ is a (possibly finite) sequence of variables and F is a function from D^γ to M with $\gamma = \mathsf{Card}(\bar{y})$. Let FT be the set of all forms. We say that a form $\langle F, \bar{y}\rangle$ *represents* a function $f \in M^V$ iff $F(v\bar{y}) = f_v$ for every $v \in V$.

A binary relation $\approx$ on the set FT is defined as follows.

$$\langle F, \bar{y}\rangle \approx \langle G, \bar{z}\rangle \overset{\text{def}}{\Longleftrightarrow} F(v\bar{y}) = G(v\bar{z}) \text{ for all } v \in V.$$

Clearly, $\approx$ is an equivalence on FT. Thus, two forms are *equivalent* iff they represent same function from M^V. Given a form $\langle F, \bar{y}\rangle$ from FT an equivalence class of $\langle F, \bar{y}\rangle$ with respect to the relation $\approx$ is denoted by $[\langle F, \bar{y}\rangle]$. Let $FT/\approx \overset{\text{def}}{=} \{[\langle F, \bar{y}\rangle] \mid \langle F, \bar{y}\rangle \in FT\}$.

The following lemma is true.

Lemma 4 *There is a one-to-one correspondence between functions from M^V and elements of the set $FT/\approx$.*

Proof Clearly, any form $\langle F, \bar{y}\rangle$, induces a function f from M^V as follows:

$$f_v \overset{\text{def}}{=} F(v\bar{y}) \text{ for all } v \in V.$$

Furthermore, for each function $f \in M^V$ there is a form $\langle F, \bar{x}\rangle$, where $\bar{x} \overset{\text{def}}{=} \langle x_i \mid i < \omega\rangle$, and F is a function $F : D^\omega \to M$, which is defined as follows.

$$F(d_i \mid i < \omega) \overset{\text{def}}{=} f_v, \text{ where } vx_i = d_i \text{for all} i < \omega.$$

It is easy to define operations s^i_j on the set FT as follows. $s^i_j\langle F, \bar{y}\rangle \overset{\text{def}}{=} \langle F, \bar{y}'\rangle$ where $\bar{y}'$ is obtained from $\bar{y}$ by replacement of all occurrences of the variable x_i for x_j.

It is not difficult to prove the following lemma.

Lemma 5 *Let* $\langle F, \bar{y}\rangle \approx \langle G, \bar{z}\rangle$. *Then* $s^i_j\langle F, \bar{y}\rangle \approx s^i_j\langle G, \bar{z}\rangle$.

Now, with help of Lemma 5, the definition of the operations s^i_j on the set M^V is straightforward. Given $f \in M^V$, let $\langle F, \bar{y}\rangle$ be a form which represents f. The result of application of s^i_j to f is a function represented by the form $s^i_j\langle F, \bar{y}\rangle$.

We have defined all algebraic operations of the similarity type σ on the set M^V. Our current aim is to prove that the algebra $\langle M^V, \sigma\rangle$ is quasicylindric. For this purpose we inroduce the following definitions.

A variable x_i is an *eigenvariable* of a function f from M^V iff there are variable interpretations v and w such that $f_v \neq f_w$ and $vx_j = wx_j$ for all $j \neq i$.

Proposition 2 *Let* $\bar{y}$ *be a sequence of different variables and* $f \in M^V$. *The function* f *can be represented by a form* $\langle F, \bar{y}\rangle$ *(for some* F*) iff* $\bar{y}$ *contains all eigenvariables of* f.

Proof Assume that a variable x_i is not in the sequence $\bar{y}$ and $\langle F, \bar{y}\rangle$ represents f. Then for all variable interpretations v and w which coincide on all variables different from x_i we have $f_v = F(v\bar{y}) = F(w\bar{y}) = f_w$. Thus, x_i is not an eigenvariable of f.

For the converse assume that $\bar{y} = \langle y_n \mid n < \gamma\rangle$ contains all eigenvariables of f. For arbitrary d_n from D we define $F(d_n \mid n < \gamma) \overset{\text{def}}{=} f_v$, where $vy_n = d_n$ $(n < \gamma)$. The definition of F does not depend on the choice of v since $\bar{y}$ contains all eigenvariables of f. It is clear that $\langle F, \bar{y}\rangle$ represents f.

Lemma 6 *For every* $n < \omega$ *and every function* f *in* M^V *the variable* x_i *is not eigenvariable of any of functions* $\forall_i f$ *and* $\exists_i f$.

Proof Let $v, w \in V$. Assume $vx_j = wx_j$ for all $j \neq i$. Then $(\forall_i f)_v = f_w \wedge f_v \wedge \bigwedge\{f_u \mid (\forall j \neq i)(ux_j \neq vx_j \wedge ux_i \neq wx_i)\} = (\forall_i f)_w$. Proof for $\exists_i f$ is similar.

It is now a routine to check all the equalities of quasicylindric algebras and prove the following proposition.

Proposition 3 *The algebra* $\langle M^V, \sigma\rangle$ *is quasicylindric.*

We call every such algebra $\langle M^V, \sigma\rangle$ which is built as above from an pseudo-boolean model $\langle M, V\rangle$ a *quasicylindric Rasiowa-Sikorski algebra.*

The following proposition is true.

Proposition 4 *For every function* f *from* M^V *the set* Δf *coincide with the set of indices of all the eigenvariables of* f.

Proof It follows from Lemma 6 that the indices of all the eigenvariables of f are in Δf. For the converse assume that $\bar{y}$ is a set of all eigenvariables of f. By Proposition 2 f can be represented by a form $\langle F, \bar{y}\rangle$ (for some F). By the definition of operations s^i_j we can obtain that if x_i does not occur in $\bar{y}$ then $s^i_j f = f$. Hence, $i \notin \Delta f$ if x_i is not an eigenvariable of f.

Let φ be a formula valuation in a pseudo-boolean model $\langle M, D\rangle$. We define a formula valuation h_φ in the algebra $\langle M^V, \sigma\rangle$ as follows. For every atomic formula $p(x_0, \ldots, x_{\#p-1})$ we let $h_\varphi p(x_0, \ldots, x_{\#p-1}) \stackrel{\text{def}}{=} [\varphi(p)(x_0, \ldots, x_{\#p-1})]$. By Proposition 4 $\Delta h_\varphi p(x_0, \ldots, x_{\#p-1}) \subseteq \{0, \ldots, \#p-1\}$. Further, by Lemma 3 the mapping h_φ can be expanded to a valuation of formulae in the algebra $\langle M^V, \sigma\rangle$. The following lemma is true for such h_φ and φ.

Lemma 7 $(h_\varphi A)_v = \varphi(A[v])$ *for every formula* A *and every variable interpretation* v.

Proof We prove this by induction on the length of the formula A. Let p be an n-ary predicate symbol. Thus, for the atomic formula $p(x_{i_0}, \ldots, x_{i_{n-1}})$ we have the following sequence of equalities.

$$\begin{aligned}&(h_\varphi p(x_{i_0}, \ldots, x_{i_{n-1}}))_v = \\ &= (h_\varphi s^{0\cdots n-1}_{i_0\cdots i_{n-1}} p(x_0, \ldots, x_{n-1}))_v = (s^{0\cdots n-1}_{i_0\cdots i_{n-1}} h_\varphi p(x_0, \ldots, x_{n-1}))_v = \\ &\qquad = \varphi(p)(vx_{i_0}, \ldots, vx_{i_{n-1}}) = \varphi(p(x_{i_0}, \ldots, x_{i_{n-1}})[v]).\end{aligned}$$

The induction step is trivial.

Let h be a formula valuation in the algebra $\langle M^V, \sigma\rangle$. We define a valuation of formulae φ_h in the pseudo-boolean model $\langle M, D\rangle$. For arbitrary atomic formula $p(x_0, \ldots, x_{n-1})$ we have $\Delta hp(x_0, \ldots, x_{n-1}) \subseteq \{0, \ldots, n-1\}$ by Lemma 2. From Propositions 2 and 4 the function $hp(x_0, \ldots, x_{n-1})$ can be represented by a form $\langle F, (x_0, \ldots, x_{n-1})\rangle$. We let $\varphi_h(p) \stackrel{\text{def}}{=} F$. Similarly to Lemma 7 the following lemma holds for such h and φ_h.

Lemma 8 $(hA)_v = \varphi_h(A[v])$ *for every formula* A *and any variable interpretation* v.

Remark 1 By Lemma 8 φ_h is uniquely defined by h and does not depend on the choice of variable sequences in atomic formulae.

Lemmata 7 and 8 immediately imply the following theorem.

Theorem 2 *A pseudo-boolean model* $\langle M, D\rangle$ *and the corresponding quasicylindric algebra* $\langle M^V, \sigma\rangle$ *have same set of valid formulae.*

The theorem entails that the semantics of pseudo-boolean models is not weaker than the semantics of quasicylindric algebras.

Theorem 3 *If a logic* L *is complete with respect to a class of pseudo-boolean models then it is complete with respect to a class of quasicylindric Rasiowa-Sikorski algebras.*

13.4.2 Kripke Semantics

The Kripke semantics of possible worlds provides one of the most intuitive interpretations for superintuitionistic logics. It is known (Ono 1973) that the cardinality of the set of all superintuitionistic predicate logics which are not complete with respect to any class of Kripke frames is at least $2^{\aleph_0}$.

An *(intuitionistic) Kripke frame* is a tuple $\langle W, \leqslant, D\rangle$, where $\langle W, \leqslant\rangle$ is a non-empty partially ordered set and D is a function which maps every $w \in W$ to some non-empty set D_w such that $w \leqslant v$ implies $D_w \subseteq D_v$ for any $w, v \in W$. Elements of W are called *worlds* and each set D_w for a $w \in W$ is called *domain* of w.

A tuple $\langle W, \leqslant, D, \nu\rangle$ is called *(intuitionistic) Kripke model* if ν is a function such that $\nu(w, p) \subseteq (D_w)^n$ for every $w \in W$ and n-ary predicate symbol p and $w \leqslant v$ implies $\nu(w, p) \subseteq \nu(v, p)$ for all $w, v \in W$.

Let $\langle W, \leqslant, D\rangle$ be a Kripke frame and $w \in W$. A *variable interpretation* in w is any mapping of Var into D_w. For every world w, let $\mathsf{V}(w)$ be the set of all variable interpretations in the world w.

For a variable interpretation γ, a variable interpretation δ in a world w is called *i-variation* of γ in w iff $\delta x_j = \gamma x_j$ for every $j \neq i$. Let $\overset{i,w}{\longleftrightarrow}$ be a binary relation on the set of $\mathsf{V}(w)$ such that $\gamma \overset{i,w}{\longleftrightarrow} \delta$ iff δ is an i-variation of γ in w. Clearly, each $\overset{i,w}{\longleftrightarrow}$ is an equivalence relation on $\mathsf{V}(w)$.

A *truth relation* $\models_\nu$ in a Kripke model $\langle W, \leqslant, D, \nu\rangle$ is defined as follows. For any world w and a variable interpretation γ in w we define $w \models_\nu A[\gamma]$ by induction of the length of given formula A.

(1) $w \not\models_\nu \bot[\gamma]$ and $w \models_\nu \top[\gamma]$;
(2) $w \models_\nu p(x_{i_0}, \ldots, x_{i_n}) \overset{\text{def}}{\iff} \langle \gamma x_{i_0}, \ldots, \gamma x_{i_n}\rangle \in \nu(w, p)$;
(3) $w \models_\nu (A \wedge B)[\gamma] \overset{\text{def}}{\iff} (w \models_\nu A[\gamma] \wedge w \models_\nu B[\gamma])$;
(4) $w \models_\nu (A \vee B)[\gamma] \overset{\text{def}}{\iff} (w \models_\nu A[\gamma] \vee w \models_\nu B[\gamma])$;
(5) $w \models_\nu \exists x_i A[\gamma]$ iff $w \models_\nu A[\delta]$ for some i-variation δ of γ in w;
(6) $w \models_\nu \neg A[\gamma]$ iff $v \not\models_\nu A[\gamma]$ for every $v \in W$ such that $w \leqslant v$;
(7) $w \models_\nu (A \to B)[\gamma]$ iff $v \models_\nu A[\gamma]$ implies $v \models_\nu B[\gamma]$ for every $v \in W$ such that $w \leqslant v$;
(8) $w \models_\nu \forall x_i A[\gamma]$ iff $v \models_\nu A[\delta]$ for every $v \in W$ such that $w \leqslant v$ and any i-variation δ of γ in the world v.

We say that A is *true* in w iff $w \models_\nu A[\gamma]$ for every variable interpretation γ in w. A formula is *true* in a Kripke model iff it is true in every world of the model. We usually omit the subscript ν in $\models_\nu$ if ν is known from the context.

For a Kripke frame $\mathfrak{F} \overset{\text{def}}{=} \langle W, \leqslant, D\rangle$, let $\mathsf{S}(\mathfrak{F})$ be the set of all functions X mapping every world w to a subset X_w of $\mathsf{V}(w)$ and such that $w \leqslant v$ implies $X_w \subseteq X_v$ for all $w, v \in W$. For a variable interpretation $\gamma \in \mathsf{V}(w)$, $s^i_j\gamma$ denotes a (unique) i-variation of γ such that $(s^i_j\gamma)x_i = \gamma x_j$.

We define operations of the similarity type σ on the set $\mathsf{S}(\mathfrak{F})$ as follows. For every $w \in W$ let:

(1) $\bot_w \overset{\text{def}}{=} \varnothing$ and $\top_w \overset{\text{def}}{=} \mathsf{V}(w)$;
(2) $(X \wedge Y)_w \overset{\text{def}}{=} X_w \cap Y_w$;
(3) $(X \vee Y)_w \overset{\text{def}}{=} X_w \cup Y_w$;
(4) $(\exists_i X)_w \overset{\text{def}}{=} \{\gamma \in \mathsf{V}(w) \mid \exists \delta \in X_w\ \gamma \overset{i,w}{\longleftrightarrow} \delta\}$;
(5) $(\neg X)_w \overset{\text{def}}{=} \{\gamma \in \mathsf{V}(w) \mid \forall v \in W\ (w \leqslant v \Rightarrow \gamma \notin X_v)\}$;
(6) $(X \to Y)_w \overset{\text{def}}{=} \{\gamma \in \mathsf{V}(w) \mid \forall v \in W\ (w \leqslant v \wedge \gamma \in X_v \Rightarrow \gamma \in Y_v)\}$;
(7) $(\forall_i X)_w \overset{\text{def}}{=} \{\gamma \in \mathsf{V}(w) \mid \forall v \in W\ \forall \delta \in \mathsf{V}(v)\ (w \leqslant v \wedge \gamma \overset{i,v}{\longleftrightarrow} \delta \Rightarrow \delta \in X_v)\}$;
(8) $(s^i_j X)_w \overset{\text{def}}{=} \{\gamma \in \mathsf{V}(w) \mid s^i_j \gamma \in X_w\}$.

It is now a routine to check that all the equalities of quasicylindric algebras are true in the algebra $\langle \mathsf{S}(\mathfrak{F}), \sigma \rangle$.

Proposition 5 *$\langle \mathsf{S}(\mathfrak{F}), \sigma \rangle$ is a quasicylindric algebra.*

Let now γ be a variable interpretation in a Kripke model $\langle W, \leqslant, D, \nu \rangle$. We define a formula valuation h_ν in the algebra $\langle \mathsf{S}(\mathfrak{F}), \sigma \rangle$ as follows. For every atomic formula $p(x_0, \ldots, x_{n-1})$ where $n = \#p$ we let

$$(h_\nu p(x_0, \ldots, x_{n-1}))_w \overset{\text{def}}{=} \{\gamma \in \mathsf{V}(w) \mid \langle \gamma x_0, \ldots, \gamma x_{n-1} \rangle \in \nu(w, p)\}.$$

Lemma 9 *$\Delta h_\nu p(x_0, \ldots, x_{n-1}) \subseteq \{0, \ldots, n-1\}$ for every n-ary predicate symbol p.*

Thus, by Lemma 3, h_ν can be extended to a formula valuation in the algebra $\langle \mathsf{S}(\mathfrak{F}), \sigma \rangle$. It is not difficult to prove the following lemma.

Lemma 10 *Let $\models$ be the truth relation in a Kripke model $\langle W, \leqslant, D, \nu \rangle$. Then $(h_\nu A)_w = \{\gamma \in \mathsf{V}(w) \mid w \models A[\gamma]\}$ for every formula A.*

An element X from $\mathsf{S}(\mathfrak{F})$ is *regular* iff for every $w \in W$, $\xi \in X_w$, and $\eta \in \mathsf{V}(w)$, the variable interpretation η belongs to X_w whenever $\xi x_i = \eta x_i$ for all $i \in \Delta X$. A subalgebra of $\langle \mathsf{S}(\mathfrak{F}), \sigma \rangle$ is *regular* iff all its elements are regular.

It is easy to see that, given a Kripke model $\langle W, \leqslant, D, \nu \rangle$, element $h_\nu A \in \mathsf{S}(\mathfrak{F})$ is regular for every formula A.

Let h be a valuation of formulae in the algebra $\langle \mathsf{S}(\mathfrak{F}), \sigma \rangle$. For every $w \in W$ and n-ary predicate symbol p we define

$$\nu_h(w, p) \overset{\text{def}}{=} \{\langle \gamma x_0, \ldots, \gamma x_{n-1} \rangle \in (D_w)^n \mid \gamma \in (hp(x_0, \ldots, x_{n-1}))_w\}$$

Lemma 11 *Let $\mathfrak{F} \overset{\text{def}}{=} \langle W, \leqslant, D \rangle$ be a Kripke frame and h be a valuation of formulae in the algebra $\langle \mathsf{S}(\mathfrak{F}), \sigma \rangle$. Then $\gamma \in (hA)_w \iff w \models_{\nu_h} A[\gamma]$ for every $w \in W$ and $\gamma \in \mathsf{V}(w)$.*

Proof The proof of the equivalence is by induction on the length of a formula A. We give only a sketch of the proof. The case when $A = \top$ or $A = \bot$ is trivial. Let $A = p(x_{i_0}, \ldots, x_{i_n})$ where $n = \#p - 1$ and $w \in W$. We have

$$(hA)_w = (h(s^{k_0}_{i_0} \cdots s^{k_n}_{i_n} s^0_{k_0} \cdots s^n_{k_n} p(x_0, \ldots, x_n)))_w =$$
$$(s^{k_0}_{i_0} \cdots s^{k_n}_{i_n} s^0_{k_0} \cdots s^n_{k_n} hp(x_0, \ldots, x_n))_w,$$

for some different $k_0, \ldots, k_n$ such that $\{k_0, \ldots, k_n\} \cap \{0, \ldots, n, i_0, \ldots, i_n\} = \varnothing$. Such a construction is required to express the operation of *simultaneous* substitution of all i_j instead of corresponding $j = 0, \ldots, n$.

Let $\gamma \in (hA)_w$. This means that $s^{k_0}_{i_0} \cdots s^{k_n}_{i_n} s^0_{k_0} \cdots s^n_{k_n} \gamma$ belongs to $(hp(x_0, \ldots, x_n))_w$. We skip a few details related to $k_0, \ldots, k_n$ while unravelling the above construction and obtain that $\gamma \in (hA)_w$ is equivalent to that $\delta \in (hp(x_0, \ldots, x_n))_w$ where δ is defined by

$$\delta x_l \overset{\text{def}}{=} \begin{cases} \gamma x_l, & l \neq 0, \ldots, n, \\ \gamma x_{i_l}, & l = 0, \ldots, n. \end{cases}$$

By the definition of ν_h we have that $\langle \delta x_0, \ldots, \delta x_n \rangle \in \nu_h(w, p)$. This is equivalent to $\langle \gamma x_{i_0}, \ldots, \gamma x_{i_n} \rangle \in \nu_h(w, p)$. Thus, $w \models_{\nu_h} A[\gamma]$.

For the converse, we assume that $w \models_{\nu_h} A[\gamma]$. That is, $\langle \gamma x_{i_0}, \ldots, \gamma x_{i_n} \rangle \in \nu_h(w, p)$. This means that $\langle \delta' x_0, \ldots, \delta' x_n \rangle \in \nu_h(w, p)$ for some δ' from the set $(hp(x_0, \ldots, x_n))_w$ such that $\delta' x_l = \gamma_{x_l}$ for all $l = 0, \ldots, n$. Since $\Delta hp(x_0, \ldots, x_n) \subseteq \{0, \ldots, n\}$, using the above definition of δ, we obtain $\delta x_l = \delta' x_l$ for all $l \in \Delta hp(x_0, \ldots, x_n)$. Hence, because $hp(x_0, \ldots, x_n)$ is regular, δ also belongs to $(hp(x_0, \ldots, x_n))_w$. Therefore, $\gamma \in (hA)_w$.

The induction step is straightforward.

Lemma 12 *Let $X \in \mathsf{S}(\mathfrak{F})$, $w \in W$, $\gamma \in X_w$, and $\delta \in \mathsf{V}(w)$. Consider the set $I \overset{\text{def}}{=} \{i < \omega \mid \gamma x_i \neq \delta x_i\}$. If $I \cap \Delta X = \varnothing$ and I is finite then $\delta \in X_w$.*

Proof Assume that $I = \{i_0, \ldots, i_n\}$. Since all i_k do not belong to ΔX, we have $\forall_{i_0} \ldots \forall_{i_n} X = X$. Therefore, $\gamma \in X_w = (\forall_{i_0} \ldots \forall_{i_n} X)_w$. By the definition of operations $\forall_i$, for any $\delta \in \mathsf{V}(w)$, if $\gamma x_j = \delta x_j$ for all $j \neq i_0, \ldots, i_n$ then $\delta \in X_w$.

Proposition 6 *Let $\langle A, \sigma \rangle$ be a locally finitary subalgebra of the algebra $\langle \mathsf{S}(\mathfrak{F}), \sigma \rangle$ and B be a set of all regular elements of A. Then B is closed under all the operations of the algebra $\langle A, \sigma \rangle$, i.e. $\langle B, \sigma \rangle$ is a subalgebra of $\langle A, \sigma \rangle$.*

Proof We only show that $s^i_j X$ is regular whenever X is regular because this is the most interesting case. All the other operations are examined similarly. If $i = j$ then $s^i_j X = X$ and, the statement is trivial. Assume that $i \neq j$ and let $w \in W$, $\gamma \in (s^i_j X)_w$, and $\gamma x_k = \delta x_k$ for all $k \in \Delta s^i_j X$. By Lemma 1, $\Delta s^i_j X \subseteq (\Delta X \setminus \{i\}) \cup \{j\}$. We define

$$\gamma' x_k \overset{\text{def}}{=} \begin{cases} \delta x_k, & k \in (\Delta X \setminus \{i\}) \cup \{j\}, \\ \gamma x_k, & \text{otherwise}. \end{cases}$$

ΔX is finite since $\langle A, \sigma \rangle$ is locally finitary. Therefore the set $I \overset{\text{def}}{=} \{k \mid \gamma x_k \neq \gamma' x_k\}$ is also finite. Furthermore, if $k \in \Delta s^i_j X$ then $\gamma' x_k = \delta x_k = \gamma x_k$ and, consequently,

$k \notin I$. Hence, $I \cap \Delta s^i_j X = \varnothing$. By Lemma 12, $\gamma' \in (s^i_j X)_w$. Thus, by the definition of the operation s^i_j, $\xi' \stackrel{\text{def}}{=} s^i_j \gamma'$ is in X_w. Let us consider ξ defined by

$$\xi x_k \stackrel{\text{def}}{=} \begin{cases} \delta x_k, & k \neq i, \\ \xi' x_k, & k = i. \end{cases}$$

It is not difficult to see that $\xi x_k = \xi' x_k$ for every $k \in \Delta X$. Thus, because X is regular, $\xi \in X_w$. It is now an easy check that $\xi = s^i_j \delta$ and, hence, $\delta \in (s^i_j X)_w$.

Theorem 4 *Let $\mathfrak{F} \stackrel{\text{def}}{=} \langle W, \leqslant, D\rangle$ be a Kripke frame. Then A is valid in the frame iff it is valid in the largest locally finitary regular subalgebra of $\langle \mathsf{S}(\mathfrak{F}), \sigma\rangle$.*

The theorem shows that the Kripke semantics is equivalent to the semantics of locally finitary regular quasicylindric algebras. A direct consequence of this theorem is the following statement.

Theorem 5 *Any superintuitionistic logic which is complete with respect to some class of Kripke frames is complete with respect to a corresponding class of locally finitary regular quasicylindric algebras.*

In the rest of the section we establish a few results on regular algebras which show a connection with Heyting algebras build from propositional Kripke frames.

The proof of the following lemma is not difficult.

Lemma 13 *Let $\mathfrak{F} \stackrel{\text{def}}{=} \langle W, \leqslant, D\rangle$ be a Kripke frame and $X \in \mathsf{S}(\mathfrak{F})$. Then X is regular and $\Delta X = \varnothing$ iff, for each $w \in W$, either $X_w = \varnothing$ or $X_w = \mathsf{V}(w)$.*

An *algebra of the upward closed sets* of a propositional Kripke frame $\mathfrak{F} \stackrel{\text{def}}{=} \langle W, \leqslant\rangle$ is a Heyting algebra $\langle \mathscr{C}(\mathfrak{F}), \top, \bot, \neg, \vee, \wedge, \rightarrow\rangle$ such that $\mathscr{C}(\mathfrak{F})$ is the set of all upward closed sets of $\mathfrak{F}$ and the algebraic operations are defined by the following equalities for all $U, V \in \mathscr{C}(\mathfrak{F})$: $\top \stackrel{\text{def}}{=} \mathscr{C}(\mathfrak{F})$, $\bot \stackrel{\text{def}}{=} \varnothing$, $U \vee V \stackrel{\text{def}}{=} U \cup V$, $U \wedge V \stackrel{\text{def}}{=} U \cap V$, $U \rightarrow V \stackrel{\text{def}}{=} \{w \in W \mid (\forall v \in W)(w \leqslant v \wedge v \in U \Rightarrow v \in V)\}$, and $\neg U \stackrel{\text{def}}{=} U \rightarrow \bot$.

Theorem 6 *Let Let $\mathfrak{F} \stackrel{\text{def}}{=} \langle W, \leqslant, D\rangle$ be a Kripke frame and $A \stackrel{\text{def}}{=} \{X \in \mathsf{S}(\mathfrak{F}) \mid \Delta X = \varnothing$ and X is regular$\}$. Then $\langle A, \sigma\rangle$ is a subalgebra of the algebra $\langle \mathsf{S}(\mathfrak{F}), \sigma\rangle$ whose reduct to the similarity type of Heyting algebras is isomorphic to the algebra of all the upward closed sets of the propositional Kripke frame $\langle W, \leqslant\rangle$.*

Proof It is clear that $\langle A, \sigma\rangle$ is a subalgebra of the algebra $\langle \mathsf{S}(\mathfrak{F}), \sigma\rangle$. We define $\varphi(X) \stackrel{\text{def}}{=} \{w \in W \mid X_w = \mathsf{V}(w)\}$ for all $X \in A$. It is not difficult to show that φ is an isomorphism (of Heyting algebras).

13.5 Algebraic Equivalents of Deductive Properties

This section considers several important properties of superintuitionistic logics: Craig interpolation property, interpolation principle for deducibility, Beth definability property, projective Beth definability property, disjunctive and existential properties. We formulate an algebraic counterparts of these properties. The proofs of the corresponding equivalence theorems can be found in (Tishkovsky 2001).

13.5.1 Interpolation Property

In this subsection we consider two variants of interpolation property for the superintuitionistic predicate logics and their algebraic counterparts. The main equivalence results are established in (Tishkovsky 2001) which, in its turn, follows the definitions and proofs in (Maksimova 1979, 1992).

A superintuitionistic predicate logic L has the *(Craig) interpolation property* (CIP), iff the condition $L \vdash A \to B$ implies both $L \vdash A \to C$ and $L \vdash C \to B$ for some formula C such that the formula C contains only those predicate symbols which occur in both formulas A and B. The formula C is called *interpolant* of the formula $A \to B$.

A superintuitionistic predicate logic L has *interpolation principle for deducibility* (IPD), iff the condition $A \vdash_L B$ implies both $A \vdash_L C$ and $C \vdash_L B$ for some formula C such that $P(C) \subseteq P(A) \cap P(B)$. The formula C is called *interpolant on deducibility* of the formula $A \to B$.

A class K of algebras is called *amalgamable* (Maksimova 1977) iff for any algebras $\mathscr{A}_0, \mathscr{A}_1, \mathscr{A}_2$ from K and embeddings $i_1 : \mathscr{A}_0 \to \mathscr{A}_1$ and $i_2 : \mathscr{A}_0 \to \mathscr{A}_2$ there exists an algebra $\mathscr{A}$ from K and embeddings $e_1 : \mathscr{A}_1 \to \mathscr{A}$ and $e_2 : \mathscr{A}_2 \to \mathscr{A}$ such that $e_1 i_1 = e_2 i_2$. A class K of algebras is called *superamalgamable* (Maksimova 1977), if the following additional condition holds for all $a_1 \in |\mathscr{A}_1|$ and $a_2 \in |\mathscr{A}_2|$:

$$\begin{cases} e_1 a_1 \leqslant e_2 a_2 \iff \exists a_0 \in |\mathscr{A}_0| \ (a_1 \leqslant i_1 a_0 \wedge i_2 a_0 \leqslant a_2), \\ e_2 a_2 \leqslant e_1 a_1 \iff \exists a_0 \in |\mathscr{A}_0| \ (a_2 \leqslant i_2 a_0 \wedge i_1 a_0 \leqslant a_1). \end{cases}$$

If class K is amalgamable (superamalgamable), then we also say that K has amalgamation property, AP (respectively: superamalgamation property or SupAP in short).

Let L be a superintuitionistic predicate logic, which is complete with respect to a variety V of quiasicylindric algebras.

Lemma 14 *If L has* IPD *(*CIP*), then the class $V^{(\omega)}$ is amalgamable (respectively: superamalgamable).*

Lemma 15 (cf. Czelakowski 1982) *Let the class $V^{(\omega)}$ be amalgamable. Let A and B be formulae, $\Gamma \overset{\text{def}}{=} \{C \in \mathrm{For}(P(A) \cap P(B)) \mid A \vdash_L C\}$. If $\Gamma \nvdash_L B$ then $A \nvdash_L B$.*

Lemma 16 *If the class $V^{(\omega)}$ is superamalgamable then L has* CIP.

Theorem 7 *Let L be a superintuitionistic predicate logic which is complete with respect to some variety V of quasicylindric algebras. The following conditions are equivalent:*

(1) L has IPD *(*CIP*);*
(2) the class $V^{(\omega)}$ *is amalgamable (superamalgamable).*

13.5.2 Beth Definability Property

A superintuitionistic predicate logic L has *Beth definability property* iff for each formula $A[p, p_0, \ldots, p_m]$ with an n-ary predicate symbol p the condition

$$A[p, p_0, \ldots, p_m] \wedge A[q, p_0, \ldots, p_m] \vdash_L p(x_0, \ldots, x_{n-1}) \equiv q(x_0, \ldots, x_{n-1})$$

implies that $A[p, p_0, \ldots, p_m] \vdash_L p(x_0, \ldots, x_{n-1}) \equiv B$ for some formula $B[p_0, \ldots, p_n]$.

Let K be arbitrary class of quasicylindric algebras closed under subalgebras. K has ES*-*property* (Maksimova 1998), iff for any algebra $\mathscr{B}$ from K and for any subalgebra $\mathscr{A}$ of the algebra $\mathscr{B}$ the following condition ES*$(\mathscr{B}, \mathscr{A})$ holds:

for any b from $|\mathscr{B}| \setminus |\mathscr{A}|$ if the set $|\mathscr{A}| \cup \{b\}$ generates $\mathscr{B}$ then there are $\mathscr{C}$ from K and monomorphisms $g, h : \mathscr{B} \to \mathscr{C}$ such that $ga = ha$ for all a in $\mathscr{A}$ and $gb \neq hb$.

K has ES*-*property for finitely generated algebras* iff the condition ES*$(\mathscr{B}, \mathscr{A})$ is satisfied for any finitely generated algebra $\mathscr{B}$ from K and its finitely generated subalgebra $\mathscr{A}$.

The proof of the following theorem is a generalisation of the proof of similar theorem in Maksimova (1998) to predicate logics.

Theorem 8 *Let L be a superintuitionistic predicate logic which is complete with respect to a variety V of quasicylindric algebras. The following conditions are equivalent:*

(1) L has Beth property;
(2) $V^{(\omega)}$ *has* ES*-*property;*
(3) $V^{(\omega)}$ *has* ES*-*property for finitely generated algebras.*

13.5.3 Projective Beth Definability Property

The projective Beth definability property was introduced by Maksimova (1998) as a natural generalisation of the Beth definability property.

A superintuitionistic predicate logic L has *projective Beth definability property* iff for each fromula $A[p, p_0, \ldots, p_m, q_0, \ldots, q_t]$ with an n-ary predicate symbol p the condition

$$A[p, p_0, \ldots, p_m, q_0, \ldots, q_t] \wedge A[q, p_0, \ldots, p_m, q_0', \ldots, q_t'] \vdash_L p(x_0, \ldots, x_{n-1}) \equiv q(x_0, \ldots, x_{n-1})$$

implies $A[p, p_0, \ldots, p_m, q_0, \ldots, q_t] \vdash_L p(x_0, \ldots, x_{n-1}) \equiv B$ for some formula $B[p_0, \ldots, p_m]$.

Let K be an arbitrary class of quasicylindric algebras closed under subalgebras. K has SES-*property* (Maksimova 1998), iff for any algebra $\mathscr{B}$ from K and for any subalgebra $\mathscr{A}$ of the algebra $\mathscr{B}$ the following condition SES($\mathscr{B}$, $\mathscr{A}$) holds:

for any b from $|\mathscr{B}| \setminus |\mathscr{A}|$ there are $\mathscr{C}$ from K and monomorphisms $g, h : \mathscr{B} \to \mathscr{C}$ such that $ga = ha$ for all a in $\mathscr{A}$ and $gb \neq hb$.

K has SES-*property for finitely generated algebras* iff the condition SES($\mathscr{B}$, $\mathscr{A}$) is true for any finitely generated algebra $\mathscr{B}$ from K and its finitely generated subalgebra $\mathscr{A}$.

The proof of the following theorem is a predicate generalisation of the proof of the similar theorem in (Maksimova 1998).

Theorem 9 *Let L be a superintuitionistic predicate logic which is complete with respect to some variety V of quasicylindric algebras. The following conditions are equivalent:*

(1) L has projective Beth property;
(2) $V^{(\omega)}$ has SES-*property;*
(3) $V^{(\omega)}$ has SES-*property for finitely generated algebras.*

13.5.4 Disjunctive Property

A superintuitionistic logic L has the *disjunctive property* (Rasiowa and Sikorski 1963) iff for every formula $A \vee B$ from L either A belongs to L or B is in L.

A quasicylindric algebra is *well connected* iff for all elements a and b of the algebra, $a \vee b = \top$ implies that either $a = \top$ or $b = \top$.

It is easy to prove the following lemma (cf. Rasiowa and Sikorski 1963).

Lemma 17 *If L has the disjunctive property then the algebra of Lindenbaum-Traski of the logic L is well connected.*

Theorem 10 *Let L be a superintuitionistic predicate logic which is complete with respect to a class K of quasicylindric algebras and K be closed under subalgebras. The following conditions are equivalent.*

(1) L has the disjunctive property,
(2) for any algebras $\mathscr{A}$ and $\mathscr{B}$ from $K^{(\omega)}$ there are a well connected locally-finitary algebra $\mathscr{C}$ such that all the formulae from L are valid in $\mathscr{C}$ and a homomorphism from $\mathscr{C}$ onto $\mathscr{A} \times \mathscr{B}$.

13.5.5 Existential Property

A superintuitionistic predicate logic L has *existential property* (Dragalin 1988) iff $L \ni \exists x_i A$ implies $L \ni s^i_j B$ for some variable x_j and some formula B which is congruent to the formula A and such that the substitution of the variable x_j for the variable x_i is admissible.

A quasicylindric algebra is $\exists$-*connected* iff for each element a of the algebra and every $i < \omega$, $\exists_i a = \top$ implies $s^i_j a = \top$ for some $j \in \omega$.

The following lemma can be proved similarly to Lemma 17.

Lemma 18 *If L has existential property then the Lindenbaum–Tarski algebra of the logic L is* $\exists$-*connected.*

Theorem 11 *Let L is a superintuitionistic predicate logic which is complete with respect to a class K of quasicylindric algebras and K be closed under subalgebras. The following conditions are equivalent.*

(1) L has an existential property,
(2) for every countable family of locally-finitary algebras $\langle \mathscr{A}_i \mid i \in \omega \rangle$ *from K there are a locally-finitary* $\exists$-*connected algebra* $\mathscr{C}$ *such that all the formulae from L are valid in* $\mathscr{C}$ *and a homomorphism from* $\mathscr{C}$ *onto the largest locally-finitary subalgebra of the algebra* $\prod_{i \in \omega} \mathscr{A}_i$.

13.6 Conclusion and Future Work

In the paper we showed how to obtain an algebraic semantics for all superintuitionistic predicate logics. We embedded the two most known semantics for these logics: Rasiowa-Sikorski algebraic models and Kripke models into the semantics of quasicylindric algebras. It turns out that, using the semantics, the results on algebraic counterparts of several important properties of superintuitionistic logics can be transferred from the propositional case rather comfortably.

Future work can proceed in several directions. Regarding generalisation of the results there are two paths. One is to generalise all the results of the paper to an arbitrary non-classical predicate logics. Given a class of propositional logics in a fixed logical language which are algebraisable in the sense of Blok and Pigozzi (1989), it looks possible to produce an appropriate algebraic semantics for the class of first-order extensions of these logics. In general, the language of the propositional logics can be very different from the language of superintuitionistic logics. This pose a problem of reformulating the equalities (Q1)–(Q18) in the corresponding algebraic language and, thus, the transfer of the results is not straightforward. Once the algebraic semantics for the first-order extensions is defined, generalising the considered equivalence theorems for logical and algebraic properties to the first-order language is not difficult.

The presented results on the embeddings of semantics clearly showed that the semantics of quasicylindric algebras is more general than two of the known semantics considered in the paper. This, however, is formally establishing a mathematical fact which is clear from general reasoning. Indeed, the quasicylindric algebraic semantics is free from domain dependencies which are inherent to both the Rasiowa–Sikorski algebraic semantics and the Kripke semantics of possible worlds. Within these semantics, interpretations of individual variables are tied to particular interpretation domains. We used this fact effectively when we built quasicylindric algebras from such interpretations. The proof provides a more interesting insight into interrelations of algebraic semantics with other sorts of semantics. The key observation in this respect is that an algebraic semantics for a (predicate) logic, if this is possible at all, can be built from *interpretations* of the logic in some structures of other semantical nature. In this case, the obtained algebraic semantics will be more generic, that is, it will suit a wider class of logics than the original semantics. Thus, the other possible generalisation is a transfer of the results on the embeddings of semantics. It is feasible to generalise the notion of semantics for a logic and prove a general result that algebraic semantics subsumes all other semantics.

As it can be seen from results of Sect. 13.4.2, regular quasicylindric algebras play same role in theory of quasicylindric algebras as Boolean set algebras in the theory Boolean algebras. It would be interesting to provide a further support for this claim by an analogue of Stone's representation theorem. Whether any quasicylindric algebra is a subalgebra of a suitable regular algebra is an open question.

Further development of the theory of quasicylindric algebras can also proceed in the direction of establishing interrelations between algebraic properties, or, equivalently, between corresponding logical properties. This direction is in line with the results in (Maksimova 1999). In this respect, considering the algebraic properties from the point of Category Theory can also lead to new exciting results. On this more general route (if it will be taken), we expect and hope to get very deep insights into the nature of logical consequence relations and that can lead the research to a 'language-independent' classification of logics and corresponding classes of algebras.

All studies on Algebraic Logic establish that the areas of Algebra and Logic are very tightly connected. However, despite great advance in both the areas, there is still a huge gap between them. I hope the results presented in this paper will provide a new view point on algebraisation of first-order logics and contribute to bridging this gap. Algebraic Logic and, especially, its field on algebraisation of first-order logics are very young and evolving areas and we should expect that more exciting and beautiful results in these areas are to come.

References

Andréka, H., Németi, I. & Sain, I. (1984). *Abstract model theoretic approach to algebraic logic*, manuscript.

Blok, W. J., & Pigozzi, D. (1989). *Algebraizable logics*, Memoirs of the American Mathematical Society. Providence: American Mathematical Society.

Czelakowski, J. (1982). Logical matrices and the amalgamation property. *Studia Logica*, *41*(4), 329–341.

Czelakowski, J. (2001). *Protoalgebraic logics*. Dordrecht: Kluwer Academic Publishers.

Czelakowski, J., & Jansana, R. (2014). Weakly algebraizable logics. *Journal of Symbolic Logic*, *65*, 641–668.

Czelakowski, J., & Pigozzi, D. (2004). Fregean logics. *Annals of Pure and Applied Logic*, *127*, 17–76.

Dragalin, A. G. (1988). *Mathematical intuitionism: Introduction to proof theory*, Translations of Mathematical Monographs. Providence: American Mathematical Society.

Ershov, Y. L., & Palyutin, E. A. (1986). *Mathematical logic*. Moscow: Mir Publishers.

Gabbay, D. M., & Maksimova, L. (2005). *Interpolation and definability: modal and intuitionistic logics*. Oxford: Oxford University Press,Clarendon Press.

Henkin, L., Monk, J. D., & Tarski, A. (1971, 1985). *Cylindric algebras, Parts 1, 2*. Amsterdam: North Holland.

Hoogland, E. (1997). Algebraic characterizations of two Beth definability properties. *Workshop on Abstract Algebraic Logic, Abstracts of talks* (pp. 48–52). Bellaterra, Spain: Centre de Recerca Matemática.

Jankov, V. A. (1968). The construction of a sequence of strongly independent superintuitionistic propositional calculi. *Soviet Mathematics Doklady*, *9*, 806–807.

Madarász, J. (1995). The Craig Interpolation theorem in multi-modal logics. *Bulletin of the Section of Logic*, *24*(3), 147–154.

Maksimova, L. (1977). Craig's theorem in superintuitionistic logics and amalgamable varieties of pseudo-boolean algebras. *Algebra and Logic*, *16*(6), 427–455.

Maksimova, L. (1986). On maximal intermediate logics with disjunction property. *Studia Logica*, *45*(1), 69–75.

Maksimova, L. (1992). Modal logics and varieties of modal algebras: The Beth properties, interpolation, and amalgamation. *Algebra and Logic*, *31*(2), 90–105.

Maksimova, L. (1995). On variable separation in modal and superintuitionistic logics. *Studia Logica*, *55*(1), 99–102.

Maksimova, L. (1998). Explicit and implicit definability in modal and related logics. *Bulletin of the Section of Logic*, *27*(1/2), 36–39.

Maksimova, L. (1999). Interrelations of algebraic, semantical and logical properties for superintuitionistic and modal logics. In *Logic, algebra and computer science* (Vol. 46, pp. 159–168), Banach center publications. Warszawa: Institute of Mathematics, Polish Academy of Sciences.

Maksimova, L. L. (1979). Interpolation theorems in modal logics and amalgamable varieties of topoboolean algebras. *Algebra and Logic*, *18*(5), 556–586.

Németi, I. (1991). Algebraization of quantifier logics, an introductory overview. *Studia Logica*, *50*(3/4), 485–570.

Ono, H. (1973). A study of intermediate predicate logics. *Publications of the Research Institute for Mathematical Sciences*, *8*, 619–649.

Pinter, C. (1973). A simple algebra of first order logic. *Notre Dame Journal of Formal Logic*, *14*(3), 361–366.

Pinter, C. (1975). Algebraic logic with generalized quantifiers. *Notre Dame Journal of Formal Logic*, *16*, 511–516.

Rasiowa, H. (1974). *An algebraic approach to non-classical logics* (Vol. 78), Studies in logic and the foundation of mathematics. Amsterdam: North-Holland.

Rasiowa, H., & Sikorski, R. (1963). *The mathematics of metamathematics*, Monografie matematiczne. Warszawa: Państwowe Wydawnictwo Naukowe.
Rybakov, V. (1997). *Admissibility of logical inference rules*. Amsterdam: Elsevier.
Tarski, A. (1935). Grundzüge des Systemenkalküls. *Erster Teil, Fundamenta Mathematicae*, *25*, 503–526.
Tishkovsky, D. E. (1999). On algebraic semantics for superintuitionistic predicate logics. *Algebra and Logic*, *38*(1), 36–50.
Tishkovsky, D. E. (2001). Algebraic equivalents of some properties of superintuitionistic predicate logics. *Algebra and Logic*, *40*(2), 218–242.

Chapter 14
Dummett Logic, Irreflexive Modality and Novikov Completeness

Alexander Yashin

Abstract We study an extension of superintuitionistic logic *LC* (*Dummett logic*, *Logic of Chains*) in the language containing an additional unary logical connective. This connective is interpreted on every finite chain by the so called *irreflexive modality*. We show that the resulting logic is conservative over *LC*, determines a *new logical connective* in *LC* w.r.t. P. Novikov's approach to the notion of a new logical connective. Moreover, we show that is an explicit example of *Novikov complete extension* of *LC*, i.e. every proper extension of it is not conservative over *LC*.

14.1 Introduction

In 50-th of XX century P. Novikov suggested his own approach to the notion of a new intuitionistic propositional connective. Let us give the exact definitions.

The class Fm of formulas is built from propositional variables by means of standard logical connectives $\vee$, $\wedge$, $\rightarrow$, $\leftrightarrow$, $\neg$. Sometimes it is convenient to add to this language standard logical constants $\mathsf{0}$ (false) and $\mathsf{1}$ (true).

The *intuitionistic propositional logic* will be denoted by Int.

A *superintuitionistic logic* (or *s.i. logic*) is defined as an arbitrary subset $L \subset Fm$, containing Int and closed under the modus ponens rule and the substitution rule (if s is a substitution and A is a formula then $s(A)$ is called an *instance* of A).

Adding to the initial language an unary connective $\varphi(\cdot)$ we obtain the class $Fm(\varphi)$ of formulas of the enriched language in a natural way. Formulas without φ are said to be *pure*.

By a *φ-logic*[1] we mean any set $\mathscr{L}$ of formulas of the φ-language containing Int and closed under the following three rules: modus ponens, substitution, and the *Replacement Rule for φ*:

[1]The term comes from Skvortsov (1983).

A. Yashin (✉)
Moscow State University of Psychology and Education, Moscow, Russia
e-mail: yashin.alexandr@ya.ru

S. Odintsov (ed.), *Larisa Maksimova on Implication, Interpolation, and Definability*,
Outstanding Contributions to Logic 15, https://doi.org/10.1007/978-3-319-69917-2_14

$$\frac{A \leftrightarrow B}{\varphi(A) \leftrightarrow \varphi(B)} \qquad RR(\varphi)$$

For a given φ-logic $\mathscr{L}$ and a set $\Gamma \subset Fm(\varphi)$ the expression $\mathscr{L} + \Gamma$ denotes the smallest φ-logic containing $\mathscr{L}$ and Γ. A φ-logic $\mathscr{L}$ is *conservative* over s.i. logic L if $B \in \mathscr{L}$ implies $B \in L$ for any pure formula B. An *explicit expression for* φ is a formula of the form $\varphi(p) \leftrightarrow B$ where p is a variable and B is a pure formula.

Definition 1 (*Novikov, see Smetanich* (1960, 1961)) A φ-logic $\mathscr{L}$ *determines a new logical connective in s.i. logic* L if the following holds:

(1) $\mathscr{L}$ is conservative over L;
(2) $\mathscr{L}$ contains $(p \leftrightarrow q) \to (\varphi(p) \leftrightarrow \varphi(q))$ (the *Replacement Axiom for* φ);
(3) for every pure formula B the φ-logic $\mathscr{L} + \varphi(p) \leftrightarrow B$ is not conservative over L (in other words it is impossible to add an explicit expression to $\mathscr{L}$ without destroying the conservativeness).

Definition 2 (*Novikov, see Smetanich* (1960, 1961)) A φ-logic $\mathscr{L}$ is said to be *complete over* L if it is conservative over L and for every formula $A \notin \mathscr{L}$ φ-logic $\mathscr{L} + A$ is not conservative over L.[2]

The first condition of the Definition 1 is considered by several authors as the initial requirement for a given φ-logic to determine a new connective. In other words a φ-logic $\mathscr{L}$ should include L as the *pure basis*. In this case we may treat an additional connective as *belonging to* L.

The second "minimal" condition is that φ-logic $\mathscr{L}$ should not contain an explicit expression for φ. P. Novikov's approach accepts some strengthening of it: there should be no conservative extensions of $\mathscr{L}$ containing an explicit expression.

The Definition 2 may be considered as an attempt to answer the following question: which φ-logics, determining a new connective, should be studied (recall that in general there are continuum many logics determining a new connective: for Int it was established in Bessonov (1977))? Of course, some useful examples should appear in connection with applications, but in theirs absence it seems natural to study *maximal* conservative φ-logics.

Remark 1 Since the term "completeness" is used in mathematical logic in various contexts we shall use *Novikov completeness* for what is mentioned in the Definition 2.

The first example of a new connective for Int w.r.t. Novikov's approach was found by Smetanich (1960, 1961):

$$T^{\varphi} = Int + \begin{cases} \varphi(p) \leftrightarrow \varphi(q) \\ \neg\neg\varphi(p) \\ \varphi(p) \to (q \vee \neg q). \end{cases}$$

[2]The author does not know whether these notions were explicitly mentioned in any of Novikov's papers.

However the first additional axiom of T^{φ} shows that φ in the Smetanich logic may be regarded as a logical constant. According to this remark we build upon the Definition 1 the following way. Say that $\mathscr{L}$ determines an *essentially nonconstant* connective φ for L, if, in addition, $\mathscr{L} + \varphi(p) \leftrightarrow \varphi(q)$ is not conservative over L (variables p and q are distinct). An example of new essentially nonconstant unary connective for Int was described by Bessonov (1977).

The so-called *Novikov problem for s.i.logic L* was formulated as follows:

(a) to give an explicit example of Novikov complete conservative extension of L with new unary essentially nonconstant connective;
(b) to describe the family of Novikov complete extensions of L and to find out which of them determine a new unary connective.

In (1960, 1961) Smetanich considered the s.i. *logic of two-element chain* (also known as the *logic* $L(J_1)$ *of the first Jaskowsky matrix* J_1). His main result concerning this logic is that there are exactly 9 Novikov complete extensions of $L(J_1)$ determining new connectives (three of them determine new constants).

In this paper we are interested in the *Dummett logic* LC (Dummett 1959) and the so-called *irreflexive modality* φ interpreted in Kripke models as

$$x \Vdash \varphi(A) \Leftrightarrow \forall y > x\, (y \Vdash A),$$

in connection with Novikov's approach.

The irreflexive modality allows us to describe a countable family of pairwise incompatible conservative over LC φ-logics such that each of them determines a new essentially nonconstant connective for LC (Yashin 2014); however the question about an explicit example of a Novikov complete extension of LC remained open.

In this paper we describe an explicit example of Novikov complete φ-logic over LC with new unary essentially nonconstant connective. Moreover, this φ-logic is incompatible with every φ-logic mentioned in Yashin (2014).

It is pre-supposed that the reader is familiar with the basic notions of metamathematics of s.i. logics (e.g. Chagrov and Zakharyashchev 1997): the intuitionistic propositional logic Int, Kripke frames, pseudo-Boolean algebras, general frames, generated subframes, p-morphism, the logic of a class of frames etc.

14.2 Preliminaries

In this section we will recall some notions and theorems from metamathematics of s.i. logics.

14.2.1 Frames

A *Kripke frame* (or simply a *frame*) is a nonempty partially ordered set W. Elements of W are called *worlds, nodes, points* etc. A frame W is called rooted if it has the smallest node which is called a *root* and is denoted by $root\ W$. A subset $X \subseteq W$ is called a *cone* if the following holds

$$\forall x, y \in W\ (x \in X \,\&\, x \leqslant y \Rightarrow y \in X).$$

The set of all cones in W will be denoted by $Con\ W$. $W^x = \{y \in W \mid x \leqslant y\}$ is called a cone *generated* by node x.

Kripke operations $\supset$ and $\neg$ on $Con\ W$ are defined as follows:

$$X \supset Y := \{x \in W \mid X \cap W^x \subseteq Y \cap W^x\},$$
$$\neg X := \{x \in W \mid X \cap W^x = \varnothing\}.$$

It is easy to see that $\neg X = X \supset \varnothing$.

A structure of the form $(Con\ W, \cap, \cup, \supset, \neg)$, where $\cap$ and $\cup$ are the usual set-theoretical operations, is an example of a *pseudo-Boolean algebra (p.b.algebra)* or *Heyting algebra* (Rasiowa and Sikorsky 1968). P.b.algebras model the intuitionistic propositional logic in the same sense that Boolean algebras model the classical propositional logic.

The following representation theorem is well known.

Theorem 1

(1) Every p.b.algebra is isomorphic to a subalgebra of algebra $Con\ W$ for some frame W.
(2) Every finite p.b.algebra is isomorphic to $Con\ W$ for some finite frame W.

In connection with this theorem the notion of a general frame is widely and successfully used. A *general frame* is a pair $\mu = (W, S)$ where W is a frame and S is a subset of $Con\ W$ containing $\varnothing$ and closed under operations $\cap, \cup, \supset$ (see Chagrov and Zakharyashchev 1997). In case $S = Con\ W$ general frame (W, S) is identified with the (ordinary) frame W. Keeping in mind the second part of the representation theorem for p.b. algebras, in the sequel all finite general frames will be identified with finite (ordinary) frames.

14.2.2 Models

Formulas are constructed from the propositional variables $VAR = \{p_i, q_j, \ldots\}$ and propositional constants $\mathsf{0}$ and $\mathsf{1}$ by means of logical connectives $\rightarrow, \wedge, \vee, \neg, \leftrightarrow$ in the usual way. The set of all formulas will be denoted by Fm. $Sub\ A$ denotes the set of all subformulas of A. In particular $A \in Sub\ A$.

A *substitution* on Fm is a function $s : Fm \to Fm$ which preserves constants and commutates with logical connectives. It is clear that substitution is fully determined by its values on variables $p \in Var$. For a substitution s and a formula A $s(A)$ is called a *substitutional instance* of A.

An *evaluation* on a general frame $\mu = (W, S)$ is a function $v : Var \to S$. A general frame together with an evaluation is called a *model*. For arbitrary model (W, S, v) the *value* $v(A)$ of a formula A is defined by induction:

$$\begin{array}{llll} v(A \vee B) & := v(A) \cup v(B), & v(0) & := \varnothing, \\ v(A \wedge B) & := v(A) \cap v(B), & v(1) & := W, \\ v(A \to B) & := v(A) \supset v(B), & v(\neg A) & := \neg\, v(A). \end{array}$$

It is clear that $v(A) \in S$.

The *forcing relation* in a given model (W, S, v) is defined by

$$x \Vdash_v A :\Leftrightarrow x \in v(A).$$

A formula A is said to be *true* in a model $M = (W, S, v)$ $(M \models_v A))$, if $v(A) = W$. A formula A is *valid* in a general frame μ $(\mu \models A)$ if for any evaluation v on μ we have $(\mu, v) \models A$. For a given class $\mathbf{M}$ of general frames the set $L(\mathbf{M})$ of formulas is defined as

$$L(\mathbf{M}) := \{A \in Fm \mid \forall \mu \in \mathbf{M} : \mu \models A\}.$$

The *intuitionistic propositional logic* Int may be defined as the set of formulas which are valid in the class of all frames. A *superintuitionistic logic* is an arbitrary set L of formulas including Int and closed under the modus ponens and the substitution rules:

$$A, A \to B \in L \Rightarrow B \in L,$$

$$A \in L,\ s : Fm \to Fm \text{ (substitution)} \Rightarrow s(A) \in L.$$

Notice, that the set Fm satisfies these conditions too, i.e. it is a s.i.logic, but a *contradictory* one. Below we will consider only noncontradictory logics.

It is clear that $L(\mathbf{M})$ is a s.i.logic called the *logic of class* $\mathbf{M}$.

Theorem 2 (Modeling theorem Rasiowa and Sikorsky 1968) *Let L be a s.i. logic, then there exists a class* $\mathbf{M}$ *of general frames such that* $L = L(\mathbf{M})$.

14.2.3 Generated Subframes

Let $\mu = (W, S)$ be a general frame and $W' \in Con\ W$. Put

$$\mu' = (W', S'),$$

where $S' := \{X \cap W' \mid X \in S\}$, then μ' is called the frame *generated* by cone W'. This notion is a generalization of the notion of W^x.

For an evaluation v on μ the corresponding evaluation v' on μ' is defined by

$$v'(p) := v(p) \cap W' \text{ for } p \in Var.$$

Theorem 3 (On generated submodels) *Under the conditions above for any formula A we have*

$$v'(A) = v(A) \cap W'.$$

14.2.4 p-Morphisms

Let $\mu = (W, S)$ and $\mu' = (W', S')$ be general frames. A function $h\colon W \longrightarrow W'$ is called a *p-morphism from μ onto μ'* if the following conditions hold:

$$\begin{array}{lll} (i) & y \in W' \Rightarrow \exists x \in W : h(x) = y & \text{(surjectivity);} \\ (ii) & x \leqslant y \Rightarrow h(x) \leqslant' h(y) & \text{(monotonicity);} \\ (iii) & h(x) \leqslant' b \Rightarrow \exists y \geqslant x : h(y) = b & \text{(conicity);} \\ (iv) & \forall X \in S' : h^{-1}(X) \in X & \text{(continuity).} \end{array}$$

We use notation $h : \mu \twoheadrightarrow \mu'$ (or $h : (W, S) \twoheadrightarrow (W', S')$) for p-morphisms; the expression $\mu \twoheadrightarrow \mu'$ means that there exists a p-morphism $h : \mu \twoheadrightarrow \mu'$.

Example 1 Let $\mu = (W, S)$, $\mu' = (W, S')$ (with the same W) be general frames and $S' \subseteq S$. Then the identity mapping $id : W \to W$ is a p-morphism from μ onto μ'.

Let $h : \mu \twoheadrightarrow \mu'$ and v be an evaluation on μ'. The evaluation $(h^{-1}v)$ on μ is defined by $(h^{-1}v)(p) := h^{-1}(v(p))$ for $p \in Var$.

Theorem 4 (On p-morphisms) *Under the conditions above for any formula A*

$$(h^{-1}v)(A) = h^{-1}(v(A)).$$

A general frame μ is called a *reduct* of a general frame η if there exist a generated subframe η' of η and a p-morphism $\eta' \twoheadrightarrow \mu$. Notation: $\mu \preceq \eta$. The notion of *majorization* on classes of general frames is defined by

$$\mathbf{M} \preceq \mathbf{N} :\Leftrightarrow \forall \mu \in \mathbf{M}\, \exists \eta \in \mathbf{N} : \mu \preceq \eta.$$

Theorem 5 (Direct comparison theorem) *Suppose* $\mathbf{M}$ *and* $\mathbf{N}$ *are two classes of general frames, then*

$$\mathbf{M} \preceq \mathbf{N} \Longrightarrow L(\mathbf{N}) \subseteq L(\mathbf{M})$$

Theorem 6 (Inverse comparison theorem). *Let* $\mathbf{F}$ *be a class of finite rooted frames and* $\mathbf{M}$ *be a class of general frames. Then*

$$L(\mathbf{M}) \subseteq L(\mathbf{F}) \Longrightarrow \mathbf{F} \preceq \mathbf{M}.$$

A class $\mathbf{M}$ of general frames is called *characteristic* (for L) if $L(\mathbf{M}) = L$.

A s.i.logic L has the *finite model property*, if there exists a class $\mathbf{H}$ of finite frames such that $L = L(\mathbf{H})$.

Proposition 1 *Suppose* $L = L(\mathbf{H})$, *where* $\mathbf{H}$ *is a class of finite frames,* $\mathbf{M}$ *is a class of general frames such that* $L \subseteq L(\mathbf{M})$. *Then* $\mathbf{M}$ *is characteristic for* L *iff* $\mathbf{H} \preceq \mathbf{M}$.

14.3 φ-Logics and φ-frames

In this section we generalize some metamathematical notions to language $\mathscr{L}(\varphi)$ obtained from ordinary propositional language by adding a new unary logical connective (we partly draw upon Skvortsov 1983).

The class of formulas of enriched language will be denoted by $Fm(\varphi)$. Formulas not containing φ are said to be *pure*.

A *general* φ*-frame* is a triple $\mu = (W, S; \Phi)$ where (W, S) is an (ordinary) general frame and $\Phi : S \to S$ is an operator on S.

Remark 2 Notice that the expression $\Phi(X)$ only makes sense for $X \in S$ and thus every use of an expressions of the form $\Phi(\cdot)$ requires us to check whether the argument belongs to S.

The notion of an *evaluation* is defined as in Sect. 14.2. Given an evaluation v the value $v(A)$ is defined the same way with an additional clause:

$$v(\varphi(A)) := \Phi(v(A)).$$

We denote by $\mathscr{L}(\mathbf{M})$ the set of formulas valid in the class $\mathbf{M}$ of general φ-frames.

The validity clause for φ immediately implies that $\mathscr{L}(\mathbf{M})$ is closed under the Replacement Rule for φ, i.e. $\mathscr{L}(\mathbf{M})$ is a φ-logic (see Sect. 14.1).

14.3.1 *Modeling of* φ*-logics*

Theorem 7 (Modeling of φ-logics) *For any* φ*-logic* $\mathscr{L}$ *(conservative over* L*) there exists a (characteristic for* L*) class* $\mathbf{M}$ *of general* φ*-frames such that* $\mathscr{L} = \mathscr{L}(\mathbf{M})$.

Some proofs of modeling theorem are based on the notion of Lindenbaum algebra and on the representation theorem for p.b.algebras. Here we describe a more direct proof.

Let us fix an noncontradictory φ-logic $\mathscr{L}$ (i.e. $\mathscr{L} \neq Fm(\varphi)$).

We consider pairs of the form $(\Gamma; \Delta)$ where $\Gamma, \Delta \subseteq Fm(\varphi)$ (Γ, Δ may be infinite). We write Γ, A instead of $\Gamma \cup \{A\}$ for brevity. A pair $(\Gamma; \Delta)$ is called *inconsistent* (w.r.t. $\mathscr{L}$) if there exist $A_1, \ldots, A_m \in \Gamma$ and $B_1, \ldots, B_k \in \Delta$ such that $A_1 \wedge \ldots \wedge A_m \to B_1 \vee \ldots \vee B_k \in \mathscr{L}$; otherwise, $(\Gamma; \Delta)$ is called *consistent*. A pair $(\Gamma; \Delta)$ is called *maximal* if $\Gamma \cup \Delta = Fm(\varphi)$.

Lemma 1 *Let $(\Gamma; \Delta)$ be a consistent pair and $C \in Fm(\varphi)$ be an arbitrary formula. Then at least one of the pairs $(\Gamma, C; \Delta)$ and $(\Gamma; \Delta, C)$ is consistent.*

Proof Follows from the famous Cut-rule, which is admissible in Int. □

Corollary 1 *For every consistent pair $(\Gamma; \Delta)$ there exists a maximal consistent pair $(\Gamma'; \Delta')$ such that $\Gamma \subseteq \Gamma'$ and $\Delta \subseteq \Delta'$.*

Maximal consistent pairs will be called *types*. Types will be denoted by small Greek letters $\sigma, \tau, \varkappa, \ldots$, in addition $\sigma = (\sigma^0; \sigma^1)$. Denote $M := \{\sigma \mid \sigma \text{ is a type}\}$. Since $\mathscr{L}$ is assumed to be noncontradictory, there exists a formula $A \notin \mathscr{L}$. The pair $(\varnothing; A)$ is consistent w.r.t. $\mathscr{L}$. By Corollary 1 there exists at least one type.

The set of types is equipped with the following relation: $\sigma \leqslant \tau :\Leftrightarrow \sigma^0 \subseteq \tau^0$. It is easy to check that $\leqslant$ is a partial ordering on M. So we have a frame $(M, \leqslant)$. It is called the *canonical frame of $\mathscr{L}$*.

Lemma 2 (Consistency machinery) *For any type δ:*

(1) $1 \in \sigma^0, 0 \in \sigma^1$;
(2) $A \wedge B \in \sigma^0 \Rightarrow A \in \sigma^0$ *and* $B \in \sigma^0$;
(3) $A \wedge B \in \sigma^1 \Rightarrow A \in \sigma^1$ *or* $B \in \sigma^1$;
(4) $A \vee B \in \sigma^0 \Rightarrow A \in \sigma^0$ *or* $B \in \sigma^0$;
(5) $A \vee B \in \sigma^1 \Rightarrow A \in \sigma^1$ *and* $B \in \sigma^1$;
(6) $A \to B \in \sigma^0 \Rightarrow B \in \sigma^0$ *or* $A \in \sigma^1$;
(7) $A \to B \in \sigma^1 \Rightarrow \exists \tau \geqslant \sigma : A \in \tau^0$ *and* $B \in \tau^1$;
(8) $\neg A \in \sigma^0 \Rightarrow A \in \sigma^1$;
(9) $\neg A \in \sigma^1 \Rightarrow \exists \tau \geqslant \sigma : A \in \tau^0$.

For every formula A define a subset $|A| \subseteq M$ the following way:

$$|A| := \{\sigma \in M \mid A \in \sigma^0\}.$$

Obviously $|A| \in Con\, M$; $|A|$ may be called *the cone corresponding to formula A*. In terms of $|\cdot|$ the Consistency machinery lemma may be reformulated as follows:

Corollary 2 *Let $A, B \in Fm(\varphi)$ be arbitrary formulas, then*

(1) $|0| = \varnothing$, $|1| = M$;

(2) $|A \wedge B| = |A| \cap |B|$;
(3) $|A \vee B| = |A| \cup |B|$;
(4) $|A \to B| = |A| \supset |B|$;
(5) $|\neg A| = \neg\, |A|$;
(6) $A \in \mathscr{L} \Rightarrow |A| = M$;
(7) $A \leftrightarrow B \in \mathscr{L} \Rightarrow |A| = |B|$.

Put $S := \{|A| \mid A \in Fm(\varphi)\}$. Clearly S is a subalgebra of $Con\ M$. So we have a general frame $\mu = (M, S)$. Let us define $\Phi\colon S \to S$ on it the following way:

$$\Phi(|A|) := |\varphi(A)|.$$

It follows from the Replacement Rule for φ that $\Phi(|A|)$ does not depend on the choice of a formula A.

Thus we obtain a general φ-frame $\mu = (M, S; \Phi)$ called the *canonical φ-frame of $\mathscr{L}$*. Let us define the *canonical evaluation* $v_0\colon Var \to S$ by putting $v_0(p) := |p|$ for $p \in Var$.

Lemma 3 (Commutation) *For every formula $A \in Fm(\varphi)$ we have $v_0(A) = |A|$.*

Corollary 3 *If $A \notin \mathscr{L}$ then $\mu \not\models A$.*

In remains to see that the canonical φ-frame μ is a model of $\mathscr{L}$.

Lemma 4 *Let $v\colon Var \to S$ be an arbitrary evaluation on the canonical φ-frame $\mu = (W, S; \Phi)$ and $A \in \mathscr{L}$. Then $v(A) = M$.*

Proof Suppose $A = A(p_1, \ldots, p_m)$ and $v(p_i) = X_i \in S$. Choose formulas B_i such that $X_i = |B_i|$. Since $A \in \mathscr{L}$, we have $A(B_1, \ldots, B_m) \in \mathscr{L}$ by the substitution rule. Moreover, since $\mathscr{L}$ is closed under the substitution rule we have $v(A) = \tilde{A}(v(p_1), \ldots, v(p_m)) = \tilde{A}(|B_1|, \ldots, |B_m|) = |A(B_1, \ldots, B_m))| = M$, where $\tilde{A}$ is the pseudo-Boolean term corresponding to formula A. □

Thus, Theorem 7 is proved.

14.3.2 The Replacement Axiom and Generated φ-Subframes

Recall that the *Replacement Axiom* for φ is a formula

$$(p \leftrightarrow q) \to (\varphi(p) \leftrightarrow \varphi(q)) \qquad RA(\varphi).$$

The Replacement Axiom for φ reflects a fundamental intuitionistic idea that the truth-value of a sentence on a given node of an intuitionistic Kripke model should depend only on the *future cone* of this node (Skvortsov 1983):

Lemma 5 *Suppose* $(W, S; \Phi) \models RA(\varphi)$, $W' \in Con\ W$ *and* $X, Y \in S$, *then*

$$X \cap W' = Y \cap W' \Longrightarrow \Phi(X) \cap W' = \Phi(Y) \cap W'.$$

Proof Assume that $X \cap W' = Y \cap W'$ but $\Phi(X) \cap W' \neq \Phi(Y) \cap W'$. It immediately follows from this that $RA(\varphi)$ is refuted in (W, S, Φ) w.r.t. an evaluation v such that $v(p) = X$ and $v(q) = Y$. □

Let $\mu = (W, S; \Phi) \models RA(\varphi)$ and $W' \in Con\ W$. We define a general φ-frame $\mu' = (W', S'; \Phi')$, where (W', S') is generated from (W, S) by W' and for every $X \in S$

$$\Phi'(X \cap W') := \Phi(X) \cap W'.$$

This definition is sound by previous lemma and μ' is called the *general* φ*-subframe generated from* μ *by* W'.

Theorem 8 *Suppose a general* φ*-frame* μ' *is generated from a general* φ*-frame* μ *by a cone* W', v *is an evaluation on* μ *and* $v'(p) := v(p) \cap W'$ *for all* $p \in Var$. *Then for every formula* $A \in Fm(\varphi)$ *we have*

$$v'(A) = v(A) \cap W'.$$

14.3.3 p_φ-Morphisms

Suppose $\mu = (W, S; \Phi)$, $\mu' = (W', S'; \Phi')$ are general φ-frames and $h : (W, S) \twoheadrightarrow (W', S')$. p-Morphism h is said to be a p_φ*-morphism* from μ onto μ' if

$$\forall X \in S' : h^{-1}(\Phi'(X)) = \Phi(h^{-1}(X)).$$

Notation: $h : (W, S; \Phi) \overset{\varphi}{\twoheadrightarrow} (W', S'; \Phi')$ or $h : \mu \overset{\varphi}{\twoheadrightarrow} \mu'$; the expression $\mu \overset{\varphi}{\twoheadrightarrow} \mu'$ means that there exists $h : \mu \overset{\varphi}{\twoheadrightarrow} \mu'$.

Theorem 9 (On p_φ-morphisms) *Let* $h : (W, S, \Phi) \overset{\varphi}{\twoheadrightarrow} (W', S', \Phi')$, v *be an evaluation on* μ' *and let* $(h^{-1}v)$ *be the corresponding evaluation on* μ *defined by* $(h^{-1}v)(p) := h^{-1}(v'(p))$ *for* $p \in Var$. *Then for every formula* $A \in Fm(\varphi)$ *we have*

$$(h^{-1}v)(A) = h^{-1}(v(A)).$$

Proof By easy induction on the length of formula A. The induction step for φ follows from the p_φ-morphism condition. □

In the sequel we assume that all general φ-frames under consideration model the Replacement Axiom for φ.

We say that μ is a *φ-reduct* of η, if there exist a generated φ-subframe $\eta' \subseteq \eta$ and p_φ-morphism $\eta' \overset{\varphi}{\twoheadrightarrow} \mu$. Notation: $\mu \overset{\varphi}{\preceq} \eta$.

Let **M** and **N** be classes of general φ-frames. We define

$$\mathbf{M} \overset{\varphi}{\preceq} \mathbf{N} :\Leftrightarrow \forall \mu \in \mathbf{M}\, \exists \eta \in \mathbf{N} : \ \mu \overset{\varphi}{\preceq} \eta.$$

Theorem 10 (Direct comparison theorem for φ-logics) *Suppose* **M** *and* **N** *are two classes of general φ-frames. Then*

$$\mathbf{M} \overset{\varphi}{\preceq} \mathbf{N} \Longrightarrow \mathscr{L}(\mathbf{N}) \subseteq \mathscr{L}(\mathbf{M}).$$

14.4 Some Information About Dummett Logic

Dummett logic appeared in Dummett (1959) as

$$LC := Int + Lin,$$

where $Lin := (A \to B) \vee (B \to A)$ is called the *Linearity Axiom.*

It is known that LC is the logic of the class **C** of finite linearly ordered frames (thus LC stands for the *Logic of Chains*). Moreover, every infinite chain is a characteristic model of LC. Dummett logic is one of the three *pretabular* s.i.logics, i.e. it is not tabular, but every proper extension of it is tabular (see Maksimova 1972).

Lemma 6 *For any rooted general frame (W, S) the following are equivalent:*

(a) $(W, S) \models LC$;
(b) *S is linearly ordered by the inclusion relation.*

Theorem 11 (Characteristic class criterion for LC) *For any class* **M** *of general φ-frames we have*

$$L(\mathbf{M}) = LC \Leftrightarrow \mathbf{C} \preceq \mathbf{M}.$$

Formulas bd_n are defined inductively:

$$bd_1 := p_1 \vee \neg p_1, \qquad bd_{n+1} := p_{n+1} \vee (p_{n+1} \to bd_n).$$

Proposition 2 (Boundaries of the depth) *For given n the formula bd_n is valid in every chain of the depth $\leqslant n$ and is refuted in every chain of the depth $> n$.*

Formulas bd_n may be viewed as typical pure formulas not belonging to LC.

14.5 Dummett Logic and the Irreflexive Modality

The *irreflexive modality* φ in Kripke models is interpreted as

$$x \Vdash \varphi(A) :\Leftrightarrow \forall y > x : y \Vdash A.$$

Consider the following three formulas:

1° $A \to \varphi(A)$,
2° $(\varphi(A) \to A) \to A$,
3° $\varphi(A) \to (B \vee (B \to A))$.

These formulas were introduced by A. Kuznetsov in (1985) in connection with the intuitionistic provability operator.[3]

The *Kuznetsov Logic* $Kz(\varphi) := Int(\varphi) + 1° + 2° + 3°$ was used as a basic point to construct an explicit example Novikov complete extension for Int determining a new unary connective (see Yashin 1999).

Proposition 3 *Formulas* 1° *and* 3° *are valid in every Kripke frame; formula* 2° *is valid in every Kripke frame satisfying the ascending chain condition (in particular, in every finite chain).*

Denote by $\mathscr{LC}$ the φ-logic of the class of finite chains equipped with the irreflexive modality operator. In other words $\mathscr{LC}$ may be considered as a hybrid between Dummett logic and Kuznetsov φ-logic.

Aside from 1°, 2° and 3° the following formulas belong to $\mathscr{LC}$ (see Yashin 1999):

4° $(A \to B) \to (\varphi(A) \to \varphi(B))$ (monotonicity);
5° $(A \leftrightarrow B) \to (\varphi(A) \leftrightarrow \varphi(B))$ (the Replacement Axiom for φ);
6° $(\varphi(A) \wedge \varphi(B)) \to \varphi(A \wedge B)$;
7° $\varphi(0) \to (A \vee \neg A)$;
8° $\neg\neg\varphi(A)$;
9° $\varphi^n(0) \to bd_n$.

Proposition 4 $\mathscr{LC}$ *determines a new essentially nonconstant connective for* LC.

Proof $\mathscr{LC}$ is conservative over LC because it is defined by a characteristic for LC class of frames.

Let us show that $\mathscr{LC}$ does not admit any explicit expression $\varphi(p) \leftrightarrow B$, where $p \in Var$ and B is pure. Put $\mathscr{L}' := \mathscr{LC} + \varphi(p) \leftrightarrow B$ and assume that $\mathscr{L}'$ is conservative over LC. Without loss of generality we assume that B contains a single variable p. A simple classification for pure formulas containing a single variable is known (Smetanich 1961), that is for every formula $B(p)$ one of the following five cases holds: i) $B \leftrightarrow 0 \in Int$; ii) $B \leftrightarrow p \in Int$; iii) $B \leftrightarrow \neg p \in Int$; iv) $B \leftrightarrow \neg\neg p \in Int$; v) B is a classical tautology.

[3] We will not be addressing this topic here.

Let us consider these cases separately.

(i) We have $\varphi(p) \leftrightarrow 0 \in \mathscr{L}'$. Since $\neg 0 \in LC$, then $\neg\varphi(p) \in \mathscr{L}'$. Using 8° we obtain $0 \in \mathscr{L}\mathscr{C}$. Quod non.

(ii) We have $\varphi(p) \leftrightarrow p \in \mathscr{L}'$. Then from 2° we obtain $(p \to p) \to p \in \mathscr{L}'$. Since $p \to p \in Int$, we get $p \in \mathscr{L}'$. Quod non.

(iii) We have $\varphi(p) \leftrightarrow \neg p \in \mathscr{L}'$. Again from 2° we obtain $(\neg p \to p) \to p \in \mathscr{L}'$. Over Int this is equivalent to $\neg\neg p \to p$ which does not belong to LC.

(iv) We have $\varphi(p) \leftrightarrow \neg\neg p \in \mathscr{L}'$. From 3° we obtain $\neg\neg p \to (q \vee (q \to p)) \in \mathscr{L}'$. This pure formula does not belong to LC, since it is refuted on a three-element chain:

$$\bullet - - - - - - \bullet_q - - - - - - \bullet_{pq}$$

(v) We have $\varphi(p) \leftrightarrow B(p) \in \mathscr{L}'$, where B is a classical tautology. In this case $B(0) \in Int$ (see Smetanich 1961). We obtain $\varphi(0) \in \mathscr{L}'$ and, by 7°, we have $p \vee \neg p \in \mathscr{L}'$ which does not belong to LC.

Finally, add $\varphi(p) \leftrightarrow \varphi(q)$ where p and q are distinct. By 3° we obtain $\varphi(p) \to (r \vee (r \to q))$ and by 1° we obtain $p \to (r \vee (r \to q))$. This formula does not belong to LC, since it is refuted in the following model:

$$\bullet_p - - - - - - \bullet_{rp},$$

which completes our proof. □

The following statement is also notable:

Proposition 5 (Implicit definability Yashin 1999)

$$Kz(\varphi) + Kz(\psi) \vdash \varphi(A) \leftrightarrow \psi(A).$$

This proposition means that at most one operator satisfying $1° \wedge 2° \wedge 3°$ can be defined on a given pseudo-Boolean algebra **B**.

For arbitrary general frame (W, S) the *separation relation* $\prec$ is defined the following way (see Goldblatt 1976a, b):

$$x \prec y :\Leftrightarrow x < y \text{ and } \exists X \in S : (x \notin X \,\&\, y \in X).$$

We write $x \prec\!\!\cdot\, y$ if $x \prec y$ and there is no z such that $x \prec z \prec y$.

Proposition 6 *Suppose* $(W, S; \Phi) \models 1° \wedge 2° \wedge 3°$. *Then for every* $X \in S$

$$x \in \Phi(X) \iff \forall y \succ x : y \in X.$$

Proof See Proposition 3.4 in Yashin (1999). □

Corollary 4 *Suppose* $(W, S; \Phi) \models 1° \wedge 2° \wedge 3°$. *Then* $x \in \Phi^n(X)$ *iff for every* $\prec$-*chain* $x \prec x_1 \prec \cdots \prec x_n$ *of the length* $n + 1$ *we have* $x_n \in X$.

This corollary justifies the notion of the *depth of a node* x *w.r.t. a cone* $X \in S$. Put $dp(x, X) \leqslant n :\Leftrightarrow x \in \Phi^n(X)$ and for $n \geqslant 1$

$$dp(x, X) = n :\Leftrightarrow dp(x, X) \leqslant n \text{ and } dp(x, X) \nleqslant n - 1.$$

If for every natural n we have $x \notin \Phi^n(X)$, put $dp(x, X) := \infty$. In addition, for nonempty X one may put $dp(x, X) = 0 :\Leftrightarrow x \in X$. Notice that the expression $dp(x, \varnothing) = 0$ does not make sense.

14.6 Chains with Reflexive Nodes and p-Morphism Reconstruction

Let F be a finite chain. Choose an arbitrary subset $R \subseteq F$. Nodes from R will be called *reflexive* (the motivation for this term will be explained below).

A p-morphism $h\colon (W, S) \twoheadrightarrow F$ is called *regular* if

$$\forall x, y \in W(x \prec y \,\&\, h(x) = h(y) \Rightarrow h(x) = h(y) \in R).$$

Example 2 Suppose $R = F$. Then every p-morphism $h\colon (W, S) \twoheadrightarrow F$ is regular.

Example 3 Suppose $R = \varnothing$. Then a p-morphism $h\colon (W, S) \twoheadrightarrow F$ is regular iff h is a strongly monotonic function, i.e. $\forall x, y \in W(x \prec y \Rightarrow h(x) < h(y))$.

Proposition 7 (On p-morphism reconstruction, 1-st case) *Suppose* $\mu = (W, S; \Phi)$ *is a rooted general* φ*-frame such that* $\mu \models Lin \wedge 1° \wedge 2° \wedge 3°$, F *is a finite chain with a subset* R *of reflexive nodes, root* $F \in R$ *and* $h\colon (W, S) \twoheadrightarrow F$. *Then there exists a regular* p*-morphism* $f\colon (W, S) \twoheadrightarrow F$.

In this section letters x, y, z, w will be used for nodes in W and a, b, c will be used for nodes in F.

If $R = F$, i.e. there are no nonreflexive nodes in F, put $f := h$. Therefore assume that F contains at least one nonreflexive node.

Consider

$$R_0 := \{x \in R \mid \exists y \in F(x \lessdot y \,\&\, y \notin R)\} \subseteq R.$$

Suppose $R_0 = \{b_1, \dots, b_n\}$, then for each b_i there is a maximal nonreflexive interval $F_i := \{a_{is_i}, \dots, a_{i1}\} \subseteq F \setminus R$ such that

$$b_i \lessdot a_{is_i} \lessdot \cdots \lessdot a_{i1}$$

and if $a_{i1} \lessdot c$ then $c \in R$.

Put $X_i := h^{-1}(\{b_i\})$, $Y_i := h^{-1}(F_i)$ and $Z_i := h^{-1}(\{x \in F \mid x > a_{i1}\})$ (if a_{i1} is the largest element of F then put $Z_i := \varnothing$).

Lemma 7 *Suppose $x \in X_i$, then there exists a chain*

$$x \prec x_{s_i} \prec\!\cdot\; x_{s_{i-1}} \prec\!\cdot \dots \prec\!\cdot\; x_1,$$

such that $x_j \in Y_i$ and $dp(x_j, Z_i) = j$ for any $j \in \{1, \dots, s_i\}$.

Proof By conicity of h there exists a chain

$$x \prec y_{s_i} \prec y_{s_i-1} \prec \dots \prec y_1$$

in W such that $h(y_j) = a_{ij}$ for $j \in \{1, \dots, s_i\}$. Thus, we have $dp(y_{s_i}, Z_i) \geq s_i$.

If $dp(y_{s_i}, Z_i) = s_i$ then put $x_j := y_j$ for $j = 1, \dots, s_i$.

If $dp(y_{s_i}, Z_i) = s_i + k$ for some $k > 0$, then there exists $x_{s_i} \succ y_{s_i}$ such that $dp(x_{s_i}, Z_i) = s_i$. Thus we can build our chain starting with x_{s_i}.

Finally, consider the case $dp(y_{s_i}, Z_i) = \infty$. In this case $y_{s_i} \notin \Phi^{s_i-1}(Z_i)$. The axiom 2° implies $y_{s_i} \notin \Phi_i^s(Z_i) \supset \Phi^{s_i-1}(Z_i)$ (recall that $s_i > 0$). By definition of p.b. operation $\supset$ there exists $z > y_{s_i}$ such that $z \in \Phi^{s_i}(Z_i)$ and $z \notin \Phi^{s_i-1}(Z_i)$. Consequently $dp(z, Z_i) = s_i$. So we can take z as the least element of the desired sequence. □

Define a function $f\colon W \longrightarrow F$ the following way (see the notation preceding Lemma 7):

$$f(x) := \begin{cases} h(x), & \text{if } h(x) \in R; \\ a_{ij}, & \text{if } h(x) \in F_i \text{ and } dp(x, Z_i) = j \leqslant s_i; \\ b_i, & \text{if } h(x) \in F_i \text{ and } dp(x, Z_i) > s_i. \end{cases}$$

Let us check the p-morphism conditions for f.

(i) *Surjectivity of f.*

• If $b \in R$ then $b \in h(X) = f(X)$.

• Consider a nonreflexive node a_{ij}. By Lemma 7 there exists $x \in Y_j$ such that $dp(x, Z_i) = j$. By definition of f we have $f(x) = a_{ij}$.

(ii) *Monotonicity of f.* Suppose $x, y \in W$ and $x < y$.

• If $x, y \in h^{-1}(R)$ then $f(x) = h(x) \leqslant h(y) = f(y)$.

• If $x, y \in Y_i$ then $dp(y, Z_i) \leqslant dp(x, Z_i)$. Therefore $f(x) \leqslant f(y)$.

• If $x \in X_i$ and $y \in Y_i$, then $f(x) = h(x) = b_i$ and either $f(y) = b_i$ or $f(y) = a_{ij}$ for some $j \in \{1, \dots, s_i\}$. In both cases $f(x) \leqslant f(y)$.

• If $x \in Y_i$ and $y \in Z_i$, then $Z_i \neq \varnothing$ and there exists a reflexive node $c \in F$ such that $a_{i1} \prec\!\cdot\; c$. In this case $f(x) < c$ and $f(y) \geqslant c$.

The other cases are considered similarly.

(iii) *Conicity of f.* Since F is finite, it is sufficient to prove

$$f(x) \lessdot c \Longrightarrow \exists y > x : f(y) = c. \qquad (*)$$

There are four cases.

• $f(x), c \notin R$. We have $f(x) = a_{ij+1}$ and $c = a_{ij}$, so (∗) follows from definitions of f and dp: if $f(x) = a_{ij+1}$ then $dp(x, Z_i) = i + 1$ by definition of f and by definition of dp there is $y > x$ such that $dp(y, Z_i) = j$, that is $f(y) = a_{ij}$.

• $f(x) \notin R$, $c \in R$. We have $f(x) = a_{i1}$, $h(x) < c$. There exists a node $y \in Z_i$ such that $y > x$ and $h(y) = c$. In this case $f(y) = h(y) = c$.

• $f(x) \in R$, $c \notin R$. We have $f(x) = b_i \lessdot a_{is_i}$. By Lemma 7 there exists $x_{s_i} \in Y_i$ such that $dp(x_{s_i}, Z_i) = s_i$. By definition $f(x_{s_i}) = a_{is_i}$.

• $f(x), c \in R$. We have $f(x) = h(x) \lessdot c$. There exists $y > x$ such that $h(y) = c$. By definition $f(y) = h(y) = c$.

(iv) *Continuity of* f. We need to show that $f^{-1}(F^c) \in S$ for $c \in F$.

• If $c \in R$ then $f^{-1}(F^c) = h^{-1}(F^c) \in S$.

• If $c = a_{ij}$ then $f^{-1}(F^c) = \Phi^j(h^{-1}(\{x \in F \mid x > a_{i1}\})) \in S$.

Finally, let us show the *regularity* of f. Suppose $x, y \in W$ such that $x \prec y$ and $f(x) = f(y) = a$. We will show this by contradiction. Thus, assume $a \in F \setminus R$. Then a belongs to some nonreflexive interval $a_{is_i} \lessdot \ldots \lessdot a_{ij}(= a) \lessdot \ldots \lessdot a_{i1}$. So $dp(x, Z_i) = dp(y, Z_i) = j$ and there exists a chain $x \prec y \prec z_{j-1} \prec \ldots \prec z_1$ where $z_1 \notin Z_i$. There is also $U \in S$ such that $x \notin U$ and $y \in U$. Contradiction with $dp(x, Z_i) = j$. So the node a is reflexive.

Therefore, Proposition 7 is proved.

Proposition 8 (On p-morphism reconstruction, 2-nd case) *Suppose* $\mu = (W, S; \Phi)$ *is a rooted general* φ*-frame,* $\mu \models Lin \wedge 1° \wedge 2° \wedge 3°$, F *is a finite chain with a subset* R *of reflexive nodes, root* $F \notin R$ *and* $h\colon (W, S) \twoheadrightarrow F$. *Then there exist a generated* φ*-subframe* $\mu' = (W', S'; \Phi')$ *and a regular p-morphism* $f\colon (W', S') \twoheadrightarrow F$.

Proof Let $root F = a \in F \setminus R$. Consider the lowest maximal nonreflexive interval beginning with $a = a_s$:

$$a(= a_s) \lessdot \ldots \lessdot a_1.$$

There exists $x \in W$ such that $h(x) = a_s$. Let $\mu' = (W', S'; \Phi')$ be a general φ-subframe generated by W^x, $root\, \mu' = x$ and let $h'\colon (W', S') \twoheadrightarrow F$ be the restriction of h on W'. Put $Y := h^{-1}(\{a_1, \ldots, a_s\})$ and $Z := h^{-1}(\{x \in F \mid x > a_1\})$. Then we have $dp(x, Z) = s$.

Let us define a new function $f\colon W' \to F$. For every $w \in Y$ put $f(w) := a_{dp(w,Z)}$. For $w \in W' \setminus Y$ function f is defined exactly as in the proof of Proposition 7. It is easy to show that f is the desired regular p-morphism from μ' onto F. □

14.7 Novikov Completeness for $\mathscr{LC}$

Theorem 12 $\mathscr{LC}$ *is a Novikov complete extension of LC.*

In order to prove this, let **M** be a characteristic class of general φ-frames and $\mathbf{M} \models Lin \wedge 1° \wedge 2° \wedge 3°$. Under these assumptions we have the following:

Proposition 9 $\mathscr{L}(\mathbf{M}) \subseteq \mathscr{L}\mathscr{C}$.

Fix a formula $A \notin \mathscr{L}\mathscr{C}$ containing variables $\bar{p} = \{p_1, \ldots, p_m\}$. There exists a finite φ-chain $(F; \Phi)$ (Φ is the irreflexive modality) and an evaluation v on F such that $root\ F \nVdash_v A$. Let a node $b \in F$ be *reflexive* if

$$\forall \varphi(B) \in Sub(A)\,(b \Vdash_v \varphi(B) \Rightarrow b \Vdash_v B).$$

Example 4 If A does not contain symbol φ, then every node in F is reflexive.

Example 5 If F consists of n nodes and A contains a subformula of the form $\varphi^n(0)$ then every node of F is nonreflexive.

Since $\mathbf{M}$ is characteristic for LC, there exists $\mu_1 = (W_1, S_1; \Phi_1) \in \mathbf{M}$ and a p-morphism $h\colon (W_1, S_1) \twoheadrightarrow F$.

By Propositions 7 and 8 there exist a generated subframe (W, S) of (W_1, S_1) and a regular p-morphism $f\colon (W, S) \twoheadrightarrow F$.

Let u be an evaluation on (W, S) concordant with v w. r. t. f, that is

$$x \Vdash_u p :\Leftrightarrow f(x) \Vdash_v p$$

for every $x \in W$ and every $p \in Sub(A)$. Then we have the following

Lemma 8 *For every formula* $C \in Sub(A)$ *we have*

$$\forall x \in W\,(x \Vdash_u C \Leftrightarrow f(x) \Vdash_v C). \qquad (\sharp)$$

Proof This lemma is proved in the usual manner — by induction on the construction of subformula C.

For $C = p \in Sub(A)$ the statement $(\sharp)$ coincides with the definition of the evaluation u. For constants 0 and 1 the statement $(\sharp)$ is obvious.

Consider a compound subformula $C \in Sub(A)$. Then the induction assumption is that *for every proper subformula of the formula C the statement* $(\sharp)$ *holds.*

Steps for standard connectives (i.e. when C has one of the forms $B_1 \wedge B_2$, $B_1 \vee B_2$, $B_1 \to B_2$ or $\neg B$) are proved in the usual way with the use of the initial properties of the p-morphism f.

Consider the case $C = \varphi(B)$.

Suppose $f(x) \nVdash_v \varphi(B)$, then there exists $c \in F$ such that $c > f(x)$ and $c \nVdash_v B$. By the conicity of f there exists $y > x$ such that $f(y) = c$. Then $y \in f^{-1}(F^c)$ and $x \notin f^{-1}(F^c)$. Since $f^{-1}(F^c) \in S$, we have $y \succ x$. By inductive hypothesis we obtain $y \nVdash_u B$. Therefore, $x \nVdash_u \varphi(B)$.

Suppose now that $x \nVdash_u \varphi(B)$, then there exists $y \succ x$ such that $y \nVdash_u B$. By inductive hypothesis we have $f(y) \nVdash_v B$ and $f(y) \geq f(x)$. One of two cases holds:

(1) $f(y) > f(x)$. Then $f(x) \nVdash_v \varphi(B)$.
(2) $f(y) = f(x)$. Since f is a regular p-morphism, $f(x)$ is a reflexive node in F. We have $f(x) \nVdash_v B$, from which $f(x) \nVdash_v \varphi(B)$ follows.

□

Let us complete the proof of Proposition 9. We have obtained $(W, S; \Phi) \not\models A$. Therefore, $A \notin \mathscr{L}(\mathbf{M})$.

Finally, the main result, that is Theorem 12, immediately follows from Proposition 9.

14.8 Conclusion

In Yashin (2014) a denumerable family of φ-logics was considered:

$$\mathscr{L}^n := \mathscr{L}(\omega + n),$$

where ω is the order type of natural numbers and n is an arbitrary natural number. The connective φ is interpreted as above by the irreflexive modality.

Each of $\mathscr{L}^n$ is conservative over LC and determines new unary nonconstant connective. All of $\mathscr{L}^n$ are *pairwise incompatible*, that is $\mathscr{L}^n + \mathscr{L}^k$ is not conservative over LC for $n \neq k$.

Proposition 10 *For every natural n logic $\mathscr{L}\mathscr{C}$ in not compatible with $\mathscr{L}^n$.*

Proof $\mathscr{L}^n$ contains the formula $\varphi^{n+1}(0) \to \varphi^n(0)$, while $\mathscr{L}\mathscr{C}$ contains the formula $(\varphi^{n+1}(0) \to \varphi^n(0)) \to \varphi^n(0)$, which is an instance of $2°$. These two formulas imply that $\varphi^n(0) \in \mathscr{L}\mathscr{C} + \mathscr{L}^n$. Finally, the last formula implies (in $\mathscr{L}\mathscr{C}$) the formula bd_n, which does not belong to LC. □

Acknowledgements The Author owes special thanks to Professor L. Maksimova for her diligent attention to problems concerning nonstandard logical connectives in nonclassical logics. Also the Author is grateful to unknown referee for useful technical comments and remarks.

References

Bessonov, A. V. (1977). New operations in intuitionistic calculus. *Mathematical notes of the Academy of Sciences of the USSR*, *22*, 503–506.
Chagrov, A., & Zakharyaschev, M. (1997). *Modal Logics.*, Oxford Logics Guides Oxford: Clarendon Press.
Dummett, M. (1959). A propositional calculus with denumarable matrix. *Journal of Symbolic Logic*, *24*, 97–106.
Goldblatt, R. I. (1976a). Metamathematics of modal logic. *I, Reports on Mathematical Logic*, *6*, 41–78.

Goldblatt, R. I. (1976b). Metamathematics of modal logic. *II, Reports on Mathematical Logic, 7,* 21–52.
Kuznetsov, A. V. (1985). On intuitionistic propositional provability calculus. *Doklady Akademii Nauk SSSR, 283,* 27–30.
Maksimova, L. L. (1972). Pretabular superintuitionistic logics. *Algebra and Logic, 11,* 308–314.
Rasiowa, H., & Sikorsky, R. (1968). *The Mathematics of Metamathematics.* Warsaw: PWN.
Smetanich, Ya. S. (1960). On completeness of a propositional calculus with an additional unary operation. In *Proceedings of Moscow mathematical society, 9,* 357–371.
Smetanich, Ya. S. (1961). O propozicionalnom ischislenii s dopolnitelnoy logicheskoĭ svyazkoĭ, *Doklady Akademii Nauk SSSR139,* 309–312. [On propositional calculi with an additional logical connective].
Skvortsov, D. P. (1983). On intuitionistic propositional calculus with an additional logical connectives. *Investigations in nonclassical logics and formal systems* (pp. 154–173). Moscow: Nauka.
Yashin, A. D. (1999). Irreflexive modality in the intuitionistic propositional logic and Novikov completeness. *Journal of Philosophical logic, 28,* 175–197.
Yashin, A. D. (2014). Irreflexive modality as a new logical connective in the Dummett logic. *Siberian Mathematical Journal, 55,* 185–190.

Chapter 15
On Linear Logic of Knowledge and Time

Veta F. Yun

The author expresses a sincere gratitude to L.L. Maksimova for her comprehensive help.

Abstract The paper is a survey of studies devoted to polymodal logics based on a class of frames with discrete linear time with current time point clusters. We consider some classes of frames connected to this class. The paper suggests several axiomatic calculi in various polymodal languages. We show that these axiomatic calculi are complete with respect to corresponding classes of frames. Offered calculi, as we show, posses many interesting properties, e.g. the finite model property and decidability.

15.1 Introduction

A local survey of studies devoted to polymodal logics based on a class of frames with discrete linear time with current time point clusters is presented in this paper.

It is well known that modal logic has a vast variety of syntaxes and semantics. On the one hand this explains the wide application of modal logics in areas of computer science and artificial intelligence. On the other hand there is rapid development in the models that describe calculus, proof, reasoning (cf. e.g. Calardo and Rybakov 2007; Fagin et al. 1995; Halpern et al. 2004; Rybakov 2008). A common approach to research is a combination of knowledge and time modalities.

There are many works devoted to this subject. Different logics of knowledge and time including linear ones were explored for example in Calardo and Rybakov (2007), Halpern et al. (2004), Lukyanchuk (2013), Rybakov (2008). For example, Rybakov (2008) examines the polymodal logic of linear discrete time with time point clusters. Here a class of special linearly ordered frames $\langle W, R \rangle$ with clusters

V. F. Yun (✉)
Sobolev Institute of Mathematics, 630090 Novosibirsk, Russia
e-mail: veta_v@mail.ru

S. Odintsov (ed.), *Larisa Maksimova on Implication, Interpolation, and Definability*, Outstanding Contributions to Logic 15, https://doi.org/10.1007/978-3-319-69917-2_15

(uncertain situations) is considered. Such frames naturally occur in computer science and artificial intelligence—clusters could be people with their knowledge, all internet resources and other information clusters available at the moment.

When defining logic with the help of frames, the most important problems are the choice of a modal language and the question of axiomatization of this logic. Rybakov (2008) extends the language of temporal logics by modal operations imitating the possibility of discovery in the time flow (i.e. two modalities - "week necessary in future" $\Box_w^+$ and "week necessary in past" $\Box_w^-$ are added). The logic of linear discrete time with time point clusters and these additional modalities examined in this work is of particular interest, due to the fact that the result is not a classical modal logic. Furthermore, this logic expresses the properties of a second order logic and possesses features which are difficult to express in usual modal languages.

V.F. Yun introduced a new relation R_1 between consecutive clusters. If temporal modalities $\Box_1$, $\Box_1^-$ connected with R_1 are added to the language then it is proved in Yun (2009) that weak modalities are expressed by the other. This inspired the author to consider wider classes of frames $\langle W, R, R_1 \rangle$ with various properties. Yun (2009) introduces the class of temporal linear $(S \subseteq R)Ind$ frames, and in Yun (2010) a class of Ind-frames with linear time is considered.

To axiomatize the class of temporal linear $(S \subseteq R)Ind$ frames it is natural to choose a temporal language with four modalities $\Box$, $\Box^-$ and $\Box_1$, $\Box_1^-$. In this language there is a calculus which is complete with respect to temporal linear $(S \subseteq R)Ind$ frames. It is proved (Yun 2009) that it is finitely approximable and therefore it is decidable.

Yun (2010) considers another class of frames (called Ind-frames with linear time) associated with linearly ordered frames $\langle W, R \rangle$ with current time point clusters. The paper suggests a calculus in the language with four modalities $\Box$, $\Box^-$, $\Box_1$ and $\Box_1^-$, complete with respect to the class Ind-frames with linear time. The author proves that the calculus is finitely approximable by finite Ind-frames with linear time and therefore it is decidable.

Rybakov (2008) examines LTK-frames $\langle \bigcup_{i \in N} C(i), R, R_\sim \rangle$ with additional relations $S_1, \ldots, S_k$ that imitate knowledge of agents and a relation $R_\sim$ describing common knowledge. Calardo and Rybakov (2007) presents an axiomatization for the polymodal logic LTK of the class of such frames. Polymodal logic LTK was formerly introduced in Calardo (2006), Calardo and Rybakov (2005).

Yun (2015b) introduces an additional relation R_1 between elements of consecutive clusters. A wider class of frames $\langle W, R, R_\sim, R_1 \rangle$ that are called inductive nearly LTK-frames are examined. The author also presents an axiomatization of the class of inductive nearly LTK-frames in the language with three modalities. It is obvious that if relations $S_1, \ldots, S_k$ and their corresponding modalities are added then axioms of $S5$ and axioms for $S_i \subseteq S_\sim$ have to be added. The author proves that the found calculus is decidable and has fmp.

Lukyanchuk (2013) presents a linear polymodal logic of knowledge and time with intransitive time relation as a set of formulae valid in frames of a special kind. Luk'yanchuk and Rimatskii (2013) introduce a calculus AS_{LTK_r} which is correct with respect to the class of these frames. In Yun (2015a), the author demonstrates a formula of the logic of knowledge and linear time with intransitive time relation that is not derivable in AS_{LTK_r}.

15.2 Preliminary Notes

In Rybakov (2008) V.V. Rybakov examined a polymodal logic of linear discrete time with time point clusters. More specifically, the author considers $\langle \bigcup_{i \in \mathbb{N}} C(i),\ R \rangle$ frames with linear ordered R-clusters of states $C(i)$ and examines the logic generated by these frames.

Let us consider the frames $\langle W,\ R_T,\ R_\sim \rangle$ where

(a) $W = \bigcup_{i \in \mathbb{N}} C(i)$ is the disjoint union of nonempty sets $C(i)$;

(b) $xRy \iff \exists i, j \in \mathbb{N}(x \in C(i), y \in C(j)$ and $i \le j)$.

Thus every set $C(i)$ is an R-cluster, so that for $x \in C(i)$: $C(i) = \{y \mid xRy$ and $yRx\}$.

We will introduce an additional relation R_1 between elements of consecutive clusters and consider frames $\langle X,\ R,\ R_1 \rangle$. The binary relation R_1 is as follows:

$$xR_1y \iff \exists i \in \mathbb{N}(x \in C(i) \text{ and } y \in C(i+1)).$$

In Rybakov (2008) V.V. Rybakov considered a polymodal language with time modal operators $\Box$, $\Box^-$ associated with the relation R, and weak modalities $\Box_w$, $\Box_w^-$:

$$x \models \Box A \iff \forall y(xRy \Longrightarrow y \models A),$$

$$x \models \Box^- A \iff \forall y(yRx \Longrightarrow y \models A),$$

$$x \models \Box_w A \iff \forall i[x \in C(i) \Longrightarrow (\forall j \ge i \exists y \in C(j)(y \models A))],$$

$$x \models \Box_w^- A \iff \forall i[x \in C(i) \Longrightarrow (\forall j \le i \exists y \in C(j)(y \models A))].$$

Let us add modal operators $\Box_1$, $\Box_1^-$ associated with relation R_1 to the language. Then it is possible to prove that weak modalities can be expressed through other modalities (Yun 2009):

Proposition 1 *The following formulae are valid in frames under consideration:*

$$\Box_w A \longleftrightarrow \Diamond_1 \Diamond_1^- A \& \Box \Diamond_1 A;$$

$$\Box_w^- A \longleftrightarrow \Diamond_1 \Diamond_1^- A \& \Box^- (\Diamond_1^- A \vee \Box_1^- \bot \& \Diamond^- A).$$

This is another reason why studying polymodal logics of frames $\langle \bigcup_{i \in \mathbb{N}} C(i),\ R \rangle$ with an additional relation between consecutive clusters is interesting. In Lukyanchuk

(2013) A.N. Lukyanchuk examined logics of frames $\langle \bigcup_{i\in N} C(i),\ R\rangle$ with an additional relation R_T between consecutive clusters satisfying different properties (reflexivity, irreflexivity, intransitivity).

The papers (Yun 2009, 2010, 2015b) deal with classes of frames with an additional relation R_1 between consecutive clusters associated with the class of special linearly ordered frames $(\bigcup_{i\in N} C(i),\ R)$ with time point clusters which satisfy special conditions.

The details on logics of these classes are given below.

15.3 Temporal Logic of Linear Time $(S \subseteq R)Ind$-Frames

We will consider a language with modal operators of the future $\Box$, $\Box_1$ and modal operators of the past $\Box^-$, $\Box_1^-$.

Let us consider the frames $\langle W,\ R,\ R_1\rangle$ where W is a nonempty set and $R,\ R_1$ are binary relations on W.

A model M on $\mathbb{F}$ is a pair $\langle \mathbb{F}, V\rangle$ where $\mathbb{F} = \langle W,\ R,\ R_1\rangle$ is a frame and V is a valuation which associates a set of worlds from W to each propositional letter $p \in P$. The valuation V can be extended in the standard way (Chagrov and Zakharyaschev 1997; Gabbay et al. 1994) onto all formulae of the considered language.

In particular for all $x \in W$:

$$x \models_V \Box A \iff \forall y(xRy \Longrightarrow y \models_V A),$$

$$x \models_V \Box_1 A \iff \forall y(xR_1y \Longrightarrow y \models_V A),$$

$$x \models_V \Box^- A \iff \forall y(yRx \Longrightarrow y \models_V A),$$

$$x \models_V \Box_1^- A \iff \forall y(yR_1x \Longrightarrow y \models_V A).$$

If $M = \langle \mathbb{F}, V\rangle$ is a model on the frame $\mathbb{F} = \langle W,\ R,\ R_1\rangle$, a formula A is said to be *true in the model M at the world x* if $x \models_V A$. A formula A is *true in the model* $M = \langle \mathbb{F}, V\rangle$ if $\forall x \in W(x \models_V A)$. A is *valid in the frame* $\mathbb{F}$ if the formula A is true in any model M on $\mathbb{F}$.

We will consider the frames with the following conditions:

(1) for all $x,y \in W$: xRy or yRx;
(2) R is a transitive relation;
(3) R is a reflexive relation;
(4) $xR_1y \Longrightarrow xRy$;
(5) if zR_1x and zR_1y then yRx and xRy;
(6) if xR_1z and yR_1z then yRx and xRy;
(7) for any $x \in X$ there is $y \in X$ such that xR_1y;
(8) $xRy \iff (x = y$ or there are $x_1, \ldots, x_n \in X$ such that $x = x_1S\ldots Sx_nSy)$.

Here S is an auxiliary relation defined as follows:

$$xSy \iff (a)\ xR_1y \text{ or } (b)\ \exists z(xR_1z \text{ and } yR_1z) \text{ or } (c)\ \exists z(zR_1x \text{ and } zR_1y).$$

Definition 1 A frame $\langle W,\ R,\ R_1\rangle$ is a linear time $(S \subseteq R)Ind$-frame if the conditions (1)–(8) are met. A model $M = \langle \mathbb{F}, V\rangle$ on a frame $\mathbb{F}$ is a linear time $(S \subseteq R)Ind$-model if $\mathbb{F}$ is a linear time $(S \subseteq R)Ind$-frame.

Note that frames $\langle \bigcup_{i\in\mathbb{N}} C(i),\ R\rangle$ (Rybakov 2008) are linear time $(S \subseteq R)Ind$-frames if a relation R_1 between consecutive clusters is defined as follows:

$$xR_1y \iff \exists i \in \mathbb{N}(x \in C(i), y \in C(i+1)).$$

Let us define **LInd** as the logic calculus obtained by adding the following axioms to the minimal temporal logic $K_t(R, R_1)$ Gabbay et al. (1994):

(1) $\Box(\Box A_1 \longrightarrow A_2) \vee \Box(A_2\&\Box A_2 \longrightarrow A_1)$;
(1^-) $\Box^-(\Box^- A_1 \longrightarrow A_2) \vee \Box^-(A_2\&\Box^- A_2 \longrightarrow A_1)$;
(2) $\Box A \longrightarrow \Box\Box A$;
(2^-) $\Box^- A \longrightarrow \Box^-\Box^- A$;
(3) $\Box A \longrightarrow A$;
(3^-) $\Box^- A \longrightarrow A$;
(4) $\Box A \longrightarrow \Box_1 A$;
(4^-) $\Box^- A \longrightarrow \Box_1^- A$;
(5) $\Diamond_1 A \longrightarrow \Box_1(\Diamond A\&\Diamond^- A)$;
(6) $\Diamond_1^- A \longrightarrow \Box_1^-(\Diamond A\&\Diamond^- A)$;
(7) $\Diamond_1 \top$;
(8) $A\&\Box(A \longrightarrow (\Box_1(A\&\Box_1^- A)\&\Box_1^-\Box_1 A)) \longrightarrow \Box A$.

Here and further $\Diamond_F$ is the notation for $\neg\Box_F\neg$ and $\Diamond_F^-$ is the notation for $\neg\Box_F^-\neg$ $(F \in \{R,\ R_1\})$.

Note that the axiom (8) is similar to the axiom of induction and the property (8) closely resembles the inductive property (Vakarelov 1988, 1992).

Theorem 1 (Soundness, Yun 2009) *For any formula A the following is true: if A is derivable in* **LInd** *then A is valid in any linear time* $(S \subseteq R)Ind$*-frame.*

Thus the calculus **LInd** is correct with respect to the class of linear frames with current time point clusters with additional relation R_1.

Also **LInd** is finitely approximable by linear time $(S \subseteq R)Ind$-models:

Theorem 2 (Yun 2009) *For any formula A the following is true: A is derivable in* **LInd** *iff A is true in any finite linear time* $(S \subseteq R)Ind$*-models.*

From the Theorem 2 it follows that **LInd** is complete with respect to linear time $(S \subseteq R)Ind$-frames. In conclusion, note that since the calculus **LInd** is finitely approximable and finitely axiomatizable, it is decidable.

15.4 The Temporal Logic of Inductive Frames with Linear Time

Definition 2 An *Ind-frame with linear time* is a frame $\langle W, R, R_1\rangle$, where the following conditions are true:

(1) for all $x, y \in W$: xRy or yRx;
(2) for any $x \in W$ there is $y \in W$ such that xR_1y;
(3) $\exists t_0(\forall x(t_0Rx)$ and $\forall y(yRt_0 \Longrightarrow \neg\exists z(zR_1y)))$;
(4) if xR_1y, zR_1y and zR_1u, then xR_1u;
(5) $xRy \iff (x = y$ or there are $x_1, \ldots, x_n \in X$ such that $x = x_1R_1 \ldots R_1x_nR_1y$ or there are $x_1, \ldots, x_m \in X$ such that $xR_1x_1R_1 \ldots R_1x_mR_1{}^{-1}y)$.

Note that if we define the relation of R_1 between consecutive clusters as before,

$$xR_1y \iff \exists i \in \mathbb{N}(x \in C(i), y \in C(i+1)),$$

then the frames $\langle\bigcup_{i\in\mathbb{N}} C(i), R\rangle$ with relation R_1 are *Ind*-frames with linear time.

Let us define **Ind**(**R**, **R**$_1$). The calculus **Ind**(**R**, **R**$_1$) is the logic calculus obtained by adding the following axioms (1)–(9) to the minimal temporal logic $K_t(R, R_1)$ (Gabbay et al. 1994):

(1) $\Box(\Box A_1 \longrightarrow A_2) \vee \Box(\Box A_2 \longrightarrow A_1)$,
$\Box(\Box^- A_1 \longrightarrow A_2) \vee \Box^-(\Box^- A_2 \longrightarrow A_1)$;

(2) $\Diamond_1\top$;

(3) $\Diamond^-\Box^-\Box_1^-\bot$;

(4) $\Diamond_1^-\Diamond_1 A \longrightarrow \Box_1^-\Diamond_1 A$, $\Diamond_1\Diamond_1^- A \longrightarrow \Box_1\Diamond_1^- A$;

(5) $A\&\Box(A \longrightarrow \Box_1(A\&\Box_1^- A)) \longrightarrow \Box A$;

(6) $\Box A \longrightarrow \Box_1 A$, $\Box^- A \longrightarrow \Box_1^- A$;

(7) $\Box A \longrightarrow A$, $\Box^- A \longrightarrow A$;

(8) $\Box A \longrightarrow \Box\Box A$, $\Box^- A \longrightarrow \Box^-\Box^- A$;

(9) $\Diamond_1^- A \longrightarrow \Box_1^-(\Diamond A\&\Diamond^- A)$.

Here the axiom (5) is similar to the axiom of induction (Vakarelov 1988, 1992).

The work Yun (2010) provides proof that the introduced calculus is correct with respect to this class of frames:

Theorem 3 (Soundness) *For any formula A: if A is derivable in* **Ind**(**R**, **R**$_1$) *then A is valid in any Ind-frame with linear time.*

Thus the calculus $\mathbf{Ind}(\mathbf{R}, \mathbf{R_1})$ is also correct with respect to the class of linear frames with current time point clusters with added relation R_1.

Moreover Yun (2010) gives proof that the calculus $\mathbf{Ind}(\mathbf{R}, \mathbf{R_1})$ is finitely approximable by finite Ind-models with linear time:

Theorem 4 *For any formula A: A is derivable in* $\mathbf{Ind}(\mathbf{R}, \mathbf{R_1})$ *iff A is true in any finite* Ind*-models with linear time.*

The completeness of the calculus $\mathbf{Ind}(\mathbf{R}, \mathbf{R_1})$ with respect to the class of all Ind-models with linear time follows immediately from Theorem 4. Also note that since the calculus $\mathbf{Ind}(\mathbf{R}, \mathbf{R_1})$ is finitely axiomatizable then from Theorem 4 also follows its decidability.

15.5 Polymodal Logic of Inductive Linear Time Frames

Let us consider a modal language with modal operators $\Box, \Box_\sim, \Box_1$. More precisely let us consider the language consisting of a countable set of propositional letters P, standard logical connectives and modal operators $\Box, \Box_\sim, \Box_1$. Formulae are defined as usual (Chagrov and Zakharyaschev 1997).

We will consider the frames $\langle W, R, R_\sim, R_1\rangle$ and the models $\langle W, R, R_\sim, R_1, V\rangle$ where W is a nonempty set, $R, R_\sim, R_1$ are binary relations on W and V is a valuation which associates a set of worlds from W to each propositional letter $p \in P$. The valuation V can be extended in the standard way (Chagrov and Zakharyaschev 1997) onto all formulae of the considered language. In particular for all $x \in W$:

$x \models_V p \iff x \in V(p)$ for any $p \in P$,

$x \models_V \Box A \iff \forall y(xRy \Longrightarrow y \models_V A)$,

$x \models_V \Box_\sim A \iff \forall y(xR_\sim y \Longrightarrow y \models_V A)$,

$x \models_V \Box_1 A \iff \forall y(xR_1 y \Longrightarrow y \models_V A)$.

A formula A is *true in the model* $M = \langle W, R, R_\sim, R_1, V\rangle$ if $x \models_V A$ for all $x \in W$. A formula A is said to be *valid in the frame* if it is true in any model on this frame.

Definition 3 A frame $\langle W, R_T, R_\sim\rangle$ with given equivalence relations $S_1, \ldots, S_k$ on W is called an LTK*-frame* (Calardo and Rybakov 2007) if the following holds:

(a) $W = \bigcup_{n\in\mathbb{N}} C^n$ is the disjoint union of nonempty sets C^n;
(b) $xRy \iff \exists n, m \in \mathbb{N}(x \in C^n, y \in C^m$ and $n \leq m)$;
(c) $xR_\sim y \iff \exists n \in (x \in C^n$ and $y \in C^n)$;
(d) S_i is some equivalence relation on each C^n, $1 \leq i \leq k$.

Thus each C^n is a R-cluster (and $R_\sim$-cluster), so that for $x \in C^n$: $C^n = \{y \mid xRy$ and $yRx\}$.

We define a new class of frames associated with the class of LTK-frames.

Definition 4 A frame $\langle W,\ R,\ R_\sim,\ R_1\rangle$ is called an *inductive nearly LTK-frame* if the following conditions are met:

(1) for all $x,\ y \in W$: xRy or yRx;
(2) $R_\sim$ is a reflexive relation;
(3) $R_\sim$ is a transitive relation;
(4) $R_\sim$ is a symmetric relation;
(5) if xR_1y and xR_1z then $yR_\sim z$;
(6) if xR_1y and $yR_\sim z$ then xR_1z;
(7) $xRy \iff (xR_\sim y$ or there are $x_1, \ldots, x_n$ such that $xR_\sim x_1R_1 \ldots R_1x_nR_1y)$.

Note that if the relation R_1 is defined as

$$xR_1y \iff \exists m \in \mathbb{N}(x \in C^m,\ y \in C^{m+1})$$

in some LTK-frame then it will satisfy the conditions (1)–(7).

LKInd is defined as the logic calculus obtained by adding the following axioms to the minimal modal logic $K_t(R,\ R_\sim,\ R_1)$ (Chagrov and Zakharyaschev 1997):

(1) $\Box(\Box A_1 \longrightarrow A_2) \vee \Box(\Box A_2 \longrightarrow A_1)$,
(2) $\Box_\sim A \longrightarrow A$;
(3) $\Box_\sim A \longrightarrow \Box_\sim\Box_\sim A$;
(4) $\Diamond_\sim A \longrightarrow \Box_\sim\Diamond_\sim A$;
(5) $\Diamond_1 A \longrightarrow \Box_1\Diamond_\sim A$;
(6) $\Box_1 A \longrightarrow \Box_1\Box_\sim A$;
(7) $\Box_\sim A \& \Box(A \longrightarrow \Box_1 A) \longrightarrow \Box A$;
(8) $\Box A \longrightarrow A$;
(9) $\Box A \longrightarrow \Box\Box A$;
(10) $\Box A \longrightarrow \Box_1 A$;
(11) $\Box A \longrightarrow \Box_\sim A$.

Yun (2015b) gives proof that **LKInd** is correct with respect to the class of all inductive nearly LTK-frames:

Theorem 5 *For any formula A: if A is derivable in* **LKInd** *then A is valid in any inductive nearly LTK-frame.*

Another fact from Yun (2015b):

Theorem 6 *For any formula A: A is derivable in* **LKInd** *iff A is true in any finite inductive nearly LTK-model.*

From Theorem 6 it follows immediately that **LKInd** is complete with respect to inductive nearly LTK-frames. Since **LKInd** is finitely axiomatizable then it is decidable.

15.6 On Linear Logic with Reflexive Intransitive Time Relation

In Lukyanchuk (2013) A.N.Lukyanchuk introduced a linear polymodal logic of knowledge and time LTK_r with reflexive intransitive time relation as a set of formulae valid in frames of a special kind. More precisely in Lukyanchuk (2013) LTK_r-frames are defined as follows:

Definition 5 A frame $\langle W, R_T, R_\sim, S_1, \ldots, S_k \rangle$ is called the LTK_r*-frame* if:

(a) $W = \bigcup_{n \in J} C^n$ is a disjunctive union of nonempty sets C^n, $J = \{1, \ldots, L\}$, $L \in \mathbb{N}$ or $J = \mathbb{N}$;
(b) $x R_T y \iff \exists n \in J((x \in C^n \text{ and } y \in C^n) \text{ or } (x \in C^n \text{ and } y \in C^{n+1}))$;
(c) $x R_\sim y \iff \exists n \in J(x \in C^n \text{ and } y \in C^n)$;
(d) S_i is some equivalence relation on each C^n.

In Luk'yanchuk and Rimatskii (2013) the study of the axiomatization problem for the linear polymodal logic of knowledge and time LTK_r with reflexive intransitive time relation was continued. The authors found a calculus AS_{LTK_r} which is correct with respect to the class of all LTK_r-frames. Also Luk'yanchuk and Rimatskii (2013) contains proof of completeness of the proposed calculus with respect to a class of frames connected with LTK_r-frames.

More precisely in Lukyanchuk (2013), Luk'yanchuk and Rimatskii (2013) the polymodal language with modal operators $\Box_T, \Box_\sim, \Box_1, \ldots, \Box_k$ is considered. Formulae and models on frames are defined as usually (Chagrov and Zakharyaschev 1997).

Given a model $M = \langle W, R_T, R_\sim, S_1, \ldots, S_k, V \rangle$, the valuation $V : P \to \mathbb{P}(W)$ can be extended in the standard way from the set P of propositional letters onto all formulae of the proposed language. In particular, for all $x \in W$:

$$x \models_V p \iff x \in V(p), p \in P$$
$$x \models_V \Box_T A \iff \forall y(x R_T y \Longrightarrow y \models_V A),$$
$$x \models_V \Box_\sim A \iff \forall y(x R_\sim y \Longrightarrow y \models_V A),$$
$$x \models_V \Box_i A \iff \forall y(x S_i y \Longrightarrow y \models_V A),\ i \in \{1, \ldots, k\}.$$

If $M = \langle \mathbb{F}, V \rangle$ is a model on frame $\mathbb{F} = \langle W, R_T, R_\sim, S_1, \ldots, S_k \rangle$, a formula A is said to be *true in the model* M *at the world* x if $x \models_V A$. A formula A is *true in the model* $M = \langle \mathbb{F}, V \rangle$ if $\forall x \in W(x \models_V A)$. A is *valid in the frame* $\mathbb{F}$ if the formula A is true in any model M on $\mathbb{F}$.

Consider the calculus of AS_{LTK_r} introduced in Luk'yanchuk and Rimatskii (2013):

Axioms of AS_{LTK_r}**:**

Axioms of CPC (classical propositional calculus);
$L_{\Box_T} : \Box_T(\Box_T A \longrightarrow B) \vee \Box_T(\Box_T B \longrightarrow A)$;
$K_{\Box_\xi} : \Box_\xi(A \longrightarrow B) \longrightarrow (\Box_\xi A \longrightarrow \Box_\xi B),\ \xi \in \{T, \sim, 1, \ldots, k\}$;

$T_{\Box_\xi} : \Box_\xi A \longrightarrow A,\ \xi \in \{T, \sim, 1, \ldots, k\};$
$4_{\Box_\xi} : \Box_\xi A \longrightarrow \Box_\xi \Box_\xi A,\ \xi \in \{\sim, 1, \ldots, k\};$
$5_{\Box_\xi} : \neg\Box_\xi A \longrightarrow \Box_\xi \neg\Box_\xi A,\ \xi \in \{\sim, 1, \ldots, k\};$
$M.1 : \Box_T A \longrightarrow \Box_\sim A;$
$M.2 : \Box_\sim A \longrightarrow \Box_i A,\ 1 \leqslant i \leqslant k;$
$AL : (\Box_\sim A \& \Box_\sim B \& \Diamond_T(\neg A \& \Box_\sim B)) \longrightarrow \Box_T B.$

Inference Rules of AS_{LTK_r}**:**
$MP : \frac{A,\ A \longrightarrow B}{B},\ Nec : \frac{A}{\Box_T A}.$
Here and further $\Diamond_\xi$ is a notation for $\neg\Box_\xi\neg$ ($\xi \in \{T, \sim, 1, \ldots, k\}$).

In Yun (2015a) a formula is proposed that is valid in all LTK_r-frames and is not derivable in AS_{LTK_r}. Thus it is proved that AS_{LTK_r} is not complete with respect to the class of all LTK_r-frames:

Theorem 7 *(1) The formula* $\Diamond_T \Diamond_\sim p \longrightarrow \Diamond_T p$ *is valid in any* LTK_r*-frame.*
(2) The formula $\Diamond_T \Diamond_\sim p \longrightarrow \Diamond_T p$ *is not derivable in* AS_{LTK_r}*.*

References

Calardo, E. (2006). Admissible inference rules in the linear logic of knowledge and time LTK. *Logic Journal of the IGPL, 14*(1), 15–34.
Calardo, E., & Rybakov, V. V. (2005). Combining time and knowlege, semantic approach. *Bulletin of the Section of Logic*, *34*(1), 13–21.
Calardo, E., & Rybakov, V. V. (2007). An axiomatization for the multi-modal logic of knowledge and time LTK. *Logic Journal of the IGPL*, *15*(3), 239–254.
Chagrov, A., & Zakharyaschev, M. (1997). *Modal logics*. Oxford logic guides 35 Oxford: Clarendon Press.
Fagin, R., Halpern, J. Y., Moses, Y., & Vardi, M. Y. (1995). *Reasoning about knowledge*. Cambridge: MIT Press.
Gabbay, D. M., Hodkinson, I. M., & Reynolds, M. A. (1994). *Temporal logic: Mathematical foundations and computational aspects* (Vol. 1). Oxford: Clarendon Press.
Halpern, J. Y., Van der Meyden, R., & Vardi, M. Y. (2004). Complete axiomatizations for reasoning about knowledge and time. *SIAM Journal on Computing*, *33*(2), 674–703.
Lukyanchuk, A. (2013). Decidability of multi-modal logic LTK of linear time and knowledge. *Journal of Siberian Federal University*, *6*(2), 220–226.
Luk'yanchuk, A. N., & Rimatskii, V. V. (2013). An axiomatization for the linear logic of knowledge and time LTK_r with intransitive time relation. *Siberian Mathematical Journal*, *54*(6), 1037–1045.
Rybakov, V. (2008). Discrete linear temporal logic with current time point clasters, deciding algorithms. *Logic and Logic Philosophy*, *17*(1–2), 143–161.
Vakarelov, D. (1988). Modal logics for Knowledge Representation Systems. Preprint No 7, Sofia University.
Vakarelov, D. (1992). Inductive modal logics. *Fundamenta Informaticae*, *16*, 383–405.
Yun, V. F. (2009). Temporal logic of linear time frames with inductions axiom. *Siberian Electronic Mathematical Reports*, *6*, 312–325.
Yun, V. F. (2010). The temporal logic of inductive frames with linear time. *Siberian Electronic Mathematical Reports*, *7*, 445–457.

Yun, V. F. (2015a). On linear logic of knowledge and time with intransitive time relation. *Siberian Mathematical Journal*, *56*(3), 565–568.
Yun, V. F. (2015b). Polymodal logic of the class of inductive linear time frames. *Siberian Electronic Mathematical Reports*, *12*, 421–431.

Appendix A
Larisa Maksimova. List of Publications

A.1 1963–1972

1. Maksimova, L. L. (1964). On system of axioms of the calculus of rigorous implication. *Algebra i Logika*, 3 (3), 59–68. Russian.
2. Maksimova, L. L. (1966). Formal deductions in the calculus of rigorous implication. *Algebra i Logika*, 5 (6), 33–39. Russian.
3. Maksimova, L. (1966). Some problems of Ackermann's calculus. International Congress of Mathematicians, Abstracts of Sect. 1, Moscow, p. 19.
4. Maksimova, L. L. (1966). Some problems of the Ackermann calculus. *Doklady Academii Nauk SSSR*, 175 (6), 1222–1224. Russian.
   Some problems of the Ackermann calculus. (English) *Soviet Mathematics*, 8 (4), 997–999 (1967).
5. Maksimova, L. (1967). Topological spaces and quasi-ordered sets. *Algebra i Logika*, 6 (4), 51–59. Russian.
6. Maksimova, L. (1967). Topological spaces and quasi-ordered sets. In *8th All-Union Colloquium on general algebra (abstracts)*, Ryga, p. 82.
7. Maksimova, L. L. (1967). On models of the calculus E. *Algebra i Logika*, 6 (6), 5–20. Russian.
8. Maksimova, L. L. On a calculus of rigorous implication. *Algebra i Logika*, 7 (2), 55–75. Russian.
   The calculus of strict entailment. (English) *Algebra and Logic*, 7 (2), 102–116 (1968).
9. Maksimova, L. (1968). Logical calculi of rigorous implication. Ph.D. thesis, Novosibirsk.
10. Maksimova, L. (1969). An interpretation of systems with rigorous implication. In *10th All-Union Algebraic Colloquium (abstracts)*, Novosibirsk, p. 113.
11. Maksimova, L. L. On E-theories. *Algebra i Logika*, 9 (5), 530–538. Russian.
    E-theories. (English) *Algebra and Logic*, 9 (5), 320–325 (1970).

S. Odintsov (ed.), *Larisa Maksimova on Implication, Interpolation, and Definability*, Outstanding Contributions to Logic 15, https://doi.org/10.1007/978-3-319-69917-2

12. Maksimova, L. (1971). Interpretation, extension theorems and separation theorems for the calculi E and R. In *4th Intern. Congress of LMPS (abstracts)*, Bucharest, pp. 34–35.
13. Lavrov, I. A. & Maksimova, L. L. (1970). Problems in Logic. Novosibirsk State University, Novosibirsk, pp. 106.
14. Maksimova, L. L. Interpretation and separation theorems for the calculi E and R. *Algebra i Logika*, 10 (4), 376–392. Russian.
    An interpretation and separation theorems for the logical systems E and R. (English) *Algebra and Logic*, 10 (4), 232–241 (1971).
15. Maksimova, L. (1971). Quasi-finite superintuitionistic logics. In *11th All-Union Algebraic Colloquium (abstracts)*, Kishinev, pp. 258–259.
16. Maksimova, L. L. (1972). Pretabular superintuitionistic logics. *Algebra i Logika*, 11 (5), 558–570. Russian.
    Pretabular superintuitionist logic (English) *Algebra and Logic*, 11 (5), 308–314 (1972).

A.2 1973–1982

17. Maksimova, L. L. Implication lattices. *Algebra i Logika*, 12(4), 445–467. Russian.
    Implication lattices. (English) *Algebra and Logic*, 12(4), 249–261 (1973).
18. Maksimova, L. (1973). A semantics for the calculus E of entailment. *Bulletin of Section Logic*, 2 (1), 18–23.
19. Maksimova, L. (1973) On the variety of R-models. In *12 All-Union Algebraic Colloquium (abstracts)*, vol. 2, p. 287, Sverdlovsk.
20. Maksimova L. & Vakarelov, D. (1974). Representation theorems for generalized Post algebras of order ω^+. *Bulletin de l'Acad. Polonaise des Sciences (Ser. math.)*, 22 (8), 757–764.
21. Maksimova, L. & Vakarelov, D. (1974). Semantics for ω^+-valued predicate calculi. *Bulletin de l'Acad. Polonaise des Sciences (Ser. math.)*, 22 (8), 765–771.
22. Maksimova, L. (1974). Implication strict. In *Encyclopedia of Cybernetics*, Kiev, pp. 352–353.
23. Maksimova, L. (1974). Logics non-classical. In *Encyclopedia of Cybernetics*, Kiev, pp. 352–353.
24. Maksimova, L. L. & Rybakov, V. V. (1974). On lattice of normal modal logics. *Algebra i Logika*, 13 (2), 182–216. Russian.
    A lattice of normal modal logics. (English) *Algebra and Logic*, 13 (2), 105–122 (1974).
25. Maksimova, L. L. & Rybakov, V. V. (1974). On lattice of modal logics. 3th All-Union Conference on Mathematical Logic (abstracts), pp. 131–132, Novosibirsk.

26. Lavrov, I. A. & Maksimova, L. L. (1975). Problems in Set Theory, Mathematical Logic and Theory of Algorithms. 1st Edition - Moscow, Nauka (1st Ed. - 1975; 2nd Ed. - 1984; 3nd Ed. - 1995; 4th Ed. -2001), 5th Edition – Moscow, Fizmatlit, 2002, 2003, 2004, 2006. Translations: English - New York, Plenum/Kluwer 2003; Hungarian - Budapest, Muszaki Konyvkiado 1987, Polish - Warszawa, Wydawnictwo Naukowe PWN 2004.
27. Maksimova. L. L. (1975). On tautologies of ω^+-valued logic. *Matematicheskie Zametki*, 17 (6), 947–955. Russian.
On tautologies in ω^+-valued logic. (English) *Mathematical Notes*, 17 (6), 568–573 (1975).
28. Maksimova, L. L. (1975). Pretabular extensions of the Lewis logic S4. *Algebra i Logika*, 14 (1), 28–55. Russian.
Pretabular extensions of Lewis S4. (English) *Algebra and Logic*, 14 (1), 16–33 (1975).
29. Maksimova, L. (1975). Locally finite varieties of topological boolean algebras. In *13th All-Union Algebraic Symposium (abstracts)*, pp. 425–426, Gomel'.
30. Maksimova, L. L. (1975). Modal logics of finite slices. *Algebra i Logika*, 14 (3), 304–319. Russian.
Finite-level modal logics. (English) *Algebra and Logic*, 14 (3), 188–197 (1975).
31. Maksimova, L. L. Principles of separation of variables in propositional logics. *Algebra i Logika*, 15 (2), 168–184. Russian.
The principle of separation of variables in propositional logics. (English) *Algebra and Logic*, 15 (2), 105–114 (1975).
32. Maksimova, L. (1976). Boolean-valued models of modal logic. In *7th All-Union Symposium of Logic and Methodology of Science (abstracts)*, pp. 133–135, Kiev.
33. Maksimova, L. (1976). Separation of variables in relevant calculi. In *4th All-Union Conference on Mathematical Logic (abstracts)*, p. 81, Kishinev.
34. Maksimova, L. (1977). Amalgamated varieties of pseudoboolean algebras. In *14th All-Union Algebraic Conference (abstracts)*, vol. 2, pp. 107–108, Novosibirsk.
35. Maksimova, L. L. Craig's interpolation theorem and amalgamable varieties. *Doklady Academii Nauk SSSR*, 237 (6), 1281–1284. Russian.
Craig's interpolation theorem and amalgamable manifolds. (English) *Soviet Mathematics Doklady*, 18, 1511–1514 (1977).
36. Maksimova, L. L. Craig's theorem in superintuitionistic logics and amalgamable varieties of pseudo-boolean algebras. *Algebra i Logika*, 16 (6), 643–681. Russian.
Craig's theorem in superintuitionistic logics and amalgamable varieties of pseudo-boolean algebras. (English) *Algebra and Logic*, 16 (6), 427–455 (1977).
37. Maksimova, L. (1978). Craig's interpolation theorem in modal logics. Modal and intensional logics (abstracts), pp. 89–91, Moscow.
38. Maksimova, L. L., Scvortsov, D. P. & Shehtman, V.B. Impossibility of finite axiomatization of Medvedev's logic of finite problems. *Doklady Academii Nauk SSSR*, 245 (6), 1051–1053;
Maksimova, L. L., Shehtman, V. B. & Skvortsov, D.P. (1979). The impossibility

of a finite axiomatization of Medvedev's logic of finitary problems. (English) *Soviet Mathematics Doklady*, 20, 394–398.

39. Maksimova, L. (1979). Interpolation properties of superintuitionistic logics. *Studia Logica*, 38 (4), 419–428.
40. Maksimova, L. L. On a classification of modal logics. *Algebra i Logika*, 18 (3), 328–340. Russian.
    A classification of modal logics. (English) *Algebra and Logic*, 18 (3), 202–210 (1979).
41. Maksimova, L. L. Interpolation theorems in modal logics and amalgamable varieties of topological boolean algebras. *Algebra i Logika*, 18 (5), 556–586. Russian.
    Interpolation theorems in modal logics and amalgamable varieties of topological Boolean algebras. (English) *Algebra and Logic*, 18 (5), 348–370 (1979).
42. Maksimova, L. L. Interpolation theorems in modal logics: Sufficient conditions. *Algebra i Logika*, 19 (2), 194–213. Russian.
    Interpolation theorems in modal logics. Sufficient conditions. (English) *Algebra and Logic*, 19 (2), 120–132 (1980).
43. Maksimova, L. (1979). Interpolation properties of superintuitionistic and modal logics. Modal and tense logics (Proceedings. of 2nd Soviet-Finnish Colloquium on Logic - Moscow 1979), pp. 54–69.
44. Maksimova, L. (1979). Almost all modal logics do not have Craig's interpolation property. In *6th International Congress of LMPS (abstracts, Sec.1–4)*, pp. 120–123.
45. Maksimova, L. (1981). On locally finite varieties of pseudoboolean algebras. In *16th All-Union Algebraic Conference (abstracts)*, part 1, pp. 99–100, Leningrad.
46. Maksimova, L. (1982). On Craig's theorem in modal logics. In *Modal and intensional logics*(8th All-Union Conference on Logic and Methodology of Science), pp. 63–65, Vilnius.
47. Maksimova, L. (1982). Interpolation in modal logics. In *6th All-Union Conference on Math. Logic (abstracts)*, p. 102, Tbilissi.
48. Maksimova, L. L. Failure of the interpolation property in modal counterparts of Dummett's logic. *Algebra i Logika*, 21 (6), 690–694. Russian.
    Absence of the interpolation property in the consistent normal modal extensions of the dummett logic. (English) *Algebra and Logic*, 21 (6), 460–463 (1982).
49. Maksimova, L. (1982). Interpolation properties of superintuitionistic, positive and modal logics. In *Intensional Logics: Theory and Applications*, Soc. Philos. Fennica, Helsinki, pp. 70–78.
50. Maksimova, L. (1982). The Lyndon interpolation theorem in modal logics. *Mathematical logic and theory of algorithms*, Nauka, Novosibirsk, pp. 45–55. Russian.

A.3 1983–1992

51. Maksimova, L. (1983). On interpolation in modal logics containing S4. In *7th International Congress of LPMS (abstracts)*, vol. 1, pp. 84–87, Salzburg.
52. Maksimova, L. (1983). Remarks on the disjunction property in superintuitionistic logics. *Semiotic aspects of intellegent activity*, VINITI, Moscow, pp. 83–85. Russian.
53. Maksimova, L. (1983). On number of amalgamated varieties of topoboolean algebras. In *17th All-Union Algebraic Congerence (abstracts)*, part 2, pp. 135–136, Minsk.
54. Maksimova, L. (1984). On number of maximal intermediate logics with the disjunction property. In *7th All-Union Conference on Math. Logic (abstracts)*, p. 95, Novosibirsk.
55. Maksimova, L. (1986). On maximal intermediate logics with the disjunction property. *Studia Logica*, 45 (1), 69–75.
56. Maksimova, L. (1986). Infinite-slice extensions of the modal logic K4 possessing interpolation property. In *8th All-Union Conference on Math. Logic (abstracts)*, p. 107, Moscow.
57. Maksimova, L. (1986). Modal K4-logics of infinite slice. In *Logic and system methods of analysis of scientific knowledge to 9th All-Union Meeting for Logic, Methodology and Philosophy of Science*, p. 35, Kharkov.
58. Maksimova, L. (1987). On interpolation in normal modal logics. In *Non- classical logics*, Stiintsa, Kishinev, pp. 40–56.
59. Maksimova, L. (1987). On a variety of diagonalizable algebras. In *19th All-Union Algebraic Conference (abstracts)*, part 1, p.170, Lvov.
60. Maksimova, L. (1987). Craig's interpolation property in propositional modal logics. In *9th International Congress of LPMS (Abstracts)*, vol. 5, pp. 35–38, Moscow.
61. Maksimova, L. Interpolation in infinite-slice extensions of the provability logic. *Algebra i Logika*, 27 (5), 581–604. Russian.
    Interpolation in infinitely layered extensions of provability logic. (English) *Algebra and Logic*, 27 (5), 361–376 (1988).
62. Maksimova, L. (1988). Definability theorems in normal extensions of the provability logic. Preprint/Institute of Mathematics, Siberian Division of AS of USSR, Novosibirsk.
63. Maksimova, L. (1988). Interpolation in extensions of modal provability logic. Summer School and Conference on Math. Logic Heyting 1988 (abstracts), p. 39, Varna.
64. Maksimova, L. (1988). Interpolation and the Beth properties in extensions of the provability logic. In *9th All-Union Conference on Math. Logic (abstracts)*, p. 97, Leningrad.
65. Maksimova, L. (1989). Interpolation in modal logics of infinite slice containing K4. *Matematiceskaya logika i algoritmiceskie problemy*, Novosibirsk, Nauka (Sib. Div.), pp. 73–91. Russian.

66. Maksimova, L. (1989). Definability theorems in normal extensions of the provability logic. *Studia Logica*, 48 (4), 495–507.
67. Maksimova, L. A continuum of normal extensions of the modal provability logic with the interpolaton porperty. *Sibirskij Matematicheskij Zhurnal*, 30 (6), 122–131. Russian.
Continuum of normal extensions of the modal logic of provability with the interpolation property. (English) *Siberian Mathematical Journal*, 30 (6), 935–944 (1989).
68. Maksimova, L. (1989). Interpolation, the Beth property and tense logic of tomorrow. Preprint of the Institute of Mathematics, Siberian Branch of the Russian Academy of Sciences Novosibirsk.
69. Maksimova, L. (1989). Relevant principles and formal deducibility. In *Directions in relevant logic*, Kluwer Academic. Publishers., Dordrecht, pp. 95–97.
70. Maksimova, L. (1990). Temporal logics with discrete moments of time and the Beth property. In *4th Asian Logic Conference (Proceedings)*, pp. 31–34, Tokyo.
71. Maksimova, L. (1990). Interpolation and definability theorems in modal logics. In *10th All-Union Conference on Math. Logic (abstracts)*, p. 105, Alma-Ata.
72. Maksimova, L. (1990). Temporal logics with discrete moments of time and the Beth property. International Congress of Mthematicians (abstracts), p. 1, Kyoto, Japan.
73. Maksimova, L. (1991). The tense logic of the next do not have the Beth property. *Journal of Applied Non-Classical Logic*, 1 (1), 73–76.
74. Maksimova, L. Temporal logics with the operator«the next» do not have interpolation or the Beth property. *Sibirskij Matematicheskij Zhurnal*, 32 (6), 109–113. Russian.
Temporal logics with the operator«the next» do not have interpolation or the Beth property. (English) *Siberian Mathematical Journal*, 32 (6), 989–993.
75. Maksimova, L. (1991). Definability and interpolation in varieties of modal algebras. In *International Conference of Algebra (abstracts)*, Barnaul, p. 81.
76. Maksimova, L. L. (1991). Beth's properties, interpolation and amalgamation in varieties of modal algebras. *Doklady Akademii Nauk SSSR*, 319 (6), 1309–1312. Russian.
Beth properties, interpolation and amalgamation in varieties of modal algebras. (English) *Soviet Mathematics Doklady*, 44 (1), 327–331 (1992).
77. Maksimova, L. (1991). On the Beth definability properties in modal logics. In *9th International Congress of LMPS (abstracts)*, vol. 1, p. 144, Uppsala 1991, Sweden.
78. Maksimova, L. (1991). Amalgamation and interpolation in normal modal logics. *Studia Logica*, 50 (3/4), 457–471.
79. Maksimova, L. Modal logics and varieties of modal algebras: The Beth properties, interpolation, and amalgamation. *Algebra i Logika*, 31 (2), 145–166. Russian.
Modal logics and varieties of modal algebras: The Beth properties, interpolation, and amalgamation. (English) *Algebra and Logic*, 31 (2), 90–105 (1992).

80. Maksimova, L. (1992). Interpolation and definability theorems in normal extensions of the modal provability logic. Logic Colloquium 1989 (abstracts). *Journal of Symbolic Logic*, 57 (1), 312–313. Berlin.
81. Maksimova, L. An analog of Beth's theorem in normal extensions of the modal logic K4. *Sibirskii Matematicheskii Zhurnal*, 33 (6), 118–130. Russian.
An analog of Beth's theorem in normal extensions of the modal logic K4. (English) *Siberian Mathematical Journal*, 33 (6), 1052–1065 (1992).
82. Maksimova, L. (1992). Definability and interpolation in classical modal logics. In *Proceedings of the International Conference on Algebra, ded. A.I Malcev, Novosibirsk'89*, Contemporary Mathematics, 131, Part 3, pp. 583–599.
83. Maksimova, L. (1993). Hallden-completeness in normal modal logics. The 1992 European Summer Meeting of the ASL (Veszprem, August 9-15, 1992), abstracts, 59; *Journal of Symbolic Logic*, 58 (3), 1134–1135.
84. Maksimova, L. (1992). Implicit and explicit definability in modal and temporal logics. In: M. Masuch & L. Pilos (eds.), Logic at Works, Proceedings of the First International Conference on Applied Logic, University of Amsterdam, Dec. 17–19.

A.4 1993–2002

85. Maksimova, L. (1995). Explicit definition of implicitly definable in some modal logics. Logic Colloquium'93. University of Keele, 1993; *Bulletin of Symbolic Logic*, 1 (1), 97–98.
86. Maksimova, L. (1995). On some aspects of the history of modal logics. Logic Colloquium '94 (Abstracts of tutorial and main lectures), Clermont-Ferrand, France, 21-30 July 1994), 37; *Bulletin of Symbolic Logic*, 1 (2), 212–213.
87. Maksimova, L. (1994). On variable separation in modal logics. Abstracts of Papers NSL'94, p.7, Kanazawa, Japan.
88. Maksimova, L. (1995). On variable separation in modal logics. *Bulletin of Section of Logic*, 24 (1), 21–25.
89. Maksimova, L. (1995). Implicit and explicit definability in modal and temporal logics. In L. Polos & M.Masuch, *Applied Logic: How, What and Why* (pp. 153–159). Kluwer Acad. Publ.
90. Maksimova, L. (1995). Interpolation in predicate logics with equality. In *10th international Congress of Logic, Methodology and Philosophy of Science, abstracts*, p.154 (1995); *Bulletin of Symbolic Logic*, 3 (2), 269–270.
91. Maksimova, L. (1995). On variable separation in modal and super- intuitionistic logics. *Studia Logica*, 55, 99–112.
92. Maksimova, L. (1996). *Review* of A.R.Anderson, N.D.Belnap, J.M.Dunn. Entailment. The logic of relevance and necessity. Volume II. Princeton Univ. Press. J. Symbolic Logic, 61, 338–341.

93. Maksimova, L. (1997). Formulas preserving interpolation and Beth's properties. Logic Colloquium'96, Abstracts, Donostia - San Sebastian, Spain, July 1996, 128; *Bulletin of Symbolic Logic*, 3 (2), 269–270.
94. Maksimova, L. (1996). Joint embedding properties in varieties of modal algebras. In *Conference on Universal Algebra and Lattice Theory, Abstracts*, Szeged, Hungary, July 1996, p.34.
95. Maksimova, L. (1996). Interrelation of Logical, Algebraic and Semantical Properties for Intermediate and Normal Modal Logics. In *Logic, Algebra and Computer Science*, Technical Report 96–11(232), 1996, Institute of Informatics, Warsaw University, p. 30.
96. Maksimova, L. (1997). On the Beth definability properties in varieties of modal algebras. Workshop on Abstract Algebraic Logic (abstracts), CRM, Bellaterra (Spain), pp. 71–75.
97. Maksimova, L. L. Explicit Definitions of Implicitly Definable Objects in some Modal Logics. *Sibirskii Matematicheskii Zhurnal*, 38 (3), 598–602. Russian. Explicit Definitions of Implicitly Definable Objects in some Modal Logics. (English) *Siberian Mathematical Journal*, 38 (3), 513–517 (1997).
98. Maksimova, L. L. Interpolation in Superintuitionistic Predicate Logics with Equality. *Algebra i Logika*, 36 (5), 536–554. Russian.
Interpolation in Superintuitionistic Predicate Logics with Equality. (English) *Algebra and Logic*, 36 (5), 319–329 (1997).
99. Maksimova, L. (1998). Projective Beth Property in varieties of modal and Heyting Algebras. In *Conference on Universal Algebra and Lattice Theory, Abstracts*, Szeged, Hungary, p. 26.
100. Maksimova, L. (1998). Explicit and Implicit Definability in Modal and Related Logics. *Bulletin of Section of Logic*, 27 (1/2), 36–39.
101. Maksimova, L. (1998). Interpolation in Superintuitionistic and Modal Predicate Logics with Equality. In: M. Kracht, M. de Rijke, H. Wansing & M. Zakharyashchev (Eds.), *Advances in Modal Logic*, (pp. 133–141). Volume I, Lecture Notes, CSLI Publications, Stanford.
102. Maksimova, L. L. (1998). Explicit and Implicit Definability in Modal, Superintuitionistic and Relevant Logics. *Logicheskie Issledovania*, Issue 5, Moscow: Nauka, pp. 53–60.
103. Maksimova, L. (1999). Projective Beth Property in Superintuitionistic Logics. Logic Colloquium '98. abstracts, Prague, 1998; *Bulletin of Symbolic Logic*, 5 (1), 121–122.
104. Maksimova, L. (1999). Interrelations of Algebraic, Semantical and Logical Properties for Superintuitionistic and Modal Logics. In *Logic, Algebra and Computer Science, Banach Center Publications*, vol. 46, Institute of Mathematics, Polish Academy of Sciences, Warszawa, pp. 159–168.
105. Maksimova L, & Voronkov, A. (2000). Complexity of Some Problems in Modal and Superintuitionistic Logics. Logic Colloquium, Utrecht, 1999, abstracts, 34; *Bulletin of Symbolic Logic*, 6 (1), 118–119.

106. Maksimova, L. (1999). Projective Beth Property and Intermediate Logics. In *11th International Congress of Logic, Methodology and Philosophy of Science, Cracow, Abstracts*, p. 72.
107. Maksimova, L. L. Projective Beth Properties in Modal and Superintuitionistic Logics. *Algebra i Logika*, 38 (3), 316–333. Russian.
Projective Beth Properties in Modal and Superintuitionistic Logics. (English) *Algebra and Logic*, 38 (3), 171–180 (1999).
108. Maksimova, L. L. Superintuitionistic Logics and Projective Beth Property. *Algebra i Logika*, 38 (6), 680–696. Russian.
Superintuitionistic logics and the projective beth property. (English) *Algebra and Logic*, 38 (6), 374–382 (1999).
109. Maksimova, L. (2000). Intuitionistic Logic and Implicit Definability. *Annals of Pure and Applied Logic*, 105, 83–102.
110. Maksimova, L. (2000). Strongly Decidable Properties of Modal and Intuitionistic Calculi. *Logic Journal of IGPL*, 8 (6), 797–819.
111. Maksimova, L. (2000). Definability and Interpolation in Extensions of Johansson's Minimal Logic. II World Congress on Paraconsistency, May 8-12, 2000, Juquehy - San Sebastiano, Brazil, Program/Abstracts, pp. 27–28.
112. Maksimova, L. (2001). Intuitionistic Logic and Projective Beth Property. LC 2000 and ELSS 2000, Abstracts of Contributed papers, Paris 2000, 244; *Bulletin of Symbolic Logic*, 7 (1), 133–134.
113. Maksimova, L. L. Decidability of the projective Beth property in varieties of Heyting algebras. *Algebra i Logika*, 40 (3), 275–286. Russian.
Decidability of the projective Beth property in varieties of Heyting algebras. (English) *Algebra and Logic*, 40 (3), 159–165 (2001).
114. Maksimova, L. (2001). Paraconsistent extensions of the minimal logic and implicit definability. Smirnov Readings - 3, Abstracts, Moscow 2001, pp. 51–52.
115. Maksimova, L. (2001). On complexity of some problems in positive calculi. *Logic, Complexity and Computer Science* (Proceedings of the International Workshop), Univ. Paris 12, France, 2001, pp. 183–193.
116. Maksimova, L. L. (2001). Implicit definability in paraconsistent extensions of the minimal logic. *Logicheskie issledovania*, Moscow:Nauka, (8), 72–81. Russian.
117. Maksimova, L. (2002).Projective Beth property in extensions of S4. Collegium Logicum, Annals of Kurt-Goedel-Society, vol. 4, Abstacts of LC 2001, Vienna, 2001, 133; *Bulletin of Symbolic Logic*, 8 (1), 149–150.
118. Maksimova, L. (2002). Complexity of interpolation and related properties in positive calculi. *Journal of Symbolic Logic*, 67 (1), 397–408.

A.5 2003–2016

119. Maksimova, L. L. Positive logics and implicit definability. *Algebra i Logika*, 42 (1), 65–93. Russian.
Implicit Definability and Positive Logics. (English) *Algebra and Logic*, 42 (1), 37–53 (2003).
120. Maksimova, L. L. (2002). Definability and Interpolation in Extensons of Johansson Minimal Logic. In: W. Carnieli, M. Coniglio & I. D'Ottavio (Eds.), Paraconsistency: The logical Way to the Inconsistent. (pp. 435–443). Marcel Dekker, Inc., New York, 2002. (Proceedings of the 2nd World Congress on Paraconsistency, Sao Paolo, Brazil, May 2000).
121. Maksimova, L. (2002). Definability in Modal and Intuitionistic Logics. *Proceeding of International Conferences on Mathematical logic*, Novosibirsk, pp. 81–91.
122. Maksimova, L. (2003). Implicit definability over Grzegorczyk modal logic. Logic Colloquium 2002, Collegium Logicum 2002, Muenster, Germany, August 2002, 43; *Bulletin of Symbolic Logic*, 9 (1), 99.
123. Maksimova, L. (2002). Projective Beth's Properties in Infinite Slice Extensions of the Modal Logic K4. *Advances in Modal Logic*, vol. 3, World Scientific, New Jersey-London-Singapore- Hong Kong, pp. 349–363.
124. Maksimova, L. (2003). Complexity of some problems in positive and related calculi. *Theoretical Computer Science*, 303 (1), 171–185.
125. Lavrov, I. & Maksimova, L. (2003). Problems in Set Theory, Mathematical Logic and the Theory of Algorithms. The University Series in Mathematics, Kluwer Academic/Plenum Publishers, New York, pp. xii+282.
126. Maksimova, L. (2004). Decidable properties of logical calculi and of varieties of algebras. Logic Colloquium 2003, Helsinki, Finland, Abstracts; *Bulletin of Symbolic Logic*, 10 (2), 238–239.
127. Maksimova, L. (2003). Restricted interpolation and the projective Beth property in equational logic. In *12th International Congress of Logic, Methodology and Philosophy of Science*. 2003, Oviedo (Spain), pp. 109–110.
128. Maksimova, L. & Voronkov, A. (2003). Complexity of Some Problems in Modal and Intuitionistic Calculi. In: M. Baaz & J. A. Makowsky (Eds.). *Computer Science Logic* Proceedings of 17th International Workshop, CSL 2003, 12th Annual Conference of the EACSL, and 8th Kurt Goedel Colloquium, KGC 2003, Vienna, Austria August 25-30, 2003. *Lecture Notes in Computer Science*, Springer, vol. 2803, pp. 397–412.
129. Maksimova, L. (2003). Restricted interpolation in modal logics. *Advances in Modal Logic*, vol. 4, King's College Publications, London, pp. 297–311.
130. Maksimova, L. L. Restricted interpolation and projective Beth's property in equational Logic. *Algebra i Logika*, 42 (6), 712–726. Russian.
Restricted Interpolation and the Projective Beth Property in Equational Logic. (English) *Algebra and Logic*, 42 (6), 398–406 (2003).

131. Maksimova, L. L. Definability in normal extensions of the modal S4 logic. *Algebra i Logika*, 43 (4), 387–410. Russian.
Definability in Normal Extensions of S4. (English) *Algebra and Logic*, 43 (4), 217–229 (2004).
132. Lavrov, I. A. & Maksimova, L.L. (2004). Zadania z teorii mnogosci, logiki matematicznej i teorii algoritmow. Wydawnictwo Naukowe PWN, Warszawa, pp. 316.
133. Gabbay, D. & Maksimova, L. (2005). Interpolation and Definability: Modal and intuitionistic logics." 509 pp, Oxford University Press, Oxford.
134. Maksimova, L. (2005). Interpolation and Joint Consistency. In *We will show them! Essays in honour Dov Gabbay*. Volume 2. S.Artemov et all, eds., King's College Publications, pp. 293–305.
135. Maksimova, L. L. (2005). Interpolation and definability in extensions of the minimal logic. *Algebra i Logika*, 44 (6), 726–750. Russian.
Interpolation and definability in extensions of the minimal logic. (English) *Algebra and Logic*, 44 (6), 407–421 (2005).
136. Maksimova, L. (2005). Interpolation and projective Beth property in extensions of minimal logic. Logic Colloquium 2004, Turin, Italy, 192; *Bulletin of Symbolic Logic*, 11 (2), 309.
137. Maksimova, L. (2005). Implicit definability and interpolation in non-classical logics. UNILOG'05, *Handbook of First World Congress and School on Universal Logic*, Montreaux - Switzerland, p. 78.
138. Maksimova, L. (2006). Decidable properties of logical calculi and of varieties of algebras. In: V. Stoltenberg-Hansen & J. Vaananen (Eds.). Logic Colloquium '03, Ser. Lecture Notes in Logic 24, Association for Symbolic Logic, pp. 146–166.
139. Maksimova, L. (2006). On a form of interpolation in modal logic. Logic Colloquium 2005, Athens, Greece, 94; *Bulletin of Symbolic Logic*, 12 (2), 340.
140. Maksimova, L. (2006). Projective Beth property in extensions of Grzegorczyk logic. *Studia Logica*, 83 (1-3), 365–391.
141. Maksimova, L. (2006). Definability and Interpolation in Non-Classical Logics. *Studia Logica*, 82 (2), 271–291.
142. Maksimova, L. L. & Shreiner, P. A. (2006). Algorithms of recognition of the tabularity and pretabularity in extensions of the intuitionistic calculus. *Vestnik, Quart.J. of Novosibirsk State University*, Ser. math., mech. and informatics, 6 (3), 49–58.
143. Maksimova, L. L. (2006). Projective Beth property and interpolation in positive and related logics. *Algebra i Logika*, 45 (1), 85–113. Russian.
The Projective Beth Property and Interpolation in Positive and Related Logics. (English) *Algebra and Logic*, 45 (1), 49–66 (2006).
144. Maksimova, L. (2006). On interpolation problem in paraconsistent extensions of minimal logic. In: G. Mints & R. de Queiroz (eds.). Proceedings of WoLLIC'2006, Volume 165 of the Electronic Notes in Theoretical Computer Science (ISSN 1571-0661), Elsevier Science Pub., pp. 107–119.

145. Maksimova, L. (2006). On interpolation problem in paraconsistent extensions of minimal logic. Abstracts of WoLLIC'2006, *Logic Journal of IGPL*, 14 (3), 528–529.
146. Maksimova, L. (2007). Weak interpolation in equational logic. Logic Colloquium 2006, Abstracts. Nijmegen, The Netherlands, 22–23; *Bulletin of Symbolic Logic*, 13 (2), 278–279.
147. Maksimova, L. L. A method of proving interpolation in paraconsistent extensions of the minimal logic. *Algebra i Logika*, 46 (5), 627–648. Russian.
A method of proving interpolation in paraconsistent extensions of the minimal logic. (English) *Algebra and Logic*, 46 (5), 341–353 (2007).
148. Maksimova, L. (2007). On modal Grzegorczyk logic. *Fundamenta Informaticae*, 81 (1–3), 203–210.
150. Maksimova, L. L. (2007). Decidability of some properties in tabular logics. In *13th International Congress of Logic, Methodology and Philosophy of SCience, Volume of Abstracts*, Beijing, China, pp. 124–125.
151. Maksimova, L. L. (2008). A weak form of interpolation in equational logic. *Algebra i Logika*, 47 (1), 94–107. Russian.
A weak form of interpolation in equational logic. (English) *Algebra and Logic*, 47 (1), 56–64 (2008).
152. Maksimova, L. L. & Orlowska, E. The Beth property and interpolation in lattice-based algebras and logics. *Algebra i Logika*, 47 (3), 307–334. Russian. The Beth property and interpolation in lattice-based algebras and logics. (English) *Algebra and Logic*, 47 (3), 176–192 (2008).
153. Maksimova, L. (2008). Interpolation and implicit definability in extensions of the provability logic. *Logic and Logical Philosophy*, 17, 129–142.
154. Maksimova, L. (2009). Interpolation and related properties in semilattice based varieties. Logic Colloquium 2008 (Bern, Switzerland), Abstracts, 2008, 40–41, *Bulletin of Symbolic Logic*, 15 (1), 122. http://www.lc08.iam.unibe.ch/slideUpload/talks/Maksimova06260849.pdf;
155. Maksimova, L. (2008). Restricted interpolation in modal and superintuitionistic logics. *Advances in Modal Logic 2008*, Submitted Abstracts of AiML 2008, Nancy, France, pp. 29–33. http://aiml08.loria.fr/talks/maksimova.pdf.
156. Alaev, P. A. & Maksimova, L. L. (2008). Mathematical Logic. Part 1. *Textbook*. Novosibirsk State University, Novosibirsk, p. 30.
157. Maksimova, L. L. Restricted interpolation property in superintuitionistic logics. *Algebra i Logika*, 48 (1), 54–89. Russian.
Restricted interpolation property in superintuitionistic logics. (English) *Algebra and Logic*, 48 (1), 33–53 (2009).
158. Maksimova, L. L. Decidability of the interpolation problem and of related properties in tabular logics. *Algebra i Logika*, 48 (6), 754-792. Russian. Decidability of the interpolation problem and of related properties in tabular logics. *Algebra and Logic*, 48 (6), 426–448 (2009).
159. Karpenko, A. V. & Maksimova, L. L. Simple weakly transitive modal algebras. *Algebra i Logika*, 49 (3), 347–366. Russian.

Simple weakly transitive modal algebras. (English) *Algebra and Logic*, 49 (3), 233–245 (2010).

160. Maksimova, L. (2010). Weak interpolation in extensions of minimal logic. In S. Feferman & W. Sieg *Proofs, Categories and Computations Essays in honor of Grigori Mints*. (With the collaboration of V. Kreinovich, V. Lipschitz and Ruy de Queiroz), College Publications, London, pp. 159–170. 978-1-84890-012-7.
161. Maksimova, L. (2010). Problem of restricted interpolation in superintuitionistic and some modal logics. *Logic Journal of IGPL*, 18 (3), 367–380. (2009; https://doi.org/10.1093/jigpal/jzp040)
162. Maksimova, L. (2010). Weak interpolation in extensions of Johansson's minimal logic. Logic Colloquium 2009 (Sofia, Bulgaria), Abstracts, 2009, 66–67; *Bulletin of Symbolic Logic*, 16 (1), 125–126. http://lc2009.fmi.uni-sofia.bg/contributed/c0c621dcaad8f8be0f1e4b55126b778a.pdf
163. Maksimova, L. (2009). An algebraic approach to non-classical logics. Conference Mal'tsev Meeting. (Abstracts), p. 20. Novosibirsk, August 24–28.
164. Maksimova, L. L. Joint consistency in extensions of the minimal logic. *Sibirskii Matematicheskii Zhurnal*, 51 (3), 604–619. Russian.
Joint consistency in extensions of the minimal logic. (English) *Siberian Mathematical Journal*, 51 (3), 479–490 (2010).
165. Maksimova, L. (2010). Joint consistency and weak interpolation over the minimal logic. Logic Colloquium 2010 (Paris, July 25-31), Abstracts of contributed talks, pp. 35–36.
166. Maksimova, L. (2010). Weak interpolation property over the minimal logic. *Advances in Modal Logic 2010*, Extended Abstracts. Steklov Mathematical Institute, Moscow, pp. 48–52.
167. Maksimova, L. L. (2010). Non-classical logics and algebra. *H.A.Vasilyev's Imaginary logic and modern non-classical logics*. Proceedings of the international conference, p. 32, Kazan, October 11–15. Russian.
168. Maksimova, L. L. Decidability of the weak interpolation property over the minimal logic. *Algebra i Logika*, 50 (2), 152–188. Russian.
Decidability of the weak interpolation property over the minimal logic. (English) *Algebra and Logic*, 50 (2), 106–132 (2011).
169. Maksimova, L. (2011). Interpolation and Definability over the Logic Gl. *Studia Logica*, 99 (1-3), 249–267. http://www.springerlink.com/openurl.asp?genre=article&id=doi:10.1007/s11225-011-9351-1
170. Maksimova, L. L. (2011). Amalgamation, interpolation and implicit definability in varieties of algebras. Conference Mal'tsev Meeting, *Sovremennye Problemy Matematiki*, 15, Steklov Mathematical Institute of Russian Academy of Sciences, M., 15–39. Russian.
Amalgamation, interpolation and implicit definability in varieties of algebras. (English) *Proceedings of the Steklov Institute of Mathematics*, 278 (1 Supplement), 66–90 (2012).
171. Maksimova, L. (2011). Well-composed J-logics and interpolation. In *5th International Conference on Topology, Algebra and Categories in Logic*, TACL 2011, pp. 203–206, Marseilles, August 26–30.

172. Maksimova, L. L. (2011). Weak interpolation and negative equivalence in extensions of minimal logic. In *Conference Mal'tsev Meeting. (abstracts)*, p. 141, Novosibirsk, October 11–14.
173. Maksimova, L. L. Interpolation and the projective Beth property in well-composed logics. *Algebra i Logika*, 51 (2), 244–275. Russian.
Interpolation and the projective Beth property in well-composed logics. (English) *Algebra and Logic*, 51 (2), 163–184 (2012).
174. Maksimova, L. (2012). Interpolation and Beth definability over the minimal logic. In: T. Bolander, T. Brauner, S. Ghilardi & L. Moss, (Eds.). *Advances in Modal Logic* College Publications, ISBN 978-1-84890-068-4, vol. 9, pp. 459–463.
175. Maksimova, L. L. The decidability of Craig's interpolation property in well-composed J-logics. *Sibirskii Matematicheskii Zhurnal*, 53 (5), 1048–1064. Russian.
The decidability of Craig's interpolation property in well-composed J-logics. (English) *Siberian Mathematical Journal*, 53 (5), 839–852 (2012).
176. Maksimova, L. L. The projective Beth property in well-composed logics. *Algebra i Logika*, 52 (2), 172–202. Russian.
The projective Beth property in well-composed logics. (English) *Algebra and Logic*, 52 (2), 116–136 (2013).
177. Maksimova, L. L. Classification of extensions of the modal logic S4. *Sibirskii Matematicheskii Zhurnal*, 54 (6), 1337–1352. Russian.
Classification of extensions of the modal logic S4. (English) *Siberian Mathematical Journal*, 54 (6), 1064–1075 (2013).
178. Maksimova, L. L. Restricted interpolation over modal logic S4. *Algebra i Logika*, 52 (4), 461–501. Russian.
Restricted interpolation over modal logic S4. (English) *Algebra and Logic*, 52 (4), 308–335 (2013).
179. Maksimova, L. L. The Lyndon property and uniform interpolation over the Grzegorczyk logic. *Sibirskii Matematicheskii Zhurnal*, 55 (1), 147–155. Russian.
The Lyndon property and uniform interpolation over the Grzegorczyk logic. (English) *Siberian Mathematical Journal*, 55 (1), 118–124 (2014).
180. Maksimova, L. (2014). Interpolation and recognition. Logic Colloquium 2013, Evora, Portugal, July 22–27, Scientific Program and Abstracts, 114; *Bulletin of Symbolic Logic*, 20 (2), 251.
181. Maksimova, L. L. (2014) Negative equivalence over the minimal logic and interpolation. *Siberian Electronic Mathematical Reports*, 11, 1–17. Russian.
182. Alaev, P. A. & Maksimova, L. L. (2014). Mathematical Logic. Part 1. *Educational textbook.* Editorial publishing center, NSU, Novosibirsk, p. 106.
183. Alaev, P. A & Maksimova, L. L. (2014). Mathematical Logic. Part 2. *Educational textbook.* Editorial publishing center, NSU, Novosibirsk, p. 96.
184. Maksimova, L. L. & Yun, V. F. (2014). Recognizability over the minimal logic. In *Conference Mal'tsev Meeting. (abstracts)*, p. 153, Novosibirsk, November 10–13.

185. Maksimova, L. L. & Yun, V. F. (2015). Recognizable logics. *Algebra i Logika*, 54 (2), 252–274. Russian.
Recognizable logics. (English) *Algebra and Logic*, 54 (2), 167–182 (2015).
186. Maksimova, L. L. & Yun, V. F. Interpolation over the minimal logic and Odintsov intervals. *Sibirskii Matematicheskii Zhurnal*, 56 (3), 600–616. Russian.
Interpolation over the minimal logic and Odintsov intervals. (English) *Siberian Mathematical Journal*, 56 (3), 476–489 (2015).
187. Maksimova, L. L. & Yun, V. F. (2015). WIP-minimal logics and interpolation. *Siberian Electronic Mathematical Reports*, 12, 7–20. Russian.
188. Maksimova, L. L. & Yun, V. F. (2015). Odintsov intervals and interpolation over the minimal logic. In *Conference Mal'tsev Meeting. (Abstracts)* p. 215, Novosibirsk, May 3–7.
189. Maksimova, L. L. & Yun, V. F. (2016). Slices over minimal logic. *Algebra i Logika* (to appear) Russian.
190. Maksimova, L. L. & Yun, V. F. (2016). The tabularity problem over minimal logic. *Sibirskii Matematicheskii Zhurnal* (to appear) Russian.
191. Maksimova, L. (2016). LC and its pretabular relatives. In: K. Bimbo (Ed.) J. Michael Dunn on information based logics. Ser. Outstanding Contributors to Logic 8, Springer International Publishing Switzerland, pp. 81–92.
192. Maksimova, L. L. & Yun, V. F. Strong decidability and strong recognizability. *Algebra i Logika* (ssubmitted). Russian.
193. Maksimova, L. L. & Yun, V. F. (2013). Calculi over minimal logic and nonembeddability of algebras. *Siberian Electronic Mathematical Reports*, 13, 704–715. Russian.
194. Maksimova, L. L. & Yun, V. F. Slices and levels of extensions of minimal logic. *Sibirskii Matematicheskii Zhurnal* (submitted). Russian.
195. Maksimova, L. L. (2016). The structure of slices over minimal logic. *Sibirskii Matematicheskii Zhurnal*, 57 (5), 1078-1087. Russian.

Index

S. Odintsov (ed.), *Larisa Maksimova on Implication, Interpolation, and Definability*, Outstanding Contributions to Logic 15, https://doi.org/10.1007/978-3-319-69917-2